T0139305

SURVIVAL ANALYSIS *with* INTERVAL-CENSORED DATA

A Practical Approach with Examples in R, SAS, and BUGS

CHAPMAN & HALL/CRC

Interdisciplinary Statistics Series

Series editors: N. Keiding, B.J.T. Morgan, C.K. Wikle, P. van der Heijden

Published titles

AGE-PERIOD-COHORT ANALYSIS: NEW MODELS, METHODS, AND EMPIRICAL APPLICATIONS Y. Yang and K. C. Land

ANALYSIS OF CAPTURE-RECAPTURE DATA R. S. McCrea and B. J.T. Morgan

AN INVARIANT APPROACH TO STATISTICAL ANALYSIS OF SHAPES S. Lele and J. Richtsmeier

ASTROSTATISTICS G. Babu and E. Feigelson

BAYESIAN ANALYSIS FOR POPULATION ECOLOGY R. King, B. J.T. Morgan, O. Gimenez, and S. P. Brooks

BAYESIAN DISEASE MAPPING: HIERARCHICAL MODELING IN SPATIAL EPIDEMIOLOGY, SECOND EDITION A. B. Lawson

BIOEQUIVALENCE AND STATISTICS IN CLINICAL PHARMACOLOGY S. Patterson and B. Jones

CAPTURE-RECAPTURE METHODS FOR THE SOCIAL AND MEDICAL SCIENCES D. Böhning, P. G. M. van der Heijden, and J. Bunge

CLINICAL TRIALS IN ONCOLOGY, THIRD EDITION S. Green, J. Benedetti, A. Smith, and J. Crowley

CLUSTER RANDOMISED TRIALS R.J. Hayes and L.H. Moulton

CORRESPONDENCE ANALYSIS IN PRACTICE, THIRD EDITION M. Greenacre

THE DATA BOOK: COLLECTION AND MANAGEMENT OF RESEARCH DATA M. Zozus

DESIGN AND ANALYSIS OF QUALITY OF LIFE STUDIES IN CLINICAL TRIALS, SECOND EDITION D.L. Fairclough

DYNAMICAL SEARCH L. Pronzato, H. Wynn, and A. Zhigljavsky

FLEXIBLE IMPUTATION OF MISSING DATA S. van Buuren

GENERALIZED LATENT VARIABLE MODELING: MULTILEVEL, LONGITUDINAL, AND STRUCTURAL EQUATION MODELS A. Skrondal and S. Rabe-Hesketh

GRAPHICAL ANALYSIS OF MULTI-RESPONSE DATA K. Basford and J. Tukey

INTRODUCTION TO COMPUTATIONAL BIOLOGY: MAPS, SEQUENCES, AND GENOMES M. Waterman

MARKOV CHAIN MONTE CARLO IN PRACTICE W. Gilks, S. Richardson, and D. Spiegelhalter

Published titles

MEASUREMENT ERROR ANDMISCLASSIFICATION IN STATISTICS AND EPIDEMIOLOGY: IMPACTS AND BAYESIAN ADJUSTMENTS P. Gustafson

MEASUREMENT ERROR: MODELS, METHODS, AND APPLICATIONS
J. P. Buonaccorsi

MEASUREMENT ERROR: MODELS, METHODS, AND APPLICATIONS
J. P. Buonaccorsi

MENDELIAN RANDOMIZATION: METHODS FOR USING GENETIC VARIANTS IN CAUSAL ESTIMATION S.Burgess and S.G.Thompson

META-ANALYSIS OF BINARY DATA USINGPROFILE LIKELIHOOD D. Böhning, R. Kuhnert, and S. Rattanasiri

MISSING DATA ANALYSIS IN PRACTICE T. Raghunathan

MODERN DIRECTIONAL STATISTICS C. Ley and T. Verdebout

POWER ANALYSIS OF TRIALS WITH MULTILEVEL DATA M. Moerbeek and S. Teerenstra

SPATIAL POINT PATTERNS: METHODOLOGY AND APPLICATIONS WITH R
A. Baddeley, E Rubak, and R. Turner

STATISTICAL ANALYSIS OF GENE EXPRESSION MICROARRAY DATA T. Speed

STATISTICAL ANALYSIS OF QUESTIONNAIRES: A UNIFIED APPROACH BASED ON R AND STATA F. Bartolucci, S. Bacci, and M. Gnaldi

STATISTICAL AND COMPUTATIONAL PHARMACOGENOMICS R. Wu and M. Lin

STATISTICS IN MUSICOLOGY J. Beran

STATISTICS OF MEDICAL IMAGING T. Lei

STATISTICAL CONCEPTS AND APPLICATIONS IN CLINICAL MEDICINE
J. Aitchison, J.W. Kay, and I.J. Lauder

STATISTICAL AND PROBABILISTIC METHODS IN ACTUARIAL SCIENCE
P.J. Boland

STATISTICAL DETECTION AND SURVEILLANCE OF GEOGRAPHIC CLUSTERS
P. Rogerson and I. Yamada

STATISTICAL METHODS IN PSYCHIATRY AND RELATED FIELDS: LONGITUDINAL, CLUSTERED, AND OTHER REPEATED MEASURES DATA
R. Gueorguieva

STATISTICS FOR ENVIRONMENTAL BIOLOGY AND TOXICOLOGY A. Bailer and W. Piegorsch

STATISTICS FOR FISSION TRACK ANALYSIS R.F. Galbraith

SURVIVAL ANALYSIS WITH INTERVAL-CENSORED DATA: A PRACTICAL APPROACH WITH EXAMPLES IN R, SAS, AND BUGS
K. Bogaerts, A. Komárek, and E. Lesaffre

VISUALIZING DATA PATTERNS WITH MICROMAPS D.B. Carr and L.W. Pickle

Chapman & Hall/CRC
Interdisciplinary Statistics Series

SURVIVAL ANALYSIS *with* INTERVAL-CENSORED DATA

A Practical Approach with Examples in R, SAS, and BUGS

Kris Bogaerts
Arnošt Komárek
Emmanuel Lesaffre

CRC Press is an imprint of the
Taylor & Francis Group, an informa business
A CHAPMAN & HALL BOOK

CRC Press
Taylor & Francis Group
6000 Broken Sound Parkway NW, Suite 300
Boca Raton, FL 33487-2742

Printed on acid-free paper
Version Date: 20171012

International Standard Book Number-13: 978-1-4200-7747-6 (Hardback)

Visit the Taylor & Francis Web site at
http://www.taylorandfrancis.com

and the CRC Press Web site at
http://www.crcpress.com

Printed and bound in the United States of America by
Edwards Brothers Malloy on sustainably sourced paper

Contents

List of Tables

List of Figures

Notation

$S(t)$	survival function S
$F(t)$	cumulative distribution function F
$f(t)$	density function $f(t)$
$\hbar(t)$	hazard function $\hbar(t)$
$H(t)$	cumulative hazard function $H(t)$
$[l, u]$	closed interval with lower limit l and upper limit u
$(l, u]$	half-open interval with lower limit l and upper limit u
$\lfloor l, u \rfloor$	open, half-open or closed interval with lower limit l and upper limit u
δ	censoring indicator
$L(\cdot)$	likelihood
$\ell(\cdot)$	$\log\{L(\cdot)\}$, log-likelihood
$\mathcal{D}$	collected data
$\mathcal{DP}$	Dirichlet process
$\mathcal{D}ir_p(\delta_1, \ldots, \delta_p)$	p-dimensional Dirichlet distribution with parameters $\delta_1, \ldots, \delta_p$
$\mathcal{N}(\mu,\, \sigma^2)$	normal distribution with mean μ and variance σ^2
φ	density of the standard normal Gaussian distribution $\mathcal{N}(0, 1)$
φ_{μ,σ^2}	density of the Gaussian distribution $\mathcal{N}(\mu,\, \sigma^2)$
Φ	standard cumulative distribution function of the Gaussian distribution $\mathcal{N}(0, 1)$
Φ_{μ,σ^2}	cumulative distribution function of the Gaussian distribution $\mathcal{N}(\mu,\, \sigma^2)$
Φ_ρ	standard bivariate cumulative distribution function of the Gaussian distribution with correlation ρ
$\mathcal{G}(\zeta,\, \gamma)$	gamma distribution with a shape parameter ζ and a rate parameter γ (with the mean ζ/γ)

$\mathcal{N}_p(\boldsymbol{\mu}, \boldsymbol{\Sigma})$	p-dimensional normal distribution with mean vector $\boldsymbol{\mu}$ and covariance matrix $\boldsymbol{\Sigma}$
$I(\cdot)$	indicator function equal to 1 if the expression between parentheses is true, and 0 otherwise
l_p	penalized log-likelihood
$\Delta^k(\cdot)$	k-order forward difference function
$\mathbb{I}$	identity matrix
$\boldsymbol{I}$	Hessian matrix
$RanF$	range of the function F
$C(u, v)$	copula
$\breve{C}(u, v)$	survival copula
$\breve{C}_\theta^C(u, v)$	Clayton copula with parameter θ
$\breve{C}_\rho^G(u, v)$	Gaussian copula with parameter ρ
$\breve{C}_\theta^P(u, v)$	Plackett copula with parameter θ
$\lVert\cdot\rVert$	Euclidean length

Preface

We speak of interval censoring when a continuous measurement can only be observed to lie in an interval. A trivial example is a rounded continuous recording, such as age recorded in years. In statistical analyses, rounding is often ignored without much harm done except when the rounding is too coarse, such as reducing age into age classes. In this book we focus on the treatment of interval-censored data in the context of survival analysis. Over the last forty years much research has been done in developing appropriate methods for dealing with such data in a variety of practical settings. However, despite the statistical developments, interval censoring is too often ignored in applications. So the authors of this book questioned why the problem of interval censoring has been kept under the radar for such a long time.

For many years there is an excellent book on the treatment of interval-censored data, see Sun (2006). However, despite the book of Sun and the numerous statistical developments, appropriate methods for interval-censored data are not commonly used in practice. For example, the Food and Drug Administration does not require the use of appropriate methods for dealing with interval censoring in the analysis of clinical trial data. Reasons for this lack of use can be plenty. Firstly, it is assumed that ignoring interval censoring has only a minimal impact on the conclusions of a statistical analysis. Another possible reason is the (perceived) absence of appropriate statistical software. While lately much (free) software has appeared, it struck us that this is not recognized by the statistical community at large. For this reason, we thought that a more practically oriented book focusing on the application of statistical software for interval-censored data might be complementary to the book of Sun.

The origin of this book lies, however, much earlier in time when the third author was contacted by his dental colleague, Dominique Declerck. On a particular day she knocked on the door and asked whether there was interest in collaborating on a large longitudinal study in oral health. The aim of this study was, she said, to obtain a good picture of the oral health among children attending primary school in Flanders. In addition the project would give valuable information on risk factors for all kinds of aspects of oral health. Furthermore, the study would enable to get a good estimate of the emergence times of permanent teeth in Flanders. The study became internationally known as the Signal Tandmobiel study, and provided a rich amount of oral health output but also statistical output. This study was also the start of our interest into the statistical treatment of interval-censored survival times. Both the first and second authors performed their doctoral research on this study.

As the title of the book suggests, our intention is to provide the reader with a practical introduction into the analysis of interval-censored survival times. Apart from the introductory chapters, each chapter focuses on a particular type of statistical analysis. Often there are more than one statistical approach to address a particular problem. First the approach will be briefly explained, then follows the statistical analysis of one of the motivating data sets. This consists in first reporting the results of the analysis and then reporting in detail how the analysis could be done with the statistical software. We note that often the results are reported with too much accuracy (too many digits behind the decimal point), but this is done to make it easier for the reader to trace better the discussed result in the statistical output.

A variety of examples, mainly medical but also some other types of examples, serve to illustrate extensively the proposed methods in the literature. Because of space limitations, but also because of the particular aim of the book, we have largely limited ourselves to those approaches for which statistical software was existing. In some cases, when feasible, some published developments were programmed by the authors. While we have included a great variety of approaches, we have given some emphasis to methods and software developed in our team. As software, we have chosen for R and SAS to illustrate frequentist approaches and R, SAS and BUGS-like software for Bayesian methods. It is clear that this book would not have been possible without the efforts of the many statisticians who devoted energy and time in developing statistical software. Luckily, this attitude has gained much popularity over the last decades.

The book is divided into four parts. Part I is the introductory part. In Chapter 1 survival concepts are reviewed and the motivating data sets are introduced. In the same chapter we give a brief introduction to the use of the frequentist software, i.e., R and SAS. In addition, we describe the R package icensBKL, which contains the data of the examples and self-written functions by the authors as well as routines that were shared by others. Chapter 2 reviews the most common statistical methods for the analysis of right-censored data. This chapter can be skipped by those who are familiar with survival analysis techniques.

Part II represents the main body of frequentist methods for analyzing interval-censored data. Chapter 3 deals with estimating the survival curve and includes the well-known nonparametric estimate of Turnbull. But also parametric as well as smoothing approaches are described. The comparison of two or more survival distributions is discussed in Chapter 4. In Chapter 5, the different approaches to analyze a proportional hazards (PH) model for interval-censored survival times are discussed. Although for interval-censored data the PH model does not have the central position in survival analysis anymore, it still attracted most of the attention among statisticians. The accelerated failure time (AFT) model is treated in Chapter 6. This model has been explored to a much lesser extent than the PH model, yet it has many interesting properties. Information loss due to interval censoring must be taken into account at the design of a study that may suffer from interval censoring. In this chapter a method is reviewed that can be used for the calculation of the necessary study size when dealing with an AFT model. This method can be used, to some extent, also for a PH model. The statistical treatment of bivariate survival times is dealt with in Chapter 7. In this chapter, we describe parametric models but also smooth and copula modelling with the aim to estimate the association of survival times in the presence of interval censoring. In the last chapter of Part II some more complex problems are treated, namely doubly interval-censored data, clustered interval-censored data and a graphical approach to display high-dimensional interval-censored observations.

In Part III Bayesian methods for interval-censored observations are reviewed. In Chapter 9 a general introduction to Bayesian concepts and Bayesian software is given. Examples of Bayesian analyses, both parametric and nonparametric, on right-censored data introduce the reader to the use of BUGS-like software and its interface with R. Bayesian estimation of the survival distribution is treated in Chapter 10. As for frequentist methods, parametric approaches as well as flexible and nonparametric approaches are reviewed. The Bayesian PH model is discussed in Chapter 11. In contrast to frequentist methods, now the semiparametric model does not play any role at all. Indeed, it turned out that the current Bayesian nonparametric techniques cannot handle the PH model appropriately. Therefore, we only reviewed parametric PH models and two approaches that model the baseline hazard in a flexible way. In Chapter 12, the Bayesian approach to fit an AFT model to interval-censored observations is illustrated. Apart from the parametric approach, two approaches that smooth the density of the AFT model are reviewed. We end this chapter with a semiparametric approach. The last chapter in this part treats the Bayesian way to analyze interval-censored observations with an hierarchical structure, multivariate models and doubly interval censoring. For each of these three topics the parametric, flexible and semiparametric approaches are introduced and illustrated with practical examples.

The concluding part of the book consists of only one chapter. Basically each statistical method can be generalized to deal with interval censoring. In Chapter 14 a quick review is given of the techniques that were omitted in the main part of the book. This chapter provides a proof of the growing interest among statisticians to deal with interval censoring in an appropriate manner.

Finally, several appendices provide extra detailed information on either the data sets, the software or some statistical background useful for the understanding of certain techniques explained in the main part of the book. The link to the supplemental materials is available on the website of the publisher (`https://www.crcpress.com/9781420077476`).

In the ten-year period of writing this book, many colleagues helped us in a variety of ways. In alphabetical order, we thank Clifford Anderson-Bergman for help with the icenReg package, Ann Belmans for support in the English language, Aysun Çetinyürek Yavuz for sending us her PhD thesis, Fang Chen for help when dealing with SAS problems, David Dejardin for adapting the program for doubly interval-censored data, Marc Delord for help with the MIICD package, Steffen Fieuws for several discussions on a variety of topics, Ruth Goodall and Volkmar Schenkel for interesting discussions, Patrick Groenen for input in the biplot program, Alejandro Jara for input when using DPpackage, Juha Karvanen for providing the mobile phone data set, Hae-Young Kim for input on the sample size calculation method, Ruth Nysen, Sun Xing and Min Zhang for providing their SAS macros, Chun Pan for help with the ICBayes package, Virginie Rondeau for assistance with the frailtypack package, Sylvie Scolas for sending her PhD thesis, Ying So for solving questions on the %ICSTEST and %EMICM macros, Koen Vanbrabant for tips and tricks in R and Wenjie Wang for help with the dynsurv package.

A special thanks goes to Lupe Gómez and her team (Malu Calle, Ramon Oller, and Klaus Langohr) for the support and several discussions on interval censoring at various times. Also thanks to David Grubbs for his patience and guidance in the process of writing this book.

The authors strongly believe that statistical theory without the necessary software is not appealing to the statistical practitioner. We hope that this book fills in the gap and that it will help statisticians, epidemiologists, etc. in applying the appropriate methodology when interval censoring is present in the data.

We also thank our wives Inge, Lenka and Lieve, for their support and understanding for the many hours of working on the book during the weekends and evenings.

To the future generation, let us hope that they are allowed to live in a (more) peaceful society, to Daan, Lien, Jindra, Zbyňka, Radka, Ondřejka, Štěpán, Annemie, Kristof, Lisa, Jesse, and Lore.

Part I

Introduction

Chapter 1

Introduction

This chapter introduces the reader to the basic concepts in survival analysis and the notation used in this book. We also familiarize the reader with the data sets that serve to illustrate the theoretical concepts. Further, we provide a brief introduction to R with a focus on how censoring is dealt with and the R package icensBKL wherein we have collected the data sets used in this book and various R functions to analyze interval-censored observations. Then follows an initiation to how the SAS software will be used in this book for the analysis of interval-censored observations. We will discuss the Bayesian software packages, such as OpenBUGS, WinBUGS, JAGS, etc. in Chapter 9, where we also start with the Bayesian treatment of interval censoring. In the remainder of the book, programs written in either of the packages will be extensively discussed.

1.1 Survival concepts

We are interested in methods to analyze survival data, i.e., data that measure the time to some event *(event time)*, not necessarily death. More precisely, the *event time* is a nonnegative real valued random variable. Data on event times are obtained in a study by monitoring subjects over (calendar) time, recording the moments of the specified event of interest and determining the time span between the event and some initial *onset time*. Examples are admission to a study and disease progression, contagion by HIV virus and onset of AIDS, tooth emergence and time to onset of caries.

Let T be the random variable representing the event time of interest. The distribution of T is uniquely determined by a nonincreasing right-continuous *survival function* $S(t)$ defined as the probability that T exceeds a value t in its domain, i.e.,

$$S(t) = \mathsf{P}(T > t), \qquad t > 0. \tag{1.1}$$

Similarly, a nondecreasing right-continuous *cumulative distribution function* (cdf) $F(t)$ is defined as the probability that T does not exceed a value t, i.e.,

$$F(t) = \mathsf{P}(T \leq t) = 1 - S(t), \qquad t > 0. \tag{1.2}$$

Another possibility is to specify the *hazard function* $\hbar(t)$ which gives the instantaneous rate at which an event occurs for an item that is still at risk for the event at time t, i.e.,

$$\hbar(t) = \lim_{\Delta t \to 0_+} \frac{\mathsf{P}(t \leq T < t + \Delta t \,\big|\, T \geq t)}{\Delta t}, \qquad t > 0. \tag{1.3}$$

We will also shortly write

$$\hbar(t)\mathrm{d}t = \mathsf{P}\big(T \in N_t(\mathrm{d}t) \,|\, T \geq t\big), \quad \text{where } N_t(\mathrm{d}t) = [t,\, t{+}\mathrm{d}t), \qquad t > 0. \tag{1.4}$$

Finally,

$$H(t) = \int_0^t \hbar(s)\,\mathrm{d}s, \qquad t > 0 \tag{1.5}$$

is called the *cumulative* (or *integrated*) *hazard function*.

Most often, and we will also assume it throughout the book unless explicitly specified otherwise, the event time T has an absolutely continuous distribution with a density $f(t)$, i.e.,

$$f(t) = -\frac{\mathrm{d}S(t)}{\mathrm{d}t} = \frac{\mathrm{d}F(t)}{\mathrm{d}t}, \qquad t > 0. \tag{1.6}$$

The (cumulative) hazard function, the survival function and the density are then related through

$$S(t) = \exp\big\{-H(t)\big\} = \frac{f(t)}{\hbar(t)}, \qquad t > 0. \tag{1.7}$$

1.2 Types of censoring

1.2.1 Right censoring

A typical feature of survival data is the fact that the time to an event is not always observed accurately and observations are subject to *censoring*. Often the study is over before all recruited subjects have shown the event of interest or the subject has left the study prior to experiencing an event. In both situations, only the lower limit l for the true event time T is known such that $T > l$. Then, we say that there is *right censoring* in the data and that such an event time is *right censored*.

1.2.2 Interval and left censoring

When the event of interest has already happened before the subject is included in the study but it is not known when it occurred, one speaks of *left*

TABLE 1.1: Taxonomy of interval-censored observations.

Observation	Limits of the interval $\lfloor l,\, u \rfloor$	Censoring indicator
Right-censored in time l	$0 < l < u = \infty$	$\delta = 0$
Exactly observed time t	$0 < l = u = t < \infty$	$\delta = 1$
Left-censored in time u	$0 = l < u < \infty$	$\delta = 2$
Interval-censored in $\lfloor l,\, u \rfloor$	$0 < l < u < \infty$	$\delta = 3$

censoring. In case when we only can say that the event happened between two examinations, we speak of *interval censoring.* Left censoring occurs for example when a permanent tooth has already emerged prior to the start of a dental study that aims to estimate its emergence distribution. In the same study, an emergence time is interval-censored when the permanent tooth is present in the mouth at the current examination but not yet at the previous examination. In HIV/AIDS studies interval censoring often occurs. Indeed, time to HIV seroconversion can only be determined by a laboratory assessment which is usually initiated after a visit to the physician. Then one can only conclude that HIV seroconversion has happened between two examinations. The same is true for the diagnosis of AIDS, which is based on clinical symptoms and needs to be confirmed by a medical examination.

Left censoring and interval censoring are quite similar for survival times. Indeed, for interval censoring it is only known that the event time T occurred after, say l, and before, say u, with $l > 0$ and $u < \infty$. For left censoring, we have the same situation, but now $l = 0$. Depending on the context, we either know that $l < T \leq u$, $l \leq T < u$, $l \leq T \leq u$, or $l < T < u$. To indicate this, we will write $T \in \lfloor l,\, u \rfloor$. We note that many methods in this book lead to the same results irrespective of whether the observed interval is closed, open or half-open.

If we allow $l = 0$ and $u = \infty$ in the definition of an interval-censored observation, interval censoring becomes a natural generalization of the commonly encountered right censoring and the less common left censoring. Even an exactly observed event time t could be considered as an interval-censored observation with $l = u = t$. Right, left and (genuine) interval censoring are exemplified in Figure 1.1. In practice, we do need to distinguish between right-censored, exactly observed, left-censored, or interval-censored event times. For this reason, we will use a censoring indicator δ which is equal to 0, 1, 2, or 3, respectively, as further exemplified in Table 1.1.

1.2.3 Some special cases of interval censoring

Several special cases of interval censoring have been studied in the literature and were assigned specific terms. *Case I* interval-censored data (Groeneboom and Wellner, 1992; Sun, 2006), often referred to as *current status* data (Diamond et al., 1986; Grummer-Strawn, 1993) pertain to the situation where

(a) Right censoring

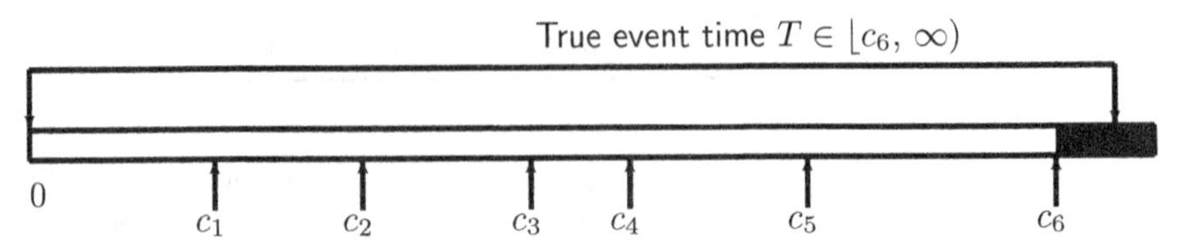

(b) Left censoring

True event time $T \in (0, c_1\rfloor$

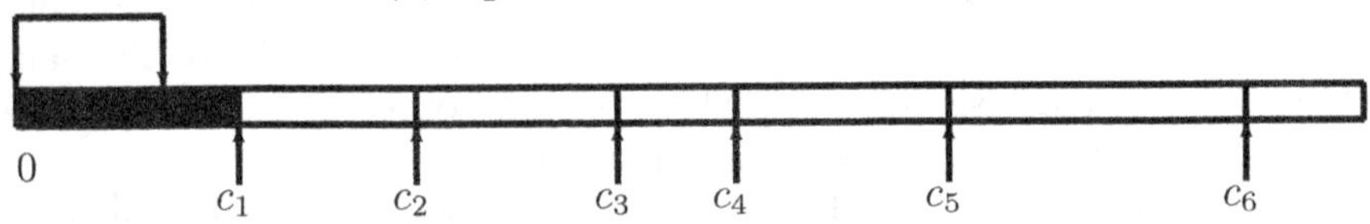

(c) Interval censoring

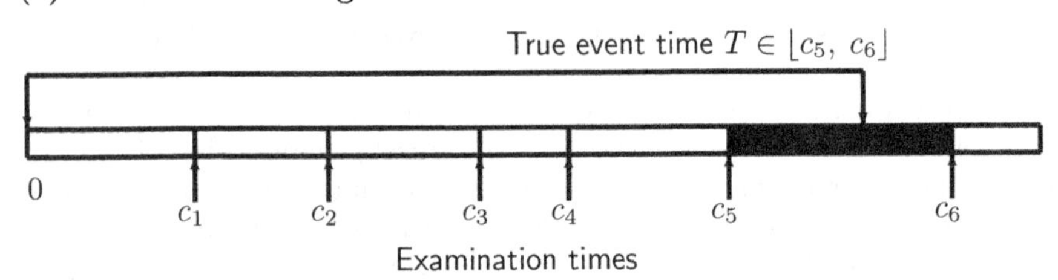

FIGURE 1.1: Right, left and interval censoring. A scheme of censored observations obtained by performing examinations to check the event status at times $c_1, \ldots, c_6$: (a) observation is right-censored at c_6 (i.e., $l = c_6$), (b) observation is left-censored at c_1 (i.e., $u = c_1$), (c) observation is interval-censored with observed interval $\lfloor c_5, c_6\rfloor$ (i.e., $l = c_5$, $u = c_6$).

the event status of each subject is examined only once at a fixed or random time. For current status data, all observations are of the form $(0, c_1\rfloor$ or $\lfloor c_1, \infty)$, where c_1 is the examination time for a specific subject. That is, it is only recorded whether the event has already happened or yet has to happen at time c_1 and hence all observations are either left- or right-censored.

When for each subject exactly two examinations are performed the single observation takes either of the forms $(0, c_1\rfloor$, $\lfloor c_1, c_2\rfloor$, $\lfloor c_2, \infty)$, with c_1 and c_2 the two examination times (fixed or random). In that case, we talk about *case II* interval censoring. Analogously, data arising from a fixed number k examination times are called *case k* interval-censored (Gómez et al., 2004).

When there are several fixed or random inspection times, the resulting data set contains only intervals of the form $\lfloor l, u\rfloor$ with $l < u$ and for at least one interval $0 < l < u < \infty$. This is referred to as *general interval censoring* (Sun, 2006). When both intervals $\lfloor l, u\rfloor$ with $0 \le l < u \le \infty$ as well as exact observations, i.e., intervals $[l, u]$ with $l = u$ are present, one speaks of *mixed* interval-censored data (Yu et al., 2000). Unless stated otherwise, we do not distinguish in this book between the above two types of interval censoring.

1.2.4 Doubly interval censoring

In survival analysis, a time to event T is computed from a specific onset time T^O to the time of event T^E, i.e., $T = T^E - T^O$, which is also referred to as the *gap time* also denoted as G. It may happen that not only T^E is interval-censored but also T^O (De Gruttola and Lagakos, 1989; Gómez and Calle, 1999; Gómez and Lagakos, 1994; Komárek and Lesaffre, 2006, 2008; Sun, 1995, 2006). Examples are (a) time to caries development (T) on a tooth, where the (chronological) time of the tooth emergence is the onset time T^O

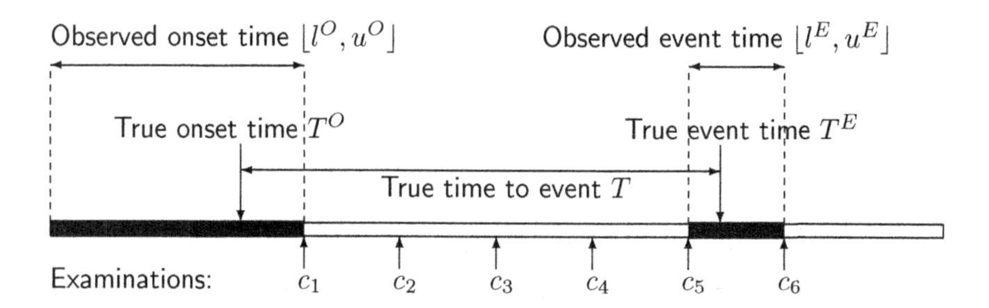

FIGURE 1.2: Doubly interval censoring. A scheme of a doubly interval-censored observation obtained by performing examinations to check the event status at times $c_1, \ldots, c_6$. The onset time is left-censored at time $u^O = c_1$ (i.e., interval-censored in the interval $(0, c_1\rfloor)$, the event time is interval-censored in the interval $\lfloor l^E, u^E\rfloor = \lfloor c_5, c_6\rfloor$.

and the (chronological) time that a cavity appears is the event time T^E; (b) time to the onset of AIDS (T), where the time of HIV seroconversion is the onset time T^O and the onset time of AIDS is the event time T^E. In both examples it is practically not feasible to know the onset time precisely, in that case we say that the time to event T is *doubly interval-censored*, see also Figure 1.2.

Note that the term *doubly censoring* has been used in the literature for another case. Gehan (1965a), Turnbull (1974), Chen and Zhou (2003) and Cai and Cheng (2004) used the term *doubly censored* data when case I interval-censored (current status) observations are mixed with exactly observed event times. This may occur when there exists a time window in which the event can be exactly observed but outside this window, the observations are either left- or right-censored.

1.2.5 Truncation

In addition to censoring, also *truncation* (Klein and Moeschberger, 2003) may occur. Truncation of survival data occurs when only those subjects whose event time lies within a certain observational window are included in the study. The most common case is *left truncation* where the truncation time is often called a "delayed entry time" since we only observe subjects from this time until they have the event of interest or are censored. As an example, left-truncated data occur in a cohort study in which subjects are included in the study only if they experienced some initial event prior to the event of interest. In this case, for all subjects in the study, the event time of interest is greater than the occurrence time of the initial event. Subjects who had the event of interest already before the recruitment started, are then not included. Ignoring left truncation leads to length-biased sampling, which may introduce substantial bias in the estimation of the survival time distribution (Klein and Moeschberger, 2003). Here, we focus on studies with interval-censored data only and truncation is not covered further.

1.3 Ignoring interval censoring

In practice, interval censoring is often addressed inappropriately. A common approach to deal with interval-censored observations is *mid-point* and *right-point imputation*. These are examples of single imputation techniques. In that case, each finite interval is replaced by its mid-point or right-point or, more generally, by an arbitrary value from within the interval. Right-censored observations are often left unchanged. This approach allows then to use classical survival software assuming only right-censored or exactly observed survival times (see Chapter 2). Nevertheless, the single imputation approach may lead to invalid inference already discussed by Law and Brookmeyer (1992).

More recently, Panageas et al. (2007) discussed the impact of single imputation in cancer studies. A popular endpoint to evaluate the effect of a cancer treatment is *progression-free survival* (PFS). PFS is defined as the time from treatment administration to the time that progression of the disease is detected often with radiological testing. Hence, while true progression occurs in-between two examinations, the recorded PFS is based on the examination when progression is detected by a scan. Panageas et al. (2007) showed that descriptive statistics like the median survival time might be severely affected by ignoring the interval-censored character of PFS. They performed a simulation study mimicking a typical Phase II clinical trial with 25 patients for which the true median PFS was 13, 26 or 52 weeks and the frequency of radiological examinations for the patients ranged from 4 to 12 weeks. The percentage bias of the estimated median PFS using the standard Kaplan-Meier methodology on the left-, mid- and right-point imputed data ranged from −7% to −39% for left-point imputation, −54% to 26% for mid-point imputation and from −8% to 68% for right-point imputation, respectively.

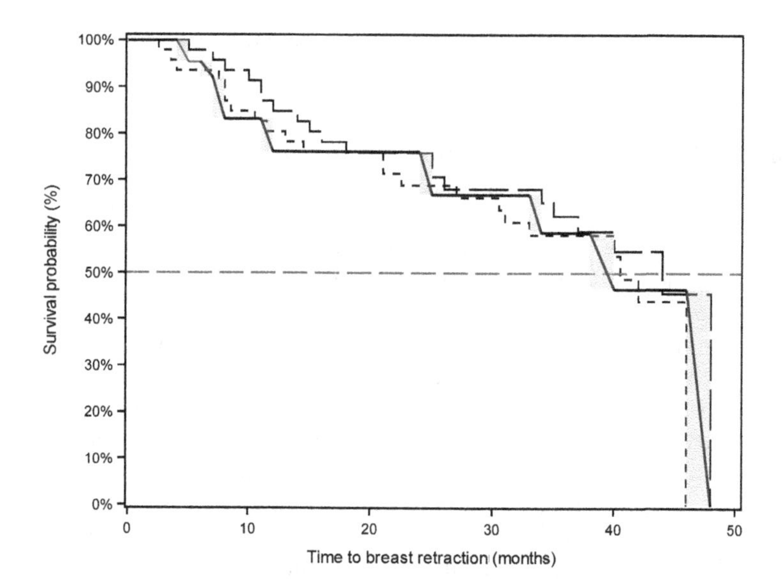

FIGURE 1.3: Breast cancer study of the radiotherapy-only group. Median time to breast retraction determined from Kaplan-Meier curve using mid-point imputed (short dashed line) and right-point imputed (long dashed line) data and the nonparametric maximum likelihood estimate assuming a linear survival curve in the regions of support (solid line).

Left-point imputation consistently underestimated the true PFS, right-point imputation method consistently overestimated the true PFS, while the results were mixed for mid-point imputation. Furthermore, the bias does not necessarily increase with wider assessment intervals but rather depends on the timing of the assessment intervals relative to the true median.

Example 1.1 Breast cancer study

We compared the median time to breast retraction for the radiotherapy-only arm of a breast cancer study introduced in Section 1.6. Because the 46 patients were only examined at prespecified visits, the data are interval-censored. In Figure 1.3 the survival curve was calculated in three different ways. Two curves are calculated ignoring the interval-censored character of the data. Standard Kaplan-Meier methodology was applied to mid- and right-point imputed data. The third curve displays the nonparametric maximum likelihood estimator assuming a linear trend in the regions of support (see Section 3.1). Interest lies in estimating the median survival time. The median determined by mid-point (40.5 months) and right-point (44 months) imputation is much larger than the median obtained from the nonparametric maximum likelihood estimator (38.6 months).

□

We also evaluated the impact of ignoring interval censoring in a bivariate context using a small simulation study.

Example 1.2 Impact of mid-point imputation

We generated 100 data sets, each with 100 data points (corresponding to 100 subjects) (T_1, T_2) from a bivariate normal distribution with the margins having means $\mu_1 = \mu_2 = 10$ and standard deviations $\sigma_1 = \sigma_2 = 1.50$ (setting 1) and $\sigma_1 = \sigma_2 = 0.75$ (setting 2), respectively. Further, the correlation coefficient ϱ between the two margins is in both settings equal to 0.6. For each setting, one simulated data set is shown in Figure 1.4. Suppose now that T_1 and T_2 are the two events of interest, e.g., emergence times of two permanent teeth and we are interested in estimating the means, the standard deviations and the correlation coefficient. Suppose further that checking whether the event defining T_1 and T_2 respectively has happened was done at visits performed at times $C_1, \ldots, C_M$ leading to interval-censored data.

We generated for each data set the first visit time C_1 (different for each subject) from the uniform distribution on the interval (3, 4) and then assumed that all subsequent visits followed equidistantly each Δ unit of time. The number of visits was sufficiently high to avoid right-censored observations. That is, all observed time intervals in all data sets are of length Δ. Maximum likelihood estimates of μ_1, μ_2, σ_1, σ_2 and ϱ were computed for each data set (a) from mid-point imputed data using standard methods for uncensored data, (b) from observed time intervals using methods for interval-censored data. Furthermore, we computed corresponding maximum likelihood estimates from the original uncensored data. Average estimates over 100 simulated data sets are shown in Table 1.2. It is seen that mid-point imputation has negligible impact on the estimation of the means. Nevertheless, as the length of the observed intervals increases, the mid-point based estimates of the standard deviations start to overestimate the true standard deviations. The impact of mid-point imputation is most profoundly seen on the estimates of the correlation coefficient. For all values of Δ, the method that takes interval censoring properly into account leads to the estimated value of ϱ being practically the same as the true value which is clearly not the case when using mid-point imputation. □

The above example showed typical features of mid-point imputation. That is, the impact of ignoring interval censoring increases with the length of observed intervals. At the same time, it is seen that the impact of the length of the observed intervals depends on the variability of the event time. Nevertheless, as in the above example, variability is often overestimated when the mid-point imputation is used and hence, one cannot really infer the impact of mid-point estimation from the lengths of observed intervals. In real data, the lengths of intervals often vary considerably rendering it even more difficult to relate a characteristic of the interval lengths to the variability of the event time.

Further illustrations of the impact of ignoring interval censoring can be found in this book. In Examples 4.1, 4.2 and 5.9, we contrast a mid- and right-point imputation analysis with a proper interval-censored data analysis.

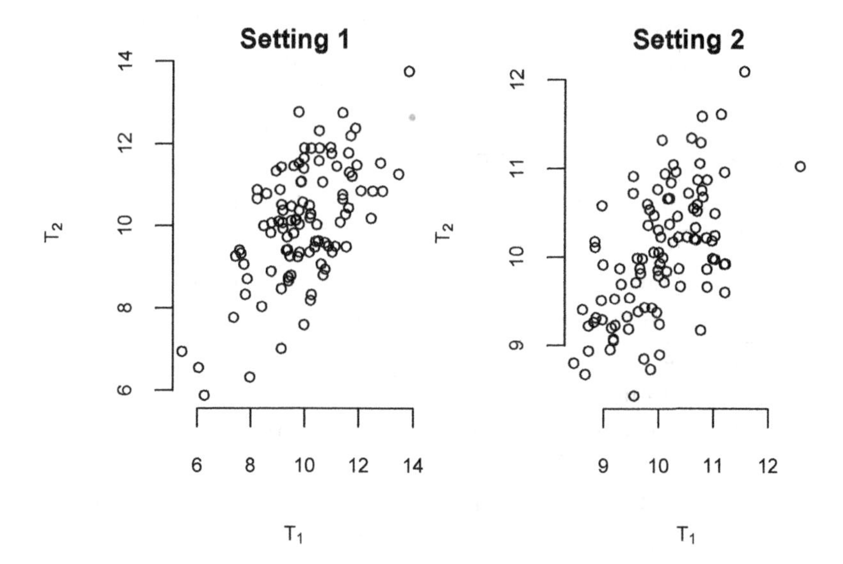

FIGURE 1.4: One (true) simulated data set from either setting used for the illustration of mid-point imputation (see Example 1.2).

TABLE 1.2: Simulation study. Average estimates (over 100 simulated data sets) from original (uncensored) data drawn from a bivariate normal distribution with means $\mu_1 = \mu_2$, standard deviations $\sigma_1 = \sigma_2$ and correlation ϱ. Interval-censored data are fabricated with different lengths Δ of observed intervals and are analyzed with (a) mid-point single imputation and (b) appropriate methodology for interval-censored data.

	μ_1	μ_2	σ_1	σ_2	ϱ
		Setting 1			
True	10.00	10.00	1.500	1.500	0.600
Original data	10.00	10.00	1.485	1.500	0.594
$\Delta = 1$					
Mid-point	10.00	10.00	1.512	1.529	0.571
Interval-censored	10.00	10.00	1.484	1.502	0.593
$\Delta = 2$					
Mid-point	10.00	10.01	1.594	1.606	0.519
Interval-censored	10.00	10.01	1.485	1.497	0.598
$\Delta = 3$					
Mid-point	10.00	10.01	1.720	1.740	0.453
Interval-censored	9.99	10.00	1.468	1.490	0.605
		Setting 2			
True	10.00	10.00	0.750	0.750	0.600
Original data	10.00	10.00	0.740	0.740	0.603
$\Delta = 1$					
Mid-point	10.00	10.00	0.795	0.794	0.522
Interval-censored	10.00	10.00	0.740	0.739	0.602
$\Delta = 2$					
Mid-point	10.01	10.02	0.940	0.938	0.401
Interval-censored	9.98	9.99	0.731	0.729	0.602
$\Delta = 3$					
Mid-point	10.19	10.20	1.280	1.279	0.389
Interval-censored	9.95	9.97	0.679	0.700	0.598

1.4 Independent noninformative censoring

Throughout the book we assume that the censoring mechanism is *independent and noninformative.* The terms "independent" and "noninformative" are used similarly as in the classical book of Kalbfleisch and Prentice (2002). We explain these concepts first for right-censored data and then for interval-censored data.

1.4.1 Independent noninformative right censoring

Kalbfleisch and Prentice (2002) introduced the notion of independent noninformative censoring for right-censored data as follows. Let C denote the random variable causing censoring such that we do not observe the true event time T, but only $Y = \min(T, C)$. The indicator variable $\delta = I[T \leq C]$ equals 0 when T is censored and 1 otherwise.

Independent censoring

The censoring mechanism is called *independent* of the random variable T when the hazard which applies to the censored population is at each time point the same as the hazard which applies if there were no censoring. That is,

$$\mathsf{P}\big(T \in N_t(\mathrm{d}t) \,\big|\, C \geq t,\, T \geq t\big) = \mathsf{P}\big(T \in N_t(\mathrm{d}t) \,\big|\, T \geq t\big) \tag{1.8}$$

for any $t > 0$. Note that independence of random variables T and C implies that the condition (1.8) is satisfied, but condition (1.8) does not necessarily imply independence of T and C.

Noninformative censoring

Kalbfleisch and Prentice (2002) called the censoring mechanism *noninformative* if the censoring random variable C does not depend on any parameters used to model the distribution of the event time T.

Summarized, for independent noninformative censoring, the censoring procedure or rules may depend arbitrarily during the course of the study on:

- previous event times of other subjects in the study;
- previous censoring times of other subjects in the study;
- random mechanisms external to the study;
- values of covariates possibly included in the model;

but must not contain any information on the parameters used to model the event time. The independent noninformative censoring includes also *type I censoring* where censoring can only happen at a preplanned time.

1.4.2 Independent noninformative interval censoring

Consider now the case of interval-censored data where the observed intervals are generated by a triplet $(L, U, T)'$. That is, we observe an interval $\lfloor l, u \rfloor$ if $L = l$, $U = u$ and $T \in \lfloor L, U \rfloor$. Note that since the observed interval $\lfloor l, u \rfloor$ must contain the event time T, the support of the random vector $(L, U, T)'$ is equal to

$$\{(l, u, t) : 0 \le l \le t \le u \le \infty\}.$$

Using the arguments given by Kalbfleisch and MacKay (1979), Williams and Lagakos (1977), and Oller et al. (2004), it can be shown that the interval-censored counterpart of condition (1.8) is given by a so-called *constant-sum condition*

$$\iint\limits_{\{(l,\,u):\, t\in\lfloor l,\,u\rfloor\}} \frac{\mathsf{P}\big(L \in N_l(\mathrm{d}l),\, U \in N_u(\mathrm{d}u),\, T \in \lfloor l, u\rfloor\big)}{\mathsf{P}\big(T \in \lfloor l, u\rfloor\big)} = 1 \quad \text{for all } t > 0. \tag{1.9}$$

In the remainder of the book we will call interval censoring *independent* if it satisfies the constant-sum condition (1.9) and *noninformative* if the distribution of censoring random variables L and U does not depend on the parameters used to model the distribution of the event time T. Finally, it is useful to mention that Oller et al. (2004) use a slightly different definition of "noninformative."

Example 1.3 Emergence time of a permanent tooth
In Section 1.6.7, the Signal Tandmobiel study® is introduced where one of the events of interest is the emergence time of a particular permanent tooth and hence T is the age when the tooth emerges. The emergence status of the tooth was checked at preplanned visits performed at times $0 < C_1 < \ldots < C_M$ leading to interval censoring for the event time T. That is, the random variables L and U are defined as follows.

$$(L, U) = \begin{cases} (0, C_1), & \text{if } T \in (0, C_1\rfloor, \\ (C_1, C_2), & \text{if } T \in \lfloor C_1, C_2\rfloor, \\ \vdots & \\ (C_M, \infty), & \text{if } T \in \lfloor C_M, \infty). \end{cases}$$

As will be explained in Section 1.6.7, the visit times $\boldsymbol{C} = (C_1, \ldots, C_M)'$ in the Signal Tandmobiel study were pre-planned irrespective of the dental status of the children. Hence, the vector $\boldsymbol{C}$ is independent of the random variable T and the censoring mechanism is independent. In this case, one can formally show (Oller et al., 2004) that the constant-sum condition (1.9) is satisfied.

Note that the censoring mechanism would not necessarily be independent in a retrospective study involving patient records of a dental practice. In that case the timing of the visits often depends on whether the patient suffered from a tooth problem.

□

1.5 Frequentist inference

Statistical models describe the probabilistic mechanism that leads to observed data. It is usually assumed that this mechanism depends on a vector of unknown parameters, let us say $\boldsymbol{\theta}$, which represents the relevant information we wish to pick up from the observed data. The assumed probabilistic mechanism together with the observed data determines the *likelihood function*, $L(\boldsymbol{\theta}) \equiv L(\boldsymbol{\theta} \mid \text{data})$, which is the cornerstone to draw inference about the unknown parameter vector $\boldsymbol{\theta}$.

Two major paradigms exist in statistics of how to use the likelihood to draw inference on $\boldsymbol{\theta}$, namely the *frequentist* and the *Bayesian* paradigms. In both paradigms, the data are assumed to be a random sample from a population described by a distribution controlled by an unknown $\boldsymbol{\theta}$. In the classical frequentist point of view, $\boldsymbol{\theta}$ is fixed. Several methods are in use to estimate the true value of the parameter $\boldsymbol{\theta}$, *maximum likelihood* (ML) being one of the most popular ones. The maximum likelihood estimator (MLE), $\widehat{\boldsymbol{\theta}}$, maximizes the likelihood function over a set Θ of possible $\boldsymbol{\theta}$ values – called the parameter space. Hypotheses about the parameter vector $\boldsymbol{\theta}$ can be tested and the accuracy of the estimates can be assessed by a confidence interval (CI). We will elaborate on these methods in this section.

In Chapter 9, we introduce the reader to the basic concepts of the Bayesian approach.

1.5.1 Likelihood for interval-censored data

The likelihood plays a principal role in drawing inference about unknown model parameters. Let T_i $(i = 1, \ldots, n)$ be a set of independent event times for which probabilistic models $\mathsf{P}_{i,\boldsymbol{\theta}}$ are assumed, where $\boldsymbol{\theta}$ is a vector of unknown parameters and i represents the subjects. The likelihood $L(\boldsymbol{\theta})$ of an independent random sample is the product of individual likelihood contributions $L_i(\boldsymbol{\theta})$, i.e.,

$$L(\boldsymbol{\theta}) = \prod_{i=1}^{n} L_i(\boldsymbol{\theta}). \qquad (1.10)$$

In this section, we discuss the general form of the likelihood for interval-censored data.

Let $\lfloor l_i,\, u_i \rfloor$ be observed intervals and δ_i $(i = 1, \ldots, n)$ corresponding censoring indicators as defined in Table 1.1. As in Section 1.4.2, we assume that the observed intervals are generated by triplets $(L_i,\, U_i,\, T_i)'$. That is, the interval $\lfloor l_i,\, u_i \rfloor$ is observed if $L_i = l_i$, $U_i = u_i$ and $T_i \in \lfloor L_i,\, U_i \rfloor$. Strictly speaking, with interval-censored data, the likelihood contribution is given by the density of observables, i.e., by the joint distribution of $(L_i,\, U_i)'$ whose support is such that $T_i \in \lfloor L_i,\, U_i \rfloor$ with probability one. That is, the likelihood contribution of the observed interval $\lfloor l_i,\, u_i \rfloor$ is given by

$$L_{i,full}(\boldsymbol{\theta}) = \mathsf{P}_{i,\boldsymbol{\theta}}\big(L_i \in N_{l_i}(dl_i),\, U_i \in N_{u_i}(du_i),\, T_i \in \lfloor l_i,\, u_i \rfloor\big). \tag{1.11}$$

If the event time T_i is independent of $(L_i,\, U_i)'$ and if the distribution of $(L_i,\, U_i)'$ does not depend on $\boldsymbol{\theta}$ (*noninformative censoring*), it directly follows that $L_{i,full}$ is proportional to the so-called *simplified likelihood* contribution

$$L_i(\boldsymbol{\theta}) = \mathsf{P}_{i,\boldsymbol{\theta}}\big(T_i \in \lfloor l_i,\, u_i \rfloor\big), \tag{1.12}$$

where the contribution to the likelihood of possibly random L_i and U_i is ignored. Furthermore, it is shown in Oller et al. (2004) that the same is true under the assumption of independent noninformative censoring as specified in Section 1.4.2 where only the *constant-sum condition* (1.9) is assumed which is weaker than independence of T_i and $(L_i,\, U_i)'$. Consequently, inference on the event times T_i can be based on the simplified likelihood (1.12). In the remainder of the book, we will use the simplified likelihood for inference but will omit the adjective "simplified." However, in the concluding Chapter 14 we bring up the possible pitfalls of ignoring the possible relationship of the true event time with the observed interval.

When $\mathsf{P}_{i,\boldsymbol{\theta}}$ corresponds to a simple parametric model for independent observations given by a continuous distribution with density $f_i(t;\, \boldsymbol{\theta})$ and survival function $S_i(t;\, \boldsymbol{\theta})$, the individual likelihood contribution (1.12) can be written as

$$\begin{aligned} L_i(\boldsymbol{\theta}) &= \oint_{l_i}^{u_i} f_i(t;\, \boldsymbol{\theta}) \mathrm{d}t \\ &= \begin{cases} \int_{l_i}^{\infty} f_i(t;\, \boldsymbol{\theta})\, \mathrm{d}t = S_i(l_i;\, \boldsymbol{\theta}), & \delta_i = 0, \\ f_i(t_i;\, \boldsymbol{\theta}), & \delta_i = 1, \\ \int_{0}^{u_i} f_i(t;\, \boldsymbol{\theta})\, \mathrm{d}t = 1 - S_i(u_i;\, \boldsymbol{\theta}), & \delta_i = 2, \\ \int_{l_i}^{u_i} f_i(t;\, \boldsymbol{\theta})\, \mathrm{d}t = S_i(l_i;\, \boldsymbol{\theta}) - S_i(u_i;\, \boldsymbol{\theta}), & \delta_i = 3, \end{cases} \end{aligned} \tag{1.13}$$

where δ_i is defined in Table 1.1. In (1.13), we made use of the notation

$$\oint_{l}^{u} f(t)\, \mathrm{d}t = \begin{cases} \displaystyle\int_{l}^{u} f(t)\, \mathrm{d}t, & \text{if } l < u \\ f(l) = f(u), & \text{if } l = u, \end{cases} \tag{1.14}$$

i.e., the integral disappears whenever the event time is exactly observed.

1.5.2 Maximum likelihood theory

As we stated above, the method of maximum likelihood is one of the most popular approaches to estimation and inference in the frequentist context. The estimator, $\widehat{\boldsymbol{\theta}}$, maximizes the likelihood, or equivalently the log-likelihood $\ell(\boldsymbol{\theta}) = \log\{L(\boldsymbol{\theta})\}$ over a set $\Theta \subseteq \mathbb{R}^q$ – the parameter space. Large sample maximum likelihood theory is discussed in detail in, for example, Cox and Hinkley (1974), Lehmann and Casella (1998) or Casella and Berger (2002). Here we give a brief and simplified overview.

The ML theory states that with a correctly specified model and under mild regularity conditions, $\widehat{\boldsymbol{\theta}}$ is approximately normally distributed with mean equal to $\boldsymbol{\theta}_0$ – the true parameter value and a covariance matrix equal to

$$\widehat{\boldsymbol{V}} = \left\{ - \frac{\partial^2 \ell}{\partial \boldsymbol{\theta} \partial \boldsymbol{\theta}'}(\widehat{\boldsymbol{\theta}}) \right\}^{-1} = \left(\widehat{v}_{kl} \right)_{k,l} \quad (k, l = 1, \ldots, q). \tag{1.15}$$

That is, for an appropriately large sample size,

$$\widehat{\boldsymbol{\theta}} \sim \mathcal{N}_q(\boldsymbol{\theta}_0, \widehat{\boldsymbol{V}}), \tag{1.16}$$

or equivalently,

$$\left(\widehat{\boldsymbol{\theta}} - \boldsymbol{\theta}_0\right)' \widehat{\boldsymbol{V}}^{-1} \left(\widehat{\boldsymbol{\theta}} - \boldsymbol{\theta}_0\right) \sim \chi^2_q, \tag{1.17}$$

where χ^2_q is a chi-square distribution with q degrees of freedom.

Standard errors, confidence intervals and confidence regions

(Approximate) standard errors for estimated model parameters $\widehat{\boldsymbol{\theta}}$ are simply square roots of the diagonal elements of matrix $\widehat{\boldsymbol{V}}$, i.e.,

$$\widehat{\mathsf{SE}}\big(\widehat{\theta}_k\big) = \sqrt{\widehat{v}_{kk}} \qquad (k = 1, \ldots, q).$$

Wald $(1-\alpha)100\%$ confidence intervals for components θ_k $(k = 1, \ldots, q)$ of the parameter vector follow directly from normal approximation (1.16) and are given by

$$\left[\widehat{\theta}_k - \widehat{\mathsf{SE}}\big(\widehat{\theta}_k\big)\, \Phi^{-1}(1-\alpha/2),\; \widehat{\theta}_k + \widehat{\mathsf{SE}}\big(\widehat{\theta}_k\big)\, \Phi^{-1}(1-\alpha/2)\right], \tag{1.18}$$

where Φ is the cdf of a standard normal distribution. A simultaneous (approximate) $(1-\alpha)100\%$ confidence region Θ_α for a multivariate parameter vector $\boldsymbol{\theta}$ is an ellipsoid derived from chi-square approximation (1.17) and is given by

$$\Theta_\alpha = \left\{ \boldsymbol{\theta} :\, \big(\boldsymbol{\theta} - \widehat{\boldsymbol{\theta}}\big)' \widehat{\boldsymbol{V}}^{-1} \big(\boldsymbol{\theta} - \widehat{\boldsymbol{\theta}}\big) < \chi^2_q(1-\alpha) \right\},$$

where $\chi^2_q(1-\alpha)$ denotes the $(1-\alpha)100\%$ quantile of the χ^2_q distribution.

Delta method

When the maximum likelihood estimate for $\boldsymbol{\theta}$ is equal to $\widehat{\boldsymbol{\theta}}$, then it follows from ML theory that the maximum likelihood estimate of $\Psi = \Psi(\boldsymbol{\theta})$, a differentiable function, is equal to $\widehat{\Psi} = \Psi(\widehat{\boldsymbol{\theta}})$. In addition, the *delta method* can be used to calculate the (approximate) standard error of $\Psi(\widehat{\boldsymbol{\theta}})$, given by

$$\widehat{\mathsf{SE}}(\widehat{\Psi}) = \sqrt{(\nabla\widehat{\Psi})' \, \widehat{\boldsymbol{V}} \, \nabla\widehat{\Psi}},$$

where

$$\nabla\widehat{\Psi} = \frac{\partial\Psi}{\partial\boldsymbol{\theta}}(\widehat{\boldsymbol{\theta}}).$$

Wald hypothesis tests

To test the null hypothesis $\mathrm{H}_0 : \boldsymbol{\theta} = \boldsymbol{\theta}_0$, we may again use the chi-square approximation (1.17) based on test statistic $X^2 = (\widehat{\boldsymbol{\theta}} - \boldsymbol{\theta}_0)' \, \widehat{\boldsymbol{V}}^{-1} \, (\widehat{\boldsymbol{\theta}} - \boldsymbol{\theta}_0)$ and compute a P-value from the cdf $F_{\chi^2_q}$ of the χ^2_q distribution as

$$P = 1 - F_{\chi^2_q}(X^2).$$

Testing the univariate null hypothesis $\mathrm{H}_0 : \theta_k = \theta_{k,0}$, $k \in \{1, \dots, q\}$, can be done with the equivalent test statistic

$$Z = \frac{\widehat{\theta}_k - \theta_{k,0}}{\sqrt{\widehat{v}_{k,k}}} = \frac{\widehat{\theta}_k - \theta_{k,0}}{\widehat{\mathsf{SE}}(\widehat{\theta}_k)}. \tag{1.19}$$

A two-sided P-value based on normal approximation (1.16) can be calculated as

$$P = 2\Big\{1 - \Phi\big(|Z|\big)\Big\} = 2\,\Phi\big(-|Z|\big). \tag{1.20}$$

When $\theta_{k,0} \equiv 0$ in (1.19), a popular test is obtained which is part of the standard output in most statistical packages.

Likelihood ratio tests

The *likelihood ratio test* (LRT) is most popular to test the null hypothesis $\mathrm{H}_0 : \boldsymbol{\theta} \in \Theta_0$, where $\Theta_0 \subset \Theta$ is a subset of the original parameter space Θ obtained by imposing ν constraints on the elements of the parameter vector. Often, Θ_0 is obtained from Θ by constraining some of the parameter components to zero. In the context of model comparison, the likelihood ratio test can be applied to nested models, i.e., when the submodel characterized by the parameter space Θ_0 is a simplified version of the original model characterized by the parameter space Θ. Let $\widehat{\boldsymbol{\theta}}_0$ maximize the (log-)likelihood over the parameter space Θ_0, then $\widehat{\boldsymbol{\theta}}_0$ is the MLE of $\boldsymbol{\theta}$ for the simpler model. The LRT statistic

$$\Lambda = -2\Big\{\ell(\widehat{\boldsymbol{\theta}}_0) - \ell(\widehat{\boldsymbol{\theta}})\Big\} \tag{1.21}$$

follows under H_0 asymptotically a chi-square distribution with ν degrees of freedom. The P-value of the test is then calculated from

$$P = 1 - F_{\chi^2_\nu}(\Lambda).$$

Although being asymptotically equivalent, the LRT is preferred over the Wald test (see, e.g., Collett, 2003).

Score tests

One more standard possibility to test the null hypothesis $H_0 : \boldsymbol{\theta} \in \Theta_0$, where $\Theta_0 \subset \Theta$ is the parameter space obtained by imposing ν constraints on the original parameter space Θ, is the *score test*, often referred to also as the *Rao's score test* or the *Lagrange multiplier test.* Its advantage above the LRT is that it only requires calculation of $\widehat{\boldsymbol{\theta}}_0$, the MLE of $\boldsymbol{\theta}$ under the constrained parameter space Θ_0. Let

$$\widehat{\boldsymbol{U}}_0 = \frac{\partial \ell}{\partial \boldsymbol{\theta}}(\widehat{\boldsymbol{\theta}}_0), \qquad \widehat{\boldsymbol{J}}_0 = -\frac{\partial^2 \ell}{\partial \boldsymbol{\theta} \partial \boldsymbol{\theta}'}(\widehat{\boldsymbol{\theta}}_0)$$

denote the first derivative (the score) vector and minus the second derivatives matrix, respectively, of the log-likelihood evaluated at $\widehat{\boldsymbol{\theta}}_0$. Under H_0 and mild regularity conditions, the score test statistic

$$R = \widehat{\boldsymbol{U}}_0' \, \widehat{\boldsymbol{J}}_0^{-1} \, \widehat{\boldsymbol{U}}_0$$

follows asymptotically a chi-square distribution with ν degrees of freedom. Analogously to the LRT, the P-value of the score test is calculated from

$$P = 1 - F_{\chi^2_\nu}(R).$$

Model selection

Nonnested models can be evaluated by means of an information criterion. The information criteria are generally defined as

$$\mathrm{IC} = -2\ell(\widehat{\boldsymbol{\theta}}) + c\nu, \tag{1.22}$$

where ν is the number of free parameters of the model and c is a penalty term specific to the information criterion IC. For example, the most commonly used *Akaike's information criterion* (AIC, Akaike, 1974) is obtained by choosing $c = 2$. The *Bayesian information criterion* (BIC, Schwarz, 1978) is obtained with $c = \log(n)$, while the *Hannan-Quinn criterion* (HQC, Hannan and Quinn, 1979) makes use of $c = 2\log\{\log(n)\}$. Clearly, for $n \geq 8$, BIC is smaller than AIC. In all cases the model with the smallest criterion is preferred. Apart from that, there is no strict rule that can be used to claim that a model is genuinely better than another model. A rule of thumb derived from simulation studies specifies that the model with the smaller AIC is "preferred" when the difference in the criterion is more than 5 and is "strongly preferred" when the difference is greater than 10. Further, several simulation studies indicated that AIC tends to select larger models, while BIC prefers smaller models. The HQ criterion selects more intermediate models. Note that when all competing models have the same number of parameters, the best model is simply the model with the greatest maximized likelihood.

1.6 Data sets and research questions

Several examples of interval-censored data will be used throughout the book for illustrative purposes. The first data set is an example of right-censored data which will be used to introduce the survival concepts in Chapter 2. In addition, throughout the book we make use of three examples of univariate and three examples of multivariate interval-censored failure time data. The last data set also covers current status data. All data are electronically available in the R package icensBKL and on the accompanying web site of the book.

1.6.1 Homograft study

The reconstruction of the right ventricular outflow tract (a portion of the right ventricle of the human heart through which blood passes in order to enter the great arteries) is an essential part of the treatment in congenital heart disease. Cryopreserved homografts became the conduit of choice for this procedure. A homograft is a heart valve replacement using a human donor valve and has an excellent hemodynamic profile, is superior in suturing and tailoring and requires no anticoagulation. In a retrospective study, Meyns et al. (2005) analyzed the long-term outcome of homografts in the right ventricular outflow tract in order to identify the predictors of homograft failure and to identify the possible contribution of an immunological mechanism. Homograft failure was the event of interest. From 1989 to 2003, the right ventricular outflow tract was reconstructed with 301 homografts in 272 patients. All patients were routinely followed up and the last available follow-up point at time of the construction of the analysis data set was used as censoring time. Therefore, we can assume independent censoring. In this explorative study, one aimed to identify the predictors of homograft failure. We consider here only the first homograft in 272 patients (a subset of the total data set) with a few selected covariates, see Appendix A.1.

TABLE 1.3: Breast cancer study. Observed intervals in months for time to breast retraction of early breast cancer patients per treatment group.

Radiation therapy only (N=46)

$(0,5], (0,7], (0,8], (4,11], (5,11], (5,12], (6,10], (7,14], (7,16], (11,15], (11,18], (17,25], (17,25], (19,26], (19,35], (25,37], (26,40], (27,34], (36,44], (36,48], (37,44], (15,\infty), (17,\infty), (18,\infty), (22,\infty), (24,\infty), (24,\infty), (32,\infty), (33,\infty), (34,\infty), (36,\infty), (36,\infty), (37,\infty), (37,\infty), (37,\infty), (38,\infty), (40,\infty), (45,\infty), (46,\infty), (46,\infty), (46,\infty), (46,\infty), (46,\infty), (46,\infty), (46,\infty), (46,\infty)$

Radiation therapy and adjuvant chemotherapy (N=48)

$(0,5], (0,22], (4,8], (4,9], (5,8], (8,12], (8,21], (10,17], (10,35], (11,13], (11,17], (11,20], (12,20], (13,39], (14,17], (14,19], (15,22], (16,20], (16,24], (16,24], (16,60], (17,23], (17,26], (17,27], (18,24], (18,25], (19,32], (22,32], (24,30], (24,31], (30,34], (30,36], (33,40], (35,39], (44,48], (48,\infty), (11,\infty), (11,\infty), (13,\infty), (13,\infty), (13,\infty), (21,\infty), (23,\infty), (31,\infty), (32,\infty), (34,\infty), (34,\infty), (35,\infty)$

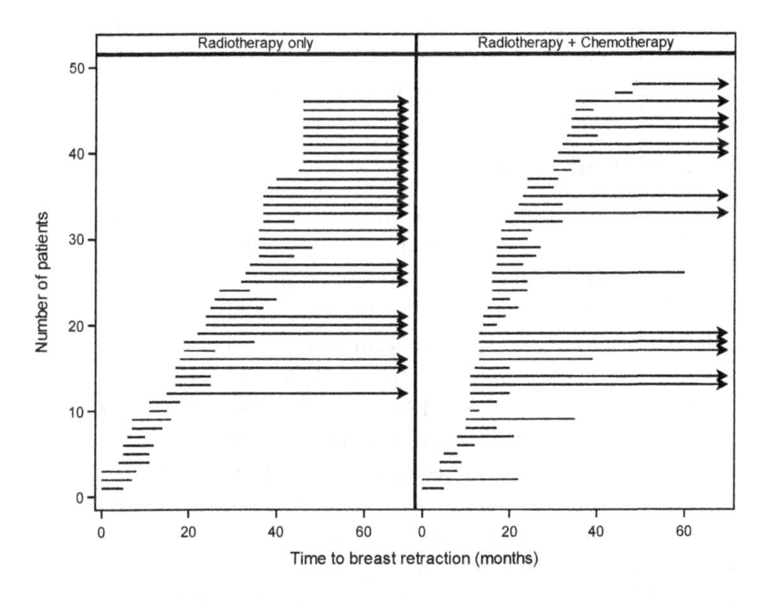

FIGURE 1.5: Breast cancer study. Observed intervals in months for time to breast retraction of early breast cancer patients per treatment group.

1.6.2 Breast cancer study

In a retrospective study on early breast cancer patients, the patients treated with primary radiation therapy and adjuvant chemotherapy were compared to those treated with radiotherapy alone for the cosmetic effects of their treatment (Beadle et al., 1984). It was of interest to evaluate the effect of the two-treatment regimen on the time until cosmetic deterioration, as determined by the appearance of breast retraction. Patients were seen at the clinic every 4 to 6 months. However, the actual time between visits increased after completion of the primary irradiation treatment, and for those patients who lived far away, follow-up intervals were often longer. The time between visits was considered unrelated to the development of retraction by the medical investigator. Consequently, independent censoring was assumed appropriate here.

The data presented in Table 1.3 are reproduced from Finkelstein and Wolfe (1985). It consists of the subset of 96 patients who were treated at the Joint Center for Radiation Therapy in Boston between 1976 and 1980. Forty-six patients were randomized to radiation therapy only regimen, while 48 patients to the radiation therapy and adjuvant chemotherapy regimen. The intervals represent the time period during which breast retraction occurred. For example, if an observation is (8, 12], then at 8 months, the patient had shown no deterioration in the initial cosmetic result, but by 12 months, retraction was present. A graphical representation of the data is shown in Figure 1.5.

The research question of interest entails the comparison of the two treatments with respect to time until cosmetic deterioration.

1.6.3 AIDS clinical trial

These data are obtained from a natural substudy of a comparative trial from the AIDS Clinical Trials Group (protocol ACTG 181). Two hundred and four patients were scheduled for clinic visits during follow-up and data were collected until either of the following two events happened: shedding of cytomegalovirus (CMV) in the urine and blood, or colonization of Mycobacterium Aviam Complex (MAC) in the sputum or stool. For some patients the survival times are left-censored because shedding or colonization already occurred before the patient was entered in the study, while for other patients the intervals are right-censored because shedding or colonization had not yet started before the end of the study. The remaining intervals are interval-censored between the last negative and first positive test of the respective samples. The data can be found in Appendix A.2 and are reproduced from Betensky and Finkelstein (1999b). A graphical representation of the data is shown in Figure 1.6.

In this study the relationship between CMV shedding and MAC colonization was of primary interest. There is no reason to expect one event time to precede the other. Knowledge of the joint distribution of these event times allows us to calculate the probability that CMV shedding precedes MAC colonization (or vice versa).

We will assume that censoring and failure times are independent and that censoring is noninformative. This assumption implies here that patients who miss a visit are neither more nor less likely to have the events of interest, which seems plausible since both events are not associated with symptoms in a patient. However, this assumption may be violated once an individual has experienced an event. The patient may then leave the trial, censoring the other event, or receive more intensive prophylactic treatments and thus be less likely to leave the trial.

1.6.4 Sensory shelf life study

According to the Institute of Food Science Technology (IFSI) Guidelines (1993), shelf life is defined as the time during which a food product will (1) remain safe, (2) retain desired sensory, chemical, physical and microbiological characteristics, and (3) comply with any label declaration of nutritional data, when stored under the recommended conditions. For the current study, the objective was to determine the sensory shelf life (SSL) of whole-fat, stirred yoghurt with strawberry pulp. A reversed storage design was used in which yoghurt pots of 150 g were kept at 4°C, and some of them were stored in a 42°C

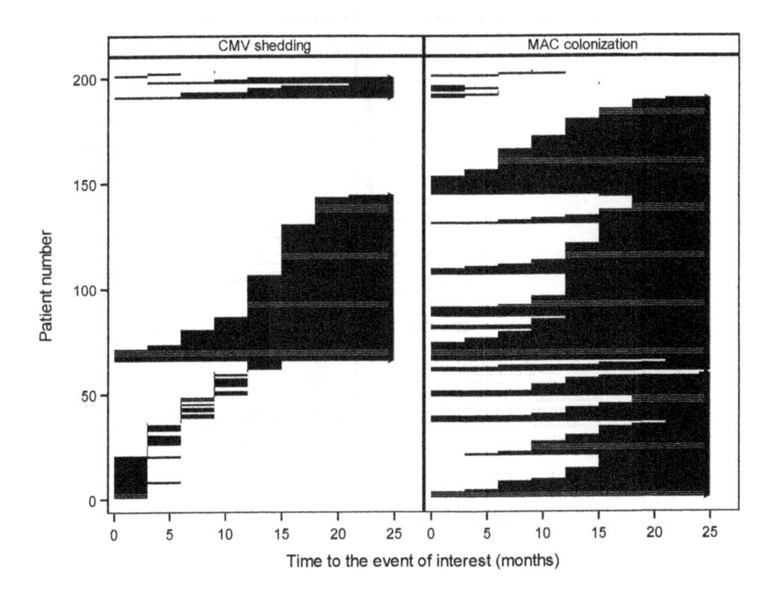

FIGURE 1.6: AIDS clinical trial. Observed intervals in months for time to CMV shedding and time to MAC colonization.

TABLE 1.4: Sensory shelf life study. Observed intervals in hours for shelf life of yoghurt stored at 42°C.

Adult (N=46)

$(0,4], (0,24], (0,24], (0,36], (0,36], (0,36], (4,8], (4,8], (4,8], (4,8], (4,8],$
$(4,24], (4,24], (4,24], (4,36], (4,48], (8,12], (8,12], (8,12], (8,12], (8,36],$
$(12,24], (12,24], (12,24], (12,24], (12,24], (12,48], (12,48], (12,48], (12,48],$
$(24,36], (24,36], (24,36], (24,36], (36,48], (36,48], (36,48], (36,48], (48,\infty),$
$(48,\infty), (48,\infty), (48,\infty), (48,\infty), (48,\infty), (48,\infty), (48,\infty)$

Child (N=47)

$(0,12], (0,12], (8,12], (8,36], (8,36], (8,48], (12,24], (12,48], (24,36], (24,36],$
$(24,36], (24,36], (24,36], (24,36], (36,48], (36,48], (36,48], (36,48], (36,48],$
$(36,48], (36,48], (48,\infty), (48,\infty), (48,\infty), (48,\infty), (48,\infty), (48,\infty), (48,\infty),$
$(48,\infty), (48,\infty), (48,\infty), (48,\infty), (48,\infty), (48,\infty), (48,\infty), (48,\infty), (48,\infty),$
$(48,\infty), (48,\infty), (48,\infty), (48,\infty), (48,\infty), (48,\infty), (48,\infty), (48,\infty), (48,\infty),$
$(48,\infty)$

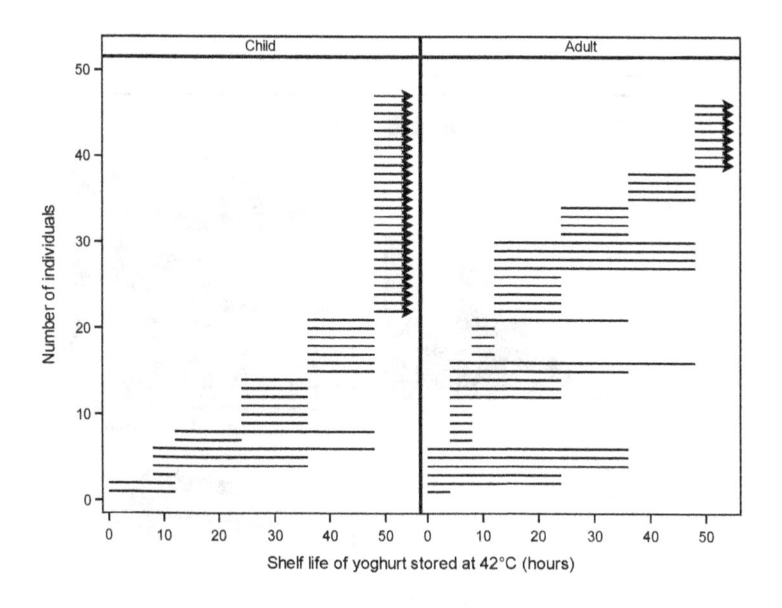

FIGURE 1.7: Sensory shelf life study. Shelf life of yoghurt stored at 42°C in hours.

oven for 0, 4, 8, 12, 24, 36 and 48 hours. These times were chosen because previous experiments showed that deterioration in flavor occurred quickly up to approximately 12 hours and then slowed down. After being stored at 42°C, the samples were refrigerated at 4°C until they were tasted. Previous microbiological analysis showed that samples were fit for consumption. More details about the experiment can be found in Hough (2010). Fifty adults between 18 and 30 years and 50 children between 10 and 12 years who consumed stirred yoghurt at least once a week were recruited from a town in Argentina. For each of the 7 samples presented in random order, the subject tasted the sample and answered the question: "Would you normally consume this product? Yes or No?". If a subject would consume the samples up to 8 hours' storage but not the samples with 12 hours' storage or longer, it is known that shelf life is somewhere between 8 and 12 hours storage. The data are thus interval-censored. Right-censored data occur when the subject accepts all samples and left-censored data if the sample with the first storage time is rejected. For subjects with inconsistent answers, several options to construct the interval are possible. Here, the widest uncertainty interval as to the storage time at which the subject rejects the yoghurt was applied. That is, from the first "yes" before a "no" until the last "no" which occurs after a "yes". Seven subjects were excluded from the analysis (4 adults and 3 children) because they preferred the stored product to the fresh product. The data presented in Table 1.4 are reproduced from Hough (2010). A graphical representation of the data is shown in Figure 1.7. It is here of interest to verify whether adults and children judge the shelf life of the yoghurt differently.

1.6.5 Survey on mobile phone purchases

In February 2013 a survey on mobile phone purchases was held in Finland among 15- to 79-year-old owners of a mobile phone. The participants were randomly sampled from a publicly available phone number directory by setting quotas in the gender, age and region of the respondents. A total of 536 completed interviews were recorded using a computer-assisted telephone interview (CATI) system. The amount of female owners but also 15- to 24-year-old owners were underrepresented in the data while male and 65- to 79-year-old owners were overrepresented in the study compared to the 2012 Finnish official statistics. The respondents were asked to answer questions about the purchase of their current and previous mobile phone and to report some family characteristics. More details about the survey may be found in Karvanen et al. (2014). The data were downloaded and adopted from `https://dvn.jyu.fi/dvn/dv/mobilephone`. We focused on the following questions:

- When did you purchase your mobile phone? (year and month; if month was not recalled, the season was asked)

- When did you purchase your previous mobile phone? (year and month; if month was not recalled, the season was asked)
- What is your gender? (male; female)
- What is your age group? (15–24; 25–34; 35–44; 45–54; 55–64; 65–79 years old)
- What is the size of your household? (1, 2, 3, 4, 5 or more persons; no answer)
- What is your household income before taxes? (30 000 or less; 30 001–50 000; 50 001–70 000; more than 70 000 euros; no answer)

The purchase times are interval-censored because only the month and not the day of purchase was asked. In addition, many respondents could not recall the time of purchase. For their current phone, 310 respondents were able to report the month and year of the purchase, an additional 115 were able to provide the season and year and 37 were not able to recall even the year. Out of 517 respondents who answered the questions about their previous phone, 117 were able to report the purchase month and year, an additional 91 were able to

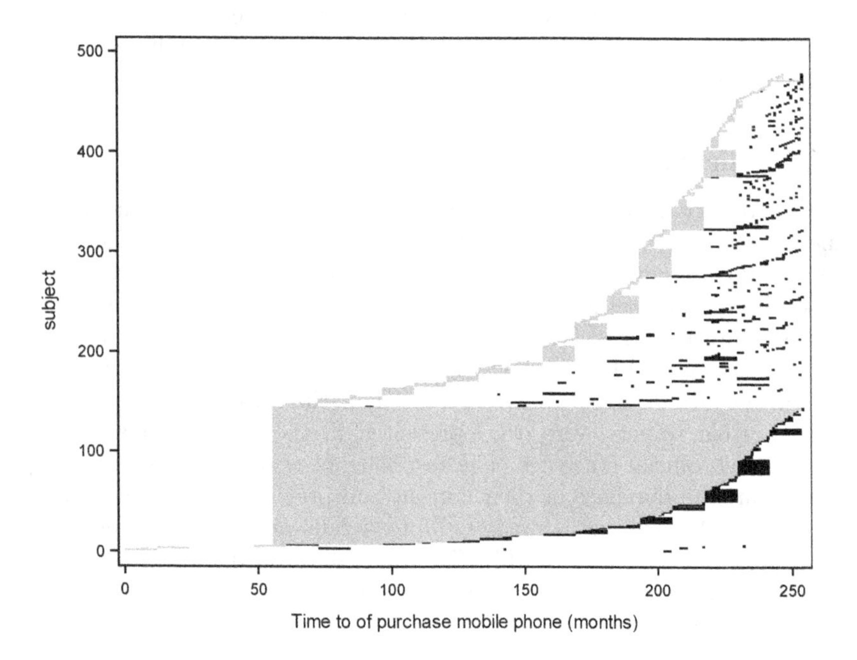

FIGURE 1.8: Mobile study. Interval-censored times of previous (gray line) and of current mobile phone purchase (black line).

report the season and year, 146 provided only the year and 163 were not able to recall even the year. A maximum purchase interval of 200 months was assumed when the purchase year is missing. The following respondents were excluded from the analysis: three respondents who reported their previous phone to have been bought after their current phone, 30 respondents for whom the purchase intervals of their previous and current phone are completely overlapping and 6 respondents who did not report their household size. As a result the data set used in our analyses contains 478 respondents, with 258 males and 220 females. The variables used in our analyses and some summary statistics of the included data can be found in Appendix A.3.

We are interested in the lifetime of the mobile phone which is defined as the difference between the purchase times of the current and previous mobile phones. Since the two purchase times are only known to happen in an interval, the time lapse between the two purchases is called doubly interval-censored. In Figure 1.8 we show the two interval-censored purchase times. Clearly, there is much more uncertainty of the timing of the first than of the second purchase. We then wished to examine whether gender, age and/or size of the household have an effect on the lifetime of a mobile phone.

1.6.6 Mastitis study

Mastitis in dairy cattle is the inflammation of the udder and the most important disease in the dairy sector of the western world. Mastitis reduces the milk production and the quality of the milk. For this study, 100 cows were included from the time of parturition (assumed to be free of infection). They were screened monthly at the udder-quarter level for bacterial infections. Since the udder quarters are separated, one quarter might be infected while other quarters remain free of infection. The cows were followed up until the end of the lactation period, which lasted approximately 300 to 350 days. Some cows were lost to follow-up, for example due to culling. Because of the approximately monthly follow-up (except during July/August for which only one visit was planned due to lack of personnel), data are interval-censored. Right-censored data are present when no infection occurred before the end of the lactation period or the cows are lost before the end of the follow-up. Two covariates were recorded. The first is the number of calvings, i.e., parity. This is a categorical cow-level covariate with the following categories: (1) one calving, (2) 2 to 4 calvings and (3) more than 4 calvings and is represented by two dummy variables (representing classes 2 and 3). The second covariate is the position of the udder quarter (front or rear). Both variables have been suggested in the literature to impact the incidence of mastitis (Adkinson et al., 1993; Weller et al., 1992). Data are adopted from Goethals et al. (2009). Some summary statistics are given in Appendix A.4. As visits were planned independently of infection times, independent noninformative censoring is a valid assumption. Under the assumption that data are recorded by side (e.g., first left, then right), a graphical representation of the data is shown in Figure 1.9.

1.6.7 Signal Tandmobiel study

The Signal Tandmobiel study was set up in 1996 to collect oral health data in Flanders (northern part of Belgium). A total of 4468 children (2315 boys and 2153 girls) born in 1989 were randomly selected through cluster (i.e., school) sampling, stratified by province and educational system such that each child had approximately equal probability to be selected. The recruited children represent about 7% of the children born in 1989 and living in Flanders.

Among other things, the Signal Tandmobiel project provides data on permanent tooth emergence for Flemish children from 7 to 12 years of age. As the children were examined annually at most six times, the emergence data are interval-censored. Tooth emergence was recorded at each examination by

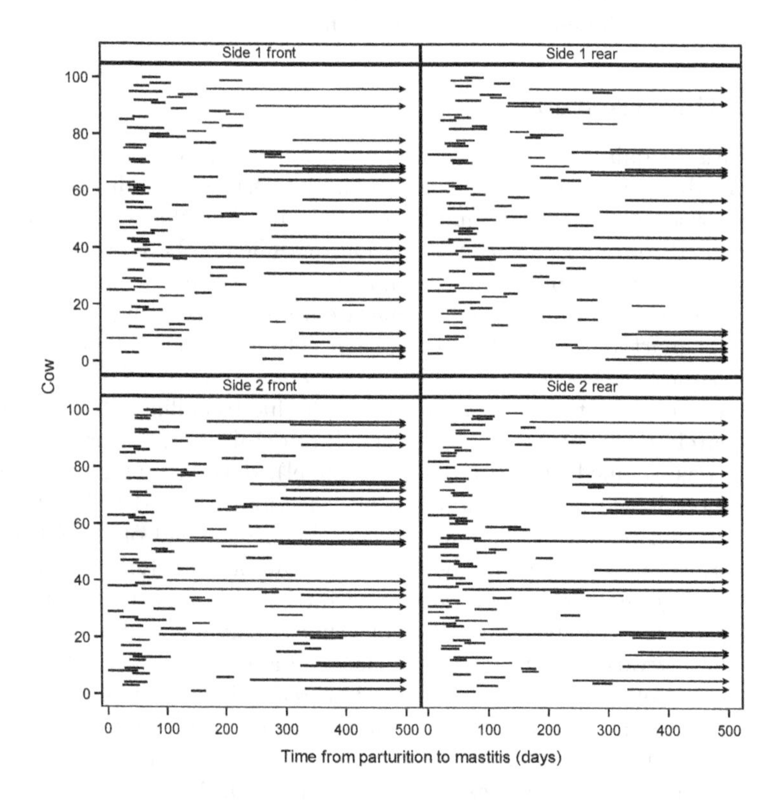

FIGURE 1.9: Mastitis study. Time from parturition to mastitis in days by location.

direct inspection. Every individual permanent tooth was scored according to its clinical eruption stage (adapted from Carvalho et al., 1989). For the current analysis, the status of tooth eruption was dichotomized into "not emerged" versus "emerged" (at least one cusp visible). Teeth extracted for orthodontic reasons were recorded as having emerged. No radiographs or additional dental records were available.

In addition, at each visit the status of each erupted tooth was recorded. Here the health status was dichotomized into "sound" versus "caries experience" (i.e., presence of caries, filled, extracted for reasons of caries), abbreviated as CE and also denoted as DMF (Decayed, Missing or Filled), being equal to 0 or 1. There was interest to find the risk factors for CE. To examine this, the period-at-risk must be determined which is the time between emergence and caries. Since both were measured in the study only on an annual basis, the begin and end measurements are interval-censored yielding then doubly interval-censored times-at-risk.

Several analyses were done on these data published in the dental and statistical literature. Much of the research was focused on individual teeth. To address the individual teeth we adopt here the Federation Dentaire Internationale (FDI) notation. This is a numbering system for deciduous (primary) and permanent teeth, illustrated in Figure 1.10.

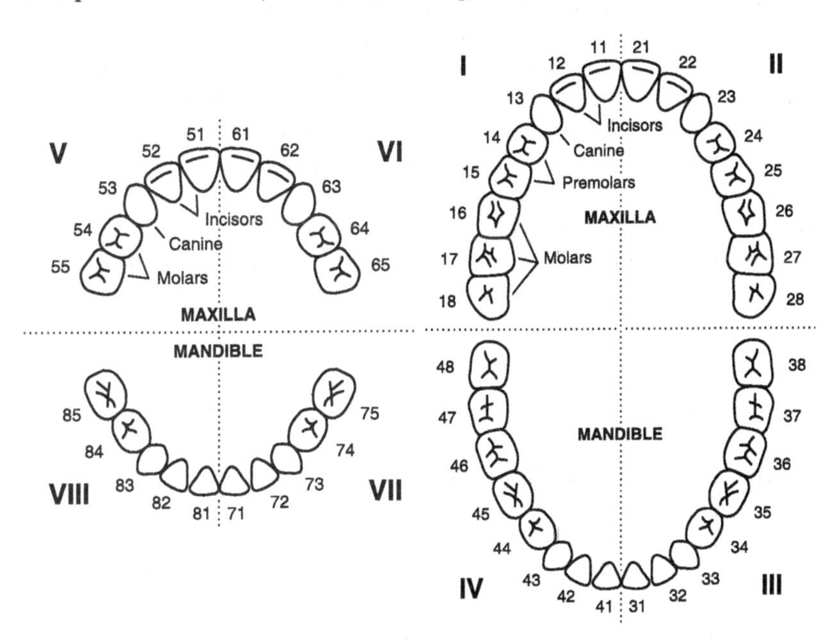

FIGURE 1.10: Signal Tandmobiel study. Federation Dentaire Internationale (FDI) numbering system of deciduous (left) and permanent (right) teeth.

TABLE 1.5: Signal Tandmobiel study. Censoring distribution for the teeth of the first quadrant for boys and girls.

Tooth	Boys (N=2315)			Girls (N=2153)		
	% censoring			% censoring		
number	left	interval	right	left	interval	right
11	49	45	6	62	34	4
12	9	77	14	21	68	11
13	0	39	61	0	56	44
14	1	56	43	0	68	32
15	1	37	62	0	47	53
16	83	15	2	89	10	1
17	0	19	81	0	29	71

Table 1.5 displays the censoring distribution for the teeth of first quadrant for boys and girls. The censoring distribution for the other quadrants are similar.

A random sample of 500 children (256 boys and 244 girls) was selected to serve as an example throughout the book. In addition to the interval-censored emergence times of teeth, a random visit was selected in order to create current status data of the emergence times for the same children. Also the time-to-caries of the four permanent first molars are included in the data set. Finally, the covariates gender, frequency of brushing, presence of sealants and occlusal plaque on the first permanent molars collected in the first year were included as potential confounders in the data set. Some summary statistics of the data are given in Appendix A.5.

As visits were planned independently of the time of emergence of the different teeth, independent noninformative censoring is a valid assumption.

1.7 Censored data in R and SAS

1.7.1 R

R (R Core Team, 2016) is a language and environment for statistical computing and graphics. It is a GNU project which is similar to the S language and environment which was developed at Bell Laboratories (formerly AT&T, now Lucent Technologies) by John Chambers and colleagues. R version 3.3.2 (2016-10-31) was used to produce the results presented in this book. For censored data, most standard survival approaches have been implemented in the by R recommended package survival (Therneau, 2017) which constitutes a part of a standard R distribution. The package is loaded into R using the command library("survival"). Additionally, methods for interval-censored data can be found in extension packages available through the *Comprehensive R Archive Network (CRAN)* on the Internet at http://cran.r-project.org or via the *Bioconductor* archive at http://www.bioconductor.org. Several of these packages are introduced throughout the book. Note that all packages have to be loaded first using the command library("NAME OF THE PACKAGE").

To supplement this book, an R package called icensBKL has been written. It contains the data introduced in Section 1.6 which are then easily accessible using the standard R command data. For example, the Signal Tandmobiel data described in Section 1.6.7 are loaded and summarized (output not shown) using the following code.

```
> library("icensBKL")
> data("tandmob", package = "icensBKL")
> summary(tandmob)
```

Further, the package icensBKL contains functions which complement the methods implemented in other R packages.

Some R example programs call other R programs. We assume that all files for a particular example are stored in the same directory. For these examples, the working directory should be set to this directory. When using RStudio (RStudio Team, 2016) as a user interface for R, this will be done automatically by running the following commands from the main program.

```
> library("rstudioapi")
> setwd(dirname(rstudioapi::getActiveDocumentContext()$path))
```

Alternatively, the working directory must be set by hand providing the correct path.

```
> setwd("C:/temp/ICbook/programs/ChapterXX/")
```

Many functions for censored data assume that the response is a Surv object. The specification of this object depends on the type of data.

Exact and right-censored data: When all event times are either exactly observed or right-censored and stored in a vector time, the Surv object is specified as

```
Surv(time, event)
```

where event is a vector of status indicators. One of the following codings can be used to indicate a right-censored or exactly observed event time, respectively: 0/1 (0=right-censored), 1/2 (1=right-censored), FALSE/TRUE (FALSE=right-censored).

This survival object will be used for the homograft data introduced in

Section 1.6.1. The variable timeFU determines exactly observed or right-censored times to the homograft failure. The censoring indicator is given by the variable homo.failure. The survival object is created and printed using the following code.

```
> library("survival")
> library("icensBKL")
> data("graft", package = "icensBKL")
> SobjGraft <- Surv(graft$timeFU, graft$homo.failure)
> print(SobjGraft)
```

```
 [1]  3.077+  3.477+  3.422+  0.016+  0.214+  1.362+
 [7]  0.137+  0.290+  0.274+  1.553   0.118+  0.701
[13]  0.866   1.674+ 11.770+  3.493+ 11.658+  2.586+
...
```

Note that the print function marks the right-censored observations by a plus sign.

Interval-censored data, possibility 1: The first possibility to specify the Surv object for general interval-censored data is

```
Surv(time, time2, event, type = "interval")
```

The vector time contains observed values for exact times, left- and right-censored times and lower limits l of observed intervals where $0 < l < u < \infty$. The vector time2 contains an arbitrary value (e.g., NA) for exact times, left- and right-censored times and contains the upper limits u of observed intervals where $0 < l < u < \infty$. The argument event indicates the type of observation: 0 for right-censored time $(0 < l < u = \infty)$, 1 for exact event time $(0 < l = u < \infty)$, 2 for left-censored time $(0 = l < u < \infty)$, 3 for observed interval $(0 < l < u < \infty)$, i.e., in the same way as indicated in Table 1.1. Table 1.6 displays this possibility on the left-hand side for the 4 different types of censoring (right, exact, left, interval).

Interval-censored data, possibility 2: Alternatively, the Surv object for general interval-censored data can be specified as

```
Surv(time, time2, type = "interval2")
```

in which case the representation introduced in Section 1.2.2 with $0 \leq l \leq u \leq \infty$ is used with zero and infinity replaced by NA. That is, the vector time contains the lower limits l of intervals (zeros replaced by NA) and the vector time2 contains the upper limits u of intervals (infinities replaced by NA). Table 1.6 displays this possibility on the right-hand side for the 4 different types of censoring (right, exact, left, interval).

TABLE 1.6: Two possibilities of specifying the 4 types of observations (right, exact, left, interval) in R.

Observation	**Surv(time,time2,event, type="interval")**			**Surv(time,time2, type="interval2")**	
	time	time2	event	time	time2
$[5, \infty)$	5	NA	0	5	NA
$[2, 2]$	2	NA	1	2	2
$(0, 3]$	3	NA	2	NA	3
$[4, 6]$	4	6	3	4	6

For example, the survival object for the interval-censored emergence time of tooth 44 in the Signal Tandmobiel data introduced in Section 1.6.7 is created in this way as indicated in the following code.

```
> library("survival")
> library("icensBKL")
> data("tandmob", package = "icensBKL")
> SobjTooth44 <- Surv(tandmob$L44, tandmob$R44,
                      type = "interval2")
> print(SobjTooth44)
```

```
 [1] 12.183436+          [ 9.040383,  9.943874]
 [3] 12.279261+          [ 9.552361, 10.354552]
 ...
[67]  7.739904-          [ 9.218344, 10.138261]
 ...
```

In this case, right-censored observations are marked by a plus sign, left-censored observations by a minus sign and intervals by closed brackets. Note that the closed intervals are produced by the print function. However, it does not necessarily mean that the observed intervals are treated as closed by all functions that deal with interval-censored data.

Models are described in R using formulae like y $\sim$ a + b. The variable y on the LHS of the tilde ('$\sim$') is the response (or dependent) variable and is often a Surv object in survival modelling. The terms on the RHS are a symbolic description of the model. Such a model consists often of a series of variables separated by + operators. Interactions between two variables are indicated with a colon, for instance a:b represents the interaction between the variables a and b. The * operator denotes factor crossing: a*b is interpreted as a + b + a:b. Also arithmetic expressions on variables like log(y) $\sim$ a + log(x) are allowed. For a complete overview of all allowed operators we refer to the R documentation.

TABLE 1.7: Specifications of interval-censored observations using SAS variables lower and upper.

Lower	Upper	Comparison	Interpretation
not missing	not missing	lower=upper	exact observation
		lower<upper	interval-censored observation
		lower>upper	observation not used
missing	not missing		left-censored observation
not missing	missing		right-censored observation
missing	missing		observation not used

1.7.2 SAS

SAS is a commercial software package which once stood for "statistical analysis software". Nowadays the statistical component is only a small part of the complete SAS system used for business intelligence and analytical software. The SAS analyses in this book are based on SAS version 9.4 TS1M3 for Windows with SAS/STAT® software version 14.1 and SAS/QC® software version 14.1. Most code will probably work under Unix or Linux as well.

For censored data, SAS has several built-in procedures and user-written macros. For exact and right-censored data, data are given by a response variable, which contains the failure/event or censoring time, and a censoring variable, which indicates whether the observation is censored or not. For interval-censored data, let two variables lower and upper contain the left and right endpoints of the censoring interval, respectively. Table 1.7 lists the different possibilities with their interpretation for the built-in procedures. Please note that user-defined macros may not necessarily adhere to these guidelines. We will indicate when this happens.

Finally, for the SAS code presented in this book we adopted the following rule: (a) SAS specific terms and options are capitalized; (b) names that you can choose (like for instance variable and data set names) are given in mixed or lower case. Note, however, that in contrast with R, SAS is not case sensitive.

All macros provided with the book should be placed into one directory. As an example we use the directory "C:/SASmacros" but it can be replaced by the directory of your choice. Note that we used a forward slash to indicate the path. This coding works both under Windows as well as Unix or Linux. The usual backward slash used in Windows may not work in other operating systems. The AUTOCALL facility allows the availability of macro definitions to the SAS session via the setting of the MAUTOSOURCE option (on by default) and the macro directory to be included in the SASAUTOS search path.

```
OPTIONS SASAUTOS = (SASAUTOS, "C:/SASmacros");
```

Alternatively, a specific macro (e.g., MacroToBeLoaded.sas) could be loaded first with an include statement.

```
%inc "C:/SASmacros/MacroToBeLoaded.sas";
```

Chapter 2

Inference for right-censored data

We now review frequentist non- and semiparametric survival methods and software for right-censored data to familiarize the reader with some methods that will be generalized to interval-censored data in the next chapters. In this chapter we only discuss the standard approaches to analyze right-censored data, and therefore the chapter can easily be omitted if the reader is familiar with survival analysis. The chapter is, though, useful as reference to methods for interval-censored observations. In this respect, we discuss the Kaplan-Meier estimate of the survival curve and statistical methods to compare nonparametrically two survival curves. Besides the immense popular Cox proportional hazards model, we also discuss the less commonly used accelerated failure time model. The survival concepts will be illustrated using the homograft data set introduced in Section 1.6.1.

2.1 Estimation of the survival function

2.1.1 Nonparametric maximum likelihood estimation

For right-censored data, the nonparametric maximum likelihood estimator (NPMLE) of a survival function is given by the well-known *Kaplan-Meier* (KM) curve (Kaplan and Meier, 1958) also called *product-limit estimator* and has been extensively studied in the literature.

The Kaplan-Meier estimator and its standard error are constructed as follows. Let $t_1 < t_2 < \ldots < t_D$ represent the D distinct observed event times. For each $k = 1, \ldots, D$, let n_k be the number of subjects still at risk just prior to t_k. Let d_k be the number of subjects that fail at t_k, and let $s_k = n_k - d_k$. The product-limit estimate of the survival distribution function at t_k is then the cumulative product

$$\widehat{S}(t_k) = \prod_{j=1}^{k} \left(1 - \frac{d_j}{n_j}\right). \tag{2.1}$$

Note that no jump is made at a censoring time. The estimate of the standard error is based on the asymptotic variability of $\widehat{S}(t_k)$ given the observed event times and is computed using *Greenwood's formula* (Greenwood, 1926) as

$$\widehat{\mathsf{SE}}\{\widehat{S}(t_k)\} = \widehat{S}(t_k)\sqrt{\sum_{j=1}^{k} \frac{d_j}{n_j s_j}}. \tag{2.2}$$

Confidence intervals based on the asymptotic normality of $\widehat{S}(t)$ might include impossible values outside the range $[0, 1]$ at extreme values of t. This problem can be avoided by applying the asymptotic normality property to a transformation of $\widehat{S}(t)$ for which the range is unrestricted. Borgan and Liestøl (1990) have shown that confidence intervals based on $g\{\widehat{S}(t)\}$ can perform better than the usual linear confidence intervals for some choices of g. Using the delta method, the standard error of $g\{\widehat{S}(t)\}$ is then estimated by

$$\widehat{\tau}(t) = \widehat{\mathsf{SE}}\{g(\widehat{S}(t)\} = g'\{\widehat{S}(t)\}\widehat{\mathsf{SE}}\{\widehat{S}(t)\}, \tag{2.3}$$

where g' is the first derivative of the function g. The 100(1 - α)% confidence interval for $S(t)$ is given by

$$g^{-1}\Big[g\{\widehat{S}(t)\} \pm z_{\frac{\alpha}{2}}\widehat{\tau}(t)\Big], \tag{2.4}$$

where g^{-1} is the inverse function of g and $z_{\frac{\alpha}{2}}$ the $100(1 - \alpha/2)\%$ critical value for the standard normal distribution. Often used transformations are log-log (Kalbfleisch and Prentice, 2002), arcsine-square root (Nair, 1984), logit (Meeker and Escobar, 1998) and log. Table 2.1 shows the different transformations and their availability in R and SAS.

Nelson and Aalen suggested another popular estimator of the survival function given by

$$\widetilde{S}(t_k) = \prod_{j=1}^{k} \exp\left(-\frac{d_j}{n_j}\right) = \exp\left(-\sum_{j=1}^{k}\frac{d_j}{n_j}\right), \tag{2.5}$$

which originates from the relationship (1.7) between the survival and cumulative hazard function. The estimator of the cumulative hazard function is given by

$$\widetilde{H}(t_k) = \sum_{j=1}^{k} \frac{d_j}{n_j}. \tag{2.6}$$

This estimator has been studied in different generality by Nelson (1969, 1972) and Aalen (1978). The Nelson-Aalen estimator (2.5) of the survival function has been examined by Altshuler (1970) and Fleming and Harrington (1984).

Example 2.1 Homograft study

We now illustrate the above described estimators on the homograft data described in Section 1.6.1. Several factors influence the survival of the homograft.

TABLE 2.1: Transformations applied to compute the confidence interval of $\widehat{S}(t)$ and their availability in R and SAS as option.

Transformation	$g(x)$	R	SAS
Linear	x	plain	linear
Log-Log	$\log\{-\log(x)\}$	log-log	loglog*
Log	$\log(x)$	log*	log
Logit	$\log\left(\dfrac{x}{1-x}\right)$	N.A.	logit
Arcsine-Square Root	$\sin^{-1}(\sqrt{x})$	N.A.	asinsqrt

Note: * denotes the default value, N.A.= Not Available.

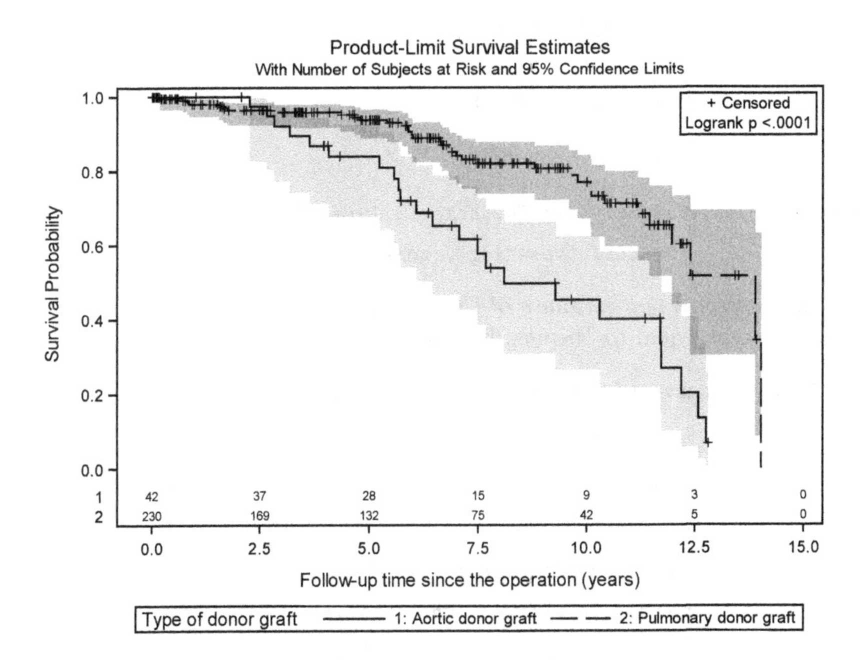

FIGURE 2.1: Homograft study. Kaplan-Meier curve (with pointwise confidence limits indicated by shaded regions) of homograft failure according to type of graft produced by SAS procedure LIFETEST.

We examine here the impact of the type of donor graft. Two different types were used: aortic homograft (AH group: 42 patients) and pulmonary homograft (PH group: 230 patients). Figure 2.1 shows the Kaplan-Meier curve for both types of homografts.

□

2.1.2 R solution

The Kaplan-Meier estimate together with its confidence limits is computed using the function survfit function from the R package survival. This function computes either a single survival curve or a survival curve for each of the subgroups defined by a factor variable as will be shown in the example below. The plot function provides basic plots of fitted Kaplan-Meier curves.

Example 2.1 in R

Having loaded the data and the survival *package, the Kaplan-Meier curves for both groups are computed using the* survfit *command as*

```
> data("graft", package="icensBKL")
> library("survival")
> Sfit <- survfit(Surv(timeFU, homo.failure) ~ Hgraft, data=graft)
```

The third command is equivalent to

```
> Sfit <- survfit(Surv(timeFU, homo.failure) ~ Hgraft, data=graft,
                  conf.type="log", conf.int=0.95)
```

The Nelson-Aalen estimate of the survival function is computed with the argument type *equal to* "fleming-harrington"*:*

```
> SfitNA <- survfit(Surv(timeFU, homo.failure) ~ Hgraft, data=graft,
                    type="fleming-harrington")
```

A survival curve for only the group with pulmonary grafts, is obtained using

```
> graft.PH <- subset(graft, Hgraft=="PH")
> Sfit.PH <- survfit(Surv(timeFU, homo.failure) ~ 1, data=graft.PH)
```

A basic summary table with the sample sizes, observed numbers of events, estimated median survival times in both groups and their 95% CI is obtained using

```
> print(Sfit)
```

```
Call: survfit(formula = Surv(timeFU, homo.failure) ~ Hgraft,
          data = graft)

             n events median 0.95LCL 0.95UCL
Hgraft=PH 230     36   13.9   11.97      NA
Hgraft=AH  42     23    8.1    7.07    12.6
```

The upper limit of the 95% CI for the survival median in the pulmonary graft group is defined as the first time t at which the upper limit of the CI for S(t) falls below 0.5. Since the output below shows there is no such time t in the pulmonary graft group, the upper limit is undefined.

The summary *command provides the Kaplan-Meier survival estimates and 95% CIs for each group.*

```
> summary(Sfit)
```

```
Call: survfit(formula = Surv(timeFU, homo.failure) ~ Hgraft,
    data = graft)

                Hgraft=PH
  time n.risk n.event survival std.err lower 95% CI upper 95% CI
  0.178    209      1    0.995 0.00477        0.986        1.000
  0.701    194      1    0.990 0.00698        0.976        1.000
  0.811    191      1    0.985 0.00866        0.968        1.000
  ...
 11.447     22      1    0.652 0.06198        0.541        0.786
 11.970     13      1    0.602 0.07481        0.472        0.768
 12.403      7      1    0.516 0.10224        0.350        0.761
 13.888      3      1    0.344 0.15613        0.141        0.837
 14.025      1      1    0.000     NaN           NA           NA

                Hgraft=AH
  time n.risk n.event survival std.err lower 95% CI upper 95% CI
  2.27     38       1    0.974  0.0260       0.9241        1.000
  2.66     37       1    0.947  0.0362       0.8790        1.000
  2.84     36       1    0.921  0.0437       0.8392        1.000
  ...
 12.59      3       1    0.134  0.0838       0.0394        0.457
 12.76      2       1    0.067  0.0633       0.0105        0.426
```

In the PH group, the first event ($d_1 = 1$) *was observed at* $t_1 = 0.178$*, at which there were* $n_1 = 209$ *subjects at risk. Thus a drop of* $1/209 \doteq 0.005$ *in the KM curve is seen, and* $\widehat{S}(t_1) = 1 - 0.005 = 0.995$.

A plot of the KM curves, without confidence limits nor symbols for censored observations (not shown) is obtained using

```
> plot(Sfit, mark.time=FALSE, lty=c(5, 1),
+       xlab="Follow-up time sine the operation (years)",
+       ylab="Survival Probability")
> legend(0, 0.6,
+          legend=c("Pulmonary donor graft", "Aortic donor graft"),
+          lty=c(5, 1), bty="n")
```

Colored curves (including confidence limits) for different strata are obtained (plot not shown) using

```
> plot(Sfit, conf.int=TRUE, lty=c(1, 2), col=c("darkgreen", "red"))
```

□

2.1.3 SAS solution

SAS software provides the Kaplan-Meier estimate with the procedures PROC LIFETEST, PROC LIFEREG and PROC RELIABILITY. The LIFETEST procedure provides a nonparametric estimate of the survival function only for right-censored data. The procedure also provides tests to compare survival curves (see Section 2.2) and rank tests for the association of the failure time variable with covariates (see Section 2.3). The LIFEREG and RELIABILITY procedures are mainly used for fitting parametric survival models for all types of censoring (see Sections 2.3 and 3.2). The two procedures also calculate the NPMLE of the survival function. The KM curve is not produced, but rather probability plots to compare the NPMLE solution with a parametric fit (see Section 2.3). We illustrate here only the LIFETEST procedure, the other two procedures are treated in Section 2.3.

Example 2.1 in SAS

The SAS *code to compute the NPMLE of $S(t)$ with the* LIFETEST *procedure is:*

```
ODS GRAPHICS ON;
PROC LIFETEST DATA=graft METHOD=KM PLOTS=(SURVIVAL) OUTSURV=ci;
TIME timeFU*homo_failure(0);
STRATA Hgraft;
RUN;
ODS GRAPHICS OFF;
```

The TIME *statement is used to indicate the failure or censoring time and the censoring status variables. In our example these are the variables* timeFU *and* homo_failure, *respectively. The zero in parentheses in* homo_failure*(0) corresponds to a right-censored observation. Note that more than one censoring code separated by commas is allowed. In our example, observations with the value 0 for* homo_failure *are censored. The* STRATA *statement identifies the variables that determine the strata levels. As we are interested in the survival curve for both types of homografts, we used the treatment variable* Hgraft *as stratum indicator. The argument* METHOD *in the* LIFETEST *statement is by default* KM *(Kaplan-Meier). To produce the Kaplan-Meier curve, we need to add* PLOTS= SURVIVAL *(or its abbreviation* S*). From* SAS *version 9.2 onwards,* PROC LIFETEST *has also* ODS GRAPHICS *capabilities which provides high quality graphics. For further details and options, we refer to the* SAS *documentation manual. Figure 2.1 shows the Kaplan-Meier curve for both types of homografts.*

The option OUTSURV=ci *produces pointwise 100(1 - α)% confidence limits for $S(t)$ in a data set named* ci. *By default, CIs are based on a log-log transformation, but other transformations can be chosen (Table 2.1 for the* CONFTYPE= *argument in the* LIFETEST *statement). The significance level α is specified in the* LIFETEST *statement by* ALPHA= α, *which is by default equal to 0.05.*

Adding the keyword NELSON *in the* LIFETEST *statement adds the Nelson-Aalen estimator to the output.*

□

2.2 Comparison of two survival distributions

2.2.1 Review of significance tests

We now compare the survival distributions S_1 and S_2 of two independent groups. This can be done by testing the null hypothesis $\mathrm{H}_0 : S_1 = S_2$. With right-censored data, the classical and most commonly used nonparametric method to compare survival distributions was derived by Mantel (1966) and is known as the *log-rank test*. This test is closely related to the *Mantel-Haenszel statistic* for stratified contingency tables and is based on the following idea.

Let $t_1 < t_2 < \ldots < t_J$ represent the J distinct observed (uncensored) event times in a combined sample obtained by merging the survival times from both groups. Let n_{j1} and n_{j2} be the numbers of subjects still at risk just prior to t_j in group 1 and group 2, respectively. Further, let d_{j1} and d_{j2} be the observed numbers of subjects that fail at t_j in the two groups. Finally, let $n_j = n_{j1} + n_{j2}$, $d_j = d_{j1} + d_{j2}$, for $j = 1, \ldots, J$. Each event time t_j is then associated with a 2 × 2 contingency table (see Table 2.2) where "survived" means that the subject survived beyond time t_j.

The random variable corresponding to d_{js}, D_{js} $(s = 1, 2)$, has a binomial

TABLE 2.2: Observed counts at time t_j to compute the log-rank test.

	Failed	Survived	
Group 1	d_{j1}	$n_{j1} - d_{j1}$	n_{j1}
Group 2	d_{j2}	$n_{j2} - d_{j2}$	n_{j2}
	d_j	$n_j - d_j$	n_j

distribution with n_{js} trials and a success probability equal to the failure rate λ_{js}. Hence $\mathrm{H}_0 : S_1 = S_2$ is equivalent to $\mathrm{H}_0 : \lambda_{j1} = \lambda_{j2}$ $(j = 1, \ldots, J)$. Under H_0, the two binomial probabilities are estimated from the combined sample as $\widehat{\lambda}_j = \widehat{\lambda}_{j1} = \widehat{\lambda}_{j2} = d_j/n_j$ $(j = 1, \ldots, J)$. Equality of two binomial probabilities can be tested with Fisher's exact test which is based on the conditional hypergeometric distribution of D_{j1} given d_j. Under the null hypothesis, this conditional distribution has mean E_{j1} and variance V_{j1} given by

$$E_{j1} = d_j \frac{n_{j1}}{n_j} = n_{j1}\,\widehat{\lambda}_j, \quad V_{j1} = d_j \frac{n_{j1}\,n_{j2}}{n_j^2}\,\frac{n_j - d_j}{n_j - 1} \quad (j = 1, \ldots, J). \tag{2.7}$$

Combining the information from the J 2×2 tables leads to the statistic

$$U_1 = \sum_{j=1}^{J} (D_{j1} - E_{j1}), \tag{2.8}$$

which has zero mean under H_0. Although the differences $D_{j1} - E_{j1}$ $(j = 1, \ldots, J)$ are not independent, it can be shown that the variance $\mathsf{var}(U_1)$ is well approximated by $\sum_{j=1}^{J} V_{j1}$ and that the statistic

$$G_1 = \frac{\sum_{j=1}^{J} (D_{j1} - E_{j1})}{\left(\sum_{j=1}^{J} V_{j1}\right)^{1/2}} \tag{2.9}$$

follows asymptotically a standard normal distribution under H_0. Equivalently, G_1^2 follows asymptotically a χ_1^2 distribution under H_0. The statistic G_1 can also be derived as the rank statistic from the logarithm of the Nelson–Aalen estimate of the survival function and is therefore known as the *log-rank test statistic*.

It is seen that the log-rank test statistic (2.9) assigns the same weight to Table 2.2 for $j = 1, \ldots, J$. That is, early and late failures contribute equally to the overall comparison of the two survival distributions. Note also that the size of the risk set, n_j, decreases with j and hence Table 2.2 based on different sample sizes contributes equally to the overall comparison with the log-rank test statistic. The *weighted log-rank test* with test statistic and weights $w_1, \ldots, w_J$

$$G_{W,1} = \frac{\sum_{j=1}^{J} w_j\,(D_{j1} - E_{j1})}{\left(\sum_{j=1}^{J} w_j^2\,V_{j1}\right)^{1/2}} \tag{2.10}$$

TABLE 2.3: Homograft study. Two-sample tests comparing aortic donor grafts with pulmonary donor grafts, obtained with R.

Test	Test statistic $G^2_{W,1}$	P-value
Log-rank	16.9	< 0.0001
Gehan-Wilcoxon	11.3	0.0008
Peto-Prentice-Wilcoxon	15.8	< 0.0001

has been proposed to improve the power of the Mantel's log-rank test for different alternatives. For example, weights $w_j = n_j/n$ lead to Gehan's generalization of the Wilcoxon test (Gehan, 1965b) which emphasizes early hazard differences. More generally, weights $w_j = w_j^\varrho = (n_j/n)^\varrho$ for $\varrho \geq 0$ lead to a class of weighted log-rank tests suggested by Tarone and Ware (1977).

With $\widehat{S}$ the Kaplan-Meier estimate of the survival distribution based on a combined sample, the weights $w_j = \widehat{S}(t_j-)$ result in the Peto-Prentice generalization of the Wilcoxon test (Peto and Peto, 1972; Prentice, 1978). A more general class, called G^ϱ class, with $w_j = w_j^\varrho = \left\{\widehat{S}(t_j-)\right\}^\varrho$ ($\varrho \geq 0$) was proposed by Harrington and Fleming (1982), and subsequently, the so-called $G^{\varrho,\gamma}$ ($\varrho \geq 0, \gamma \geq 0$) class with $w_j = w_j^{\varrho,\gamma} = \left\{\widehat{S}(t_j-)\right\}^\varrho \left\{1 - \widehat{S}(t_j-)\right\}^\gamma$ was proposed by Fleming and Harrington (1991, Chapter 7). With appropriate ϱ and γ parameters, the test is sensitive against alternatives of early, middle or late hazard differences.

An alternative broad class of methods for comparing two survival distributions was proposed by Pepe and Fleming (1989, 1991). Their test statistic is based on a weighted Kaplan-Meier estimate of the survival function in the pooled sample. For alternatives with crossing hazards their class of tests outperforms the weighted log-rank tests described above.

Example 2.2 Homograft study

In Example 2.1 we estimated the survival curve for both types of donor grafts. Now we wish to assess whether the difference between the two survival curves is statistically significant. In total, there are $J = 58$ distinct event times with $\sum_{j=1}^J d_{j1} = 36$ observed events in the pulmonary donor graft group and $\sum_{j=1}^J d_{j2} = 23$ observed events in the aortic donor graft group. The total of expected events computed using (2.7) is $\sum_{j=1}^J E_{j1} = 48.1$ in the pulmonary donor graft group and $\sum_{j=1}^J E_{j2} = 10.9$ in the aortic donor graft group. For the log-rank test statistic G_1^2 we obtained 16.9 (P < 0.0001) concluding that the two survival distributions are significantly different at $\alpha = 0.05$. The prognosis of graft survival is therefore different for patients who received a pulmonary donor graft from patients with an aortic donor graft. In Table 2.3, we show that Gehan and Peto & Prentice generalizations of the Wilcoxon test confirm the significant differences.

□

2.2.2 R solution

The G^{ϱ} class of weighted log-rank tests proposed by Harrington and Fleming (1982) is implemented in the R function survdiff (package survival) with by default, $\varrho = 0$ leading to the Mantel's log-rank test. Additionally, $G^{\varrho,\gamma}$ significance tests are implemented in the R package FHtest (Oller and Langohr, 2015).

Example 2.1 in R
First, load the data and the survival *package. Mantel's log-rank test is computed using the* survdiff *function with a* formula *as first argument with on the LHS the response as a* Surv *object (see Section 1.7.1) and on the RHS the variable which determines the groups (*Hgraft*).*

```
> data("graft", package="icensBKL")
> library("survival")
> survdiff(Surv(timeFU, homo.failure) ~ Hgraft, data=graft)
```

```
Call:
survdiff(formula = Surv(timeFU, homo.failure) ~ Hgraft, data = graft)

             N Observed Expected (O-E)^2/E (O-E)^2/V
Hgraft=PH 230       36     48.1      3.06      16.9
Hgraft=AH  42       23     10.9     13.56      16.9

 Chisq= 16.9  on 1 degrees of freedom, p= 3.86e-05
```

The function reports the group sample sizes (column N*), the observed events in both groups* ($\sum_{j=1}^{J} d_{j1}, \sum_{j=1}^{J} d_{j2}$ *in column* Observed*), the numbers of expected events in both groups* ($\sum_{j=1}^{J} E_{j1}, \sum_{j=1}^{J} E_{j2}$ *in column* Expected*). The last column is the squared log-rank statistic. The value of* G_1^2 *is given by* Chisq.

The rho *argument selects a particular test from the* G^{ϱ} *class. For example, Peto-Prentice generalization of the Wilcoxon test, equivalent to the* G^1 *test, is computed with* rho=1.

```
> survdiff(Surv(timeFU, homo.failure) ~ Hgraft, data=graft, rho=1)
```

```
Call:
survdiff(formula = Surv(timeFU, homo.failure) ~ Hgraft, data = graft,
    rho = 1)

            N Observed Expected (O-E)^2/E (O-E)^2/V
Hgraft=PH 230     28.8    38.30      2.38      15.8
Hgraft=AH  42     18.1     8.56     10.64      15.8

 Chisq= 15.8  on 1 degrees of freedom, p= 6.93e-05
```

Note that in this case, the values in columns Observed *and* Expected *are weighted observed and expected events obtained from both groups.*

□

2.2.3 SAS solution

The log-rank and Wilcoxon-Gehan test are the default tests computed by the LIFETEST procedure.

Example 2.1 in SAS
Note that the code is similar to the code used in Section 2.1.3.

```
ODS GRAPHICS ON;
PROC LIFETEST DATA=graft PLOTS=(SURVIVAL(TEST));
TIME timeFU*homo_failure(0);
STRATA Hgraft;
RUN;
ODS GRAPHICS OFF;
```

The TIME *and* STRATA *statement are the same as in Section 2.1.3 where we estimated the survival curves for both types of grafts. Strata levels are formed according to the nonmissing values of the variables specified in the* STRATA *statement. The* MISSING *option can be used to allow missing values as a valid stratum level. In our example we have two strata: subjects with a pulmonary donor graft and subjects with an aortic donor graft. Requesting the option* TEST *asks to display the P-value of the log-rank test onto the plot. In the final part of the output, the results of applying the log-rank, Wilcoxon, likelihood ratio (not covered by this section) test are shown. Other tests, such as the Peto-Prentice generalization of the Wilcoxon test, can be specified by adding the* TEST= *option in the* STRATA *statement. For the other available tests and their respective keywords, we refer to the* SAS *documentation manual.*

Test of Equality over Strata

Test	Chi-Square	DF	Pr > Chi-Square
Log-Rank	16.9375	1	<.0001
Wilcoxon	11.2688	1	0.0008
-2Log(LR)	14.9983	1	0.0001

□

2.3 Regression models

We consider here common regression models for survival analysis that express the distribution of T as a function of one or more exploratory variables denoted by vector $\boldsymbol{X}$. However, we limit ourselves to *Cox's proportional hazards (PH) model* and the *accelerated failure time (AFT) model*, which have also implementations for interval-censored data in R and SAS.

2.3.1 Proportional hazards model

2.3.1.1 Model description and estimation

The Cox proportional hazards model (Cox, 1972) has a semiparametric nature. It assumes a parametric form for the effects of the explanatory variables, but allows the hazard function of a "typical" subject to be unspecified. The PH model, or often also called the Cox model, assumes that the hazard function at t given the covariates $\boldsymbol{X}$ has the following form

$$\hbar(t \mid \boldsymbol{X}) = \hbar_0(t) \exp(\boldsymbol{X}'\boldsymbol{\beta}) \tag{2.11}$$

in which $\hbar_0(t)$ is an arbitrary unspecified baseline hazard function (where baseline refers to the "typical" subject) and $\boldsymbol{\beta}$ is a vector of regression parameters. Model (2.11) assumes a multiplicative effect of the covariates on the hazard function and hence implies that the ratio of the hazard functions for two subjects with different covariates is constant. For instance, when comparing the hazard of two groups characterized by one covariate, i.e., $X = x$ or $X = x + 1$, one has

$$\frac{\hbar(t \mid X = x + 1)}{\hbar(t \mid X = x)} = \exp(\beta). \tag{2.12}$$

Thus for a PH model, the regression coefficient β is easily interpreted as the log-hazard ratio or $\exp(\beta)$ is the hazard ratio comparing the instantaneous risk of the groups represented by $X = x + 1$ versus $X = x$. Expressed in survival functions, Model (2.11) corresponds to the following model

$$S(t \mid \boldsymbol{X}) = S_0(t)^{\exp(\boldsymbol{X}'\boldsymbol{\beta})}, \tag{2.13}$$

where S_0 represents the baseline survival function.

For right-censored data, the *partial likelihood approach* can be used for the estimation of the regression parameters (Cox, 1972, 1975). The likelihood function of the PH model can be factored into two parts: one part that depends on both $\hbar_0(t)$ and $\boldsymbol{\beta}$ and a second part that depends on $\boldsymbol{\beta}$ alone. The partial likelihood function treats the second part as though it were an ordinary likelihood function. The approach is efficient because the resulting estimator of $\boldsymbol{\beta}$ is asymptotically equivalent to the estimator of $\boldsymbol{\beta}$ obtained from the full likelihood function. However, in the partial likelihood approach, ties among observed (uncensored) event times need special care. Breslow (1974) and Efron (1977) addressed this problem. Breslow's approach is used by default in most software packages including SAS. Nevertheless, Efron's method (used by default in R) is more accurate and computationally as effective as Breslow's method.

Having estimated the parameters of the PH model, one is often interested in estimating the survival or hazard functions for specific combinations of covariates. This necessitates to estimate the baseline hazard function $\hbar_0$ as well. Several methods have been suggested in the literature. Popular choices then are a Kaplan-Meier-like estimator and the Nelson-Aalen estimator (also called Breslow estimator). For more details and an overview, we refer to Kalbfleisch and Prentice (2002, Section 4.3).

2.3.1.2 Model checking

Several tests and diagnostic tools have been developed in the literature to evaluate the validity of the PH assumption (2.11), see for example Collett (2003). We focus here on two residuals, the Lagakos and deviance residuals, that are also useful in the interval-censored setting. *Lagakos residuals*, also called *Martingale residuals*, are developed as an improvement on the Cox-Snell residuals r_i^C $(= -\log\{S(t \mid X_i)\})$, which are based on the result that when T has a survival distribution $S(t)$, then $-\log S(T)$ has an exponential distribution with mean one. The Lagakos residual is then defined as

$$\widehat{r}_i^L = \delta_i - \widehat{r}_i^C, \tag{2.14}$$

where δ_i is the censoring indicator. These residuals might vary between $-\infty$ and 1, have mean zero and are approximately uncorrelated in large samples. They may be plotted versus the linear predictor $\boldsymbol{X}'\boldsymbol{\beta}$ to validate the exponential link function or versus any of the covariates used in the model to check their functional form in the model. Lagakos residuals may not be symmetrically distributed about zero, even when the fitted model is correct. Therefore,

TABLE 2.4: Homograft study. Cox PH hazards model with MLEs of the model parameters, standard errors (SE) and hazard ratios (HR) with a 95% confidence interval (CI) using Breslow's method for ties, obtained with R.

Parameter	Estimate	SE	HR	95% CI	P-value
Aortic donor graft	0.4179	0.2878	1.519	0.864 – 2.670	0.1465
Age	−0.1060	0.0226	0.899	0.860 – 0.940	<0.0001

Therneau et al. (1990) introduced the more symmetrical distributed *deviance residuals*, defined as

$$\widehat{r}_i^D = \text{sign}(\widehat{r}_i^L)\big[-2\{\widehat{r}_i^L + \delta_i \log(\delta_i - \widehat{r}_i^L)\}\big]^{\frac{1}{2}}. \tag{2.15}$$

These residuals are the components of the deviance D, namely $D = -2\{\log \widehat{L}_c - \log \widehat{L}_f\} = \sum(\widehat{r}_i^D)^2$ where $\widehat{L}_c$ and $\widehat{L}_f$ are the maximized partial likelihoods under the current model under investigation and the full (or saturated) model, respectively. Although being symmetrically distributed around zero, they do not necessarily sum to zero. Like the Lagakos residuals, they may be plotted versus the linear predictor $\boldsymbol{X}'\boldsymbol{\beta}$ or any of the covariates in the model.

In SAS, also another model checking method suggested by Lin et al. (1993) is implemented, which is based on distributional properties of cumulative sums of residuals over follow-up time and/or covariate values. Namely their distributions under the assumed model approximate a zero-mean Gaussian process such that each observed process can be compared, both visually and analytically, with a number of simulated realizations from the approximated null distribution. Alternatively, Grambsch and Therneau (1994) describe methods based on weighted residuals. Their methodology has been implemented in R and is applied in Example 2.3.

Example 2.3 Homograft study

In Example 2.2 we found that there was a significant difference between the survival curves for both types of donor grafts. However, confounding factors were not corrected for. For example, the age distribution in both groups is significantly different ($P < 0.0001$). The median age for the pulmonary donor graft patients was 17 years (Q1–Q3: 8–26) and for the aortic donor graft patients was 7 years (Q1–Q3: 2–10). We now investigate whether the significant difference may be due to an imbalance in age between the two groups.

The estimated regression coefficients and associated statistics are shown in Table 2.4, computed using Breslow's method to deal with tied event times. As can be observed, the hazard ratio of 0.899 ($= \exp(-0.106)$) expresses the decrease in instantaneous risk when a patient gets one year older. One can also compute the decrease in hazard rate for, say an increase in 10 years. In that case the hazard ratio is equal to $\exp(-1.06) = 0.346$. Summarized, older patients have a lower hazard and are therefore less at risk of having an early graft failure (statistically significantly different from 1). After correction for age, the difference between the two types of grafts is no longer significant ($P = 0.1465$). Hence, we conclude that after correction for age, the advantage of a pulmonary donor graft over an aortic donor graft is not clear anymore. Note that we corrected only for age, correcting for other important confounding factors may still change the results.

Further, Figure 2.2 shows estimated survival curves for a 14 (=median) year old patient for both types of graft based on a fitted PH model and the Nelson-Aalen (Breslow) estimate of the baseline hazard. Compared to Figure 2.1 where no adjustment was done for age, now the difference between both groups almost vanished when adjusted for age.

Figure 2.3 shows the Martingale residuals versus the linear predictor. Also plots versus the covariates age and treatment group (not shown) do not reveal any problems.

Further, the significance tests of Lin et al. (1993) lead to P-values of 0.430 and 0.525 for type of graft and age, respectively, indicating again that there is no evidence against PH. This is confirmed by the Grambsch and Therneau's test with P-values equal to 0.646 and 0.994 for type of graft and age, respectively, and 0.889 for a global test.

□

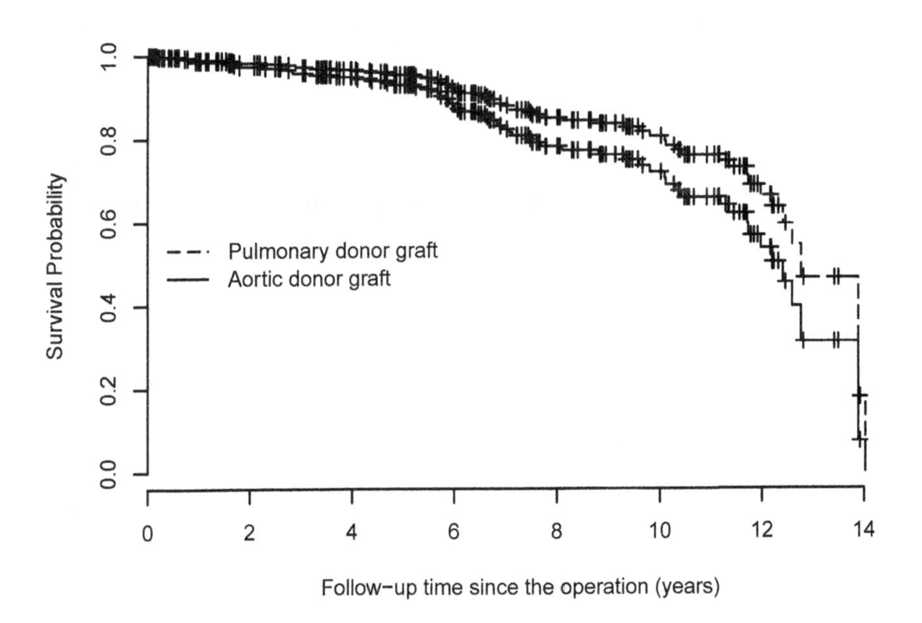

FIGURE 2.2: Homograft study. Estimated survival curves for 14-year-old patients based on the PH model corrected for age, obtained with R package survival.

2.3.1.3 R solution

The PH model for right-censored data is supported by the function coxph of the package survival. The function produces an object of class coxph which can be supplied as argument to other supportive functions, e.g., survfit (computation of the estimated survival function for specified combinations of covariates), and cox.zph (PH diagnostics of Grambsch and Therneau, 1994).

Example 2.3 in R

The PH model is fitted by the coxph *function. Note that the variable* Hgraft *is coded as* factor *in the* data.frame graft *(*PH *for pulmonary donor grafts and* AH *for aortic donor grafts) which guarantees that a proper set of 0/1 dummy variables is used.*

```
> data("graft", package="icensBKL")
> library("survival")
> fit <- coxph(Surv(timeFU, homo.failure) ~ Hgraft + age, data=graft)
> summary(fit)
```

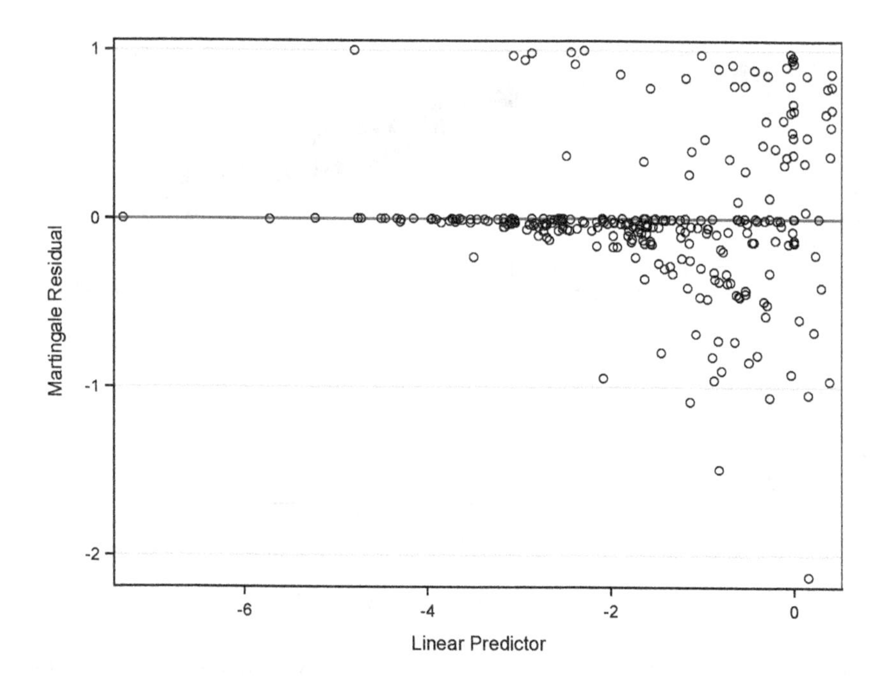

FIGURE 2.3: Homograft study. Martingale residuals vs. linear predictor obtained with SAS procedure PHREG.

```
Call:
coxph(formula = Surv(timeFU, homo.failure) ~ Hgraft + age,
      data = graft)

  n= 271, number of events= 59
   (1 observation deleted due to missingness)

              coef exp(coef) se(coef)      z Pr(>|z|)
HgraftAH  0.41726   1.51780  0.28779  1.450    0.147
age      -0.10606   0.89937  0.02259 -4.694 2.67e-06 ***
---
Signif. codes:  0 '***' 0.001 '**' 0.01 '*' 0.05 '.' 0.1 ' ' 1

         exp(coef) exp(-coef) lower .95 upper .95
HgraftAH    1.5178     0.6589    0.8635    2.6680
age         0.8994     1.1119    0.8604    0.9401

Concordance= 0.764  (se = 0.044 )
Rsquare= 0.154   (max possible= 0.851 )
Likelihood ratio test= 45.27  on 2 df,   p=1.478e-10
Wald test            = 32.1  on 2 df,   p=1.068e-07
Score (logrank) test = 37.81  on 2 df,   p=6.158e-09
```

Age is unknown for one patient (see remark at the top of the output). Since in R *Efron's method is used by default to handle ties, the results differ from those of Table 2.4. The results in Table 2.4 can be obtained by adding the argument* method*:*

```
> fit.breslow <- coxph(Surv(timeFU, homo.failure) ~ Hgraft + age,
+                       data=graft, method="breslow")
```

The column exp(-coef) *in Table 2.4 expresses the hazard ratio when we move from $x + 1 \to x$ or when we compare the reference category with another category of a categorical covariate. Finally, highly significant results of all global tests (likelihood-ratio, Wald, score) indicate that the survival of the graft depends significantly on either age or the type of graft or both.*

The estimated survival curves for two patients having the median age of 14 years but with different types of graft are shown in Figure 2.2 using:

```
> Sfit <- survfit(fit, newdata=data.frame(age=rep(14, 2),
+                                   Hgraft=factor(c("PH", "AH"))))
> plot(Sfit, lty=c(5, 1),
+      xlab="Follow-up time since the operation (years)",
```

```
+       ylab="Survival Probability")
> legend(0, 0.6,
+         legend=c("Pulmonary donor graft", "Aortic donor graft"),
+         lty=c(5, 1), bty="n")
```

The requested covariate combinations are specified by the argument newdata of the function survfit.

To test the validity of the PH assumption using the approach of Grambsch and Therneau (1994), the function cox.zph can be used.

```
> cox.zph(fit)
```

```
               rho    chisq     p
HgraftAH 0.056118 2.11e-01 0.646
age      0.000637 5.74e-05 0.994
GLOBAL         NA 2.34e-01 0.889
```

The Martingale residuals are obtained directly with fit$residuals. Other residuals, like the deviance residuals, may be obtained via the residuals function. The residuals may be plotted versus the linear predictor or covariates in the model to verify the model. The code for several example plots is available in the supplementary materials.

```
> resdev=residuals(fit,type="deviance")
```

□

2.3.1.4 SAS solution

The PH model for right-censored data can be fitted using the SAS PHREG procedure with the following code:

Example 2.3 in SAS

```
PROC PHREG DATA=icdata.graft;
  MODEL timeFU*homo_failure(0)=Hgraft age / RISKLIMITS;
RUN;
```

Hgraft is a binary variable (1=aortic donor graft; 0=pulmonary donor graft) and thus no CLASS statement is needed. The LHS in the MODEL statement specifies the survival time variable (timeFU) and the censoring indicator variable (homo_failure) with the censoring values within parentheses. The RHS specifies the explanatory variables. We compared the survival of the two types of graft (Hgraft) corrected for the age of the patient. The option RISKLIMITS provides CIs for hazard ratios in the output. By default, Breslow's method for *ties is used. Specifying the option* TIES=EFRON *requests Efron's method to be used for handling the ties.*

Several residuals can be outputted to verify the model as described above. The options RESMART *and* RESDEV *request the Martingale and deviance residuals, respectively. From* LOGSURV *one can calculate the Cox-Snell residuals.* XBETA *requests the linear predictor* $\boldsymbol{X}'\boldsymbol{\beta}$.

```
OUTPUT OUT=Outp XBETA=Xb  RESMART=Mart RESDEV=Dev LOGSURV = ls;
```

With these residuals one can then create different residual plots. For instance, Figure 2.3 shows the Martingale residuals versus the linear predictor. The code for this plot and other residual plots can be found in the supplementary materials.

The PH assumption can also be verified graphically and numerically by the methods of Lin et al. (1993) by adding the following code behind the MODEL statement.

```
ASSESS PH / RESAMPLE;
```

Supremum Test for Proportionals Hazards Assumption

Variable	Maximum Absolute Value	Replications	Seed	Pr > MaxAbsVal
Hgraft	0.8165	1000	12345	0.4300
age	1.1161	1000	12345	0.5250

For a graphical check, ODS GRAPHICS *must be turned on (figures not shown). The number of simulated residual patterns can be modified with the option* NPATHS= *of the* ASSESS *statement.*

□

2.3.2 Accelerated failure time model

2.3.2.1 Model description and estimation

The accelerated failure time (AFT) model is a useful, however less frequently used alternative to the PH model and given by

$$\log(T) = \boldsymbol{X}'\boldsymbol{\beta} + \varepsilon, \qquad (2.16)$$

with T the survival time, $\boldsymbol{X}$ a vector of covariates, $\boldsymbol{\beta}$ the vector of regression parameters and ε an error random variable. Let $\hbar_0$ and S_0 denote the baseline hazard and survival function, respectively, of the random variable $T_0 = \exp(\varepsilon)$. For a subject with covariate vector $\boldsymbol{X}$ the hazard and survival function are assumed to be

$$\hbar(t \mid \boldsymbol{X}) = \hbar_0\{\exp(-\boldsymbol{X}'\boldsymbol{\beta})\, t\}\, \exp(-\boldsymbol{X}'\boldsymbol{\beta}), \tag{2.17}$$

$$S(t \mid \boldsymbol{X}) = S_0\{\exp(-\boldsymbol{X}'\boldsymbol{\beta})\, t\}. \tag{2.18}$$

Further,

$$T = \exp(\boldsymbol{X}'\boldsymbol{\beta})\, T_0, \tag{2.19}$$

i.e., the effect of a covariate implies an acceleration or deceleration compared to the baseline event time. The effect of a covariate on the hazard for the AFT and PH model is exemplified in Figure 2.4. As for the PH model, a covariate acts multiplicatively on the baseline hazard function for an AFT model. In addition, the time scale is accelerated or decelerated for an AFT model. Finally, the AFT hazard increases with increasing values of the covariates when $\beta < 0$, whereas in the PH model this happens when $\beta > 0$.

Several approaches have been suggested in the literature to estimate the regression coefficients in an AFT model semiparametrically, i.e., without specifying the distribution of the error term ε. The first class of methods, known as the *Buckley-James approach*, is based on the generalization of the least squares method to censored data (Buckley and James, 1979; Miller, 1976). However, a drawback of the Buckley-James method is that it may fail to converge or may oscillate between several solutions. The second class of methods is based on linear-rank-tests (e.g., the weighted log-rank test introduced in Section 2.2) for censored data and is discussed in full detail in Kalbfleisch and Prentice (2002, Chapter 7). However, the linear-rank-test-based estimation of

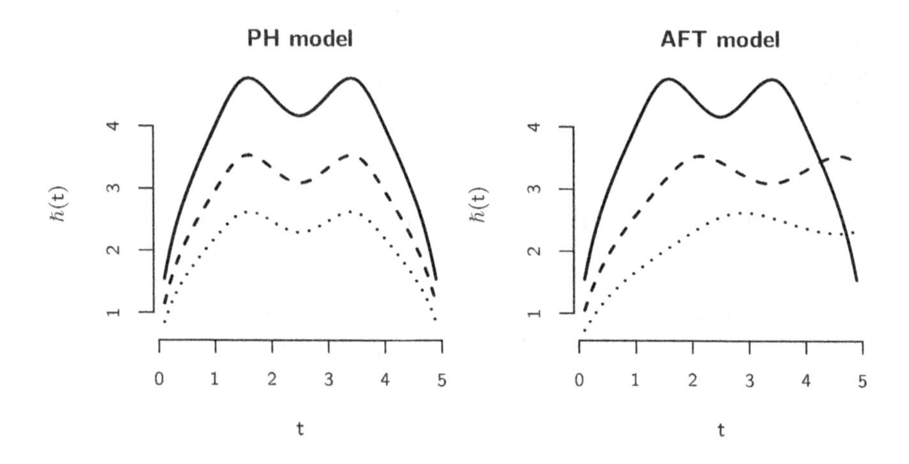

FIGURE 2.4: Impact of a covariate on the hazard of a PH and AFT model. Covariate X taking value 0 (full line), 0.6 (dashed line) and 1.2 (dotted line) based on $\beta = -0.5$ for the PH model and $\beta = 0.5$ for the AFT model.

the regression parameters of the AFT model could be computationally cumbersome. Jin et al. (2003) suggested an algorithm to compute this estimate using a linear programming technique, but is only applicable for data sets of moderate size.

In view of the computational difficulties mentioned above, a fully parametric form of the AFT model is frequently used in practice. That is, one assumes that the error term ε has a specific density $g(\varepsilon)$. The most common assumptions for $g(\varepsilon)$ are the normal density, the logistic density and the Gumbel (extreme-value) density. Hence, the parametric AFT model assumes a parametric form for the effects of the explanatory variables and also assumes a parametric form for the underlying survival function. Estimation then proceeds by a standard maximization of the log-likelihood.

Further, we point out (see Kalbfleisch and Prentice, 2002, Section 2.3.4) that the AFT model and the PH model of Section 2.3.1 are equivalent if and only if the distribution of the error term ε in (2.16) is Gumbel (extreme-value) which corresponds to the Weibull distribution for the survival time T.

Finally, the AFT model has the advantage that the regression parameters of the included covariates do not change when other (important and independent) covariates have been omitted. Of course, the neglected covariates have an impact on the distribution of the error term ε in (2.16) which is typically changed into one with larger variability. Such a property does not hold, however, for the PH model, see Hougaard (1999). This phenomenon is often called "non-collapsibility."

2.3.2.2 Model checking

When only a categorical covariate is used in the model, the Kaplan-Meier curve can be calculated for the subjects of each category separately. These Kaplan-Meier curves can then be overlayed with the fitted parametric survival curves for the specific groups. When continuous or many covariates are used, the subjects could be divided into a certain number of groups (e.g., 3, referring to low, medium and high risk patients) based on the risk score $\boldsymbol{X}'\boldsymbol{\beta}$. Comparing the Kaplan-Meier curves with the fitted curves in each group provides again an indication of goodness of fit.

The residuals introduced in Section 2.3.1.2 can also be calculated and used in a similar way.

Example 2.4 Homograft study

In Example 2.3 the significant effect of the donor type disappeared when controlling for age. We now check whether this also happens with the AFT model. For this, we need to choose first $g(\cdot)$. We considered a normal, logistic and Gumbel density. Because the models are not nested, we used AIC to compare them, see Table 2.5. The column AIC(T) is the AIC based on the event times T_i obtained with the R function survreg and the SAS procedure LIFEREG, while the entries in column AIC($\log T$) are based on $Y_i = \log(T_i)$

TABLE 2.5: Homograft study. AIC for three AFT models, obtained with SAS.

Model	AIC(T)	AIC($\log T$)
Weibull	445.708	245.689
Log-logistic	455.325	255.307
Log-normal	480.741	280.722

TABLE 2.6: Homograft study. Weibull AFT model, obtained with R.

Parameter	Estimate	SE	exp(Estimate)	P-value
Intercept	2.3080	0.1286	10.054	< 0.0001
Aortic donor graft	-0.2566	0.1425	0.774	0.0719
Age	0.0415	0.0102	1.042	< 0.0001

and only returned by SAS. Note that the log-likelihoods differ only by the additive constant $\sum_{i=1}^{n}\{\log(t_i)\}^{\delta_i}$ stemming from the Jacobian of the log-transformation and with δ_i equal to 0 or 1 for right-censored or exact data, respectively. The Weibull AFT model seems to fit the data best. For more details about the Weibull distribution, we refer to Section 3.2.

Parameter estimates of the Weibull AFT model are shown in Table 2.6. We find again that the effect of the type of graft is not significant ($P = 0.0719$) but the effect of age is highly significant ($P < 0.0001$) with a positive regression coefficient indicating that older patients are less likely to have an early graft failure. On average, the median survival time is 1.042 times longer when the patient is one year older. Note that AFT regression coefficients have opposite signs to PH model regression coefficients (see, e.g., Example 2.3).

□

2.3.2.3 R solution

The parametric AFT model is fitted using the function survreg from the R package survival. This function is discussed in more detail in Section 3.2 and in Appendix D.4. Linear-rank-test-based estimation of the semiparametric AFT model used to be implemented in the contributed package rankreg (Zhou and Jin, 2009) which, however, is no more available from *CRAN*. Here, the parametric AFT model is illustrated.

Example 2.4 in R

The code to fit the parametric AFT model is similar to the code used to fit the PH model in Example 2.3. Namely, we replace the function coxph *by the function* survreg *and add the argument* dist*, which specifies the required baseline distribution. For example, the Weibull AFT model is fitted using the following code:*

```
> data("graft", package="icensBKL")
```

```
> library("survival")
> fit <- survreg(Surv(timeFU, homo.failure) ~ Hgraft + age,
+                dist="weibull", data=graft)
> summary(fit)
```

```
Call:
survreg(formula = Surv(timeFU, homo.failure) ~ Hgraft + age,
    data = graft, dist = "weibull")
              Value Std. Error     z        p
(Intercept)  2.3080     0.1286 17.95 4.85e-72
HgraftAH    -0.2566     0.1425 -1.80 7.19e-02
age          0.0415     0.0102  4.06 4.84e-05
Log(scale)  -0.7139     0.1069 -6.68 2.40e-11

Scale= 0.49

Weibull distribution
Loglik(model)= -218.9   Loglik(intercept only)= -238.1
        Chisq= 38.48 on 2 degrees of freedom, p= 4.4e-09
Number of Newton-Raphson Iterations: 13
n=271 (1 observation deleted due to missingness)
```

The log-normal and log-logistic models are fitted by changing the value of argument dist *to* "lognormal" *and* "loglogistic", *respectively. Besides estimated regression coefficients and their standard errors, the next output shows the estimated scale of the error distribution, see Section 3.2 for more details. The highly significant* χ^2 *test indicates that the hypothesis* $H_0 : \beta(\text{Hgraft}) = 0$ *and* $\beta(\text{age}) = 0$ *is rejected and that the event time depends on either age or the type of graft or both.*

AIC is computed using

```
> AIC <- -2*fit$loglik[2] + 2*4
> print(AIC)
```

```
[1] 445.7076
```

Acceleration factors $(\exp(\boldsymbol{\beta}))$ *are obtained using*

```
> exp.beta <- exp(coef(fit))
> print(exp.beta)
```

```
(Intercept)     HgraftAH           age
 10.0539163    0.7737131     1.0423316
```

Unfortunately, there is no easy function to calculate the residuals for an AFT model. In the supplementary materials, the coding of Cox-Snell, Martingale and deviance residuals and their use in plots are shown.

□

2.3.2.4 SAS solution

The parametric AFT model can be fitted in SAS with the LIFEREG procedure. The parameter estimates and corresponding standard errors can be obtained using the following code.

Example 2.4 in SAS

```
PROC LIFEREG DATA=icdata.graft;
MODEL timeFU*homo_failure(0)=Hgraft age/DIST=weibull;
RUN;
```

The MODEL *statement is similar to that for the PH model (see Section 2.3.1.4), except that we now need to specify the distribution with the* DIST= *option. Here the code for the best fitting model, i.e., Weibull, is given but the Weibull distribution is anyway the default option. Also the log-logistic and log-normal distribution can be fitted using the keyword* LLOGISTIC *or* LNORMAL, *respectively.*

Part of the output is given below.

```
The LIFEREG Procedure

                  Fit Statistics

-2 Log Likelihood                             237.689
AIC (smaller is better)                       245.689
AICC (smaller is better)                      245.840
BIC (smaller is better)                       260.098

         Fit Statistics (Unlogged Response)

-2 Log Likelihood                             437.708
Weibull AIC (smaller is better)               445.708
Weibull AICC (smaller is better)              445.858
Weibull BIC (smaller is better)               460.116
```

(Cont.)

```
Algorithm converged.

       Type III Analysis of Effects
                          Wald
Effect        DF     Chi-Square    Pr > ChiSq

Hgraft         1         3.2398        0.0719
age            1        16.5099        <.0001

          Analysis of Maximum Likelihood Parameter Estimates

                          Standard   95% Confidence    Chi-
Parameter     DF Estimate    Error        Limits       Square Pr > ChiSq

Intercept      1   2.3080   0.1286   2.0559   2.5600 322.18       <.0001
Hgraft         1  -0.2566   0.1425  -0.5359   0.0228   3.24       0.0719
age            1   0.0415   0.0102   0.0215   0.0615  16.51       <.0001
Scale          1   0.4897   0.0523   0.3972   0.6039
Weibull Shape  1   2.0419   0.2182   1.6560   2.5177
          Analysis of Maximum Likelihood Parameter Estimates
```

Note that, in addition to the parameter estimates for the intercept and both explanatory variables (type of graft and age), also the scale and shape parameter of the Weibull distribution are reported.

The standardized residuals ($\frac{\log(T)-\boldsymbol{X}'\boldsymbol{\beta}}{\sigma}$, SRES=) and Cox-Snell residuals (CRES=) are easily obtained with an output statement. Martingale and deviance residuals must be calculated in a data step. In the supplementary materials the code for some residual plots is given.

```
OUTPUT OUT=Outp XBETA=Xb  SRES=stdres CRES = CSres ;
```

□

Part II

Frequentist methods for interval-censored data

Chapter 3

Estimating the survival distribution

In this chapter we review frequentist approaches to estimate the survival distribution $S(t)$ in the presence of interval censoring. We first introduce the nonparametric maximum likelihood estimator (NPMLE) also called Turnbull estimator of $S(t)$. Rather than giving a rigorous derivation of the estimator, we have opted for a more intuitive approach. The R software and three SAS procedures that produce the NPMLE of $S(t)$ are described. Then we move to parametric modelling and review again the available software. We end with flexible approaches to estimate the survival function based on smoothing techniques. The flexible survival models are illustrated with R functions.

We assume that i.i.d. survival times are available with cumulative distribution function $F(t)$, survival function $S(t)$ and density $f(t)$. But, rather than observing T_i $(1, \ldots, n)$ exactly, we observe a set of intervals and censoring indicators $\mathcal{D} = \big\{\lfloor l_i,\, u_i \rfloor,\, \delta_i : \, i = 1, \ldots, n\big\}$. For clarity, for an exact observation we will write $t_i = l_i = u_i$ (when $\delta_i = 1$).

3.1 Nonparametric maximum likelihood

3.1.1 Estimation

In contrast to the Kaplan-Meier estimator (see Section 2.1), the *NPMLE* of the survival function for interval-censored data has in general no closed solution and must be obtained by an iterative algorithm. An exception is case I interval-censored data (current status data) for which an analytical solution is available (Robertson et al., 1988), however is not used by any SAS or R program. In the next paragraph we compute the NPMLE of a survival distribution.

Let $\lfloor l_i,\, u_i \rfloor$ $(i = 1, \ldots, n)$ be the observed intervals from n independent subjects which contain the unknown times of the event of interest, t_i. Peto (1973) was the first to note that the maximum likelihood solution results in a set of intervals $\{[p_j, q_j]\}_{j=1}^{m}$ with the property that the estimated survival function is constant outside the intervals. Further, the mass assigned to each of the intervals is well determined but within each interval there is no information as to how that mass is assigned. The intervals are called *regions of*

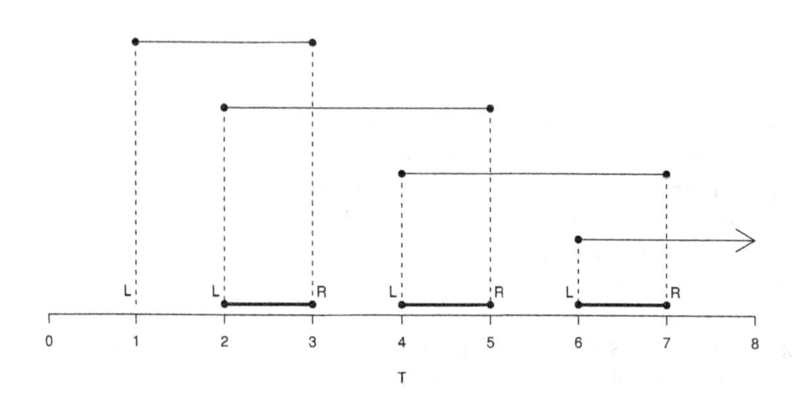

FIGURE 3.1: Determination of regions of possible support for the Turnbull estimate from observations $[1,3]$, $[2,5]$, $[4,7]$ and $[6,\infty)$. The bold lines on the bottom indicate the 3 regions of possible support.

Interval 1	(L1	R1]		[L1	R1]		
Interval 2		(L2	R2]		[L2		R2]
Ordering	(L1	R1] (L2	R2]	[L1	[L2	R1]	R2]

FIGURE 3.2: Ordering of two interval-censored observations $\lfloor L1, R1\rceil$ and $\lfloor L2, R2\rceil$ with tied endpoints $R1$ and $L2$ in the reduction algorithm to determine the regions of possible support. The ordering for half-open and closed intervals is depicted on the LHS and RHS, respectively.

possible mass or support because the ML procedure can only tell in which regions events might occur. Therefore one can only claim that the intervals $\{\lfloor p_j, q_j \rceil\}_{j=1}^{m}$ have mass ≥ 0 but not necessarily all have mass > 0. Hence these are the intervals that are assigned possible mass.

Peto (1973) and Turnbull (1976) suggested a simple reduction algorithm to identify the intervals of possible mass from the data. Namely, given the observations $\lfloor l_i, u_i \rceil$ $(i = 1, \ldots, n)$, rank the time points $\{l_i\}$ and $\{u_i\}$ in increasing order and keep track of whether the point is a left or a right endpoint. The regions of possible mass are then the intervals with a left endpoint immediately followed by a right endpoint. This finding facilitates the nonparametric estimation of the survival function considerably. The reduction process is illustrated in Figure 3.1 using 4 hypothetical observations $[1, 3]$, $[2, 5]$, $[4, 7]$ and $[6, \infty)$. When moving from left to right, the first endpoint at 1 is a left endpoint. The following endpoint at 2 is also a left endpoint. Hence, no region of possible mass is yet found. A right endpoint is found at 3. Because the interval $[2, 3]$ is formed by a left endpoint immediately followed by a right endpoint, we have found our first region of possible support. Similarly, two more regions of possible mass are found, namely $[4, 5]$ and $[6, 7]$, all indicated by bold lines in Figure 3.1.

Like in our example, Peto (1973) and Turnbull (1976) used closed intervals. However, also half-open intervals may be used. It is easily seen that the NPMLE solution depends on the closure properties of the intervals. Take the following simple example: the observations $[1, 2]$ and $[2, 3]$ give rise to only one region of support namely $[2, 2]$, but for observations $(1, 2]$ and $(2, 3]$ the reduction algorithm returns the original intervals as possible regions of support. The two situations are illustrated in Figure 3.2.

Given the regions of possible support, the mass assigned to each of these intervals must be estimated in a second step. For half-open or closed intervals the above reduction algorithm gives rise to a set of intervals $\{\lfloor p_j, q_j \rceil\}_{j=1}^{m}$. Define $s_j = S(p_j-) - S(q_j+)$ $(j = 1, \ldots, m)$. Then the vector $\boldsymbol{s} = (s_1, \ldots, s_m)'$ where $\sum_{j=1}^{m} s_j = 1$ and $s_j \geq 0$ $(j = 1, \ldots, m)$ defines equivalence classes in the space of distribution functions S which are flat outside of $\bigcup_{j=1}^{m} \lfloor p_j, q_j \rceil$. All functions in the same equivalence class will have the same likelihood because the likelihood depends only on the values in p_j and q_j $(j = 1, \ldots, m)$ and not on how the function evolves between p_j and q_j $(j = 1, \ldots, m)$. Thus, the search for the MLE of the function S can be restricted to these classes and reduces to maximizing

$$L = \prod_{i=1}^{n} \left(\sum_{j=1}^{m} \alpha_{ij} s_j \right),$$

where

$$\alpha_{ij} = \begin{cases} 1 & \text{if } \lfloor p_j, q_j \rceil \subset \lfloor l_i, u_i \rceil, \\ 0 & \text{otherwise.} \end{cases}$$

Therefore, the NPMLE of S can be estimated by constrained maximization of the likelihood L with linear constraints

$$1 - \sum_{j=1}^{m} s_j = 0,$$

$$s_j \geq 0 \quad (j = 1, \ldots, m).$$

This can be accomplished with a variety of algorithms such as the *self-consistency algorithm* of Turnbull (1976) which is in fact an example of the Expectation-Maximization (EM) algorithm (Dempster et al., 1977). In the absence of exact event times, the self-consistency algorithm requires at each time point the expected number of subjects at risk or subjects who fail. With these "pseudo data," a product limit estimator is calculated. We start with some initial estimates for s_j $(j = 1, \ldots, m)$, for instance by taking $\widehat{s}_j = 1/m$ for all j. We then estimate S with the following algorithm.

$$\begin{aligned} \widehat{S}(q_0+) &= 1 \\ \widehat{S}(q_j+) &= \widehat{\pi}_j \widehat{S}(q_{j-1}+) \quad (j = 1, \ldots, m) \\ \widehat{\pi}_j &= (n_j^{'} - d_j^{'})/n_j^{'} \end{aligned} \tag{3.1}$$

where

$$\begin{aligned} q_0 &= 0, \\ d_j^{'} &= \sum_{i=1}^{n} \left(\alpha_{ij} \widehat{s}_j / \sum_{k=1}^{m} \alpha_{ik} \widehat{s}_k \right), \\ n_j^{'} &= \sum_{k=j}^{m} \sum_{i=1}^{n} \left(\alpha_{ik} \widehat{s}_k / \sum_{r=1}^{m} \alpha_{ir} \widehat{s}_r \right). \end{aligned}$$

After this step, the $\widehat{s}_j$ $(j = 1, \ldots, m)$ are calculated again using the new estimates $\widehat{S}(q_j+)$ $(j = 1, \ldots, m)$ and this process is repeated until the required accuracy is obtained. Note that the self-consistency algorithm may stop at a local optimal solution and hence does not yield the NPMLE.

Other algorithms that may be used are the *projected gradient method* (Wu, 1978), the *intra-simplex direction method* (Lesperance and Kalbfleisch, 1992), the *iterative convex minorant (ICM) algorithm* (Groeneboom and Wellner, 1992), the *EM-ICM algorithm* (Wellner and Zahn, 1997) or mixture estimation methods such as the *vertex exchange method* (Böhning et al., 1996). We refer to the literature for further details about the different algorithms. Finally, note that the Kaplan-Meier estimate is a special case of the Turnbull estimate.

Using theory on concave programming with linear constraints, Gentleman and Geyer (1994) described how one can determine if a candidate estimate $\widehat{\boldsymbol{s}}$ of $\boldsymbol{s}$ is indeed the NPMLE. They showed that the Kuhn-Tucker conditions, which yield necessary and sufficient criteria to provide an optimal and valid solution, can be applied to this maximization problem. Let

$\ell(\boldsymbol{s}) = \sum_{i=1}^{n} \log\big(\sum_{j=1}^{m} \alpha_{ij} s_j\big)$ denote the log-likelihood and $d_k = \partial\ell/\partial s_k$ $(k = 1, \ldots, m)$. The Kuhn-Tucker conditions state that $\widehat{\boldsymbol{s}}$ is a MLE if and only if there exist values μ_j $(j = 0, \ldots, m)$ such that

$$\mu_j s_j = 0 \quad (j = 1, \ldots, m), \tag{3.2}$$

$$\mu_j \geq 0 \quad (j = 1, \ldots, m), \tag{3.3}$$

$$\frac{\partial}{\partial s_j}\left\{\ell(\boldsymbol{s}) + \sum_{k=1}^{m} s_k(\mu_k - \mu_0)\right\} = d_j + \mu_j - \mu_0 = 0 \quad (j = 1, \ldots, m). \tag{3.4}$$

The μ_j are called *Lagrange multipliers*. In general, a Lagrange multiplier is a scalar variable introduced to help solving a problem in n_v variables with n_c constraints to a solvable problem in $n_v + n_c$ variables with no constraints. From these conditions one can derive that $\mu_0 = n$. The left-hand side of (3.4), $d_j + \mu_j - \mu_0$, is called the *reduced gradient*, because it is the gradient with respect to the free variables. The Kuhn-Tucker conditions are satisfied if the Lagrange multipliers are ≥ 0 and zero for all $j = 1, \ldots, m$ where $s_j \neq 0$ and the reduced gradient is zero. The intervals $\lfloor p_j, q_j \rfloor$ with $s_j > 0$ $(j = 1, \ldots, m)$ are called the *regions of support* of the NPMLE of S.

Given that the NPMLE is indifferent to where the probability mass is placed inside the regions of support, this renders the NPMLE not unique. Gentleman and Vandal (2002) refers to this as *representational non-uniqueness*.

Simple descriptive statistics such as the median survival time can most often not be determined with the NPMLE since the median survival time will be located in one of the regions of support. Often an extra assumption is then made that in the regions of support the mass is distributed in a particular way, e.g., linearly or exponentially. This allows to obtain an estimate of, for example, the median survival time.

3.1.2 Asymptotic results

While this book does not focus on asymptotic properties of the considered methods, it is imperative to look at the behavior of the Turnbull estimator when the sample size increases. We summarize here the results that have been formulated in the review Section 3.6 of Sun (2006). This section contains also the references to the original publications. While for right-censored data one could rely on counting processes and Martingale theory, such tools are not available for interval-censored data. Most important is the result on consistency of the estimator. Provided some regularity conditions are satisfied, it can be shown that the NPMLE converges to the true survival distribution when the sample size increases. An important condition is however that the support of the censoring distribution (of the left and right endpoints of the intervals) includes the support of the survival distribution. In practice this implies that, if subjects are evaluated at approximately the same time points, it cannot be guaranteed that the NPMLE will estimate the survival distribution consistently over its support. This is important for, e.g., cancer trials

TABLE 3.1: Breast cancer study. Regions of possible support and NPMLE equivalence classes for the radiotherapy-only group.

$(p_j, q_j]$	(4, 5]	(6, 7]	(7, 8]	(11, 12]	(15, 16]	(17, 18]	(24, 25]
s_j	0.046	0.033	0.089	0.071	0	0	0.093

$(p_j, q_j]$	(25, 26]	(33, 34]	(34, 35]	(36, 37]	(38, 40]	(40, 44]	(46, 48]
s_j	0	0.082	0	0	0.121	0	0.466

where often patients are evaluated for progression of their disease at regular time points, say every three or six months. Then the Turnbull estimator is not guaranteed to give a good picture of the actual survival distribution, and the same is true for estimators of the survival distribution such as the median. Note that the same is true (and even more so) for the Kaplan-Meier estimator when based on single imputation techniques. In the same section of Sun (2006) the asymptotic distribution of the NPMLE for current status and case II interval-censored data is treated. It is shown that the asymptotic distributions are quite complex and not Gaussian.

Despite this, possibly disappointing, result we stress that comparative inference based on appropriate significance tests (taking into account the interval-censored character of the data) still provides the correct type I error rate, see Section 4.3 (and Section 6.6 for the parametric case).

Example 3.1 Breast cancer study

As an illustration we calculate the NPMLE for the early breast cancer patients treated with radiotherapy alone. Fourteen regions of possible support are found but mass > 0 has been attributed to only eight regions, see Table 3.1. The left panel of Figure 3.3 displays the NPMLE of the cumulative distribution function F. The shaded parts in the graph correspond to regions with positive mass. From above we know that within such regions the distribution of the assigned mass is unknown. The right panel of Figure 3.3 shows the NPMLE of the survival function S. Now we assumed a linear decrease in survival in each of the regions of support leading to a piecewise linear estimate $\widehat{S}$ of S.

□

3.1.3 R solution

Several algorithms have been implemented in the R package Icens (Gentleman and Vandal, 2016) to compute the NPMLE, namely: EM, PGM, VEM, ISDM and EMICM. For details, see Appendix D.2.

Example 3.1 in R

We look here at the radiotherapy-only group.

```
> data("breastCancer", package="icensBKL")
> breastR <- subset(breastCancer, treat=="radio only",
+                                 select=c("low", "upp"))
```

The maximum observed survival time is 48 months. Hence for right-censored observations, any value greater than 48 months (say 99 months) can be chosen as right limit. This is done below, namely all missing values of variable upp *(right-censored observations) are replaced by 99 and all missing values of variable* low *(left-censored observations) by zero.*

```
> breastR[is.na(breastR[,"upp"]), "upp"] <- 99
> breastR[is.na(breastR[,"low"]), "low"] <- 0
```

The NPMLE using the EM-iterative convex minorant algorithm is computed as

```
> library("Icens")
> NPMLE <- EMICM(breastR)
> print(NPMLE)
```

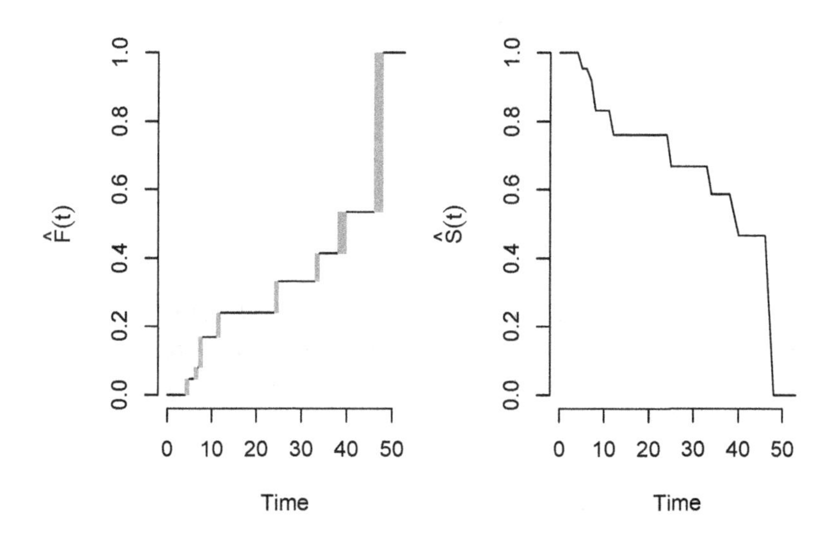

FIGURE 3.3: Breast cancer study of the radiotherapy-only group. Left: NPMLE of the cumulative distribution function obtained with R package Icens. Right: NPMLE of the survival function with the additional assumption of a piecewise linear survival curve obtained with R package icensBKL.

```
$sigma
 [1] 0.04634677 0.07971014 0.16837751 0.23913044 0.23913044
 [6] 0.23913044 0.33177627 0.33177627 0.41356204 0.41356204
[11] 0.41356204 0.53444187 0.53444187 1.00000000

$pf
 [1] 0.04634677 0.03336337 0.08866737 0.07075292 0.00000000
 [6] 0.00000000 0.09264584 0.00000000 0.08178577 0.00000000
[11] 0.00000000 0.12087983 0.00000000 0.46555813

$llk
[1] -58.06002

$weights
 [1] 626.75573 236.08264 604.37156 267.74235 274.08638
 [6] 356.04281 382.51309 294.83287 152.40891 110.02664
[11] 229.56934  28.96042 178.39798  84.24237

$lastchange
 [1]  2.176309e-09  2.622902e-15  9.322573e-10  1.199288e-09
 [5]  1.199288e-09  1.199288e-09 -1.521947e-08 -1.521947e-08
 [9] -7.066287e-09 -7.066287e-09 -7.066287e-09 -2.250526e-08
[13] -2.250526e-08  0.000000e+00

$numiter
[1] 10

$eps
[1] 1e-07

$intmap
     [,1] [,2] [,3] [,4] [,5] [,6] [,7] [,8] [,9] [,10]
[1,]    4    6    7   11   15   17   24   25   33    34
[2,]    5    7    8   12   16   18   25   26   34    35
     [,11] [,12] [,13] [,14]
[1,]    36    38    40    46
[2,]    37    40    44    48

attr(,"class")
[1] "icsurv"
```

The possible regions of support are found in the columns of the matrix intmap *(lower and upper limits in first and second rows, respectively) with corresponding equivalence classes s_j given by object* pf*. For instance, the first possible region of support is $(4, 5]$ to which 0.046 mass was assigned. The cumulative values of the s_j's and thus the NMPLE of the cdf is given by* sigma *and* llk *is the log-likelihood value at the NPMLE solution. The objects* weights, lastchange, numiter *and* eps *pertain to the optimization procedure.*

The NPMLE of the cdf in the left panel of Figure 3.3 is obtained using

```
> plot(NPMLE, ylab=expression(hat(F)(t)), main="")
```

The object NPMLE *of class* icsurv *is converted into a* data.frame NPMLE2 *(part is given in the first five rows of columns* time *and* cdf *below) using the function* icsurv2cdf *from package* icensBKL *as follows:*

```
> library("icensBKL")
> NPMLE2 <- icsurv2cdf(NPMLE)
> print(NPMLE2[1:5,])
```

```
  time        cdf
1    0 0.00000000
2    4 0.00000000
3    5 0.04634677
4    6 0.04634677
5    7 0.07971014
```

The right panel of Figure 3.3 was obtained from

```
> plot(NPMLE2$time, 1-NPMLE2$cdf, type="l",
+      xlab="Time", ylab=expression(hat(S)(t)), xlim=c(0, 53))
> lines(c(48, 53), c(0, 0))
```

The other algorithms implemented in the functions EM, PGM, VEM, ISDM *can be used in a similar way.*

□

3.1.4 SAS solution

There are four options to calculate the NPMLE of the survival function with SAS software: the SAS/STAT procedure LIFEREG, the SAS/STAT procedure ICLIFETEST (available from SAS/STAT 13.1 in SAS 9.4) the SAS/QC procedure RELIABILITY and a macro, provided by the SAS institute, called %EMICM (which is an updated version of the previously called macro %ICE). The macro %EMICM requires the SAS/IML and SAS/GRAPH software. The four options all assume half-open intervals, i.e., $(L, R]$, with $L = R$ for exact observations. As noted above, the NPMLE depends on how the interval-censored data are represented (open, half-open, closed). In practical terms, this is not a problem since one can easily change the type of interval, without effectively changing the observed data. For example, to interpret the half-open intervals $(a, b]$ as closed, one could subtract a small enough value ϵ from a and use $[a - \epsilon, b]$. Only care has to be taken not creating new intervals of possible support. To provide the final solution we again represent the intervals of possible support as $(p + \epsilon, q]$. Procedures LIFEREG and RELIABILITY estimate the NPMLE of $F(t)$. The NPMLE of the survival function is then obtained from $S(t) = 1 - F(t)$. The procedure ICLIFEREG and the macro deliver both the NPMLE of $F(t)$ and $S(t)$.

Procedure RELIABILITY

The PROBPLOT statement in procedures LIFEREG and RELIABILITY requests the NPMLE of the survival function. The RELIABILITY procedure uses the self-consistency algorithm of Turnbull (1976) to calculate the NPMLE of the cdf. The standard errors of the cumulative probability estimates are then computed from the inverse of the observed Fisher information matrix.

Example 3.1 in SAS (PROC RELIABILITY)

The WHERE *statement below selects the radiotherapy-only group for analysis. In order to take the left-censored observations correctly into account, we first need to set the zeros in the left endpoint to missing (see Section 1.7.2).*

```
DATA breast;
SET icdata.breast;
WHERE tx=1;
IF lower=0 THEN lower=.;
RUN;
```

To avoid a parametric (normal) fit, the option FIT=NONE *is added to the* PROBPLOT *statement.*

```
ODS OUTPUT ProbabilityEstimates=NPMLE1;
PROC RELIABILITY DATA=breast;
PROBPLOT ( lower upper ) / ITPRINTEM
                           PRINTPROBS
                           PPOUT
                           FIT=NONE;
RUN;
```

The variables, lower *and* upper *represent the left and right endpoint of the observed intervals. They must be specified between parentheses behind the* PROBPLOT *keyword. Initial values are provided with the* PROBLIST= *option. Convergence is obtained if the change in the log-likelihood between two successive iterations is less than delta (10^{-8} by default). The algorithm stops* *if no convergence is reached after (by default) 1 000 iterations. We experienced, however, that convergence is not obtained after 1 000 iterations, so in the option* MAXITEM= *the maximum number of iterations often needs to be increased.*

An interval probability that converged to a value smaller than the tolerance (10^{-6} by default) is set to zero and the interval probabilities are adjusted so that they add up to one. In addition, the algorithm is restarted and often converges in a few extra iterations. The tolerance value can be adapted with the TOLPROB= *option. With the option* NOPOLISH *the small interval probabilities are not set to zero.*

The iterative history of the estimated interval probabilities is produced with the ITPRINTEM *option. The option* PRINTPROBS *causes the procedure to print the regions of possible support with their corresponding probabilities. Also the corresponding reduced gradient and the Lagrange multiplier are displayed. The option* PPOUT *provides a tabular output of the estimated cdf, standard errors and corresponding confidence limits for each cumulative probability. The probability estimates are stored in a data set* NPMLE1 *with the* ODS OUTPUT *statement.*

Below we show the regions of possible support with their probability, the reduced gradient and the Lagrange multiplier. The latter two are used in checking convergence of the MLE, i.e., for the MLE all Lagrange multipliers should be nonnegative and zero for the intervals with strictly positive probabilities. The reduced gradients should be near zero. Since both conditions were satisfied here, we can claim that the solution provides the MLE.

Lower Lifetime	Upper Lifetime	Probability	Reduced Gradient	Lagrange Multiplier
4	5	0.0463	5.154988E-11	0
6	7	0.0334	-2.58218E-10	0
7	8	0.0887	8.437695E-11	0
11	12	0.0708	-1.77565E-11	0
15	16	0.0000	0	24.27929508
17	18	0.0000	0	7.6501505254
24	25	0.0926	-5.506679E-7	0
25	26	0.0000	0	9.3605205062
33	34	0.0818	-6.830398E-7	0
34	35	0.0000	0	10.521856622
36	37	0.0000	0	2.8664514442
38	40	0.1209	3.7891652E-6	0
40	44	0.0000	0	2.7862740135
46	48	0.4656	-7.542635E-7	0

The output below displays the estimated cdf, standard errors and corresponding confidence limits for each cumulative probability.

Cumulative Probability Estimates

Lower Lifetime	Upper Lifetime	Cumulative Probability	Pointwise 95% Confidence Limits Lower	Pointwise 95% Confidence Limits Upper	Standard Error
.	4	0.0000	0.0000	0.0000	0.0000
5	6	0.0463	0.0071	0.2473	0.0431
7	7	0.0797	0.0132	0.3599	0.0700
...					
44	46	0.5344	0.3314	0.7267	0.1066
48	.	1.0000	1.0000	1.0000	0.0000

The survival function is obtained from the cumulative distribution function with an additional data step.

□

The PROBPLOT statement creates a probability plot of the model for the event times using the distribution specified in the DISTRIBUTION statement (if not specified, a Gaussian model is used). This is particularly useful for comparing a fitted parametric model to the NPMLE (see Section 3.2). By default the NPMLE of the cdf is plotted on the figure. The option NPINTERVALS=POINT or SIMUL produces pointwise confidence intervals or simultaneous confidence bands valid over an interval and computed as the "Equal precision" case of Nair (1984), respectively. Note that the simultaneous confidence intervals are only plotted, they are not printed as output. The confidence intervals given in the probability estimates table are always pointwise intervals, regardless of the NPINTERVALS= specification.

Procedure LIFEREG

Also the LIFEREG procedure calculates the NPMLE of the cdf with the self-consistency algorithm of Turnbull (1976).

Example 3.1 in SAS (PROC LIFEREG)

Again, the left endpoint of a left-censored observation should be first set at missing. Then the Turnbull estimate of $F(t)$ can be obtained using the following code:

```
ODS OUTPUT ProbabilityEstimates=NPMLE1;
PROC LIFEREG DATA=breast;
MODEL (lower, uppper) = ;
PROBPLOT / ITPRINTEM
           PRINTPROBS
```

```
            PPOUT;
RUN;
```

The variables defining the endpoints of the censoring interval (in our example, lower *and* uppper*) must now be specified in a* MODEL *statement. Covariates behind the equality sign are included in the default parametric model that the* LIFEREG *procedure fits, but are ignored in the NPMLE.*

The options in the PROBPLOT *statement and the output (not shown) are similar to those of the* RELIABILITY *procedure (see above). The simultaneous confidence intervals are printed as output.*

□

Plot of estimated survival function

Neither the RELIABILITY nor the LIFEREG procedure have a built-in option to produce a plot of $\widehat{F}(t)$ or $\widehat{S}(t)$ as in Figure 3.3. However, a plot can be created using the output data set with the probability estimates of the cdf. The SAS macro %PlotTurnbull has been written by the authors to facilitate this task.

Example 3.1 in SAS (Macro %PlotTurnbull)
The survival function for the example can be drawn with the following codes:

```
%PlotTurnbull(data=NPMLE1,
              CDF=0,
              CI=1,
              join=1);
```

The data set NPMLE1 *produced by the* RELIABILITY *or* LIFEREG *procedure is used as input. The argument* CDF= *chooses the cdf (*CDF=1*) or survival function S (*CDF=0*) and the argument* CI=1 *adds 95% confidence intervals. The argument* JOIN= *determines how the natural gaps in the NPMLE are drawn. The default value is 0, leaving the gaps empty, indicating that it is unknown how the probability is distributed in the interval. The value 1 assumes a piecewise linear estimate $\widehat{S}$ as shown in the right panel of Figure 3.3. Three more options for drawing the curve are available. Full details of these and other options are found in the documentation of the macro available in the supplementary materials.*

□

Macro %EMICM

Example 3.1 in SAS (Macro %EMICM)
The %EMICM *macro assumes that left-censored data are represented by a zero or a missing value in the left endpoint, while right-censored data must*

have a missing value in the right endpoint. If $L = R$, an exact observation is assumed.

The macro is applied below to the breast data set, stores the NPLME estimates in F_therapy_only *and requests a plot (output not shown).*

```
%EMICM(data=breast,
      left=lower,
      right=upper,
      out=F_therapy_only,
      options=PLOT);
```

Only the LEFT= *and* RIGHT= *arguments are required and should contain the left and right endpoints of the time intervals, respectively.*

*Three maximization techniques are available: the self-consistency algorithm of Turnbull (*EM*), the iterative convex minorant algorithm (*ICM*) and the EM-ICM (*EMICM*) algorithm, which is the default method. With the* METHOD= *argument a particular algorithm is chosen. By default, convergence is concluded when subsequent solutions are close enough (*ERRORTYPE=1*) but other criteria are possible and are based on log-likelihood (*ERRORTYPE=2*), score (=3) or a combination of all (=4). The variance is estimated using the generalized Greenwood formula obtained by resampling (Sun, 2001). The resampling size can be chosen by the* mRS= *argument (by default 50). Adding* GROUP=group *will produce a separate NPMLE for each value of group.*

The keyword PLOT *in the* OPTIONS= *argument produces a figure of the estimated survival curve.*

Note that the previous version of the macro %EMCIM, the %ICE macro, provided Wald-based pointwise confidence intervals based on the observed information matrix. Goodall et al. (2004) showed that these confidence intervals are too wide and provided the authors the SAS *program Goodall.sas (available in the supplementary materials), which suggests two alternative ways of calculating the confidence intervals providing a better coverage.*

□

Procedure ICLIFETEST

The ICLIFETEST procedure offers a nonparametric estimate of the survival function for a group of subjects with interval-censored observations. A comparison between groups (see Section 4.1) is also available. The procedure basically replaces the above %EMICM macro which was the SAS tool to do the job until SAS/STAT 13.1 was released. The procedure differs in how exact observations ($l_i = u_i$) are handled. Namely, while the %EMICM macro subtracts 0.0001 from the left endpoint, the ICLIFETEST procedure treats exact observations as a different class following the proposal of Ng (2002).

Example 3.1 in SAS (PROC ICLIFETEST)
Left-censored observations may have 0 or a missing value for the left endpoint. The NPMLE estimate of $F(t)$ and $S(t)$ can be obtained using the following code:

```
ODS GRAPHICS ON;
PROC ICLIFETEST DATA=breast METHOD=EMICM PLOTS=(SURVIVAL(CL))
                PLOTS=(SURVIVAL(CL)) OUTSURV=outsurv  IMPUTE(SEED=123);
TIME (lower,upper);
RUN;
ODS GRAPHICS OFF;
```

*By default, the ICLIFETEST procedure uses the EMICM algorithm (Wellner and Zahn, 1997) to estimate $S(t)$ but also Turnbull's algorithm (*METHOD= TURNBULL*) or the ICM algorithm (*METHOD=ICM*) are possible. The standard errors are by default estimated using the generalized Greenwood formula obtained by imputation (Sun, 2001). The number of imputation samples can be chosen by adding the* NIMSE= *argument (by default 1 000) to the* IMPUTE *option. Alternatively, they may be obtained using bootstrap sampling (Sun, 2001) by adding the keyword* BOOTSTRAP(NBOOT=1000) *to the PROC ICLIFETEST statement. ODS GRAPHICS must be enabled to generate any plot. The* PLOTS *keyword* SURVIVAL(CL) *requests that the survival functions with 95% pointwise confidence intervals are created. Dashed lines visually connect the survival estimates across the regions of support for which the estimates are not defined. The lines may be omitted by adding the* NODASH *option between parentheses. Adding the option* FAILURE *will plot the NPMLE estimate of $F(t)$. One also has the ability to plot a kernel-smoothed hazard function (HAZARD), the log of the negative log of $\widehat{S}(t)$ versus the log of time (LOGLOGS) or the negative log of $\widehat{S}(t)$ versus time (LOGSURV). The output (not shown) is similar to the output of the* RELIABILITY *procedure (see above).*

□

3.2 Parametric modelling

Experience may suggest a particular distribution of the survival time T in which case one might opt for parametric modelling. Loosely speaking, under the correct assumptions the parametric estimated survival function will provide more precise inference than the NPMLE.

Assume that the cdf F belongs to a family of distributions indexed by $\boldsymbol{\theta} = (\theta_1, \ldots, \theta_p)'$. Then the cdf, survival function, hazard function and density are denoted as $F(t) = F(t; \boldsymbol{\theta})$, $S(t) = S(t; \boldsymbol{\theta})$, $\hbar(t) = \hbar(t; \boldsymbol{\theta})$ and $f(t) =$

$f(t;\,\boldsymbol{\theta})$, respectively. An example is the Weibull distribution with parameters $\boldsymbol{\theta} = (\gamma,\,\alpha)'$. The density function and the survival function are then given by

$$\left.\begin{aligned} f(t;\,\boldsymbol{\theta}) &= \frac{\gamma}{\alpha}\Bigl(\frac{t}{\alpha}\Bigr)^{\gamma-1}\exp\Bigl\{-\Bigl(\frac{t}{\alpha}\Bigr)^{\gamma}\Bigr\}, && \text{for } t>0,\\ S(t;\,\boldsymbol{\theta}) &= \exp\Bigl\{-\Bigl(\frac{t}{\alpha}\Bigr)^{\gamma}\Bigr\}, && \text{for } t>0, \end{aligned}\right\} \tag{3.5}$$

In Appendix B, we provide a list of other popular parametric survival distributions, i.e., log-normal, log-logistic, exponential and Rayleigh.

3.2.1 Estimation

The likelihood for n interval-censored event times is given by $L(\boldsymbol{\theta}) = \prod_{i=1}^{n} L_i(\boldsymbol{\theta})$, where $L_i(\boldsymbol{\theta})$ involves the integral from l_i to u_i as in Expression (1.13). Note that in contrast to the NPMLE, the choice of type of interval (open, closed or half-open) does not matter here. Maximization of $\ell(\boldsymbol{\theta}) = \log\{L(\boldsymbol{\theta})\}$ over a set Θ of admissible values yields the MLE $\widehat{\boldsymbol{\theta}}$.

For the two-parameter survival distributions given in Appendix B ($\boldsymbol{\theta} = (\gamma,\,\alpha)'$) the likelihood is often expressed in terms of the location μ and log-scale $\log(\sigma)$ of the distribution of $Y = \log(T)$ (see Equations (B.1) and (B.2)). For $\widehat{\mu}$ and $\widehat{\log(\sigma)}$, the parameters γ and α are computed using relationship (B.2), and their standard errors are obtained with the delta method.

3.2.2 Model selection

When there is uncertainty about the parametric assumption, several models may be fitted to choose the best fitting model. If no covariates are involved, the fit closest to the NPMLE (Section 3.1) may highlight the best model. A popular model selection is based on information criteria like AIC or BIC.

3.2.3 Goodness of fit

To verify the appropriateness of a chosen parametric distribution, a goodness of fit (GOF) test that assesses the overall fit of the model to the given data set must be performed. However, research on GOF tests for interval-censored data is rare. Lawless and Babineau (2006) developed an omnibus GOF test and Ren (2003) developed a Cramer–von Mises type GOF test based on a resampling method called the "leveraged bootstrap." For further details, we refer to the respective papers. Unfortunately a practical implementation of these methods appears still to be lacking.

A graphical check that can be performed is to overlay the NPMLE and the fitted parametric model of the survival function as illustrated in Figure 3.4 or overlay a probability plot with the NPMLE of the cdf as in Figure 3.5.

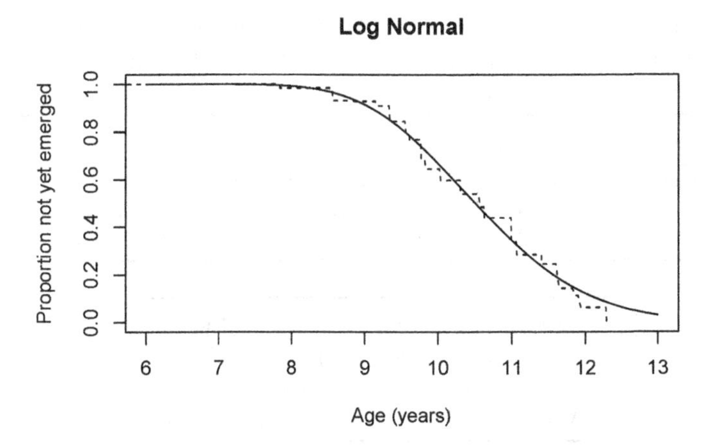

FIGURE 3.4: Signal Tandmobiel study. Log-normal model for emergence of tooth 44 of boys (solid line) overlaid with the NPMLE of the survival function (dashed line), both obtained with the R packages survival and Icens.

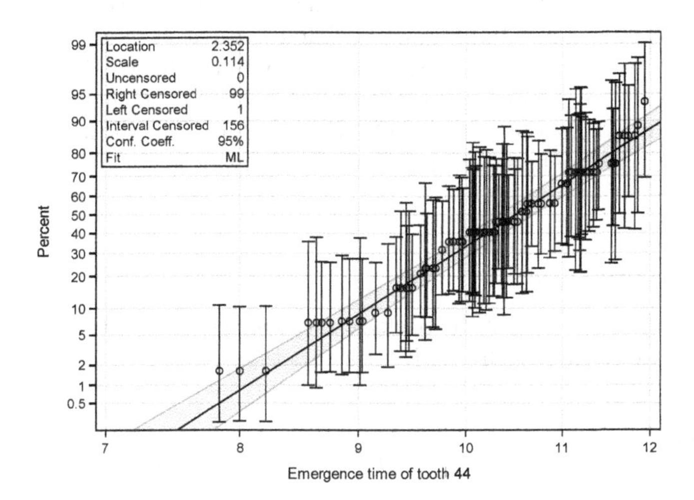

FIGURE 3.5: Signal Tandmobiel study (boys). Probability plot of log-normal model for emergence of tooth 44 overlaid with the NPMLE, solid line: estimated parametric survival function, circles: NPMLE of the survival function with pointwise 95% CI. Plot is obtained from SAS using the RELIABILITY procedure.

TABLE 3.2: Signal Tandmobiel study. Parametric modelling of emergence of tooth 44 of boys, MLEs of the model parameters, log-likelihood value at maximum, mean ($\widehat{\mathrm{E}T}$) and median ($\widehat{\mathrm{med}T}$) of the fitted distribution, obtained with R.

Model	$\widehat{\mu}$ (SE)	$\widehat{\log(\sigma)}$ (SE)	$\widehat{\gamma}$	$\widehat{\alpha}$
Log-normal	2.352 (0.0084)	−2.175 (0.060)	8.8062	10.506
Log-logistic	2.353 (0.0084)	−2.723 (0.067)	15.2216	10.518
Weibull	2.399 (0.0074)	−2.402 (0.064)	11.0495	11.014

Model	$\ell(\widehat{\boldsymbol{\theta}})$	$\widehat{\mathrm{E}T}$	$\widehat{\mathrm{med}T}$
Log-normal	−308.0	10.574	10.506
Log-logistic	−308.7	10.593	10.518
Weibull	−311.7	10.521	10.655

Example 3.2 Signal Tandmobiel study

We applied three popular two-parameter models to estimate the distribution of the emergence times of tooth 44 (permanent right mandibular first premolar) of boys. In Table 3.2 we report ML estimates for μ and $\log(\sigma)$ (location and log-scale of $\log(T)$), γ and α (parameters of the distribution of T). We also report the estimated mean and median emergence times. Further, in Figure 3.4 the best fitting log-normal is overlaid with the NPMLE. Also the other two models are very close to the NPMLE, but the log-normal model appears to be closest and has the lowest AIC and BIC. But, all in all there is not much difference in the predictive ability of the three models.

□

3.2.4 R solution

The function survreg of the package survival (see Appendix D.4 for details) can be used to establish the MLE for a variety of parametric survival models.

Example 3.2 in R

First load the data, select boys and create a survival object:

```
> data("tandmob", package="icensBKL")
> Boys  <- subset(tandmob, fGENDER=="boy")
> Sboy  <- Surv(Boys$L44, Boys$R44, type="interval2")
```

Estimate for instance the log-normal *survival distribution:*

```
> fit.LN <- survreg(Sboy ~ 1, dist="lognormal")
> summary(fit.LN)
```

```
Call:
survreg(formula = Sboy ~ 1, dist = "lognormal")
             Value Std. Error     z          p
(Intercept)  2.35     0.00844 278.7  0.00e+00
Log(scale)  -2.18     0.05998 -36.3 4.91e-288

Scale= 0.114

Log Normal distribution
Loglik(model)= -308   Loglik(intercept only)= -308
Number of Newton-Raphson Iterations: 7
n= 256
```

The row (Intercept) refers to the estimate of μ, the row Log(scale) provides the estimate of $\log(\sigma)$ (see Table 3.2). Further, Scale gives the estimate of σ. Values of $\widehat{\gamma}$ and $\widehat{\alpha}$ in Table 3.2 are obtained from

```
> gamma.LN <- 1/fit.LN$scale
> alpha.LN <- exp(coef(fit.LN))
```

The fitted survival function, as shown in Figure 3.4, is computed and plotted using

```
> tgrid <- seq(6, 13, length=100)
> S.LN <- plnorm(tgrid, meanlog=coef(fit.LN), sdlog=fit.LN$scale,
+                 lower.tail=FALSE)
> plot(tgrid, S.LN, ylim=c(0, 1),
+      xlab="Age (years)", ylab="Proportion not yet emerged",
+      main="Log Normal")
```

The log-logistic model is estimated using

```
> fit.LL <- survreg(Sboy ~ 1, dist="loglogistic")
> gamma.LL <- 1/fit.LL$scale
> alpha.LL <- exp(coef(fit.LL))
> S.LL <- pllogis(tgrid, shape=gamma.LL, scale=1/alpha.LL,
+                  lower.tail=FALSE)
```

Finally, the estimates for the Weibull model are obtained using

```
> fit.W <- survreg(Sboy ~ 1, dist="weibull")
> gamma.W <- 1/fit.W$scale
> alpha.W <- exp(coef(fit.W))
```

```
> S.W <- pweibull(tgrid, shape=gamma.W, scale=1/alpha.W,
+                  lower.tail=FALSE)
```

□

3.2.5 SAS solution

Parametric models can be fitted using the SAS/STAT procedure LIFEREG or the SAS/QC procedure RELIABILITY. The advantage of the RELIABILITY procedure is that there is a built-in option to display the confidence limits of the NPMLE of the survival curve on the probability plot. We will first discuss the LIFEREG statements.

Example 3.2 in SAS (PROC LIFEREG)

The log-normal *model of Table 3.2 can be fitted with the following* SAS *commands*

```
ODS GRAPHICS;
PROC LIFEREG DATA=icdata.tandmob OUTEST=estimates;
WHERE gender=0;                    /* select boys */
MODEL (L44,R44)= /D=LNORMAL;
PROBPLOT  MAXITEM=5000;
RUN;
ODS GRAPHICS OFF;
```

The variables L44 *and* R44 *contain the endpoints of the censoring interval of the emergence time of tooth 44. With* DISTRIBUTION=LNORMAL *(or* DIST= *or* D=*) a log-normal model is fitted. In Appendix E.1, distributions that can be fitted with the* LIFEREG *procedure are listed. The* OUTEST= *argument in the* PROC LIFEREG *statement is used to store the parameter estimates into a data set* estimates.

The PROBPLOT *statement creates a probability plot from the data. Probability plots use an inverse distribution scale so that a cdf plots as a straight line if the model is correctly specified. Two fits are displayed: the model specified in the* MODEL *statement and the Turnbull estimate of the cdf, which will plot approximately as a straight line, thus providing a visual assessment of goodness of fit. When covariates are involved, the values in the* XDATA= *data set are used to create the probability plot. If no* XDATA= *data set is specified, continuous variables are set to their overall mean values and categorical variables specified in the* CLASS *statement are set to their highest level. The* LIFEREG *procedure does not display the confidence interval of the NPMLE of the cdf and ignores covariates for the NPMLE.*

The following output for the parameter estimates is produced.

```
The LIFEREG Procedure

          Analysis of Maximum Likelihood Parameter Estimates

                          Standard   95% Confidence      Chi-
Parameter   DF Estimate     Error        Limits         Square Pr > ChiSq

Intercept    1   2.3519    0.0084    2.3354    2.3684 77673.7      <.0001
Scale        1   0.1136    0.0068    0.1010    0.1277
```

In contrast to R, *the* LIFEREG *procedure reports the scale estimate only on the original scale. The PROBPLOT statement creates a figure similar to Figure 3.5 but without the confidence intervals for the NPMLE of $F(t)$. Values of $\widehat{\gamma}$, $\widehat{\alpha}$, $\widehat{ET}$ and $\widehat{medT}$ in Table 3.2 can be calculated as follows:*

```
DATA _NULL_;
SET estimates;
gamma=1/_SCALE_;
alpha=EXP(Intercept);
median=EXP(Intercept);
mean=EXP(Intercept + 0.5*_SCALE_**2);
PUT "gamma=" gamma 6.2;
PUT "alpha=" alpha 6.4;
PUT "Median= " median 5.2;
PUT "Mean  = " mean   5.2;
RUN;
```

The log-logistic *and* Weibull *model can be fitted using the same code replacing the keyword for the distribution with* LLOGISTIC *and* WEIBULL, *respectively. For the* log-logistic *model, the values of $\widehat{\alpha}$, $\widehat{\gamma}$, $\widehat{ET}$ and $\widehat{medT}$ in Table 3.2 can be calculated as above for the* log-normal *model but replacing the calculation of the mean as follows:*

```
mean=EXP(Intercept)*CONSTANT('PI')*_SCALE_/
       (SIN(CONSTANT('PI')*_SCALE_));
```

For the Weibull *model, the values of $\widehat{\gamma}$, $\widehat{\alpha}$, $\widehat{ET}$ and $\widehat{medT}$ in Table 3.2 can be calculated as above for the* log-normal *model but replacing the calculation of the parameters as follows:*

```
gamma=1/_SCALE_;
alpha=EXP(Intercept);
median=(log(2)**_SCALE_)*EXP(Intercept);
mean=gamma(1 + _SCALE_)*EXP(Intercept);
```

□

Example 3.2 in SAS (PROC RELIABILITY)

The log-normal model can be fitted by the RELIABILITY *procedure as follows:*

```
ODS OUTPUT ParmEst=estimates2;
PROC RELIABILITY DATA=icdata.tandmob;
WHERE gender=0;                        /* select boys */
DISTRIBUTION LOGNORMAL;
PROBPLOT (L44,R44)/ MAXITEM=5000 PCONFPLT LLOWER=7 LUPPER=12;;
INSET;
LABEL L44="Emergence time of tooth 44";
RUN;
```

The MAXITEM=5000 *option increases the maximum number of iterations from 1 000 (default) to 5 000 iterations. The* PCONFPLT *option asks for the pointwise CIs around the NPMLE of the cdf. The* INSET *statement puts the model information onto the plot. Finally, the* LABEL *statement sets the label on the X-axis of the probability plot.*

The first part of the output (not shown) is similar to the output of the LIFEREG *procedure, but without the chi-square tests. Now also summary measures such as mean, median, percentiles (not shown), etc. are given.*

```
  Other Lognormal Distribution
            Parameters
Parameter                       Value

Mean                          10.5736
Mode                          10.3710
Median                        10.5056
Standard Deviation             1.2046
```

The estimates $\widehat{\gamma}$, $\widehat{\alpha}$ *can be calculated as for the* LIFEREG *procedure after modifying the output data set* estimates2 *as shown below.*

```
PROC TRANSPOSE DATA=estimates2 (KEEP=parameter estimate)
     OUT=estimates (RENAME=(location=Intercept scale=_SCALE_));
ID parameter;
RUN;
```

Figure 3.5 displays the probability plot of the log-normal model together with the NPMLE of the cdf (and 95% CIs).

□

3.3 Smoothing methods

A trade-off between fully nonparametric and strict parametric estimation is provided by smoothing methods where the survival function, density or hazard function is expressed in a flexible way. Estimation is the same as for parametric models in Section 3.2 but typically more involved. A flexible model for the survival function, density or hazard function determines the likelihood of the form (1.13) which has to be maximized with respect to unknown coefficients (ML estimation) or is incorporated in a Bayesian model (see Chapter 10). We again assume that the event times are continuously distributed, such that the estimated survival distribution is continuous as well, but now they enjoy more flexibility than the parametric models of Section 3.2.

Because of their parametric and flexible nature, smoothing methods are quite popular for modelling interval-censored data. The smoothing methods avoid the complication of closeness assumptions on the intervals, which may create confusion in purely nonparametric methods. On the other hand, their flexible nature gives them the status of an "almost" nonparametric approach. Therefore, these methods are often referred to in the literature as nonparametric or semiparametric. Various smoothing approaches have been suggested in the literature, but most popular for modelling interval-censored data are those based on *smoothing splines*. The spline-based smoothing approach typically fits the survival function, (log) density or (cumulative) hazard function as a combination of several basis functions. Several types of basis functions have been suggested in the literature: (natural) cubic splines, B-splines, P-splines, M-splines, I-splines and G-splines. A brief overview of them is given in Appendix F.3 and we will refer to this overview rather than explain the specific spline technology in detail in the respective sections.

In this section, we describe three flexible modelling approaches that can directly be applied to estimate a survival distribution. The first approach, called *logspline density estimation* and proposed by Kooperberg and Stone (1992), applies spline smoothing on the logarithm of the density. The second and third approaches fit the density with a mixture of Gaussian mixtures. The third approach can also be viewed as a spline-based smoothing technique.

3.3.1 Logspline density estimation

3.3.1.1 A smooth approximation to the density

Kooperberg and Stone (1992) express $\log f(t)$ as a *natural cubic spline*. A natural cubic spline is a cubic spline for which the second derivatives are zero at the edges of the support, see Appendix F.3. More specifically here, the support $(0,\, T_{max})$ with $0 < T_{max} \leq \infty$ of the survival distribution is divided into subintervals $(s_0 \equiv 0,\, s_1]$, $(s_1,\, s_2]$, $\ldots$, $(s_K,\, s_{K+1} \equiv T_{max})$, using K inner knots $(K \geq 3)$ $0 < s_1 < s_2 < \ldots < s_K < s_{K+1}$. Density $f(t)$ is then modelled as

$$f(t) = f(t;\, \boldsymbol{w}) = \exp\{w_1 + w_2\, t + w_3(t - s_1)_+^3 + \ldots + w_{K+2}(t - s_K)_+^3\},$$
$$0 < t < s_{K+1}, \quad (3.6)$$

where $\boldsymbol{w} = (w_1, \ldots, w_{K+2})'$ is a vector of unknown weights and $(t - s_k)_+^3$ $(k = 1, \ldots, K)$ is the positive part of $(t - s_k)^3$. The weights $\boldsymbol{w}$ satisfy

$$\int_0^{s_{K+1}} \exp\{w_1 + w_2\, t + w_3(t - s_1)_+^3 \ldots + w_{K+2}(t - s_K)_+^3\}\, \mathrm{d}t = 1, \quad (3.7)$$

to ensure that (3.6) is a density.

3.3.1.2 Maximum likelihood estimation

The vector of unknown weights $\boldsymbol{w}$ is estimated using maximum likelihood. To get the MLE of the weights $\boldsymbol{w}$, likelihood (1.10) with $f_i(t_i;\, \boldsymbol{\theta}) \equiv f(t_i;\, \boldsymbol{w})$ must be maximized over $\boldsymbol{\theta} = \boldsymbol{w}$ under restriction (3.7).

An important part of the logspline density estimation is the selection of the number and position of knots. Kooperberg and Stone (1992) suggest to select the knots in a stepwise manner and to choose the final model using information criterion (1.22) with $\nu = K$. Kooperberg and Stone recommend BIC, since it leads in general to simpler models than AIC and hence reduces the chance of finding spurious modes in the density estimate.

Example 3.3 Breast cancer study

We now use the approach of Kooperberg and Stone (1992) to estimate the distribution of the time to breast retraction for breast cancer patients who have been treated with only radiotherapy.

The procedure selects initially six knots. Subsequent stepwise knot deletion and model selection lead to the final model with three knots: $s_1 = 4$, $s_2 = 37.39$, $s_3 = 48$ with ML estimates of the weights being $\widehat{w}_1 = -4.26$, $\widehat{w}_2 = -6.65 \cdot 10^{-3}$, $\widehat{w}_3 = -2.72 \cdot 10^{-6}$, $\widehat{w}_4 = 11.28 \cdot 10^{-6}$, $\widehat{w}_5 = -8.56 \cdot 10^{-6}$. The maximized log-likelihood value is -64.80 with $\mathsf{BIC} = 141.08$.

The estimated density and survival function are shown in Figure 3.6 together with the NPMLE of the survival function.

Note that, since the maximal upper limit of the interval-censored observations (48) exceeds the maximal right-censored observation (46), the last region of support of the NPMLE has 48 as the upper endpoint and not ∞ as would be the case if at least one right-censored observation had exceeded 48. Consequently, the NPMLE of $S(\cdot)$ is zero beyond $t = 48$ months which may not be realistic.

□

3.3.1.3 R solution

Logspline density estimation can be done with the function oldlogspline of the R package logspline (Kooperberg, 2016). The function produces an object of class oldlogspline that contains information on the fitted model. The functions doldlogspline, poldlogspline, qoldlogspline provide information on the density, cumulative distribution function and quantiles of the estimated distribution, respectively. The estimated density can be visualized with the plot function. Finally, the function roldlogspline produces a random sample from the estimated distribution. A detailed description of the syntax of the function oldlogspline is given in Appendix D.5.

In the same package, the function logspline allows for uncensored observations, an improved knot deletion and an additional algorithm, as described in Stone et al. (1997).

Example 3.3 in R

As before, we start by loading the data and selecting the radiotherapy-only group. Then we create vectors of upper limits of left-censored and lower limits of right-censored observations and a two-column matrix with interval-censored observations.

```
> Left <- breastR$upp[is.na(breastR$low)]
> Right <- breastR$low[is.na(breastR$upp)]
> Interval <- subset(breastR, !is.na(low) & !is.na(upp))
```

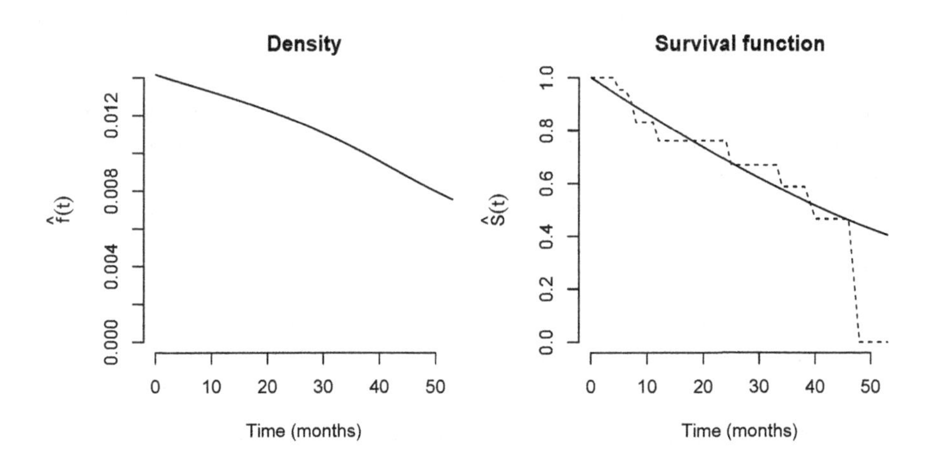

FIGURE 3.6: Breast cancer study (radiotherapy-alone group). Distribution of the time to breast retraction estimated using the logspline method (solid lines) obtained with R package logspline. Left panel: estimated density, right panel: estimated survival function compared to NPMLE (dashed line).

The output from logspline density estimation is shown below. Recall that the initial knots are chosen automatically and the final knots are selected with BIC.

```
> library("logspline")
> fit <- oldlogspline(right=Right, left=Left, interval=Interval,
+                     lbound=0)
> summary(fit)
```

```
 knots loglik    AIC minimum penalty maximum penalty
     3 -64.80 141.08            0.62             Inf
     4 -64.53 144.37              NA              NA
     5 -64.17 147.49            0.05            0.62
     6 -64.15 151.26            0.00            0.05
the present optimal number of knots is  3
penalty(AIC) was the default: BIC=log(samplesize): log( 46 )= 3.83
```

The relevant part of the output is shown above and includes the columns knots *(number of knots),* loglik *(corresponding maximal value of the log-likelihood) and* AIC*. The column* AIC *is in fact equal to BIC as also indicated by the last line of the output. We can see that the algorithm starts with six knots and ends with three knots. Columns* minimum penalty *and* maximum penalty *show the range for the penalty parameter c for which the information criterion (1.22) selects the model on the corresponding row. Hence, for a penalty parameter of 0 to 0.05 the model with 6 knots would be selected, for a penalty parameter of 0.05 to 0.62 the model with 5 knots would be selected and for a penalty parameter higher than 0.62 the model with 3 knots would be selected. The model with 4 knots would not be selected for any value of the penalty as indicated by the missing values (*NA*). Finally, the model with 3 knots is selected.*

The following print statement gives the starting values of knots $s_1, \ldots, s_6$:

```
> print(fit$knots)
```

and are

```
[1]  4.00000 14.87743 27.88053 37.39427 45.92727 48.00000
```

The final values of spline weights $\boldsymbol{w}$ are the non-zero elements in:

```
> print(fit$coef)
```

```
[1] -4.257438e+00 -6.651228e-03 -2.718045e-06  0.000000e+00
[5]  0.000000e+00  1.127636e-05  0.000000e+00 -8.558312e-06
```

Note that $\widehat{w}_1 =$ fit$coef[1] *and* $\widehat{w}_2 =$ fit$coef[2] *are the estimated intercept and slope from Equation (3.6). Further,* fit$coef[3], ..., fit$coef[8] *are the estimated weights* $\widehat{w}_3, \ldots, \widehat{w}_8$ *corresponding to the original six knots* $s_1, \ldots, s_6$*. Zeros in* fit$coef *correspond to the knots that were deleted using the stepwise knot deletion procedure. For example, from the fact that* $\widehat{w}_4 =$ fit$coef[4] *is zero, we conclude that the knot* $s_2 =$ fit$knots[4-2] $\equiv$ fit$knots[2] $=$ *14.88 has been deleted and does not contribute to the final solution.*

To display the fitted density, we use the classical plot *function:*

```
> plot(fit)
```

The fitted density and survival function shown in Figure 3.6 were drawn using

```
> tgrid <- seq(0, 53, length=100)
> f <- doldlogspline(tgrid, fit)
> plot(tgrid, f, type="l",
+      xlab="Time (months)", ylab=expression(hat(f)(t)),
+      ylim=c(0, 0.014), main="Density")
```

and

```
> F <- poldlogspline(tgrid, fit)
> plot(tgrid, 1 - F, type="l",
+      xlab="Time (months)", ylab=expression(hat(S)(t)),
+      ylim=c(0, 1), main="Survival function")
```

Selection of the number of knots on the basis of AIC (in this case leading to the same solution) is provided by:

```
> fitAIC <- oldlogspline(right=Right, left=Left, interval=Interval,
+                        lbound=0, penalty=2)
> summary(fitAIC)
```

```
 knots loglik    AIC minimum penalty maximum penalty
     3 -64.80 135.59            0.62             Inf
     4 -64.53 137.05              NA              NA
     5 -64.17 138.34            0.05            0.62
     6 -64.15 140.29            0.00            0.05
the present optimal number of knots is  3
penalty(AIC) was  2 , the default (BIC)  would have been 3.83
```

□

3.3.2 Classical Gaussian mixture model

We present here the classical Gaussian mixture approach in a frequentist framework as an introduction to the penalized Gaussian mixture approach described in the next section. For examples of this approach in a Bayesian context, we refer to Section 10.2.

The Gaussian mixture model

To model unknown distributional shapes *finite mixture distributions* have been advocated by, e.g., Titterington et al. (1985, Section 2.2) as *semiparametric* structures. In this section, we concentrate on Gaussian mixtures. Since the support of the Gaussian mixture is the real line, Gaussian mixture models for a survival distribution are developed for $Y = \log(T)$. In the following, let $g(y)$ be the density of Y, that is,

$$
\begin{aligned}
g(y) &= \exp(y)\, f\{\exp(y)\}, && y \in \mathbb{R}, \\
f(t) &= t^{-1} g\{\log(t)\}, && t > 0.
\end{aligned}
\tag{3.8}
$$

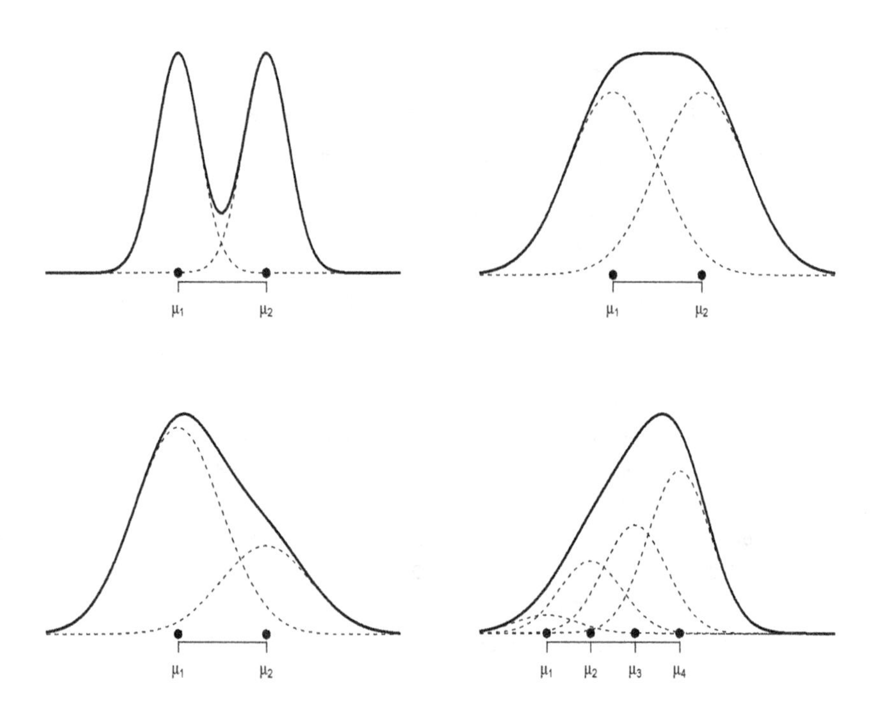

FIGURE 3.7: Several densities expressed as two- or four-component Gaussian mixtures.

Using a (shifted and scaled) finite Gaussian mixture, the density $g(y)$ is expressed as

$$g(y) = g(y;\, \boldsymbol{\theta}) = \frac{1}{\tau} \sum_{k=1}^{K} w_k \, \varphi_{\mu_k, \sigma_k^2}\Bigl(\frac{y-\alpha}{\tau}\Bigr), \tag{3.9}$$

where $\alpha \in \mathbb{R}$ is the fixed shift parameter and $\tau > 0$ is the fixed scale parameter. Theoretically, it is sufficient to consider $\alpha = 0$ and $\tau = 1$. Nevertheless, in some situations (see Section 10.2), it is useful to choose α and τ such that the shifted and scaled distribution of Y, i.e., the distribution of $\tau^{-1}(Y - \alpha)$ has approximately zero mean and unit variance.

In general, the vector of unknown parameters is composed of K mixture components $\boldsymbol{w} = (w_1, \ldots, w_K)'$, the mixture weights, $\boldsymbol{\mu} = (\mu_1, \ldots, \mu_K)'$, the mixture means, and $\boldsymbol{d} = \bigl(\sigma_1^2, \ldots, \sigma_K^2\bigr)'$ the mixture variances. That is, $\boldsymbol{\theta} = (K,\, \boldsymbol{w}',\, \boldsymbol{\mu}',\, \boldsymbol{d}')'$. Note that even if the shift parameter α and the scale parameter τ are assumed to be fixed in Expression (3.9), the location and the variability of the distribution of Y still depends on the unknown parameter $\boldsymbol{\theta}$ and as such it is estimated. We have

$$\mathsf{E}(Y) = \alpha + \tau \sum_{k=1}^{K} w_k \mu_k,$$

$$\mathsf{var}(Y) = \tau^2 \sum_{k=1}^{K} w_k \biggl\{ \sigma_k^2 + \Bigl(\mu_k - \sum_{j=1}^{K} w_j \mu_j\Bigr)^2 \biggr\}.$$

Figure 3.7 illustrates how two- or four-component Gaussian mixtures can be used to obtain densities of different shapes as a function of the (unknown) parameters.

Maximum likelihood estimation

The observed data, denoted as $\boldsymbol{\mathcal{D}} = \bigl\{\lfloor l_i,\, u_i \rfloor,\, \delta_i\bigr\}$, represents the interval-censored random sample $T_1, \ldots, T_n$. For convenience, let $\boldsymbol{\mathcal{D}}^\star = \bigl\{\lfloor l_i^\star,\, u_i^\star \rfloor,\, \delta_i\bigr\}$ be the corresponding interval-censored random sample $Y_1, \ldots, Y_n$. That is, $l_i^\star = \log(l_i)$, $u_i^\star = \log(u_i)$ with the convention that $\log(0) = -\infty$ and $\log(\infty) = \infty$ and $l_i^\star = u_i^\star = y_i$ for uncensored observations ($\delta_i = 1$). Likelihoods of $\boldsymbol{\mathcal{D}}$ and $\boldsymbol{\mathcal{D}}^\star$ differ only by a multiplicative factor stemming from the Jacobian of the logarithmic transformation. Hence it is possible to consider only the likelihood of $\boldsymbol{\mathcal{D}}^\star$ for inference. The individual likelihood contributions (1.13) are then computed as

$$L_i(\boldsymbol{\theta}) = \oint_{l_i^\star}^{u_i^\star} g(y;\,\boldsymbol{\theta})\mathrm{d}y$$

$$= \begin{cases} \int_{l_i^\star}^{\infty} g(y;\,\boldsymbol{\theta})\,\mathrm{d}y = 1 - G(l_i^\star;\,\boldsymbol{\theta}), & \delta_i = 0, \\ g(y_i;\,\boldsymbol{\theta}), & \delta_i = 1, \\ \int_{-\infty}^{u_i^\star} g(y;\,\boldsymbol{\theta})\,\mathrm{d}y = G(u_i^\star;\,\boldsymbol{\theta}), & \delta_i = 2, \\ \int_{l_i^\star}^{u_i^\star} g(y;\,\boldsymbol{\theta})\,\mathrm{d}y = G(u_i^\star;\,\boldsymbol{\theta}) - G(l_i^\star;\,\boldsymbol{\theta}) & \delta_i = 3, \end{cases} \tag{3.10}$$

where

$$G(y) = G(y;\,\boldsymbol{\theta}) = \sum_{k=1}^{K} w_k\,\Phi_{\mu_k,\sigma_k^2}\Big(\frac{y-\alpha}{\tau}\Big) \tag{3.11}$$

is the cumulative distribution function corresponding to the Density (3.9). Maximum likelihood based methods pose two main difficulties when estimating $\boldsymbol{\theta}$:

1. When the number of mixture components K is unknown, one of the basic regularity conditions for the validity of the classical maximum likelihood theory is violated. Namely, the parameter space does not have a fixed dimension since the number of unknowns (number of unknown mixture weights, means and variances) is one of the unknowns. This difficulty is discussed in detail by Titterington et al. (1985, Section 1.2.2).

2. For a fixed $K \geq 2$, the likelihood becomes unbounded when one of the mixture means, say μ_1, is equal to one of the observations y_i ($i \in \{1, \ldots, n\}$) and when the corresponding mixture variance, σ_1^2, converges to zero. Then the MLE does not exist, see, e.g., McLachlan and Basford (1988, Section 2.1).

In the frequentist approach, the first problem is addressed by varying K and choosing the model based on AIC or BIC. To accommodate the second problem, homoscedastic Gaussian mixtures, i.e., with $\sigma_k^2 = \sigma^2$ for all k can be used leading to a bounded likelihood.

We return in Section 10.2 for a Bayesian approach to fit a Gaussian mixture model to survival distributions. In that section, we also suggest other, more natural, solutions to the two above-mentioned difficulties.

R solution

We do not treat the method here in detail but the classical Gaussian mixture model may be fitted in R by using the function smoothSurvReg of the R package smoothSurv (Komárek, 2015). The function fits in principle the penalized Gaussian mixture model introduced in the next section but, by setting the smoothing parameter λ equal to zero, the classical Gaussian mixture model is obtained.

3.3.3 Penalized Gaussian mixture model

An alternative Gaussian mixture representation

The penalized Gaussian mixture (PGM) combines classical spline smoothing ideas from Section 3.3.1 and Gaussian mixtures from Section 3.3.2. In the logspline density estimation of Section 3.3.1, the unknown function $\log f(t)$ was expressed as a linear combination of prespecified (basis) polynomials (see Equation (3.6)) where only the weights had to be estimated. In this part, similar to what was done in Section 3.3.2, we model the density $g(y)$ of $Y = \log(T)$. A more general formulation of Equation (3.11) to fit the density $g(y)$ is

$$g(y) = g(y;\, \boldsymbol{\theta}) = \frac{1}{\tau} \sum_{k=-K}^{K} w_k\, B_k\Big(\frac{y-\alpha}{\tau}\Big), \tag{3.12}$$

where $\boldsymbol{w} = (w_{-K}, \ldots, w_K)'$ is a vector of unknown weights, α is the unknown location, τ is the unknown scale parameter, and $B_k(y)$, $k = -K, \ldots, K$ are prespecified basis functions. Hence the total vector of unknown parameters is now $\boldsymbol{\theta} = (\alpha,\, \tau,\, \boldsymbol{w}')'$. A suitable set of basis functions, suggested by Ghidey et al. (2004) and Komárek et al. (2005a) might consist of Gaussian densities with equidistant means $\boldsymbol{\mu} = (\mu_{-K}, \ldots, \mu_K)'$ playing the role of knots and a common basis standard deviation σ_0. That is, $B_k \equiv \varphi_{\mu_k,\,\sigma_0^2}$, and

$$g(y) = g(y;\, \boldsymbol{\theta}) = \frac{1}{\tau} \sum_{k=-K}^{K} w_k\, \varphi_{\mu_k,\,\sigma_0^2}\Big(\frac{y-\alpha}{\tau}\Big). \tag{3.13}$$

See Figure 3.8 for an illustration. The conditions

$$0 < w_k < 1 \ (k = -K, \ldots, K) \qquad \sum_{k=-K}^{K} w_k = 1 \tag{3.14}$$

ensure that (3.13) is a density and hence will be taken into account in the estimation procedure. Further, to make the shift and scale parameters α and τ identifiable, the first two moments of the standardized mixture density $g^\star(y^\star) = \sum_{k=-K}^{K} w_k\, \varphi_{\mu_k,\sigma_0^2}(y^\star)$ must be fixed, e.g., setting the mean to zero and the variance to one. Therefore, additional constraints

$$\mathsf{E}\Big(\frac{Y-\alpha}{\tau}\Big) = \sum_{k=-K}^{K} w_k \mu_k = 0, \tag{3.15}$$

$$\mathsf{var}\Big(\frac{Y-\alpha}{\tau}\Big) = \sum_{k=-K}^{K} w_k(\mu_k^2 + \sigma_0^2) = 1 \tag{3.16}$$

must be considered when estimating the model parameters.

In contrast to Section 3.3.2, the number of mixture components $(2K +$

1), the means $\boldsymbol{\mu}$ and the basis standard deviation σ_0 are all assumed to be prespecified in advance. Motivated by penalized B-spline smoothing (Eilers and Marx, 1996) with B-splines, the Gaussian densities in the above mixture construction is sometimes referred to as *G(aussian)-splines*. Also the number of components ($\equiv$ basis functions) $(2K + 1)$ is taken to be relatively high (≈ 30), knots $\boldsymbol{\mu}$ are equidistant and centered around zero, that is $\mu_k = k\delta$, $k = -K, \ldots, K$, where $\delta > 0$ is the chosen distance between two consecutive knots. Komárek et al. (2005a) recommend to take $\delta \approx 0.3$, $\sigma_0 \approx (2/3)\delta$ since the shift and scale parameters are estimated, the prespecified knots have to cover a high probability region of the zero-mean, unit-variance distribution. In most practical situations, taking μ_{-K} (μ_K) between -6 and -4.5 (4.5 and 6) provides the range of knots broad enough.

Constrained estimation is avoided when the alternative parametrization based on transformed weights is used: $\boldsymbol{a} = (a_{-K}, \ldots, a_K)'$ such that

$$a_k(\boldsymbol{w}) = \log\left(\frac{w_k}{w_0}\right) \qquad (k = -K, \ldots, K). \tag{3.17}$$

The coefficient a_0 is taken equal to zero and thus not estimated. The original

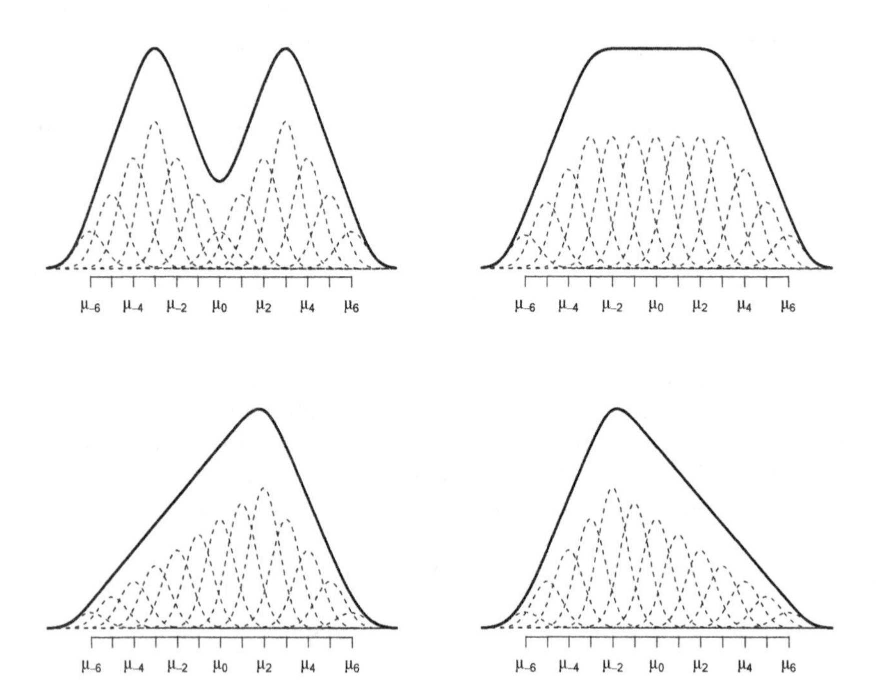

FIGURE 3.8: Several densities expressed as homoscedastic Gaussian mixtures with $(2K + 1) = 13$ and equidistant set of means – knots.

weights $\boldsymbol{w}$ are computed from the transformed weights $\boldsymbol{a}$ by

$$w_k(\boldsymbol{a}) = \frac{\exp(a_k)}{\sum_{j=-K}^{K} \exp(a_j)} \qquad (k = -K, \ldots, K) \tag{3.18}$$

and automatically satisfy the constraints (3.14). Constraints (3.15) and (3.16) can also be avoided if two coefficients, say, a_{-1} and a_1, are expressed as functions of remaining coefficients $\boldsymbol{a}_{-(-1,0,1)} = (a_{-K}, \ldots, a_{-2}, a_2, \ldots, a_K)'$, see Komárek et al. (2005a) for details. Hence, the vector of parameters of the Model (3.13) that have to be estimated is $\boldsymbol{\theta} = (\alpha,\, \tau,\, \boldsymbol{a}'_{-(-1,0,1)})'$. According to the notation of Section 3.3.2, the individual likelihood contribution $L_i(\boldsymbol{\theta})$ is again computed using Expression (3.10), now with $g(y)$ given by (3.13) and $G(y)$ defined by

$$G(y) = G(y;\, \boldsymbol{\theta}) = \sum_{k=-K}^{K} w_k(\boldsymbol{a})\, \Phi_{\mu_k, \sigma_0^2}\Big(\frac{y - \alpha}{\tau}\Big). \tag{3.19}$$

Penalized maximum likelihood estimation

The above method assumes that K is large, but when is large "large"? Also, when K is large overfitting the data comes quickly around the corner, but on the other hand a too small K causes underfitting. A solution to this problem is to take a large K (and do not worry about "too large"), but penalize the model for complexity. This then yields penalized splines or *P-splines*. The term P-spline is, however, reserved for penalized B-spline. Nevertheless, here in a similar spirit the G-splines will be penalized for model complexity.

Penalized Gaussian mixture modelling has been implemented in the two R packages: smoothSurv and bayesSurv. The first is likelihood-based and the second makes use of Bayesian estimation procedures. Here we focus on penalized maximum likelihood estimation (PMLE).

The Gaussian mixture model has $2K+1-3+2 = 2K$ free parameters, with $2K$ (number of weights minus one) relatively high. Overfitting the data and identifiability problems are avoided by applying a penalty on the $\boldsymbol{a}$ coefficients to the log-likelihood resulting in the penalized maximum likelihood estimation method. Let $\ell(\boldsymbol{\theta}) = \sum_{i=1}^{n} \log\big\{L_i(\boldsymbol{\theta})\big\}$ be the log-likelihood of the model. The PMLE of $\boldsymbol{\theta}$ is obtained by maximizing the penalized log-likelihood

$$\ell_P(\boldsymbol{\theta};\, \lambda) = \ell(\boldsymbol{\theta}) \;-\; q(\boldsymbol{a};\, \lambda), \tag{3.20}$$

where $q(\boldsymbol{a};\, \lambda)$ is the penalty term and $\lambda > 0$ is a fixed tuning parameter that controls the smoothness of the fitted error distribution. Further, the penalty term inhibits identifiability problems due to overparametrization. For a given λ, the penalty is expressed as a sum of squared finite differences of order s of the coefficients $\boldsymbol{a}$ of adjacent mixture components:

$$q(\boldsymbol{a};\, \lambda) = \frac{\lambda}{2} \sum_{k=-K+s}^{K} \big(\Delta^s a_k\big)^2, \tag{3.21}$$

where $\Delta^1 a_k = a_k - a_{k-1}$, $\Delta^s a_k = \Delta^{s-1} a_k - \Delta^{s-1} a_{k-1}$, $(s = 2, \ldots)$. Komárek et al. (2005a) recommend to use $s = 3$ which can also lead to an empirical check for normality of the density g and consequently for a log-normal distribution of the survival time T. Indeed, they show that for $\lambda \to \infty$ maximization of the penalized log-likelihood (3.20) leads to a Gaussian density. Hence high optimal values of λ (see below) indicate normality of the density g.

Selecting the smoothing parameter

With $\lambda = 0$ penalized maximum likelihood is equivalent to standard maximum likelihood with $\nu = 2K$ free parameters. As λ increases, the penalty term gets more important and the *effective number of parameters* denoted by ν decreases. To find a trade-off between a too parsimonious (low ν, high λ) and a too complex (high ν, low λ) model, Komárek et al. (2005a) chose the value of the smoothing parameter λ by minimizing Akaike's information criterion, which now takes the form

$$\mathsf{AIC}(\lambda) = -2\ell\big\{\widehat{\boldsymbol{\theta}}(\lambda)\big\} + 2\nu(\lambda), \tag{3.22}$$

where $\widehat{\boldsymbol{\theta}}(\lambda)$ maximizes the penalized log-likelihood (3.20) and $\nu(\lambda)$ is the effective number of parameters with given λ. As Komárek et al. (2005a) further explain, the effective number of parameters is given by

$$\nu(\lambda) = \text{trace}\big(\widehat{\boldsymbol{H}}^{-1}\widehat{\boldsymbol{I}}\big),$$

where

$$\widehat{\boldsymbol{H}} = -\frac{\partial^2 \ell_P\{\widehat{\boldsymbol{\theta}}(\lambda);\, \lambda\}}{\partial\boldsymbol{\theta}\partial\boldsymbol{\theta}'}, \qquad \widehat{\boldsymbol{I}} = -\frac{\partial^2 \ell\{\widehat{\boldsymbol{\theta}}(\lambda)\}}{\partial\boldsymbol{\theta}\partial\boldsymbol{\theta}'}.$$

In the R package smoothSurv (see Section 3.3.3.1), the optimal value of λ is found using a grid search. Since the log-likelihood is of order $O(n)$, it is useful to let the search grid depend on the sample size n. This allows also to use approximately the same grid for data sets of different size while maintaining the proportional importance of the penalty term in the penalized log-likelihood at the same level. In applications presented by Komárek et al. (2005a), a grid of λ values being $n\,e^2$, $n\,e^1$, $\ldots$, $n\,e^{-9}$ was used, which is the strategy used in Example 3.4.

Inference

It is of interest to derive characteristics of the fitted survival distribution. For example, the mean survival time is given by

$$\mathsf{E}(T) = \sum_{k=-K}^{K} w_k \exp\Big(\alpha + \tau\,\mu_k + \frac{\tau^2\sigma_0^2}{2}\Big). \tag{3.23}$$

$\mathsf{E}(T)$ is estimated by plugging the estimated values from the above penalized approach into Expression (3.23). The standard errors can be obtained using

TABLE 3.3: Signal Tandmobiel study (boys). Estimated parameters of the emergence distribution of tooth 44, obtained with R.

Parameter	Estimate	SE	95% CI
α	2.351	0.0084	(2.335, 2.368)
$\log(\tau)$	−2.186	0.0630	(−2.310, −2.063)
$\mathsf{E}(T)$	10.566	0.0919	(10.386, 10.746)

the delta method as soon as an estimate for $\mathsf{var}(\widehat{\boldsymbol{\theta}})$ is available. Komárek et al. (2005a) discuss two possible approaches to estimate $\mathsf{var}(\widehat{\boldsymbol{\theta}})$. The first proposal leads to the *pseudo-covariance* matrix

$$\widehat{\mathsf{var}}_P(\widehat{\boldsymbol{\theta}}) = \widehat{\boldsymbol{H}}^{-1}, \tag{3.24}$$

which is motivated by a pseudo-Bayesian technique for generating confidence bands around the cross-validated smoothing spline suggested by Wahba (1983). The second approach, suggested by Gray (1992) and based on formal properties of the ML estimates, leads to the *asymptotic covariance* matrix

$$\widehat{\mathsf{var}}_A(\widehat{\boldsymbol{\theta}}) = \widehat{\boldsymbol{H}}^{-1}\,\widehat{\boldsymbol{I}}\,\widehat{\boldsymbol{H}}^{-1}. \tag{3.25}$$

Since for finite samples the middle matrix $\widehat{\boldsymbol{I}}$ in the asymptotic covariance matrix does not have to be positive definite, Komárek et al. (2005a) prefer the pseudo-covariance matrix. Simulations also showed better coverage with the pseudo-covariance than with the asymptotic covariance matrix.

Example 3.4 Signal Tandmobiel study

To compare the results with those obtained in Example 3.2, we estimate the distribution of the emergence time of tooth 44 (permanent right mandibular first premolar) of boys once more. An equidistant set of 41 knots ranging from -6 to 6 with $\delta = 0.3$ was taken together with a basis standard deviation of $\sigma_0 = 0.2$. The third order penalty ($s = 3$ in Equation (3.21)) was used to maximize the penalized log-likelihood.

We know from Example 3.2 that a log-normal distribution is a sensible model for the distribution of the emergence time. Recall that high λ values indicate a log-normal distribution, hence it is not surprising that the optimal AIC value of 620.8 was found for the highest λ in the grid search ($\lambda = n\,e^2 = 256\,e^2 \doteq 1\,892$). This explains the effective number of parameters of $\nu = 2.63$, while there are in total 40 model parameters. Since there are two "standard" parameters in the model (shift α and scale τ), the estimation of the distributional shape consumed only 0.63 degrees of freedom. For comparison, with the smallest λ in the grid search ($\lambda = n\,e^{-9} \doteq 0.032$) AIC = 628.3 and $\nu = 9.42$.

Table 3.3 shows some estimated model parameters. Note that the shift

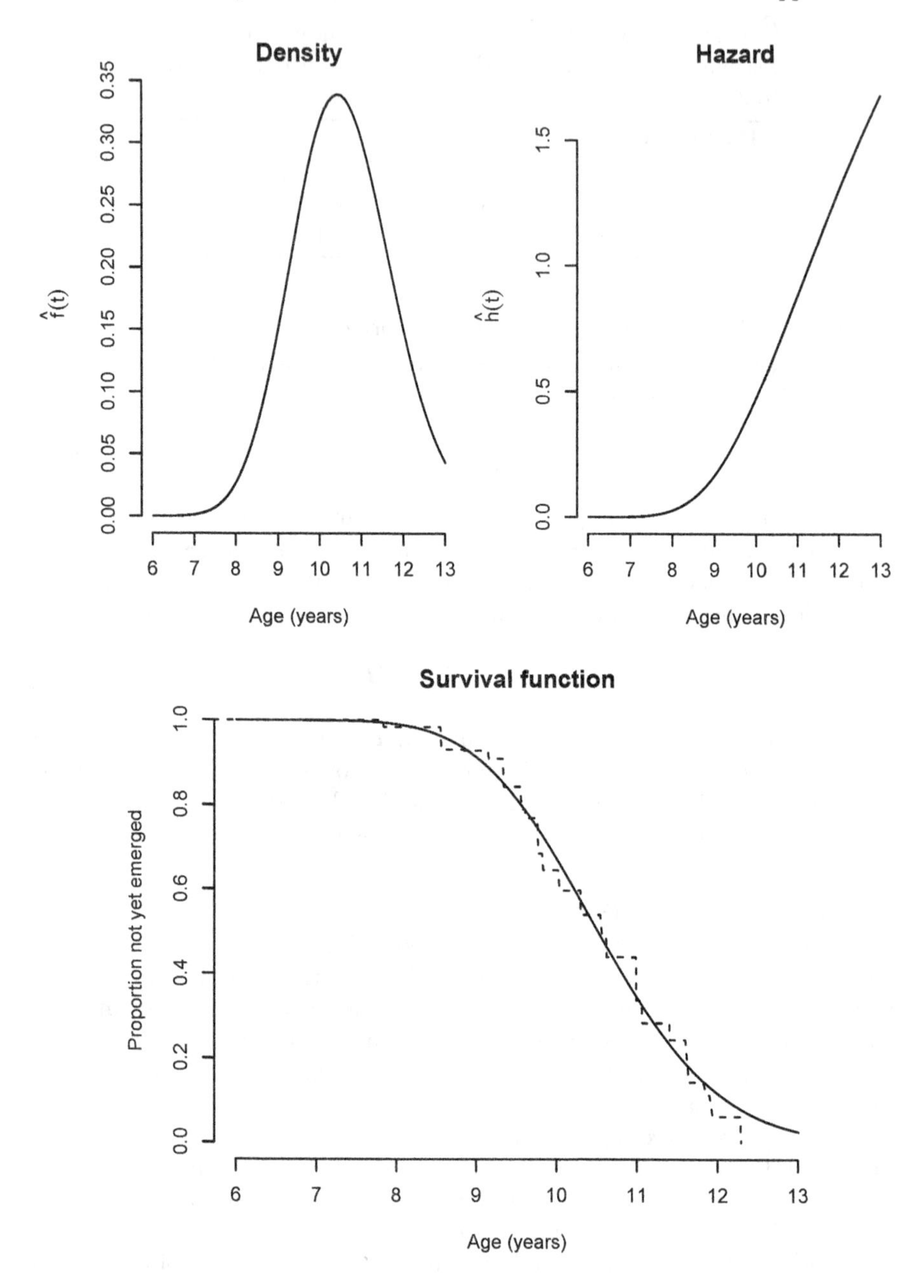

FIGURE 3.9: Signal Tandmobiel study (boys). Distribution of the time to emergence of tooth 44 estimated using the penalized Gaussian mixture. Upper panel: Estimated density and hazard function, lower panel: estimated survival function compared to NPMLE (dashed line) obtained with R package smoothSurv.

parameter α corresponds to μ in the log-normal model in Table 3.2. Similarly, the log-scale $\log(\tau)$ corresponds to $\log(\sigma)$ in the log-normal model. Comparing the standard errors in Tables 3.2 and 3.3, it is seen that a more flexible penalized Gaussian mixture practically does not lead to a decrease in efficiency compared to a parametric model when these are practically the same. However, with the penalized Gaussian mixture, we are more confident about the estimated distribution.

Finally, Figure 3.9 shows various estimates of the distribution of the emergence time T all derived from the estimated density g of $\log(T)$. The survival function is again compared to the NPMLE. An increasing hazard function is in accordance with what is known about emergence of tooth 44. Namely, there is an increasing probability that tooth 44 emerges from age 8 to 9 years, and then increases with age if emergence has not happened yet.

□

3.3.3.1 R solution

The methods described in this section have been implemented in the R package smoothSurv (Komárek, 2015). We note that the package allows for covariates in the model, which leads to the accelerated failure time (AFT) described in Section 6.2. However, fitting an AFT model with only an intercept is equivalent to simply estimating a survival distribution, as described now. The function smoothSurvReg provides the model fit, see Appendix D.6 for details.

Example 3.4 in R

First, load the data, select boys only and create a survival object for tooth 44. The model is fitted using:

```
> library("smoothSurv")
> mod <- smoothSurvReg(Sboy ~ 1, info=FALSE)
```

Setting info=FALSE *suppresses printing detailed information on the iterative optimization of the penalized log-likelihood. The same result is obtained from*

```
> mod <- smoothSurvReg(Sboy ~ 1, info=FALSE,
+                      sdspline=0.2,
+                      by.knots=0.3, dist.range=c(-6, 6),
+                      difforder=3)
```

with now the basis functions and the order of the penalty explicitly specified. A summary of the fitted model is obtained with

```
> print(mod)
```

```
Call:
smoothSurvReg(formula = Sboy ~ 1, info = FALSE)

Estimated Regression Coefficients:
             Value Std.Error Std.Error2      Z     Z2
(Intercept)  2.351   0.00838   0.008362 280.58 281.21
Log(scale)  -2.186   0.06303   0.062377 -34.69 -35.05
                     p          p2
(Intercept)  0.000e+00  0.000e+00
Log(scale)  1.266e-263 3.593e-269

Scale = 0.1123

Penalized Loglikelihood and Its Components:
     Log-likelihood: -307.759
            Penalty: -0.0874537
   Penalized Log-likelihood: -307.8464

Degree of smoothing:
   Number of parameters: 40
                  Mean parameters: 1
                 Scale parameters: 1
                Spline parameters: 38

                  Lambda: 7.389056
             Log(Lambda): 2
                      df: 2.627579

AIC (higher is better):  -310.3866

Number of Newton-Raphson Iterations:  4
n = 256
```

The section "Estimated Regression Coefficients" displays the estimated shift parameter α and the log-scale $\log(\tau)$, shown in Table 3.3. In columns Std.Error *and* Std.Error2 *the standard errors are given derived from the pseudo-covariance matrix (3.24) and the asymptotic covariance matrix (3.25), respectively. Columns* Z, Z2, p, p2 *are corresponding Wald statistics and P-values for the tests of hypotheses $H_0 : \alpha = 0$ and $H_0 : \log(\tau) = 0$, respectively (but they are of no interest here).*

In the "Degree of smoothing" section, the value of Lambda *is the optimal value of λ/n and similarly,* Log(Lambda) *is the optimal value of $\log(\lambda/n)$. The row* df *gives the effective number of parameters ν pertaining to the optimal* *value of the smoothing parameter λ. Finally, we should highlight that the AIC used in the function* smoothSurv *is defined in an alternative way as*

$$\mathsf{AIC}_{alter}(\lambda) = \ell\{\widehat{\boldsymbol{\theta}}(\lambda)\} - \nu(\lambda), \tag{3.26}$$

that is, $\mathsf{AIC}(\lambda)$ *defined in (3.22) is equal to* $-2\ \mathsf{AIC}_{alter}(\lambda)$.

Estimates of the weights $\boldsymbol{w}$ *together with their standard errors can be seen with (only a part of the output shown):*

```
> print(mod, spline=TRUE)
```

```
Details on (Fitted) Error Distribution:
         Knot SD basis    c coef. Std.Error.c Std.Error2.c
knot[1]  -6.0       0.2 7.790e-09   4.267e-08    2.915e-08
knot[2]  -5.7       0.2 3.708e-08   1.734e-07    1.213e-07
knot[3]  -5.4       0.2 1.635e-07   6.461e-07    4.631e-07
knot[4]  -5.1       0.2 6.679e-07   2.203e-06    1.620e-06
...
               Z     Z2         p        p2
knot[1]   0.1826 0.2672 8.551e-01 7.893e-01
knot[2]   0.2138 0.3056 8.307e-01 7.599e-01
knot[3]   0.2531 0.3530 8.002e-01 7.241e-01
knot[4]   0.3032 0.4123 7.618e-01 6.801e-01
...
```

The estimated weights are given in column c coef.

Confidence intervals for the model parameters shown in the first part of Table 3.3 are calculated using the following code:

```
> confint(mod, level = 0.95)
```

```
                   2.5 %     97.5 %
(Intercept)   2.33494394  2.3677932
Log(scale)   -2.30991811 -2.0629259
Scale         0.09926938  0.1270816
```

The values of AIC_{alter} *for all values of* λ *in the grid search are shown with (only a part of the output shown):*

```
> mod$searched
```

```
     Lambda Log(Lambda)       AIC       df PenalLogLik
1 7.3890561           2 -310.3866 2.627579   -307.8464
2 2.7182818           1 -310.7006 3.046868   -307.7527
3 1.0000000           0 -311.0497 3.537396   -307.6424
4 0.3678794          -1 -311.3725 4.129021   -307.4706
...
```

(Cont.)

```
     LogLik nOfParm fail
1 -307.7590      40    0
2 -307.6537      40    0
3 -307.5123      40    0
4 -307.2435      40    0
...
```

Note that the columns Lambda *and* Log(Lambda) *give, as explained above, values of* λ/n *and* $\log(\lambda/n)$. *The column* df *shows the corresponding effective number of parameters* ν *and columns* PenalLogLik *and* LogLik *give the maximized values of the penalized log-likelihood and log-likelihood, respectively.*

The mean survival time $\mathsf{E}(T)$ *and its confidence interval (see Table 3.3) is computed using*

```
> estimTdiff(mod)
```

```
Estimate of Expectation
and Its 95% Confidence Intervals
```

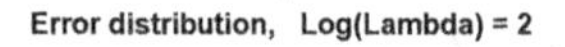

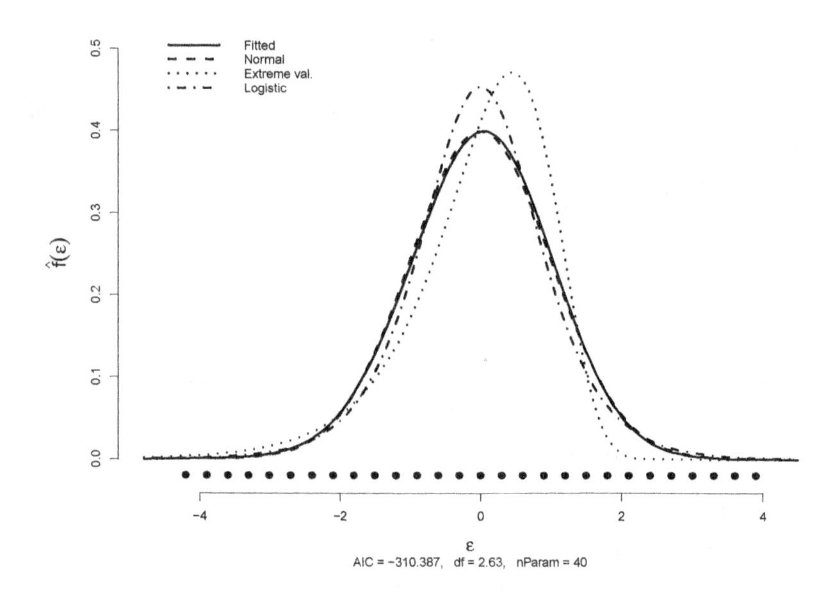

FIGURE 3.10: Signal Tandmobiel study (boys). Distribution of the standardized log-time to emergence of tooth 44 estimated using the penalized Gaussian mixture compared to parametric densities, obtained with the R package smoothSurv.

(Cont.)

```
       ET Std.Error     Lower     Upper         Z         p
 10.56613   0.09185  10.38610  10.74616 115.03500   0.00000
```

The standard error and confidence interval are derived from the pseudo-covariance matrix (3.24).

We can compute and plot the estimated survival density $\widehat{f}(t)$, hazard function $\widehat{h}(t)$ and survival function $\widehat{S}(t)$ with

```
> fdensity(mod, xlim=c(6, 13))
> hazard(mod, xlim=c(6, 13))
> survfit(mod, xlim=c(6, 13))
```

The plots of Figure 3.9 were drawn using (code to draw the NPMLE not shown):

```
> dens <- fdensity(mod, xlim=c(6, 13), by=0.1, plot=FALSE)
> plot(dens$x, dens$y1, type="l",
+      xlab="Age (years)", ylab=expression(hat(f)(t)), main="Density")
> haz <- hazard(mod, xlim=c(6, 13), by=0.1, plot=FALSE)
> plot(haz$x, haz$y1, type="l",
+      xlab="Age (years)", ylab=expression(hat(h)(t)), main="Hazard")
> S <- survfit(mod, xlim=c(6, 13), by=0.1, plot=FALSE)
> plot(S$x, S$y1, type="l", ylim=c(0, 1),
+      xlab="Age (years)", ylab="Proportion not yet emerged",
+      main="Survival function")
```

Finally, the estimate of the density

$$g^\star(y^\star) = \sum_{k=-K}^{K} w_k \varphi_{\mu_k, \sigma_0^2}(y^\star)$$

of the standardized logarithmic event time $Y^\star = \tau^{-1}\{\log(T) - \alpha\}$ can easily be plotted together with three parametric densities for comparison (see Figure 3.10) using

```
> plot(mod, resid=FALSE)
```

Note that also in this plot, the shown value of AIC is $\mathsf{AIC}_{alter}(\lambda)$ defined by Equation (3.26) and "Log(Lambda) = 2" in the title of the plot means that $\log(\lambda/n) = 2$.

□

3.4 Concluding remarks

For censored data, it is standard to always start with the NPMLE. Both SAS and R have ample possibilities to calculate the NPMLE. For R, we recommend to use the function EMICM from the package Icens. For SAS, we recommend PROC ICLIFETEST, which also uses the EMICM algorithm as default to calculate the NPMLE. When in a later stage covariates need to be included, parametric modelling or penalized smoothing methods may be used.

Chapter 4

Comparison of two or more survival distributions

We now review frequentist methods to compare two or more survival distributions with interval-censored event times. We will regularly refer to the corresponding methods in Section 2.2, where we reviewed such methods for right-censored data. Parametric or semiparametric regression models for interval-censored survival times will be treated in subsequent chapters.

Throughout the chapter, we assume that we are dealing with K independent samples $T_{k1}, \ldots, T_{k,N_k}$ $(k = 1, \ldots, K)$ of event times with cumulative distribution functions $F_1(t), \ldots, F_K(t)$, and survival functions $S_1(t), \ldots, S_K(t)$. For each sample, a set of intervals and censoring indicators $\mathcal{D}_k = \big\{ \lfloor l_{ki},\, u_{ki} \rfloor,\, \delta_{ki} : i = 1, \ldots, N_k \big\}$ $(k = 1, \ldots, K)$, is observed. It is the aim to test the null hypothesis

$$\mathrm{H}_0 : S_1 = S_2 = \cdots = S_K \tag{4.1}$$

against the alternative $\mathrm{H}_1 : S_{k_1} \neq S_{k_2}$ for at least one pair $k_1 \neq k_2$.

Furthermore, we need some notation that addresses the whole sample irrespective of the group structure. Namely, let $n = \sum_{k=1}^{K} N_k$ be the total sample size, $T_1, \ldots, T_n$ be the combined sample obtained by merging the samples from all considered groups. Similarly, let $\mathcal{D} = \big\{ \lfloor l_i,\, u_i \rfloor,\, \delta_i : \; i = 1, \ldots, n \big\}$ be the combined set of observed intervals and censoring indicators. Further, let X_{ik} be the indicator variable such that for $i = 1, \ldots, n$; $k = 1, \ldots, K$, $X_{ik} = 1$ if T_i originates from the kth sample, and $X_{ik} = 0$ otherwise and let $\boldsymbol{X}_i = (X_{i1}, \ldots, X_{iK})'$ $(i = 1, \ldots, n)$.

4.1 Nonparametric comparison of survival curves

The classical procedures for right-censored data have been extended in various ways to deal with interval censoring. We consider here the two most important types of nonparametric methods that compare survival distributions: (i) weighted log-rank tests (WLRT) introduced in Section 2.2, and (ii) tests based on the integrated weighted difference between the estimated survival functions. As for right-censored data, these two types of tests are particularly powerful against some specific cases of alternative hypotheses. The weighted log-rank tests are most powerful when the survival functions do not cross (ordered survival functions), whereas crossing survival functions are better detected by the second type of tests.

For current status data, Gehan (1965a) and Mantel (1967) proposed the Gehan-Wilcoxon version of the WLRT. With general interval-censored data, Self and Grossman (1986) developed the Peto-Prentice-Wilcoxon version of the WLRT, and Finkelstein (1986) discussed the classical log-rank test for Cox PH models. Her work was further generalized by Sun (1996) and Fay (1996, 1999). Pan (2000b), Zhao and Sun (2004), Sun et al. (2005), Zhao et al. (2008), Huang et al. (2008) and Oller and Gómez (2012) examined different versions of the WLRT for interval-censored events.

Petroni and Wolfe (1994) and Andersen and Rønn (1995) were among the first to consider tests based directly on the estimated survival functions. Later references include Zhang et al. (2001), Fang et al. (2002), Lim and Sun (2003), Zhang et al. (2003), and Yuen et al. (2006). Other sources that review methods to compare several survival distributions in the presence of interval censoring include Gómez et al. (2004), Sun (2006, Chap. 4), Gómez et al. (2009), Zhang and Sun (2010), or Fay and Shaw (2010).

Most of the above-mentioned methods rely on asymptotic properties of the test statistics. For right-censored data, these can be derived elegantly using counting process theory (Andersen et al., 1993; Fleming and Harrington, 1991). However, with interval-censored data, this theory is of no use making it much more difficult to derive asymptotic theory. Standard maximum likelihood theory was applied by, e.g., Finkelstein (1986) and Fay (1996, Sec. 3.3). Alternatives to ML approaches include permutation-based significance tests (Fay, 1996, Sec. 3.2; Fay and Shih, 1998; and Oller and Gómez, 2012) and various resampling methods such as multiple imputation (Huang et al., 2008; Pan, 2000b) and bootstrap (Yuen et al., 2006).

The above brief review illustrates that a broad range of approaches for nonparametric comparison of two or more survival distributions is available in the frequentist literature. Also several software implementations became available during the past decade. An R implementation of a class of K-sample tests for interval-censored data of Oller and Gómez (2012) was first available as a supplement to Gómez et al. (2009). We included this implementation also in the R package icensBKL. Nowadays, an R package FHtest (Oller and Langohr, 2015) provides the implementation of the methods of Oller and Gómez (2012) in full generality. Additional R packages covering nonparametric tests for comparison of two or more survival distributions under interval censoring include: the interval package (Fay, 2014; Fay and Shaw, 2010) and the glrt package (Zhao and Sun, 2015) which contains generalized log-rank tests based on papers of Zhao and Sun (2004), Sun et al. (2005) and Zhao et al. (2008). A SAS macro is also available for the first two approaches. All R implementations of the K-sample tests are based on interval-censored generalizations of the weighted log-rank test. In the remainder of this section, we briefly explain the main ideas behind them but we focus on the permutation principle and standard ML theory.

4.1.1 Weighted log-rank test: derivation

In Section 2.2, we showed that for right-censored data, the test statistic of the two-sample WLRT is based on weighted differences between observed and expected (under H_0) numbers of events in each of the samples calculated at the (uncensored) event times in the combined sample (numerator of $G_{W,1}$ in Equation (2.10)). This procedure can easily be extended to the K-sample case. With right-censored data, the test statistic of the *K-sample WLRT* is based on a vector $\boldsymbol{U} = (U_1, \ldots, U_K)'$, with

$$U_k = \sum_{j=1}^{J} w_j (D_{jk} - E_{jk}) \qquad (k = 1, \ldots, K), \tag{4.2}$$

where J is the number of distinct uncensored event times in the combined data, D_{jk} and E_{jk} $(j = 1, \ldots, J)$ are the observed and expected numbers of events, respectively, in the kth group pertaining to the jth uncensored event time in the combined sample, and $w_1, \ldots, w_J$ are suitable weights. In other words, each of the statistics $U_1, \ldots, U_K$ compares observed and expected numbers of events in the respective group at each point in which the Kaplan-Meier estimate of the combined sample survival function makes a jump. Translated to the interval-censored setting, the WLRT statistic must compare observed and expected numbers of events at each region of support of the NPMLE of the survival function of the combined sample. Consequently, Expression (4.2) can be used to define the vector of weighted log-rank test statistics with interval-censored data as well (Fay, 1999).

For reasons of simplicity, we will assume that all observed intervals are half-open taking the form of $(l, u]$. Let $\big\{(p_j, q_j] : j = 1, \ldots, J\big\}$ be the regions of support of the NPMLE $\widehat{S}$ of the survival function of the combined sample $T_1, \ldots, T_n$ and let

$$s_j = \widehat{S}(p_j-) - \widehat{S}(q_j+), \qquad (i = 1, \ldots, n;\ j = 1, \ldots, J)$$

be the corresponding equivalence classes. To test Hypothesis (4.1) the WLRT statistic is again based on $\boldsymbol{U} = (U_1, \ldots, U_K)'$ with components given by (4.2), but the "observed" numbers of events D_{jk} $(k = 1, \ldots, K;\ j = 1, \ldots, J)$ are only estimated numbers of events occurring in the kth sample in $(p_j, q_j]$. The expected numbers E_{jk} $(k = 1, \ldots, K;\ j = 1, \ldots, J)$ are again expected (under H_0) numbers of events in the jth region of support of $\widehat{S}$. We now explain how to determine the "observed" and expected numbers of events with interval-censored data.

Define the estimated overall survival function for the ith subject (hence irrespective of the group structure), $\widehat{S}^{(i)}$, as $\widehat{S}$ truncated at the ith observed interval $(l_i,\, u_i]$ as follows:

$$\widehat{S}^{(i)}(t) = \mathsf{P}_{\widehat{S}}\big(T_i > t \,\big|\, T_i \in (l_i,\, u_i]\big) = \begin{cases} 1, & t \le l_i, \\ \dfrac{\widehat{S}(t) - \widehat{S}(u_i)}{\widehat{S}(l_i) - \widehat{S}(u_i)}, & l_i < t \le u_i, \\ 0, & t > u_i. \end{cases}$$

An estimate of the survival function for the kth group, $\widehat{S}_k$, could be the average of $\widehat{S}^{(i)}$ values over the individuals from that group. That is, using the indicator covariates $\boldsymbol{X}_i$, we can write

$$\widehat{S}_k(t) = \frac{1}{N_k}\sum_{i=1}^{n} X_{ik}\widehat{S}^{(i)}(t).$$

The estimated total number of events in the combined sample pertaining to the jth interval $(p_j,\, q_j]$ is then

$$D_j = n\, s_j = n\left\{\widehat{S}(p_j-) - \widehat{S}(q_j+)\right\} \qquad (j = 1, \ldots, J).$$

While the estimated total number at risk just prior to the jth interval is

$$n_j = n\,\widehat{S}(p_j-) \qquad (j = 1, \ldots, J).$$

The estimated "observed" number of events in the kth group occurring in the jth interval $(p_j,\, q_j]$ is

$$D_{jk} = N_k\left\{\widehat{S}_k(p_j-) - \widehat{S}_k(q_j+)\right\} \qquad (k = 1, \ldots, K,\ j = 1, \ldots, J),$$

while correspondingly the number at risk just prior to that interval is

$$n_{jk} = N_k\,\widehat{S}_k(p_j-) \qquad (k = 1, \ldots, K,\ j = 1, \ldots, J),$$

and the correspondingly expected number of events (under H_0) is

$$E_{jk} = \frac{n_{jk}}{n_j}\,D_j \qquad (k = 1, \ldots, K,\ j = 1, \ldots, J).$$

The vector $\boldsymbol{U}$ upon which the test statistic of the WLRT is based, can now be written as

$$\boldsymbol{U} = \sum_{j=1}^{J} w_j\,(\boldsymbol{D}_j - \boldsymbol{E}_j), \tag{4.3}$$

where $\boldsymbol{D}_j = (D_{j1}, \ldots, D_{jK})'$, $\boldsymbol{E}_j = (E_{j1}, \ldots, E_{jK})'$ $(j = 1, \ldots, J)$.

Further insight into the weighted log-rank test is obtained by rewriting

Expression (4.2) as

$$U_k = \sum_{j=1}^{J} w_j \left(D_{jk} - \frac{n_{jk}}{n_j} D_j \right)$$
$$= \sum_{j=1}^{J} w_j \, n_{jk} \Big\{ \widehat{\hbar}^{(k)}(q_j+) - \widehat{\hbar}(q_j+) \Big\} \qquad (k = 1, \ldots, K),$$

where $\widehat{\hbar}^{(k)}(q_j+) = D_{jk}/n_{jk}$ and $\widehat{\hbar}(q_j+) = D_j/n_j$ are estimated hazards in the kth and pooled sample, respectively. That is, the kth component of the weighted log-rank vector $\boldsymbol{U}$ is an integrated weighted difference between estimated kth group and overall hazard functions.

4.1.2 Weighted log-rank test: linear form

As for right-censored data (Kalbfleisch and Prentice, 2002, Chap. 7), the interval-censored version of the WLRT vector $\boldsymbol{U}$ can be written (see Fay, 1996, 1999; Gómez et al., 2009; Oller and Gómez, 2012) as

$$\boldsymbol{U} = \sum_{i=1}^{n} c_i \, \boldsymbol{X}_i, \tag{4.4}$$

where c_i $(i = 1, \ldots, n)$ are scores uniquely related to the weights w_j $(j = 1, \ldots, J)$ in Equation (4.3), see below for some examples. That is, a particular weighted log-rank test can either be specified by the weights $w_1, \ldots, w_J$ in (4.3), or the scores $c_1, \ldots, c_n$ in (4.4). By changing the weights or scores, the WLRT can be made sensitive against alternatives of early, middle or late hazard differences. In Section 2.2, we briefly reviewed the most common choices of the weights for right-censored data and showed that they are based either on numbers at risk and events at each of distinct observed (uncensored) event times or on values of the Kaplan-Meier function of the pooled sample evaluated at observed event times. In Sections 4.1.3 and 4.1.4, we show how to extend this general concept to interval-censored data.

4.1.3 Weighted log-rank test: derived from the linear transformation model

A flexible and broad class of weighted log-rank tests can be obtained as efficient score tests of the regression coefficients in appropriate linear transformation models. This approach is sketched below but described in detail by Fay (1996), Fay (1999) and Fay and Shaw (2010).

A linear transformation model suitable for the K-sample problem is given by

$$g(T_i) = -\boldsymbol{X}_i'\boldsymbol{\beta} + \varepsilon_i \qquad (i = 1, \ldots, n), \tag{4.5}$$

where g is an unknown monotone increasing transformation (functional parameter), $\boldsymbol{\beta} = (\beta_1, \ldots, \beta_K)'$ is a vector of unknown regression coefficients (appropriately constrained for identifiability reasons, e.g., by condition $\beta_K = 0$) reflecting the differences between groups under comparison, and $\varepsilon_1, \ldots, \varepsilon_n$ are i.i.d. random variables with a known distribution function F_ε, density f_ε and quantile function F_ε^{-1} reflecting (together with g) the baseline survival distribution. Note that unknown shift (intercept) and scaling of the distribution F_ε are also implicitly included in the model via the transformation g. A model-based survival function for an observation with covariate values $\boldsymbol{X}$ is then

$$S(t \mid \boldsymbol{X}) = 1 - F_\varepsilon\big\{g(t) + \boldsymbol{X}'\boldsymbol{\beta}\big\}.$$

Under the null hypothesis of no differences between groups, $\boldsymbol{\beta} = \boldsymbol{0}$ (regression coefficients of indicator covariates $\boldsymbol{X}_i$) and $S(t \mid \boldsymbol{X}) = S(t) = 1 - F_\varepsilon\big\{g(t)\big\}$ does not depend on covariates so that $g(t) = F_\varepsilon^{-1}\big\{1 - S(t)\big\}$. Let $\widehat{S}$ be the NPMLE of the survival function of a combined sample $T_1, \ldots, T_n$ with regions of support $(p_j,\, q_j]$ $(j = 1, \ldots, J)$, then

$$\widehat{g}(t) = F_\varepsilon^{-1}\big\{1 - \widehat{S}(t)\big\}$$

is the NPMLE of the unknown functional parameter g under the null hypothesis of $\boldsymbol{\beta} = \boldsymbol{0}$.

We will further assume independent noninformative censoring, then for half-open intervals the likelihood of Model (4.5) is

$$L = \prod_{i=1}^{n} \big\{S(l_i \mid \boldsymbol{X}_i) \; - \; S(u_i \mid \boldsymbol{X}_i)\big\}. \tag{4.6}$$

One can show that (4.4) with scores given by

$$c_i = \frac{f_\varepsilon\big\{\widehat{g}(u_i)\big\} \; - \; f_\varepsilon\big\{\widehat{g}(l_i)\big\}}{\widehat{S}(l_i) \; - \; \widehat{S}(u_i)} \qquad (i = 1, \ldots, n) \tag{4.7}$$

is the efficient score vector to test the null hypothesis $\boldsymbol{\beta} = \boldsymbol{0}$. Furthermore, Fay (1999) showed that the efficient score vector can also be written as in (4.3) with weights

$$w_j = \widehat{S}(p_j-)\,\frac{\dfrac{f_\varepsilon\big\{\widehat{g}(q_j+)\big\}}{\widehat{S}(q_j+)} - \dfrac{f_\varepsilon\big\{\widehat{g}(p_j-)\big\}}{\widehat{S}(p_j-)}}{\widehat{S}(p_j-) - \widehat{S}(q_j+)} \qquad (j = 1, \ldots, J). \tag{4.8}$$

Different choices of F_ε lead to different sets of scores $c_1, \ldots, c_n$ and weights $w_1, \ldots, w_J$ and consequently also to different WLRT vectors $\boldsymbol{U}$ which lead to asymptotically efficient K-sample tests in situations when Model (4.5) with a particular choice for F_ε actually holds. Two specific choices lead to interval-censored generalizations of the classical log-rank test and Peto-Prentice-Wilcoxon test.

Firstly, for the Gumbel (extreme value) distribution $F_\varepsilon(y) = 1 - \exp\{-\exp(y)\}$, scores (4.7) and weights (4.8) become

$$c_i = \frac{\widehat{S}(l_i)\,\log\{\widehat{S}(l_i)\} - \widehat{S}(u_i)\,\log\{\widehat{S}(u_i)\}}{\widehat{S}(l_i) - \widehat{S}(u_i)} \qquad (i = 1, \ldots, n), \tag{4.9}$$

$$w_j = \widehat{S}(p_j-)\,\frac{\log\{\widehat{S}(p_j-)\} - \log\{\widehat{S}(q_j+)\}}{\widehat{S}(p_j-) - \widehat{S}(q_j+)} \qquad (j = 1, \ldots, J), \tag{4.10}$$

leading to a test which was also discussed for PH models for interval-censored data by Finkelstein (1986). Indeed, from Section 2.3.1, we know that Model (4.5) with the Gumbel distribution for ε's leads to hazard functions which satisfy the PH assumption. Furthermore, note that $w_j \to 1$ as $\widehat{S}(p_j-)/\widehat{S}(q_j+) \to 1$. That is, when $\widehat{S}$ becomes close to continuous, the classical log-rank test is obtained and hence the WLRT with scores or weights given by (4.9) and (4.10), respectively, can be viewed as an interval-censored generalization of the classical log-rank test.

For the logistic distribution $F_\varepsilon(y) = \exp(y)/\{1+\exp(y)\}$, scores (4.7) and weights (4.8) become

$$c_i = \widehat{S}(l_i) + \widehat{S}(u_i) - 1 \qquad (i = 1, \ldots, n) \tag{4.11}$$

$$w_j = \widehat{S}(p_j-) \qquad (j = 1, \ldots, J). \tag{4.12}$$

This produces the interval-censored generalization of the Peto-Prentice-Wilcoxon test.

4.1.4 Weighted log-rank test: the $G^{\varrho,\gamma}$ family

In Section 2.2, we introduced the $G^{\varrho,\gamma}$ class of weighted log-rank tests for right-censored data proposed by Fleming and Harrington (1991). Recall that the parameters ϱ and γ represent sensitivity of the test against early or late hazard differences. The extension to interval-censored data proposed by Oller and Gómez (2012) is now briefly outlined. For given $\varrho \geq 0$, $\gamma \geq 0$, they define for $j = 1, \ldots, J$ the weight function in (4.3) as follows

$$\begin{aligned} w_j &= w_j^{\varrho,\gamma} \\ &= \widehat{S}(p_j-)\,\frac{B\big(1-\widehat{S}(q_j+);\,\gamma+1,\,\varrho\big) - B\big(1-\widehat{S}(p_j-);\,\gamma+1,\,\varrho\big)}{\widehat{S}(p_j-) - \widehat{S}(q_j+)}, \end{aligned} \tag{4.13}$$

where for $y \geq 0$, $a > 0$, $b \geq 0$,

$$B(y;\,a,\,b) = \int_0^y x^{a-1}\,(1-x)^{b-1}\,\mathrm{d}x$$

is an incomplete beta function. Oller and Gómez (2012) further show that the corresponding scores for the linear form of the weighted log-rank test are given for $i = 1, \ldots, n$ by

$$c_i = c_i^{\varrho,\gamma} = \frac{\widehat{S}(u_i)\,B\bigl(1 - \widehat{S}(u_i);\ \gamma + 1,\ \varrho\bigr)\ -\ \widehat{S}(l_i)\,B\bigl(1 - \widehat{S}(l_i);\ \gamma + 1,\ \varrho\bigr)}{\widehat{S}(l_i)\ -\ \widehat{S}(u_i)}. \qquad (4.14)$$

Note that for $\gamma = 0$ and $\varrho > 0$, the weights (4.8) and scores (4.7) are obtained, which result from the linear transformation Model (4.5) where $F_\varepsilon(y) = 1 - \bigl\{1 + \varrho\exp(y)\bigr\}^{-\frac{1}{\varrho}}$. In particular, for $\varrho = 1$, this is the logistic distribution function and hence the expressions for weights $w_j^{1,0}$ $(j = 1, \ldots, J)$ and scores $c_i^{1,0}$ $(i = 1, \ldots, n)$ can be simplified into (4.12) and (4.11), respectively, leading to the interval-censored generalization of the Peto-Prentice-Wilcoxon test. Further, for $\gamma = 0$ and $\varrho \to 0$, the weights (4.13) and scores (4.14) tend to the weights (4.10) and scores (4.9), respectively, giving the interval-censored generalization of the classical log-rank test.

Oller and Gómez (2012) further note that for $\gamma > 0$, the weighted log-rank cannot be written in general as an efficient score statistic. Nevertheless, the fact that the weights (4.13) indeed generalize the original Fleming and Harrington $G^{\varrho,\gamma}$ class of the tests is seen by observing that $w_j^{\varrho,\gamma}$ tends to $\bigl\{\widehat{S}(p_j-)\bigr\}^{\varrho}\bigl\{1 - \widehat{S}(p_j-)\bigr\}^{\gamma}$ as $\widehat{S}(p_j-)/\widehat{S}(q_j+) \to 1$. That is, the weights (4.13) are close to the original Fleming and Harrington proposal when $\widehat{S}$ is nearly continuous. Consequently, analogously to right-censored case, tests sensitive to reveal early, late or median hazard differences can be constructed by appropriate choices of ϱ and γ values.

4.1.5 Weighted log-rank test: significance testing

A nonparametric test

To establish the significance of the weighted log-rank test based on the score vector $\boldsymbol{U}$ with given weights $w_1, \ldots, w_J$ or scores $c_1, \ldots, c_n$, the distribution of the score vector $\boldsymbol{U}$ under null hypothesis (4.1) must be determined. A possible way is to exploit the permutation principle (see, e.g., Fay, 1996, Sec. 3.2; Fay and Shih, 1998; and Oller and Gómez, 2012) which is based on the idea that under H_0, the labels on the scores c_i are exchangeable (see, e.g., Hájek et al., 1999). The permutation distribution of $\boldsymbol{U}$ is obtained by calculating $\boldsymbol{U}$ for all $n!$ possible permutations of labels. In practice, this can only be done exactly for small sample sizes n. For moderate to large sample sizes, the central limit theorem for exchangeable random variables leads to the normal approximation with mean

$$\mathsf{E}(\boldsymbol{U}) = \frac{1}{n}\sum_{i=1}^{n} c_i = \boldsymbol{0}$$

and covariance matrix

$$\begin{aligned} \boldsymbol{V}_P &= \frac{1}{n-1}\sum_{i=1}^{n}(\boldsymbol{U}_i - \overline{\boldsymbol{U}})(\boldsymbol{U}_i - \overline{\boldsymbol{U}})' \\ &= \frac{1}{n-1}\Big(\sum_{i=1}^{n} c_i^2\Big)\Big\{\sum_{i=1}^{n}(\boldsymbol{X}_i - \overline{\boldsymbol{X}})(\boldsymbol{X}_i - \overline{\boldsymbol{X}})'\Big\}, \end{aligned} \tag{4.15}$$

where $\overline{\boldsymbol{X}} = n^{-1}\sum_{i=1}^{n}\boldsymbol{X}_i$. Note that the components of the K-dimensional random vector $\boldsymbol{U}$ sum up to zero with probability one and the matrix $\boldsymbol{V}_P$ is of rank $K-1$. With $\boldsymbol{V}_P^-$ the generalized inverse of matrix $\boldsymbol{V}_P$, statistic

$$G_P^2 = \boldsymbol{U}'\boldsymbol{V}_P^-\boldsymbol{U} \tag{4.16}$$

follows under $\mathrm{H}_0 : \boldsymbol{\beta} = \boldsymbol{0}$ asymptotically a χ_{K-1}^2 distribution leading a significance test for equality of the K survival distributions with Model (4.5).

The permutation test, however, also assumes that the censoring mechanism does not depend on covariates and hence is the same in all groups. Nevertheless, the P-value of the permutation test remains valid even if assumed Model (4.5) does not hold but the test might not be asymptotically efficient anymore (Oller and Gómez, 2012).

Test based on classical ML theory

Standard ML theory can also be applied to test $\mathrm{H}_0 : \boldsymbol{\beta} = \boldsymbol{0}$ in (4.5) in the presence of a nuisance functional parameter g (e.g., Finkelstein, 1986; Fay, 1996, Sec. 3.3; Sun, 1996). Large sample ML theory leads to an asymptotic normal approximation of the distribution of the score vector $\boldsymbol{U}$ with mean $\mathsf{E}(\boldsymbol{U}) = \boldsymbol{0}$ and a particular covariance matrix $\boldsymbol{V}_{ML}$ calculated from the observed Fisher information matrix of likelihood (4.6). A statistical test for the above null-hypothesis is then derived from the asymptotic χ_{K-1}^2 distribution of statistic $G_{ML}^2 = \boldsymbol{U}'\boldsymbol{V}_{ML}^-\boldsymbol{U}$. Exact formula to calculate $\boldsymbol{V}_{ML}$ for common choices of scores c_i $(i = 1, \ldots, n)$ are given in Fay (1999) and in Oller and Gómez (2012).

Standard ML theory requires only independent and noninformative censoring, and hence the censoring mechanism may differ between the groups. But standard ML theory also requires that the number of parameters of Model (4.5) does not increase with the sample size, which is a problem here because the functional parameter g is implicitly estimated in a nonparametric way. This difficulty could be overcome by assuming the *grouped continuous model* (e.g., Finkelstein, 1986), which restricts the set of possible limits of observed intervals to be finite and constant, i.e., not dependent on the sample size. In a study where interval-censored observations result from checking the survival status at prespecified occasions, the grouped continuous model is obtained by setting the maximal possible observed time (in the study time scale) and establishing the precision with which time is recorded such that the limits of the observed intervals originate from a discrete distribution with a finite support. Fay (1996) argued that the application of standard maximum likelihood theory to weighted log-rank tests with interval-censored data is also hampered by the fact that the parameters which determine the nonparametric estimate of the functional parameter g often approach the boundary of the parameter space invalidating maximum likelihood theory.

Fay (1996) compared the permutation test and the ML approach by simulations and concluded that the permutation test performs best when the same censoring mechanism applies to all groups, and the ML-based test otherwise. Later, Fay and Shaw (2010) revised their recommendation and suggested to use the permutation test unless there is extreme unequal censoring and sample sizes in the K groups. They support this argument by referring to results of simulation studies with right-censored data (Heinz et al., 2003). Oller and Gómez (2012) recommended the permutation test because the ML-based test too often suffers from boundary problems.

Example 4.1 Signal Tandmobiel study. Two-sample tests

We illustrate the use of K-sample tests on the emergence times of tooth 44. Firstly, we evaluate the dependence of the emergence distribution on the dichotomized DMF (caries history) status of tooth 84 (primary mandibular right first molar) which is the predecessor of tooth 44. It is hypothesized that caries on tooth 84 may influence the emergence process of the succeeding tooth 44. Thus we wish to check whether children with sound tooth 84 (DMF = 0,

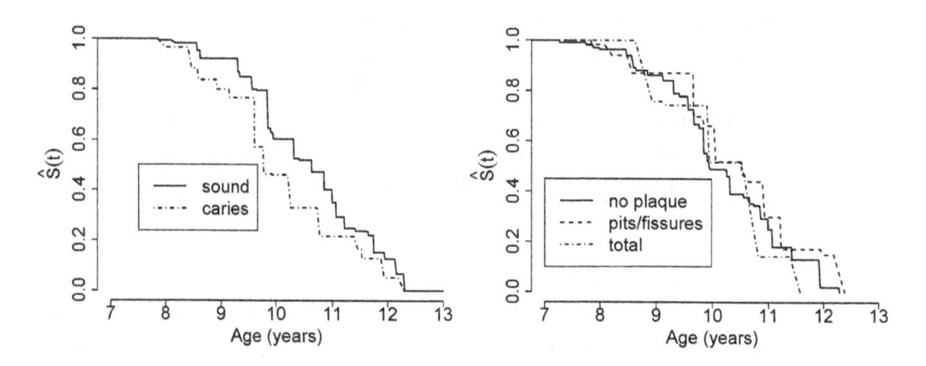

FIGURE 4.1: Signal Tandmobiel study. NPMLE of the survival functions for emergence of permanent tooth 44. Left panel: two groups according to baseline DMF status of primary tooth 84 (solid line: sound tooth 84 (DMF = 0), dashed line: caries history on tooth 84 (DMF = 1)). Right panel: three groups according to occlusal plaque status of permanent tooth 46 (solid line: no plaque (OH46o = 0), dashed line: plaque in pits/fissures (OH46o = 1), dotted-dashed line: plaque on total surface (OH46o = 2)). Both obtained with R.

TABLE 4.1: Signal Tandmobiel study. Two-sample tests from $G^{\varrho,\gamma}$ family (PPW = Peto-Prentice-Wilcoxon) comparing emergence distributions of permanent tooth 44 in two groups according to baseline DMF status of primary predecessor 84. P-values are calculated from asymptotic χ^2_1 distribution of statistic G^2_P. Results from the kSampleIcens function from the R package icensBKL.

Parameters	Test statistics G^2_P	P-value
$\varrho = 0,\ \gamma = 0$ (Log-rank)	8.62	0.0033
$\varrho = 1,\ \gamma = 0$ (PPW)	13.78	0.0002
$\varrho = 0,\ \gamma = 1$	1.41	0.2354
$\varrho = 1,\ \gamma = 1$	4.12	0.0423

$N_1 = 343$) have a different emergence distribution of tooth 44 (recorded in an interval-censored way) than children whose tooth 84 was either decayed, missing due to caries or filled (DMF = 1, $N_2 = 153$). For four children, the baseline DMF status of tooth 84 was unknown. As mentioned above, it is reasonable to assume that the missingness process is completely at random (probability of an observation being missing does not depend on observed or unobserved measurements) and hence can be ignored.

The two nonparametrically estimated survival functions are shown in the left panel of Figure 4.1. The two emergence distributions seem to differ in the sense that the emergence of teeth with a primary predecessor with caries history (DMF = 1) is somehow accelerated compared to the DMF = 0 group. However, this difference appears to vanish over time.

Since the visit times were regulated independent of the participating children, the censoring process must be independent and noninformative, and of the same type in both groups. Thus we can apply the permutation approach, thereby avoiding the ML approach with no clear theoretical properties as the number of parameters increases with the sample size.

The results from the two-sample tests derived from the $G^{\varrho,\gamma}$ family are shown in Table 4.1 for four different choices of ϱ and γ. It is not surprising that the $G^{1,0}$ (Peto-Prentice-Wilcoxon) test based on scores (4.11) and weights (4.12) leads to the most significant P-value as the groups are the most different at early emergence. That the $G^{0,1}$ test does not yield a significant result also does not come as a surprise, since it is most sensitive towards late hazard differences.

That the emergence of the (DMF = 1) group is accelerated compared to the (DMF = 0) group can be revealed by a more detailed analysis of the test statistic. For example, for the $G^{1,0}$ case, the score vector $\boldsymbol{U} = \big(U_1,\ U_2\big)'$ (Equations (4.3) or (4.4)) equals $\big(-19.31,\ 19.31\big)'$ from which we conclude that in the (DMF = 0) group, (weighted) more events happened than expected under H_0. The estimated covariance matrix $\boldsymbol{V}_P$ (Equation (4.15)) equals

$$\boldsymbol{V}_P = \begin{pmatrix} v_{P11} & v_{P12} \\ v_{P21} & v_{P22} \end{pmatrix} = \begin{pmatrix} 27.04 & -27.04 \\ -27.04 & 27.04 \end{pmatrix},$$

and the test statistic G_P^2 (Equation (4.16)) is simply the squared Z-score for either the first or second group, i.e.,

$$Z_1 = \frac{U_1}{\sqrt{v_{P11}}} = -\frac{U_2}{\sqrt{v_{P22}}} = -Z_2 = -3.71$$

and

$$G_P^2 = Z_1^2 = Z_2^2 = 13.78.$$

As a comparison, a standard log-rank test on mid- and right-point imputed data yields a P-value of 0.0015 and 0.0037, respectively. They yield the same conclusion.

□

Example 4.2 Signal Tandmobiel study. Comparison of three samples
In the second analysis, we compare the emergence distributions of tooth 44 in three groups defined by the value of another baseline covariate, i.e., amount of occlusal plaque on a spatially close tooth 46 (permanent mandibular right second molar). Three groups are compared: children with no plaque (OH46o $= 0$, $N_1 = 300$), plaque present in pits and fissures (OH46o $= 1$, $N_2 = 96$), and plaque on total occlusal surface (OH46o $= 2$, $N_3 = 20$). The nonparametrically estimated survival functions in the three groups are shown in the right panel of Figure 4.1. There are no clear differences between the groups. From the results of Table 4.2, we do not see significant differences between the three groups. For example, for the interval-censored generalization of the Peto-Prentice-Wilcoxon test ($G^{1,0}$ test), $\boldsymbol{U} = \big(5.95, -6.63, 0.68\big)'$ with the estimated covariance matrix $\boldsymbol{V}_P$ and its generalized inverse being

$$\boldsymbol{V}_P = \begin{pmatrix} 21.78 & -18.02 & -3.75 \\ -18.02 & 19.22 & -1.20 \\ -3.75 & -1.20 & 4.96 \end{pmatrix}, \quad \boldsymbol{V}_P^- = \begin{pmatrix} 0.0315 & 0.0096 & -0.0411 \\ 0.0096 & 0.0406 & -0.0502 \\ -0.0411 & -0.0502 & 0.0912 \end{pmatrix},$$

leading to the test statistic value $G_P^2 = 2.30$ and P-value of 0.3164.

As a comparison, a standard log-rank test on mid- and right-point imputed data yields a P-value of 0.2855 and 0.3200, respectively. These are both larger than the log-rank test on interval-censored data but the conclusion remains the same.

□

TABLE 4.2: Signal Tandmobiel study. Tests from the $G^{\varrho,\gamma}$ family (PPW = Peto-Prentice-Wilcoxon) comparing emergence distributions of permanent tooth 44 in three groups according to the baseline amount of occlusal plaque on a spatially close tooth 46. P-values are calculated from asymptotic χ^2_2 distribution of statistic G^2_P. Results from the kSampleIcens function from the R package icensBKL.

Parameters	Test statistics G^2_P	P-value
$\varrho = 0,\ \gamma = 0$ (Log-rank)	2.96	0.2275
$\varrho = 1,\ \gamma = 0$ (PPW)	2.30	0.3164
$\varrho = 0,\ \gamma = 1$	2.53	0.2820
$\varrho = 1,\ \gamma = 1$	2.42	0.2988

4.1.6 R solution

A nonparametric comparison of two or more survival distributions based on Model (4.5) and under interval censoring is comprehensively treated with the ictest function of the R package interval. Namely, the function allows to perform directly the generalized log-rank test with scores (4.9), the generalized Peto-Prentice-Wilcoxon test with scores (4.11), WLRTs with scores derived from assuming a normal distribution for errors in the linear transformation model, and a test proposed by Sun (1996). Furthermore, users may also define their own score generating distribution F_ε. The permutation approach (both exact and asymptotic), standard ML and the multiple imputation method of Huang et al. (2008) can be used. In addition to the manual of the package, sample illustrations are provided by Fay and Shaw (2010). See Appendix D.3 for a brief introduction to the package and additional illustrations.

The R implementation for K-sample weighted log-rank tests derived from the $G^{\varrho,\gamma}$ family is described in Gómez et al. (2009) and is included as the function kSampleIcens in the R package icensBKL. Note that P-values are calculated using the asymptotic permutation distribution of the weighted log-rank vector $\boldsymbol{U}$. See Appendix D.1 for the syntax of the function. Additionally, an R package FHtest (Oller and Langohr, 2015) is available which provides even more flexibility and fully implements related methods proposed by Oller and Gómez (2012).

To perform K-sample tests with interval-censored data, one could also use the package glrt (Zhao and Sun, 2015) which focuses on the generalized log-rank tests developed by Zhao and Sun (2004), Sun et al. (2005) and Zhao et al. (2008). Details and the methodology can be found in the above references and the glrt package. We now illustrate the use of the R package interval and the function kSampleIcens.

Example 4.1 in R (WLRTs derived from the linear transformation model using the package interval)
As always, we start by loading the data. We evaluate the dependence of the

emergence times of tooth 44 on the DMF status of tooth 84. We therefore create a data frame tandDMF84 *containing only children with known DMF status of tooth 84 represented by variable* fDMF_84.

```
> tandDMF84 <- subset(tandmob, !is.na(fDMF_84))
```

As in Example 3.1, we create a two-column matrix called ADMF84 *specifying the lower and upper limits of the observed intervals for the emergence times of tooth 44 whereby the value of 20 (years of age) plays the role of infinity.*

```
> ADMF84 <- tandDMF84[, c("L44", "R44")]
> ADMF84[is.na(ADMF84[, "L44"]), "L44"] <- 0
> ADMF84[is.na(ADMF84[, "R44"]), "R44"] <- 20
```

The WLRT based on scores (4.9) or weights (4.10) is performed by the call to ictest *with argument* scores="logrank2".

```
> ictest(L=ADMF84[, "L44"], R=ADMF84[, "R44"],
+        group=tandDMF84[, "fDMF_84"],
+        scores="logrank2", method="pclt",
+        icontrol=icfitControl(maxit=100000))
```

Argument method="pclt" *indicates that the P-value should be based on the asymptotic permutation distribution of the weighted log-rank vector* $\boldsymbol{U}$. *The first row of Table 4.1 is based on the output below.*

```
        Asymptotic Logrank two-sample test (permutation form),
        Finkelstein's scores

data:  {L,R} by group
Z = 2.9355, p-value = 0.00333
alternative hypothesis: survival distributions not equal

         n Score Statistic*
caries 153        23.68885
sound  343       -23.68885
* like Obs-Exp, positive implies earlier failures than expected
```

The values in column labeled Score Statistic* *are components of the weighted logrank vector* $\boldsymbol{U} = (U_1, U_2)'$ *and the* Z *statistic is a signed square root of the statistic* $G^2_{\rm P}$ *(Equation (4.16)) which is equal to* $U_1/\sqrt{v_{\rm P11}}$ *in the two-sample case, where* $v_{\rm P11}$ *is the estimated variance of* U_1 *taken from matrix* $\boldsymbol{V}_{\rm P}$ *(Equation (4.15)).*

For argument scores="wmw", *the interval-censored generalization of Peto-Prentice-Wilcoxon test based on scores (4.11) or weights (4.12) is obtained (second row of Table 4.1). The following code illustrates that it is also possible to specify input data using the* formula *expression with a* Surv *object.*

```
> ictest(Surv(L44, R44, type = "interval2") ~ fDMF_84,
+         data=tandDMF84, scores="wmw", method="pclt",
+         icontrol=icfitControl(maxit=100000))
```

```
        Asymptotic Wilcoxon two-sample test (permutation form)

data:  Surv(L44, R44, type = "interval2") by fDMF_84
Z = 3.7125, p-value = 0.0002052
alternative hypothesis: survival distributions not equal

                 n Score Statistic*
fDMF_84=caries 153         19.30555
fDMF_84=sound  343        -19.30555
* like Obs-Exp, positive implies earlier failures than expected
```

For method="scoretest"*, the P-value is based on standard ML theory. Below, we show the ML-based results for the Peto-Prentice-Wilcoxon WLRT.*

```
> ictest(Surv(L44, R44, type = "interval2") ~ fDMF_84,
+         data=tandDMF84, scores="wmw", method="scoretest",
+         icontrol=icfitControl(maxit=100000))
```

```
        Asymptotic Wilcoxon two-sample test (score form)

data:  Surv(L44, R44, type = "interval2") by fDMF_84
Chi Square = 14.6194, p-value = 0.0001316
alternative hypothesis: survival distributions not equal

                 n Score Statistic*
fDMF_84=caries 153         19.30555
fDMF_84=sound  343        -19.30555
* like Obs-Exp, positive implies earlier failures than expected
```

The Chi Square *statistic in the output is in fact the G^2_{ML} statistic with the χ^2_1 distribution as reference.*

□

Example 4.1 in R *($G^{\varrho,\gamma}$ **weighted log-ranks using the package icensBKL**)*
We exploit the two column matrix ADMF84 *with observed intervals and column* fDMF_84 *of data frame* tandDMF84 *with a factor variable reflecting pertinence of children into groups (see the code above) and calculate once more*

the results of the generalization of the classical log-rank test, i.e., a $G^{0,0}$ test. The arguments rho *and* gamma *determine the values of ϱ and γ, respectively.*

```
> kSampleIcens(A=ADMF84, group=tandDMF84[, "fDMF_84"],
+                  rho=0, gamma=0)
```

However, the above code did not produce the solution shown below. Note that the NPMLE did not yet converge with the function. Rather, we need to calculate first the NPMLE of the survival/distribution function based on a pooled sample based on a sufficiently large number of iterations and supply it as icsurv *argument to the* kSampleIcens *function.*

```
        K-Sample Weighted Logrank G(0, 0) with Interval-Censored Data
        (permutation form)

data:  ADMF84
G2 = 8.6174, df = 1, p-value = 0.00333
```

This can be achieved with the function PGM *of the package* Icens *and is also useful to evaluate the effect of different values of ϱ and γ as illustrated below. The code below produces the results shown in Table 4.1. The results are then exported to a data frame* wlrtDMF84*. Note that, when calling the* PGM *function, we increased the default maximal number of iterations to 10 000 to ensure convergence with the project gradient method.*

```
> cdfpoolDMF84 <- PGM(ADMF84, maxiter=10000)
> kS.DMF84 <- list()
> kS.DMF84[[1]] <- kSampleIcens(A=ADMF84, group=tandDMF84[, "fDMF_84"],
+                                  icsurv=cdfpoolDMF84, rho=0, gamma=0)
> kS.DMF84[[2]] <- kSampleIcens(A=ADMF84, group=tandDMF84[, "fDMF_84"],
+                                  icsurv=cdfpoolDMF84, rho=1, gamma=0)
> kS.DMF84[[3]] <- kSampleIcens(A=ADMF84, group=tandDMF84[, "fDMF_84"],
+                                  icsurv=cdfpoolDMF84, rho=0, gamma=1)
> kS.DMF84[[4]] <- kSampleIcens(A=ADMF84, group=tandDMF84[, "fDMF_84"],
+                                  icsurv=cdfpoolDMF84, rho=1, gamma=1)
> names(kS.DMF84) <- paste("rho = ", sapply(kS.DMF84, "[[", "rho"), ",
+                     gamma = ", sapply(kS.DMF84, "[[", "gamma"), sep="")
> wlrtDMF84 <- data.frame(rho = sapply(kS.DMF84, "[[", "rho"),
+                         gamma = sapply(kS.DMF84, "[[", "gamma"),
+                         Statistic = sapply(kS.DMF84, "[[", "statistic"),
+                         P.value = sapply(kS.DMF84, "[[", "p.value"))
```

```
> rownames(wlrtDMF84) <- 1:nrow(wlrtDMF84)
> print(wlrtDMF84)
```

```
  rho gamma Statistic       P.value
1   0     0  8.617400 0.0033296680
2   1     0 13.782687 0.0002052188
3   0     1  1.407984 0.2353914079
4   1     1  4.121250 0.0423477873
```

With $\varrho \geq 0$ *and* $\gamma = 0$*, the functions* kSampleIcens *and* ictest *from the package* interval *(with* method="pclt" *with appropriate choice of weighted log-rank scores) are two implementations of the same method and therefore must give the same output. Small differences in the test statistic and P-value may occur because the* kSampleIcens *function uses the projected gradient method while the* ictest *function uses the self-consistency EM algorithm to calculate the NPMLE of the survival function (Sections 4.1.3 and 4.1.4).*

□

Example 4.2 in R (WLRTs derived from the linear transformation model using the package interval)

We examine whether the distribution of emergence times of tooth 44 depends on the baseline amount of occlusal plaque on tooth 46 (variable fOH46o*). Similar as in Example 4.1, we again create a data frame* tandOH46 *containing only children with known* fOH46o *value and a two-column matrix called* AOH46 *comprising observed lower and upper limits of emergence times with infinity replaced by 20 (years) of age. The second row of Table 4.2 shows the interval-censored generalization of the Peto-Prentice-Wilcoxon test (WLRT based on weights (4.12) or scores (4.11)) with asymptotic inference based on the permutation distribution of the weighted log-rank vector* $\boldsymbol{U}$*. The values are obtained using*

```
> ictest(L=AOH46[, "L44"], R=AOH46[, "R44"],
+        group=tandOH46[, "fOH46o"],
+        scores="wmw", method="pclt")
```

```
        Asymptotic Wilcoxon k-sample test (permutation form)

data:  {L,R} by group
Chi Square = 2.3018, p-value = 0.3163
alternative hypothesis: survival distributions not equal

                n Score Statistic*
pits/fissures  96       -6.6302916
no plaque     300        5.9463064
```

(Cont.)

```
total              20          0.6839852
* like Obs-Exp, positive implies earlier failures than expected
```

The Chi Square value corresponds to the statistic G^2_P (Equation (4.16)), the values in column labeled Score Statistic* are again components of the weighted log-rank vector $\boldsymbol{U} = \big(U_1, U_2, U_3\big)'$. As in the previous example, the formula expression might be used to specify the response (output not shown), as follows:

```
> ictest(Surv(L44, R44, type = "interval2") ~ fOH46o, data=tandOH46,
+           scores="wmw", method="pclt")
```

Furthermore, by modification of scores and/or method values, different variants of the WLRT and/or significance tests may be obtained.

□

Example 4.2 in R ***($G^{\varrho,\gamma}$ weighted log-ranks using the package icensBKL)***

We use the two column matrix AOH46 and data frame tandOH46 with a factor variable fOH46o to show the use of the function kSampleIcens for comparing emergence distributions of tooth 44 in three groups defined by the amount of occlusal plaque on tooth 46. In the code below, we apply the generalized Peto-Prentice-Wilcoxon test, i.e., a $G^{1,0}$ test (second row of Table 4.2). The results of the test procedure are stored in object ks.OH46.

```
> cdfpoolOH46 <- PGM(AOH46, maxiter=10000)
> kS.OH46 <- kSampleIcens(A=AOH46, group=tandOH46[, "fOH46o"],
+                          icsurv=cdfpoolOH46, rho=1, gamma=0)
> print(kS.OH46)
```

```

        K-Sample Weighted Logrank G(1, 0) with Interval-Censored Data
        (permutation form)

data:  AOH46
G2 = 2.3018, df = 2, p-value = 0.3164

```

To get the value of the weighted log-rank vector $\boldsymbol{U}$:

```
> kS.OH46$u
```

```
     no plaque pits/fissures     total
[1,]  5.946285     -6.630269 0.6839886
```

The estimated covariance matrix V_P of the log-rank vector U (Equation (4.15)) can be extracted by:

```
> kS.OH46$varu
```

```
               no plaque pits/fissures     total
no plaque      21.775682    -18.021254 -3.754428
pits/fissures -18.021254     19.222671 -1.201417
total          -3.754428     -1.201417  4.955845
```

□

4.1.7 SAS solution

The macro %ICSTEST and the procedure ICLIFETEST (available from SAS/STAT 13.1 in SAS 9.4) are available to compare two or more survival functions nonparametrically. Both compute the generalized log-rank tests of Zhao and Sun (2004). The %ICSTEST macro also supports the generalized log-rank tests of Sun et al. (2005). On the other hand, the ICLIFETEST procedure has the ability to perform a stratified test, a trend test and multiple comparisons for more than 2 groups. In addition, it can output scores, which can be supplied to the NPAR1WAY procedure, to yield permutation tests.

Example 4.1 in SAS (Generalized log-rank tests using macro %ICSTEST)

We first create a data set without missing values for the grouping variable DMF_84. *Otherwise, they would be treated as a separate group.*

```
DATA tandmob;
 SET icdata.tandmob;
WHERE NOT MISSING(DMF_84);
RUN;
```

Based on the code below, the generalized log-rank tests of Zhao and Sun (2004) and Sun et al. (2005) are calculated. The test of Sun et al. (2005) has two optional arguments rho *and* gamma *with default values set to 0, leading to a generalization of the classical log-rank test. For values $\varrho \geq 0$ or $\gamma \geq 0$, the family differs from the $G^{\varrho,\gamma}$ weighted log-rank tests proposed by Oller and Gómez (2012) and does not include the Peto-Prentice-Wilcoxon test statistic.*

```
%ICSTEST(data=tandmob,
       left=L44,
       right=R44,
```

```
        group=dmf_84,
        rho=0,
        gamma=0);
```

Checking for convergence of S is controlled by the argument ERRORTYPE, which specifies the specific criterion, i.e., closeness of consecutive parameter estimates (ERRORTYPE=1), the log-likelihood function (ERRORTYPE=2), the gradient of the log-likelihood function (ERRORTYPE=3) or the maximum measure of the previous 3 options (ERRORTYPE=4). The argument RATECONV= (1e-7 by default) controls the convergence criterion. Variance estimation for the test of Zhao and Sun (2004) is based on resampling with a generalization of the Greenwood formula as suggested by Sun (2001). The number of samples is controlled by the argument mRS= and is by default set to 50.

```
          Generalized Log-rank Test I (Zhao & Sun, 2004)

               Test Statistic and Covariance Matrix

       dmf_84              U           cov(U)

            0       -23.1213         56.6668       -56.6668
            1        23.1213        -56.6668        56.6668

                 Chi-Square      DF Pr > Chi-Square

                     9.4340       1            0.0021

      Generalized Log-Rank Test II (Sun, Zhao & Zhao, 2005)
                          xi(x)=xlog(x)

               Test Statistic and Covariance Matrix

             dmf_84              U        cov(U)

                  0    -23.6889      64.9884    -64.9884
                  1     23.6889     -64.9884     64.9884

                    ChiSquare      DF Pr>ChiSquare

                       8.6348       1         0.0033
```

Both tests yield a significant result and are comparable to the $G^{0,0}$ test (see Table 4.1).

□

Example 4.1 in SAS (Generalized log-rank tests using PROC ICLIFETEST)

The ICLIFETEST *procedure has the ability to output scores that are derived from a permutation form of the generalized log-rank statistic. Using these scores in the* NPAR1WAY *procedure allows the user to calculate a permutation test and compute P-values based on normal theory approximation or Monte Carlo simulation.*

*The WLRT based on scores (4.9) or weights (4.10) is performed by using Finkelstein's weights (*WEIGHT=FINKELSTEIN*).*

```
PROC ICLIFETEST DATA=tandmob IMPUTE(SEED=123);
TIME (L44,R44);
TEST dmf_84 / WEIGHT=FINKELSTEIN OUTSCORE=OUT;
RUN;

PROC NPAR1WAY DATA=out SCORES=DATA;
CLASS dmf_84;
VAR score;
EXACT SCORES=DATA / MC SEED=1234;
run;
```

The SCORES=DATA *option ensures that the input data are used as scores. The option* MC *ensures that the exact test is performed within a reasonable amount of time. We show below the result of the asymptotic permutation test.*

```
Data Scores Two-Sample Test

Statistic (S)          23.6889
Z                       2.9355
One-Sided Pr >  Z       0.0017
Two-Sided Pr > |Z|      0.0033
```

*The procedure also supports the generalized log-rank statistic using (a) Fay's weights (*WEIGHT=FAY*, based on a proportional odds model (Fay, 1999)) and yielding the generalized Peto-Prentice-Wilcoxon test, (b) using Sun's weights (*WEIGHT=SUN*, default option (Sun, 1996)) or (c) using Harrington and Fleming's weights (*WEIGHT=FLEMING(ϱ,γ) *(Harrington and Fleming, 1982)). However, these latter weights are currently not supported for the* OUTSCORE= *option.*

□

Example 4.2 in SAS ***(Generalized log-rank tests using PROC ICLIFETEST)***

We first create a data set without missing values for the grouping variable oh46o. *Note, that the* ICLIFETEST *procedure automatically removes observations with a missing grouping variable, so leaving out explicitly these observations is not necessary. To include observations with a missing grouping variable as a separate group, the keyword* MISSING *should be added to the* ICLIFETEST *statement.*

The procedure supports several tests. In the code below, we ask again for the generalized Peto-Prentice-Wilcoxon test, i.e., a $G^{1,0}$ *test (second row of Table 4.2).*

```
PROC ICLIFETEST DATA=tandmob2 IMPUTE(SEED=123);
TIME (L44,R44);
TEST oh46o / WEIGHT=FLEMING(1,0);
RUN;
```

The output shows first basic information of the data and method used, subsequently the nonparametric survival estimates per group are provided (not shown). Next the generalized log-rank statistics, the covariance matrix for the generalized log-rank statistics and the requested test of equality over groups are shown. The result deviates slightly from the result in Table 4.2 because we used the ML approach. To obtain the permutation form reported in Table 4.2, we need to change WEIGHT=FAY *(equivalent to* FLEMING(1,0) *but allows to output the scores when adding the option* OUTSCORES=out*) and post-process the scores using the* NPAR1WAY *procedure like in Example 4.1 (output not shown). Note that the variance is calculated via a multiple imputation method according to Sun (1996). The addition of the* SEED *option ensures reproducible results.*

```
Generalized Log-Rank Statistics

oh46o               Log-Rank

no plaque           5.946306
pits/fissures       -6.63029
total               0.683986

Covariance Matrix for the Generalized Log-Rank Statistics

oh46o                 0              1              2

0               22.5224       -18.7833        -3.7391
1              -18.7833        19.8216        -1.0383
2               -3.7391        -1.0383         4.7774
```

(Cont.)

```
                Test of Equality over Group

                                                      Pr >
        Weight                  Chi-Square    DF    Chi-Square

FLEMING-HARRINGTON(1, 0)          2.2418       2      0.3260
```

For ordered groups, the keyword TREND *asks for a trend test to detect ordered alternatives such as $H_1 : S_1(t) \geq S_2(t) \geq \cdots \geq S_K(t)$ (with at least one strict inequality).*

Post-hoc pairwise testing may be requested by adding the keyword DIFF=ALL *for all pairwise tests or* CONTROL *for comparisons with a control group. By default, the unadjusted and Tukey-Kramer (Kramer, 1956) adjusted P-values are reported. Other options for adjusting are possible with the specification of the keyword* ADJUST*. The Bonferroni correction (*BON*), Scheffé correction (*SCHEFFE*) or Dunnett's two-tailed comparison of a control group with all other groups (*DUNNETT*) are available among others. For more details, we refer to Appendix E.3 or the* SAS *documentation.*

□

4.2 Sample size calculation

Given the fact that interval-censored data are best analyzed taking into account the interval-censored nature of the data, it is also recommended to do so when designing a study. In practice, the sample size is often calculated assuming the event times are exactly known. It may yield a good approximation if the assessments are done frequently. Otherwise, the power will be overestimated as illustrated by a simulation study of Sun and Chen (2010). They also proposed a sample size formula based on Finkelstein's score test (see Section 5.3.1) for a 1:1 randomized clincial trial with equal assessment intervals in both groups. The sample size formula is

$$n = \frac{4(Z_{\alpha/2} + Z_\beta)^2 v}{u^2} \tag{4.17}$$

where u and v are the expected value and variance of the score function under the alternative hypothesis. Note that u and v depend on the baseline survival distribution and the assessment schedule. Hence, it is different from the similar looking sample size formula for right-censored data (Collett, 2003). When the assessments become less frequent, more events are probably needed to preserve the same power. Unfortunately, there appears to be no practical implementation of the method. In Section 6.5, a sample size method based on the AFT model and with a SAS macro available is described. Alternatively, a simulation study must be set up. This means a large number of data sets (e.g., 1 000) of a certain sample size and with events, visits and censoring times generated from specific distributions must be generated. All data sets are then analyzed with the same interval-censored data analysis method that will be used for the future study. The proportion of significant treatment effects over all data sets, i.e., the power of the applied method, is determined for the given sample size. The process is then repeated with a smaller or large sample size until the required power is obtained.

4.3 Concluding remarks

Based on Fay and Shaw (2010) and Oller and Gómez (2012), we recommend the permutation test unless there is extreme unequal censoring and sample sizes in the K groups. This can be accomplished with the R function ictest from the package interval or the kSampleIcens function from the package icensBKL. Also, the SAS procedure ICLIFETEST combined with the procedure NPAR1WAY can perform the permutation test. The generalized log-rank test resembles best the well-known log-rank test used for right-censored data which gives the same weight regardless of the time at which an event occurs. More emphasis on early or late differences can be obtained using the Peto-Prentice-Wilcoxon test (i.e., $G^{1,0}$ test) or $G^{0,1}$ test, respectively.

To illustrate the potential danger of using mid-point imputation in the presence of interval censoring, we performed a small simulation study (100 replications) mimicking a randomized controlled clinical trial where the event times in the control group were simulated according to a Weibull distribution with scale 100 and shape 5. In both groups, 160 patients were included. Two different censoring schemes were examined: (1) examinations every 30 (± 2) days in both groups; (2) examinations every 30 (± 2) days in the experimental group and every 90 (± 7) days in the control group. We considered the case of a hazard ratio equal to 1 and a hazard ratio of 1.05. A sufficient number of visits was planned such that no event was right-censored. The analysis was performed with SAS. We analyzed the true event times and the mid-point imputed times by a standard log-rank test for right-censored data. The interval-censored data were analyzed using the generalized log-rank test of Sun (1996), the default test in PROC ICLIFETEST. For the equal censoring scheme, the type I error rate was close to 5% for all settings. However, for the unequal censoring scheme, the type I error rate was seriously inflated up to 100% for the mid-point imputed times, while again close to 5% for the true event times and the interval-censored times. For a HR of 1.05, the power for the true event times was 51%. For the equal censoring scheme, the power decreased for the interval-censored times to 39% but further decreased for the mid-point imputed times to 34%. However, for the unequal censoring scheme, the power for the interval-censored times dropped to 4.4% and for the mid-point imputed times even to 0%. In summary, great care must be taken when designing the study with an unequal censoring scheme.

Chapter 5

The proportional hazards model

The proportional hazards (PH) model is by far the most popular survival model in medical applications since its introduction in 1972 by Cox. A major selling argument of this model is the partial likelihood estimation technique which allows to estimate regression parameters hereby ignoring the baseline hazard function. However, this elegant feature only applies to right-censored survival times. For interval-censored data, the PH model is still popular but now the baseline hazard must be estimated along with the regression parameters.

Numerous approaches have been suggested to fit a PH model to interval-censored data, from purely parametric methods to semiparametric methods. We start this chapter with a purely parametric approach. Then we review the various flexible approaches and we end with the semiparametric approaches. Admitted, a strict ordering in the approaches may not be achieved or even possible.

Given the large number of approaches, each with its advantages and disadvantages, it may not be easy for the statistical practitioner to make a choice. Therefore, we have compared the results at the end of the chapter on one particular data set and provide some guidance to the reader about which method one might choose in practice. However, we realize that the choice will most likely be driven by practical arguments like the familiarity of the user with the particular software such as SAS or R or with the statistical approach involved.

Let $\lfloor l_i,\, u_i \rfloor$ denote again the observed interval in which the survival event for the ith subject occurs and $\boldsymbol{X}_i$ represent the p-dimensional vector of covariates from subject i $(i = 1, \ldots, n)$. As before, let $S(t \mid \boldsymbol{X})$ denote the survival function for a subject with covariates $\boldsymbol{X}$. The likelihood function for interval-censored observations is then proportional to

$$L = \prod_{i=1}^{n} \big\{ S(l_i \mid \boldsymbol{X}_i) - S(u_i \mid \boldsymbol{X}_i) \big\}. \tag{5.1}$$

Assume that $S(t \mid \boldsymbol{X})$ is specified by the PH Model (2.13). The likelihood function (5.1) is then

$$L(\boldsymbol{\beta},\, S_0) = \prod_{i=1}^{n} \Big\{ S_0(l_i)^{\exp(\boldsymbol{X}_i'\boldsymbol{\beta})} - S_0(u_i)^{\exp(\boldsymbol{X}_i'\boldsymbol{\beta})} \Big\}, \tag{5.2}$$

with $\boldsymbol{\beta}$ a vector of regression parameters and $S_0(t)$ the baseline survival function.

In Section 5.1, we treat PH models with a parametric baseline survival distribution. In Section 5.2, we start with the piecewise exponential baseline survival model and then go on with PH models with a baseline survival function modelled smoothly. In Section 5.3, we discuss semiparametric approaches. An imputation approach is described in Section 5.4. Model checking tools are discussed in Section 5.5. We briefly address sample size calculation in Section 5.6. Finally, we provide some practical guidance in Section 5.7.

5.1 Parametric approaches

5.1.1 Maximum likelihood estimation

We assume a parametric form for the baseline survival distribution such as Weibull, gamma, log-normal, etc. Further details on these distributions can be found in Appendix B. Both the parameters and corresponding tests are obtained using the maximum likelihood method.

Example 5.1 Breast cancer study

As an illustration we will compare the two treatments for the early breast cancer study. The Weibull baseline survival function is given by $S_0(t) = \exp\{-(t/\alpha)^\gamma\}$ with $\gamma > 0$ (shape parameter) and $\alpha > 0$ (scale parameter) such that the survival function for an individual with covariate X is

$$S(t) = \exp\{-(t/\alpha)^\gamma\}^{\exp(\beta X)},$$

where here X is an indicator variable for the radiotherapy and adjuvant chemotherapy group and β the corresponding regression parameter. The MLE of β is equal to $\widehat{\beta} = 0.9164$ which corresponds with a hazard ratio of 2.50 (95% CI: 1.43 – 4.39) for the treatment effect. The estimated baseline survival function corresponds to $\widehat{\gamma} = 1.6146$ and $\widehat{\alpha} = 49.3667$. The log-likelihood of the fitted model equals -143.321 with an AIC value of 292.6.

□

5.1.2 R solution

A PH model with a parametric baseline survival function may be fitted using the function `ic_par` of the `R` package `icenReg` (Anderson-Bergman, 2017). The function allows for an exponential, Weibull, gamma, log-normal, log-logistic or generalized gamma distributed baseline hazard. The function `diag_baseline` allows to graphically compare the fitted parametric distribution with the NPMLE. The function `diag_covar` allows to graphically assess whether the covariate effect satisfies the PH assumption (or PO assumption, if assumed). This is illustrated in Section 5.5.2.

A PH model with a Weibull baseline survival function may also be fitted with the function frailtyPenal of the R package frailtypack (Rondeau and Gonzalez, 2005; Rondeau et al., 2015, 2012). Although the function is primarily designed to fit frailty models (see Section 8.2.1), it may also fit a standard PH model with a Weibull baseline hazard. For the moment, it is the only parametric distribution available in the package. Alternatively, the likelihood may be programmed and maximized using the function optim which then allows for any other parametric survival model. A program for the Weibull baseline survival function is available in the supplementary materials.

Example 5.1 in R (package icenReg)

After loading the breast cancer data from the icensBKL *package, the lower boundary for left-censored observations must be set to* 0. *When a* cbind(low, upp) *object is used for the response (instead of a* Surv *object), the upper boundary for right-censored observations must be set to* Inf. *A PH model with a Weibull baseline survival function is fit to the data with the command:*

```
> modelWeibull <- ic_par(Surv(low, upp, type = "interval2") ~ treat,
+                 data = breastCancer, model = "ph", dist = "weibull")
```

The output of the model is:

```
Model:  Cox PH
Baseline:  weibull
Call: ic_par(formula = Surv(low, upp, type = "interval2") ~ treat,
    data = breastCancer, model = "ph", dist = "weibull")

                  Estimate Exp(Est) Std.Error z-value         p
log_shape           0.4791    1.615   0.11990   3.996 6.439e-05
log_scale           3.6090   36.950   0.08792  41.050 0.000e+00
treatradio+chemo    0.9164    2.500   0.28290   3.239 1.201e-03

final llk =  -143.3208
Iterations =  7
```

Additional output, such as a 95% confidence interval, must be calculated post-hoc using the estimates of the model.

□

Example 5.1 in R (package frailtypack)

After loading the data, some data management is needed in order to get the data ready for analysis. Left-censored observations must have a zero in the left endpoint and the right endpoint of right-censored observations must be equal to the left endpoint. Finally, an indicator variable for interval-censored data must be created. The function SurvIC *is used to take the interval-censored data into account, when fitting a PH model with a Weibull hazard to the data.*

```
> weibullfit <- frailtyPenal(SurvIC(low, upp, event) ~ treat,
+                     data = breastCancer, hazard = "Weibull")
```

```
  Cox proportional hazards model parameter estimates
  using a Parametrical approach for the hazard function

                         coef exp(coef) SE coef (H)        z        p
treatradio+chemo 0.916633    2.50086     0.282968 3.23935 0.001198


      marginal log-likelihood = -143.33
      Convergence criteria:
      parameters = 2.86e-06 likelihood = 8.49e-05 gradient = 1.89e-10

      AIC = Aikaike information Criterion     = 1.55667

The expression of the Aikaike Criterion is:
         'AIC = (1/n)[np - l(.)]'

      Scale for the weibull hazard function is : 49.36
      Shape for the weibull hazard function is : 1.62

The expression of the Weibull hazard function is:
         'lambda(t) = (shape.(t^(shape-1)))/(scale^shape)'
The expression of the Weibull survival function is:
         'S(t) = exp[- (t/scale)^shape]'

      n= 94
      n events= 56
      number of iterations:  9
```

Note that an alternative definition of AIC, which originates from the likelihood cross-validation criterion defined in (5.8) when no penalty is present. Multiplying with $2n$ yields the classical AIC value yielding the same inference.

A 95% CI for the hazard ratio is obtained with

```
> exp(weibullfit$coef - qnorm(1 - 0.05/2) * sqrt(weibullfit$varH))
> exp(weibullfit$coef + qnorm(1 - 0.05/2) * sqrt(weibullfit$varH))
```

□

5.1.3 SAS solution

The PHREG procedure cannot fit interval-censored survival times, rather we need the ICPHREG procedure (available from SAS/STAT 13.2 in SAS 9.4). It allows to fit the baseline hazard with a piecewise constant model (see Section 5.2.1) or a cubic spline model (see Section 5.2.3) with SPLINES(DF=1) corresponding to a Weibull survival model. An example program is available in the supplementary materials. Alternatively, the analysis can be done with the NLMIXED procedure, which requires the log-likelihood function to be explicitly specified. This will be illustrated now.

Example 5.1 in SAS (PROC NLMIXED)

In a preliminary data step we create all variables necessary for fitting the parametric PH model. The lower limit of left-censored observations is put to a missing value. A variable dummy *is created that takes the value 1 for all observations. Its use is explained below. Finally, we create an indicator variable* treat *for the radiotherapy and adjuvant chemotherapy group from the treatment variable* tx.

In the NLMIXED *procedure, the log-likelihood function must be provided by the user. Therefore, we create variables S_l and S_r for which we calculate the survival function in the left and right endpoint, respectively. The contribution of each individual to the likelihood function taking into account the type of censoring (left, right or interval) is expressed in the variable* lik *with* llik *as its logarithm. In the* MODEL *statement, we use the general likelihood specification.* NLMIXED *requires standard a statement "var ~ distribution". When the likelihood is specified via interval probabilities and not via a regression model, any numeric variable can be used in the model statement as long as it does not contain missing values. Here we have created a dummy variable as dependent variable.*

```
ODS OUTPUT PARAMETERESTIMATES=est;
PROC NLMIXED DATA=breast;
PARMS beta1=0;
BOUNDS gamma > 0, alpha>0;
S_r=EXP(-(1/alpha*upper)**gamma)**EXP(beta1*treat);
S_l=EXP(-(1/alpha*lower)**gamma)**EXP(beta1*treat);
     IF     MISSING(lower) AND NOT MISSING (upper) THEN lik = 1 - S_r;
ELSE IF NOT MISSING(lower) AND     MISSING (upper) THEN lik = S_l;
ELSE lik = S_l - S_r;
llik=log(lik);
MODEL dummy ~ GENERAL(llik);
RUN;
```

The output below is standard. Note, however, that we estimated the hazard ratio for treatment (and 95% CI) by exponentiating the corresponding parameter estimate and the confidence interval. We do not recommend the ESTIMATE *statement which computes the standard error with the delta method on the hazard scale. Indeed, the normal distribution is better approximated on the log hazard ratio scale (Collett, 2003). For a categorical covariate with $K > 2$ categories, a* CONTRAST *statement is needed to compare the effect of each of the $K - 1$ classes to a control class.*

```
                Fit Statistics

-2 Log Likelihood                         286.6
AIC (smaller is better)                   292.6
AICC (smaller is better)                  292.9
BIC (smaller is better)                   300.3

                          Parameter Estimates

                     Standard                                95% Confidence
Parameter Estimate      Error    DF t Value Pr > |t|            Limits

beta1        0.9164    0.2829    94    3.24    0.0017    0.3546    1.4782
gamma        1.6146    0.1936    94    8.34    <.0001    1.2303    1.9990
alpha       49.3667    6.9376    94    7.12    <.0001   35.5919   63.1415

Parameter Estimates

Parameter Gradient

beta1      8.973E-6
gamma      9.013E-7
alpha       -3.9E-7
```

□

5.2 Towards semiparametric approaches

We start this section with the piecewise exponential baseline survival model. This model assumes on the hazard scale a piecewise constant behavior with data-driven or user-defined knots. The effect of each of the covariates is assumed to satisfy the PH assumption. The choice of knots is always somewhat arbitrary, but they impact the shape of the baseline hazard and may influence the estimation of the regression parameters. Approaches have therefore been proposed in the literature to obtain a smooth shape of the baseline hazard/survival distribution and with possibly less impact on the estimation of the regression parameters. Three flexible modelling approaches implemented in statistical software will be discussed here. Two approaches are based on spline-smoothing technology. In Section 3.3 we have pointed out that several types of splines have been used in the statistical literature and this is also the case when dealing with interval-censored survival times. The spline-smoothing approach is popular in case of interval-censored observations because on top of providing a smooth solution, they largely enjoy the properties of parametric models.

5.2.1 Piecewise exponential baseline survival model

5.2.1.1 Model description and estimation

The *piecewise exponential baseline survival* (PEBS) model could be considered a first step towards nonparametric modelling of the baseline hazard function. For the right-continuous version of the model (also a left-continuous version is used), K inner knots $0 \equiv s_0 < s_1 < \ldots < s_K < s_{K+1} \equiv \infty$ divide the baseline hazard such that $\hbar_0(t) = r_k$ if $s_{k-1} \le t < s_k$, $(k = 1, \ldots, K+1)$ and with $r_k \ge 0$. The baseline cumulative hazard function then becomes a piecewise linear function given by

$$H_0(t) = \sum_{k=1}^{K+1} r_k \Delta_k(t),$$

with

$$\Delta_k(t) = \begin{cases} 0, & t < s_{k-1}, \\ t - s_{k-1}, & s_{k-1} \le t < s_k, \\ s_k - s_{k-1}, & t \ge s_k. \end{cases}$$

The positivity constraint on r_k is removed by working with v_k defined by $r_k = \exp(v_k)$ $(k = 1, \ldots, K+1)$. Covariates are introduced under the PH assumption. The knots s_k $(k = 1, \ldots, K)$ can be chosen equidistant as done by the function frailtyPenal of the R package frailtypack or could be based on the quantiles of the censoring times as suggested by Ibrahim et al. (2001). The latter approach is implemented in the SAS procedure ICPHREG. In both cases, the selection of the number of knots is guided by an information criterion such as AIC.

The function frailtyPenal allows for fitting a PEBS PH model to interval-censored data. The set of knots consists of outer knots equal to the smallest and largest data points and inner knots put equidistantly (Rondeau and Gonzalez, 2005; Rondeau et al., 2012). Note that for right-censored data, the knots can also be put according to the percentiles of the data points. However, details on how exactly the knots are selected with interval-censored data are not given, only that the number of intervals should be between 1 and 20. In addition, the output does not contain information on the choice of knots.

The ICPHREG procedure uses the *equally spaced quantile partition (ESQP) method* (Ibrahim et al., 2001) adapted to interval-censored data to select the knots. First, for all left and genuine interval-censored observations the midpoint is computed. These values are merged with the observed boundaries of these observations. Next a unique sorted list is created from all these time points. The ICPHREG procedure then partitions the time axis into the specified number of intervals with each interval containing an approximately equal number of time points. A motivation of the choice of knots is, however, not provided.

The ESQP approach appears preferable as it takes the density of observed censoring explicitly into account, while the frailtypack package may put knots where there is practically no mass. This may create computational problems as we encountered in our analysis of the breast cancer data. The user of the frailtypack package is therefore advised to inspect the estimated hazard function for irregularities.

We will now illustrate the two approaches to fit a PEBS PH model to the breast cancer data.

Example 5.2 Breast cancer study

We compare again the two treatments in the early breast cancer study.

The function frailtyPenal of the R package frailtypack fits a PEBS PH model to interval-censored data with equidistant points. The optimal model (model 1) according to a (modified) AIC is based on 4 intervals (i.e., 3 knots) with hazard ratio equal to 2.46 (95% CI: 1.40 – 4.31).

The ESQP method proposed by Ibrahim et al. (2001), adapted to interval-censored data, yields an optimal model (model 2) with 3 intervals (i.e., 2 knots) according to AIC, with hazard ratio equal to 2.45 (95% CI: 1.40 – 4.27).

To evaluate which of the two fitted models is most appropriate for the breast cancer data, we compared the respective models with AIC and BIC. As a result we obtained for model 1: AIC = 296.1, BIC = 306.3; and model 2: AIC = 296.8, BIC = 307.0. Hence, model 1 is the best fitting model according to AIC and BIC but the difference between the models is small. The models also do not fit better than the simple Weibull model in Example 5.1 which yielded an AIC of 292.6.

□

5.2.1.2 R solution

The R package frailtypack (Rondeau and Gonzalez, 2005; Rondeau et al., 2015, 2012) can fit a standard and a frailty Cox proportional hazards model with a piecewise constant baseline hazard with equidistant knots, see also Section 8.2.1.

Example 5.2 in R

The data must be prepared as described in Section 5.1.2. The code below fits a PEBS PH model with the frailtyPenal *function. The number of time intervals specified in* nb.int *may be chosen between 1 and 20. The function* SurvIC *ensures that the interval-censored character of the data is taken into account. The best fitting model is obtained by minimizing AIC (somewhat differently defined, see output). Here, the models with 9, 10 and above 11 time intervals caused convergence problems. The best fitting model is based on 4 time intervals and was obtained using the following code:*

```
> PWfit4 <- frailtyPenal(SurvIC(low, upp, event) ~ treat,
+       data = breastCancer, hazard = "Piecewise-equi", nb.int = 4)
> breastCancer.fit4
> summary(PWfit4)
> plot(PWfit4)
> plot(PWfit4, type.plot = "Survival")
```

```
  Cox proportional hazards model parameter estimates
  using a Parametrical approach for the hazard function

                      coef exp(coef) SE coef (H)       z         p
treatradio+chemo 0.899607    2.45864     0.286325 3.1419 0.0016785

      marginal log-likelihood = -144.05
      Convergence criteria:
      parameters = 5.21e-06 likelihood = 4.83e-05 gradient = 2.16e-10

      AIC = Aikaike information Criterion      = 1.58563

The expression of the Aikaike Criterion is:
        'AIC = (1/n)[np - l(.)]'

      n= 94
      n events= 56
      number of iterations:  5
      Exact number of time intervals used:  4
```

The hazard ratio with 95% CI is produced by the summary *function, while the* plot *function provides the baseline hazard (default) or baseline survival function. Apart from the convergence problems with some choices of knots, we also noticed computational difficulties in computing the hazard function from 5 knots onwards.*

□

5.2.1.3 SAS solution

Now we apply the ICPHREG procedure to fit a PEBS PH model to the breast cancer data.

Example 5.2 in SAS (PROC ICPHREG)

Because the ICPHREG *procedure has a* CLASS *statement, no preliminary data management is needed for the variable* tx.

We fit a piecewise exponential baseline survival model with the number of intervals ranging from 2 to 10. The best fitting model according to AIC with 3 intervals is shown below.

```
ODS GRAPHICS ON;
PROC ICPHREG DATA=icdata.breast
             PLOTS(OVERLAY=GROUP CL)=(S) ;
CLASS tx;
MODEL (lower,upper) = tx / B=PIECEWISE(NINTERVAL=3);
HAZARDRATIO 'Chem+radio vs radio' tx;
BASELINE COVARIATES=covs / GROUP=all;
RUN;
ODS GRAPHICS OFF;
```

In the above MODEL *statement the interval-censored response with left (*lower*) and right (*upper*) endpoints is regressed on* tx. *The PEBS PH model is chosen with the option* B=PIECEWISE *and the number of intervals is specified by* NINTERVAL=. *Alternative to the default ESQP method, the user may choose the knots by using the option* INTERVALS=. *The* ICPHREG *procedure assumes right continuity of the hazard function. With the* HAZARDRATIO *statement the hazard ratio is calculated. With the* PLOTS *option the survival curves for the treatment groups are overlaid (not shown). For this, the (previously defined)* COVARIATES= *data set must be defined in the* BASELINE *statement. The option* GROUP=, *specifying here the grouping variable* all, *determines which plots should be overlaid. Also the cumulative hazard functions may be requested by adding the keyword* CUMHAZ. *The* BASELINE *statement may also be used to export several statistics to an output data set (see Appendix E.4). The following output for the parameter estimates is produced (using a slightly different format and label for* tx*).*

Effect	Tx	DF	Estimate	Standard Error	95% Confidence Limits		Chi-Square	Pr > ChiSq
Haz1		1	0.0203	0.0055	0.0096	0.0310		
Haz2		1	0.0539	0.0135	0.0275	0.0803		
Haz3		1	0.0587	0.0193	0.0209	0.0965		
tx	R	1	-0.8946	0.2839	-1.4509	-0.3382	9.93	0.0016
tx	R+C	0	0.0000					

The knots are positioned at 14.25 and 27.25 months (not shown). There are two types of parameters: the hazard parameters Haz1 *to* Haz3 *and the regression parameter* tx*. For the treatment effect we obtained a hazard ratio of 2.45 (95% CI:* 1.40 – 4.27*).*

□

5.2.2 SemiNonParametric approach

5.2.2.1 Model description and estimation

Zhang and Davidian (2008) proposed a general framework for flexibly modelling time-to-event data in the presence of covariates and subject to arbitrary patterns of censoring. One of the investigated models is based on the PH assumption with a baseline density from a large class of densities $\mathcal{H}$. The elements of $\mathcal{H}$ may be arbitrary well approximated by the "SemiNonParametric" (SNP) density estimator of Gallant and Nychka (1987). The approximation is based on a truncated series expansion. For a fixed degree of truncation, a parametric representation is obtained rendering likelihood-based inference computationally and conceptually straightforward with arbitrarily censored data.

The class $\mathcal{H}$ contains "smooth" densities that are sufficiently differentiable to rule out "unusual" features such as jumps or oscillations but they are allowed to be skewed, multimodal, fat- or thin-tailed. In one dimension a density $h \in \mathcal{H}$ can be approximated by a truncated infinite Hermite series, yielding the SNP representation

$$h_K(z) = P_K^2(z)\psi(z), \qquad P_K(z) = a_0 + a_1 z + a_2 z^2 + \ldots + a_K z^K, \tag{5.3}$$

where the base density $\psi(z)$ is the "standardized" form of a known density with a moment generating function. Also $h_K(z)$ must be a density, i.e., $\int h_K(z)\mathrm{d}z = 1$. The method now consists in determining the best $\boldsymbol{a} = (a_0, a_1, \ldots, a_K)'$ with an appropriate K to approximate $h(z)$. With $K = 0$, $h_K(z)$ reduces to $\psi(z)$. Values $K \geq 1$ control the departure from ψ and hence also the flexibility in approximating the true $h(z)$.

Zhang and Davidian (2008) show that $\int h_K(z)\mathrm{d}z = 1$ is equivalent to $a'\boldsymbol{A}a = 1$, with $\boldsymbol{A}$ a known positive definite matrix depending on ψ. This condition can be translated into $c'c = 1$ with $c = \boldsymbol{A}^{1/2}a$ because of the positive definiteness property of $\boldsymbol{A}$. Hence c can be expressed in polar coordinates summarized in a $(K+1)$-dimensional parameter vector $\boldsymbol{\phi}$.

Assume now that the positive random variable T has survival function $S_0 = P(T > t)$ and $\log(T) = \mu + \sigma Z$, $\sigma > 0$ where Z takes values in $(-\infty, \infty)$ and has density $h \in \mathcal{H}$. The approximation suggested by (5.3) can then be applied to estimate h. Zhang and Davidian (2008) argue that the standard normal base density ($\psi(z) = \varphi(z)$) and the standard exponential base density ($\psi(z) = e^{-z}$) are most natural to use. In both cases, approximations for $f_0(t)$ and $S_0(t)$ can be constructed using the above expressions. Under the normal base density representation, for fixed K and $\boldsymbol{\theta} = (\mu, \sigma, \boldsymbol{\phi}')'$, we have for $t > 0$

$$\begin{aligned} f_{0,K}(t;\boldsymbol{\theta}) &= (t\sigma)^{-1} P_K^2\{(\log t - \mu)/\sigma; \boldsymbol{\phi}\}\varphi\{(\log t - \mu)/\sigma\}, \\ S_{0,K}(t;\boldsymbol{\theta}) &= \int_{(\log t - \mu)/\sigma}^{\infty} P_K^2(z;\boldsymbol{\phi})\varphi(z)\mathrm{d}z. \end{aligned} \tag{5.4}$$

For the exponential base density representation,

$$\begin{aligned} f_{0,K}(t;\boldsymbol{\theta}) &= (\sigma e^{\mu/\sigma})^{-1} t^{(1/\sigma - 1)} P_K^2\{(t/e^{\mu})^{1/\sigma}; \boldsymbol{\phi}\} \exp\{-(t/e^{\mu})^{1/\sigma}\}, \\ S_{0,K}(t;\boldsymbol{\theta}) &= \int_{(t/e^{\mu})^{1/\sigma}}^{\infty} P_K^2(z;\boldsymbol{\phi}) \exp(-z)\mathrm{d}z. \end{aligned} \tag{5.5}$$

For PH Model (2.13), Zhang and Davidian (2008) assume that $S_0(t)$ can be well approximated by either (5.4) or (5.5) for an appropriate K. Hence it is assumed that

$$S(t \mid \boldsymbol{X}; \boldsymbol{\beta}, \boldsymbol{\theta}) \approx S_{0,K}(t;\boldsymbol{\theta})^{\exp(\boldsymbol{X}'\boldsymbol{\beta})}.$$

Estimation of the model parameters is done with maximum likelihood over all combinations of base density (normal, exponential) and $K = 0, 1, \ldots, K_{max}$ with the HQ criterion (see Section 1.5.2) as selection tool. They argue that $K_{max} = 2$ is likely to be sufficient to achieve an excellent fit.

Example 5.3 Breast cancer study

For the comparison of the two treatments we now assume that the baseline hazard of the PH model can be well approximated by either (5.4) or (5.5) for an appropriate K. According to the HQ criterion the best fitting model has an exponential base function with $K = 0$ (HQ = 295.72). The estimate of 0.9164 for the treatment effect (hazard ratio of 2.50 with 95% CI: 1.44 – 4.35) is the same as in Example 5.1, which should not come as a surprise since this model is equivalent to a Weibull regression model. Our results are slightly different from those of Zhang and Davidian (2008) due to slightly different data sets.

□

5.2.2.2 SAS solution

Zhang and Davidian provide SAS macros that can fit to arbitrarily censored data a PH model, an AFT model (see Section 6.3) or a proportional odds model (not covered in the book).

Example 5.3 in SAS

The macros expect the value -1 as upper limit for right-censored observations, while the lower limit for left-censored observations must be 0. This is done in a first step. Then the (main) macro %SNPmacroint *is called as follows:*

```
%SNPmacroint(data=breast, xvar=treat, depl=lower, depr=upper, np=1,
             aft=0, ph=1, po=0);
```

All covariates in the model should be listed in the xvar= *argument (with categorical variables as indicator variables) and their number in the np= argument. To avoid computational problems, it is recommended to center continuous covariates at zero. Here,* treat *is the sole covariate. The lower and upper limits of the interval-censored observations are given in the* depl= *and* depr= *arguments, respectively. The model (AFT, PH or PO) is chosen by supplying 1 or 0 in* aft=, ph= *and* po=. *Here, we chose for a PH model.*

Part of the output for the best fitting model looks as follows:

```
                    estph

3.5417835 0.6193397 0.9163798 295.72357

               varph

0.0183419 -0.000373 0.0298396
-0.000373 0.0055136 -0.003927
0.0298396 -0.003927 0.0800524

  hqc_ph

296.79107
299.41478
301.39128
295.72357
298.68489
301.20051

                    E_0_PH

3.5417831 0.6193397 0.9163794 295.72357
3.5417835 0.6193397 0.9163798 295.72357
3.5417839 0.6193413 0.9163769 295.72357
```

In estph *the estimates of the parameters μ, σ, ϕ_1 (if $K \geq 1$), ϕ_2 (if $K \geq 2$), β('s) and the HQ info is given. The covariance matrix of all parameters is given by* varph. *According to HQ, the exponential approximation together with K=0 fitted best the data, see the HQ values for a normal and exponential approximation for $K = 0, 1, 2$ (N_0, N_1, N_2, E_0, E_1, E_2) in* hcq_ph. *Since*

*$K = 0$ in our example, only the estimates of μ, σ and β and the HQ value are reported. Further output (*E_O_PH*) gives the parameter estimates obtained from several models for both the normal and exponential approximation with $K = 0, 1, 2$. Only the results of the best fitting model are shown here. Computational problems are flagged when all values are put to 99999. This occurred here for the normal approximation with $K = 2$. The choice of the starting values is documented in the web appendix of Zhang and Davidian (2008).*

P-values and confidence intervals must be calculated by the user. The test statistic Z for treatment equals $0.9163798/\sqrt{0.0800524} = 3.2388$ *($P = 0.0012$) and the 95% CI is obtained using* $\exp(0.9163798 \pm 1.96 \times \sqrt{0.0800524}) = [1.44; 4.35]$.

□

5.2.3 Spline-based smoothing approaches

5.2.3.1 Two spline-based smoothing approaches

There are, generally speaking, two main approaches for spline smoothing the baseline survival distribution or hazard. In the first approach one takes a relatively few, well-placed, inner knots, while in the second approach one takes a possibly large number of inner knots but with a penalization to avoid too different adjacent spline coefficients. In both cases maximum likelihood estimation yields the spline coefficients.

Royston and Parmar (2002) proposed to model the log of the baseline cumulative hazard in terms of natural cubic splines with K inner knots and outer knots $s_0 \equiv s_{min}$ and $s_{K+1} \equiv s_{max}$, which may be different from the extreme observations (Appendix F.3). Since natural cubic splines are cubic splines constrained to be linear beyond the boundary knots, Royston and Parmar (2002) reformulated the expression of a cumulative hazard function in terms of a natural cubic spline of degree $K + 1$ as:

$$\log\{H_0(t)\} = \gamma_0 + \gamma_1 y + \gamma_2 v_1(y) + \cdots + \gamma_{K+1} v_K(y), \tag{5.6}$$

where $y = \log(t)$ represents the time on the log scale. The v_k $(k = 1, \ldots, K)$ are the basis functions, which are computed as

$$v_k(y) = (y - s_k)^3_+ - e_k(y - s_{min})^3_+ - (1 - e_k)(y - s_{max})^3_+,$$

where $e_k = \dfrac{s_{max} - s_k}{s_{max} - s_{min}}$ and $(y - a)_+ = \max(0,\, y - a)$. When $K = 0$, the log of $H_0(t)$ becomes $\gamma_0 + \gamma_1 y$, which corresponds to a common form of the Weibull model. When $\gamma_1 = 1$, the Weibull model further reduces to the exponential model. Royston and Parmar (2002) recommended to check the fit of models with degrees of freedom between 1 and 4 and to select the best fitting model based on an information criterion.

Joly et al. (1998) proposed a penalized likelihood approach to model arbitrarily censored (and truncated) data. A smooth baseline hazard $\hbar_0$ is obtained by applying a roughness penalty to the log-likelihood proportional to the integrated second order derivative of $\hbar_0$. The penalized log-likelihood is defined as

$$\ell_P(\boldsymbol{\beta}, \hbar_0) = \ell(\boldsymbol{\beta}, \hbar_0) - \kappa \int \hbar_0''^2(u)\mathrm{d}u, \tag{5.7}$$

where $\ell(\boldsymbol{\beta}, \hbar_0)$ is the log-likelihood and a given penalty κ (> 0). The smooth solution is obtained by maximizing $\ell_P(\boldsymbol{\beta}, \hbar_0)$ to yield $\widehat{\boldsymbol{\theta}}^P = \{\widehat{\boldsymbol{\beta}}^P, \widehat{\hbar}_0^P\}$.

To find the smooth solution, the authors first approximate the baseline hazard function by M-splines $M_k(t)\,(k = 1, \ldots, K)$ and the cumulative hazard function by integrated M-splines, called I-splines $(k = 1, \ldots, K)$ (see Appendix F.3). More specifically, the baseline cumulative hazard function is written as $H_0(t) = \sum_{k=1}^K w_k I_k(t)$ with w_k positive weights and by differentiation $\hbar_0(t) = \sum_{k=1}^K w_k M_k(t)$. The weights $w_1, \ldots, w_K$ and the regression coefficients $\beta_1, \ldots, \beta_p$ are then determined by maximizing the penalized log-likelihood given by Expression (5.7).

It is common to choose κ using *cross-validation*. When $\boldsymbol{\theta}$ represents the set of weights and regression coefficients, the standard cross-validation score is equal to $V(\kappa) = \sum_{i=1}^n \ell_i\big(\widehat{\boldsymbol{\theta}}^P_{(-i)}\big)$, where $\widehat{\boldsymbol{\theta}}^P_{(-i)}$ is the maximum penalized likelihood estimator of (5.7) for the sample in which the ith individual is removed and ℓ_i is the log-likelihood contribution of this individual. Maximizing $V(\kappa)$ gives $\widehat{\kappa}$, but this procedure is too time consuming. Joly et al. (1998) proposes an approximation based on a one-step Newton-Raphson approximation of $\widehat{\boldsymbol{\theta}}^P_{(-i)}$, starting from $\widehat{\boldsymbol{\theta}}^P$. Then additional first order approximations yield the following likelihood cross-validation (LCV) criterion to be minimized:

$$\overline{V}(\kappa) = \frac{1}{n} tr\Big\{(\widehat{\boldsymbol{H}}^P)^{-1}\widehat{\boldsymbol{H}} - \ell\big(\widehat{\boldsymbol{\beta}}^P, \widehat{\hbar}_0^P\big)\Big\} \tag{5.8}$$

where $\widehat{\hbar}_0^P \equiv (\widehat{w}_1^P, \ldots, \widehat{w}_K^P)'$ and $\widehat{\boldsymbol{\beta}}^P$ are the maximum penalized likelihood estimators, $\widehat{\boldsymbol{H}}^P$ is minus the converged Hessian of the penalized log-likelihood, $\widehat{\boldsymbol{H}}$ is minus the converged Hessian matrix of the log-likelihood. Lower values of LCV indicate a better fitting model. Note that originally Joly et al. (1998) applied a two-step procedure where in the first step only the baseline hazard, i.e., $w_1, \ldots, w_K$ is estimated using the above cross-validation approach. In the second step $\widehat{\kappa}$ is fixed and $\hbar_0$ is re-estimated together with the regression coefficients. In the function frailtyPenal of the R package frailtypack the penalized estimation is done in one step.

Example 5.4 Signal Tandmobiel study

We revisit Example 4.2, where we compared the emergence distributions of tooth 44 in three plaque groups observed at a spatially close tooth 46. No significant difference in emergence times between the three groups was observed. We now want to verify whether this result still stands after correction for gender given the impact of gender on the emergence distribution of permanent teeth (Leroy et al., 2003). For both smoothing techniques, the nonsignificant difference in the three emergence distributions remained. The cubic spline approach of Royston and Parmar (2002) yielded a best fitting model (AIC = 1 117.9) with 2 degrees of freedom and $P = 0.3638$ for the difference in the three groups. With the penalized spline approach of Joly et al. (1998), the best fitting model (LCV = 1.354) has 19 degrees of freedom with $\kappa = 10$ and $P = 0.3960$ for the difference in the three groups.

□

5.2.3.2 R solution

The penalized likelihood approach is implemented in the R package frailtypack (Rondeau and Gonzalez, 2005; Rondeau et al., 2015, 2012).

Example 5.4 in R

The data must be prepared in a similar way as described in Section 5.1.2. The following code fits a PH model with a baseline hazard estimated similarly. Specifying "Splines" in the hazard *argument calls for a baseline hazard estimated using a spline with equidistant intervals. The number of knots is specified by the argument n.knots and must lie between 4 and 20. Rondeau et al. suggest to start with 7 knots and decrease/augment until convergence is obtained. We determined the number of knots by minimizing LCV while holding the smoothing parameter fixed to 10 000 as recommended by Rondeau et al. Once the number of knots is determined, the (positive) smoothing parameter is chosen by refitting the model with several kappa values. The best fitting model corresponds to the smallest LCV. With the number of knots varying between 4 and 20 and smoothing parameters* kappa *= 1, 10, 100, 1 000 or 10 000, the best fitting model was obtained using*

```
> tandmob.fit19.k10 <- frailtyPenal(
+            SurvIC(L44, R44, event) ~ fOH46o + fGENDER,
+            data = tandmob, hazard = "Splines",
+            n.knots = 19, kappa = 10)
> print(tandmob.fit19.k10)
> summary(tandmob.fit19.k10)
> plot(tandmob.fit19.k10)
```

This yielded the following output.

```
  Cox proportional hazards model parameter estimates
  using a Penalized Likelihood on the hazard function

                          coef exp(coef) SE coef (H) SE coef (HIH)
fOH46opits/fissures -0.1938419  0.823788    0.145629      0.145478
```

(Cont.)

```
fOH46ototal          0.0364736  1.037147   0.282193     0.282126
fGENDERgirl          0.2390644  1.270060   0.120709     0.120494
                             z          p
fOH46opits/fissures -1.331065 0.183170
fOH46ototal          0.129251 0.897160
fGENDERgirl          1.980506 0.047647

         chisq df global p
fOH46o 1.85077  2     0.396

      penalized marginal log-likelihood = -552.3
      Convergence criteria:
      parameters = 0.00043 likelihood = 9.58e-06 gradient = 1.23e-07

      LCV = the approximate likelihood cross-validation criterion
            in the semi parametrical case     = 1.35385

      n= 416  ( 84  observation deleted due to missing)
      n events= 291
      number of iterations:  4

      Exact number of knots used:  19
      Value of the smoothing parameter:  10, DoF:  6.52
```

The summary of the model provides the hazard ratio and corresponding 95% CI. The estimated baseline hazard function with 95% CI can be plotted (not shown) using the plot statement. Adding the argument type.plot = "survival" *plots the baseline survival function.*

□

5.2.3.3 SAS solution

The cubic spline based PH model can be fitted using the ICPHREG procedure. The inner knots are determined as for the piecewise constant model (see Section 5.2.1.3), while the outer knots are placed at the minimum and maximum values of the same sequence. By default, a model with 2 degrees of freedom is fitted.

Example 5.4 in SAS

No preprocessing is needed, the following code fits the best fitting model according to the AIC criterion.

```
ODS GRAPHICS ON;
PROC ICPHREG DATA=icdata.tandmob PLOTS(OVERLAY=GROUP)=(S) ;
```

```
CLASS oh46o gender/DESC;
MODEL (L44,R44) = oh46o gender/ B=SPLINES(DF=2);
TEST oh46o gender;
HAZARDRATIO 'oh46o' oh46o / DIFF=REF;
BASELINE OUT=BaseHazard COVARIATES=covs
         TIMELIST=4 to 12 by 1 SURVIVAL=S /group=gender ROWID=oh46o;
RUN;
ODS GRAPHICS OFF;
```

Categorical variables are defined in the CLASS *statement. The option* DESC *selects the 'no plaque group' as reference group, but another group could have been taken too, see* SAS *documentation. The* MODEL *statement has option* B=SPLINES *to model the baseline cumulative hazard function by natural cubic splines with the option* DF= *to choose the degrees of freedom. Alternatively, one could specify the knots by using the option* INTERVALS=. *The* TEST *statement enables to perform type III (by default) Wald tests for the model effects. Hazard ratios with 95% CIs are calculated with the* HAZARDRATIO *statement. The same options as in Section 5.2.1.3 are available here. The* BASELINE *statement calculates the predicted values for a specific set of covariates. For more details, we refer to Appendix E.4 or the* SAS *documentation.*

The following output for the parameter estimates is produced.

Analysis of Maximum Likelihood Parameter Estimates

Effect	Oral hygiene occl46	Gender	DF	Estimate	Standard Error	95% Confidence Limits	
Coef1			1	-55.7879	8.8497	-73.1330	-38.4428
Coef2			1	26.2324	4.4726	17.4661	34.9986
Coef3			1	28.2251	7.6603	13.2112	43.2391
oh46o	total		1	0.0367	0.2822	-0.5163	0.5897
oh46o	pits/fissures		1	-0.2029	0.1457	-0.4886	0.0827
oh46o	no plaque		0	0.0000			
gender		girl	1	0.2283	0.1207	-0.0084	0.4649
gender		boy	0	0.0000			

Analysis of Maximum Likelihood Parameter Estimates

Effect	Oral hygiene occl46	Gender	Chi-Square	Pr > ChiSq
Coef1				
Coef2				

(Cont.)

Coef3				
oh46o	total		0.02	0.8965
oh46o	pits/fissures		1.94	0.1638
oh46o	no plaque			
gender		girl	3.58	0.0587
gender		boy		

The parameters Coef1 *to* Coef3 *correspond to* γ_0 *to* γ_2 *in (5.6), respectively. Note that the Chi-square statistics and P-values are not calculated for testing these parameters equal to be zero. The overall effect of* oh46o, *obtained with the TEST statement (output not shown), was not significant* $(P = 0.3638)$ *indicating no significant difference in the emergence curves between the three groups.*

□

5.3 Semiparametric approaches

Turnbull's nonparametric estimator of the survival distribution for interval-censored observations has been first generalized to the Cox PH model by Finkelstein (1986). In this section we first describe Finkelstein's approach, and then move to methods that are based on the same generalization to the Cox PH model but differ in computational aspects. Numerous developments can be found in the literature, in this book we focus (almost exclusively) on approaches that are supported by (generally available) statistical software.

5.3.1 Finkelstein's approach

Finkelstein (1986) extended the nonparametric approach of Turnbull to the proportional hazards model with interval-censored data. For this she assumed the model expressed in (5.2). Note that likelihood (5.2) depends only on the baseline hazard $\hbar_0$ through its values at the different observation time points. Let $s_0 = 0 < s_1 < \ldots < s_{K+1} = \infty$ denote the ordered distinct time points of all observed time intervals $\lfloor l_i, u_i \rfloor$ $(i = 1, \ldots, n)$. Further, let $\alpha_{ik} = I\{s_k \in \lfloor l_i, u_i \rfloor\}$ $(k = 1, \ldots, K+1,\ i = 1, \ldots, n)$ with I the indicator function. To remove the range restrictions on the parameters for S_0, the likelihood is parametrized by $\gamma_k = \log\{-\log S_0(s_k)\}$ $(k = 1, \ldots, K+1)$. Note that because $S_0(s_0) = 1$ and $S_0(s_{K+1}) = 0$, $\gamma_0 = -\infty$ and $\gamma_{K+1} = \infty$. In terms of $\boldsymbol{\beta}$ and $\boldsymbol{\gamma} = (\gamma_1, \ldots, \gamma_K)'$, the log-likelihood function $\ell(\boldsymbol{\beta}, S_0)$ can be written as

$$\ell(\boldsymbol{\beta}, \boldsymbol{\gamma}) = \sum_{i=1}^{n} \log \left[\sum_{k=1}^{K+1} \alpha_{ik} \left\{ \mathrm{e}^{-\zeta_{k-1} \exp(\boldsymbol{X}_i'\boldsymbol{\beta})} - \mathrm{e}^{-\zeta_k \exp(\boldsymbol{X}_i'\boldsymbol{\beta})} \right\} \right],$$

where $\zeta_k = \sum_{m=0}^{k} \exp(\gamma_m)$.

To estimate the mass of the regions of possible support and the regression parameters, Finkelstein (1986) proposed a Newton-Raphson algorithm. It turns out that the score equations are a generalization of the self-consistency algorithm suggested by Turnbull (see Equations (3.1)). In addition, Finkelstein derived a score test for testing $\boldsymbol{\beta} = \mathbf{0}$.

Finkelstein's approach was intended for discrete survival times but can also be used to survival times with a (right-)continuous survival distribution. For such data, the times $s_1 < \ldots < s_K$ are determined by the steps in the empirical survival function (see Section 3.1). However, asymptotic results may be inappropriate when the number of parameters increase with the sample size. When the continuous data are grouped, preferably in data independent classes, then it is appropriate to treat time as a discrete random variable. Another way to limit the number of steps with continuous survival distributions is given by the approach of Farrington, which is described in the next section.

We postpone the illustration of Finkelstein's approach until we have reviewed the related approaches.

5.3.2 Farrington's approach

Farrington's (1996) approach allows to fit interval-censored data with a generalized linear model. At the time of publication, this was a definite advantage because the model could then be implemented with the GLIM package (not used these days anymore). Farrington suggested three possibilities to parametrize the survival distribution with the first two proposals assuming a piecewise constant baseline hazard yielding a piecewise exponential survival model. His third proposal directly models the survival function, but puts mass only at a subset of the observed censoring times. If based on all censoring times, this approach is equivalent to Finkelstein's generalization. But the fact that a technique is offered to base the computation on a much reduced set of observation times and its relationship to generalized linear models, makes the approach worthwhile to be discussed. The method goes as follows.

Suppose that the data have been ordered with first l left-censored observations ($l_i = 0$), then r right-censored observations ($u_i = \infty$) and finally g genuine interval-censored observations ($0 < l_i < u_i < \infty$), with $n = l + r + g$. Since $S_i(0) = 1$ and $S_i(\infty) = 0$, likelihood function (5.1) can be written as

$$\prod_{i=1}^{l} \{1 - S_i(u_i)\} \prod_{i=l+1}^{l+r} S_i(l_i) \prod_{i=l+r+1}^{n} S_i(l_i)\{1 - S_i(u_i)/S_i(l_i)\}. \qquad (5.9)$$

This likelihood is equivalent to that of $n + g$ independent Bernoulli trials with probability p_i, response y_i (and covariate vector $\boldsymbol{X}_i$), i.e.,

$$\prod_{i=1}^{n+g} p_i^{y_i}(1-p_i)^{1-y_i}. \tag{5.10}$$

This can be seen as follows. Each left-censored observation is represented as a binary observation with $y_i = 1$ and $p_i = 1 - S_i(u_i)$ $(i = 1, \ldots, l)$, while for a right-censored observation $y_i = 0$ and $p_i = 1 - S_i(l_i)$ $(i = l+1, \ldots, l+r)$. An interval-censored observation $(l_i, u_i]$ is represented by two binary observations: (1) $y_i = 0$ and $p_i = 1 - S_i(l_i)$ and (2) $y_{i+g} = 1$ and $p_{i+g} = 1 - S_i(u_i)/S_i(l_i)$ for $(i = l+r+1, \ldots, n+g)$. Thus, by defining a set of $n+g$ binary observations as above, Expression (5.10) becomes equivalent to Expression (5.9). We now parametrize the survival probabilities, such that only at a subset of $s_0 = 0 < s_1 < \ldots < s_{K+1} = \infty$ there is mass on the survival distribution. We denote this subset of J strictly positive ordered times as $s_{(1)}, s_{(2)}, \ldots s_{(J)}$, with $J \leq K$. We will detail below how this subset needs to be chosen. First, we parametrize the survival distribution. Define

$$\theta_j = \log\left\{\frac{S_0(s_{(j-1)})}{S_0(s_{(j)})}\right\},$$

where $s_{(0)} = 0$. Then $\theta_j \geq 0$ for $j = 1, \ldots, J$ and

$$S_0(s_{(j)}) = \mathrm{e}^{-\theta_j}\, S_0(s_{(j-1)}).$$

The baseline survival function at time t is therefore given by

$$S_0(t) = \exp\left(-\sum_{j=1}^{J} \theta_j d_j\right)$$

with

$$d_j = \begin{cases} 1 & \text{if } s_{(j)} \leq t, \\ 0 & \text{if } s_{(j)} > t, \end{cases}$$

for $j = 1, \ldots, J$.

The survival function for the ith individual at time t then becomes

$$S_i(t) = S_0(t)^{\exp(\boldsymbol{X}_i'\boldsymbol{\beta})} = \exp\left\{-\exp(\boldsymbol{X}_i'\boldsymbol{\beta})\sum_{j=1}^{J}\theta_j d_j\right\}. \tag{5.11}$$

Thus, the response probabilities p_i in (5.10) are expressed in the unknown parameters $\theta_1, \ldots, \theta_J$ and the unknown regression coefficients $\beta_1, \ldots, \beta_p$, interpretable as log-hazard ratios.

We can now evaluate the survival function for the ith individual ($i =$

TABLE 5.1: Method of Farrington (1996). Definition of binary response variables y_i and auxiliary intervals B_i for constructing the indicator variables d_{ij}, $i = 1, \ldots, n+g$ and $j = 1, \ldots, J$.

Type of observation	$\mathbf{y_i}$	$\mathbf{B_i}$	
$(0, u_i]$	1	$(0, u_i]$	$i = 1, \ldots, l$
$(l_i, \infty]$	0	$(0, l_i]$	$i = l+1, \ldots, l+r$
$(l_i, u_i]$	0	$(0, l_i]$	$i = l+r+1, \ldots, n$
	1	$(l_i, u_i]$	$i = n+1, \ldots, n+g$

$1, \ldots, n)$, so that for the three types of interval-censored data, p_i can be written as

$$p_i = 1 - \exp\left\{ - \exp(\boldsymbol{X}_i'\boldsymbol{\beta}) \sum_{j=1}^{J} \theta_j d_{ij} \right\}, \tag{5.12}$$

where

$$d_{ij} = \begin{cases} 1, & \text{if } s_{(j)} \text{ in interval } B_i \\ 0, & \text{otherwise} \end{cases}$$

for $j = 1, \ldots, J$ and B_i as defined in Table 5.1.

Key is the actual choice of the J step times $s_{(j)}$ $(j = 1, \ldots, J)$. Taking all unique values of l_i and u_i for the $s_{(j)}$ will generally lead to too many θ-parameters. Farrington (1996) suggested a sequential procedure for selecting the step times. To start with, a minimal set is chosen such that each of the B_i intervals $(i = 1, \ldots, n+g)$ includes at least one of the times $s_{(j)}$. Then at least one of the values of d_{ij} in Equation (5.12) will be equal to one and hence $\sum_{j=1}^{J} \theta_j d_{ij}$ will be greater than zero. From the remaining censoring times (if any), one picks the one that significantly (at a given level) increases the log-likelihood the most. The process is stopped when no step time can be found to increase the log-likelihood significantly.

The difference of this approach with Finkelstein's approach lies in the parametrization of the steps, i.e., $\gamma_m = \sum_{j=1}^{m} \theta_j$ and, more importantly, in the mass allocated to the regions of support. In Finkelstein's approach, the mass follows immediately from the maximization algorithm. In Farrington's approach, additional step times are selected sequentially and as long as they contribute significantly to the likelihood. Hence, even when the inclusion of an extra step time adds a mass that is small but not negligible, it might be neglected. Of course, the sequential procedure of searching for extra step times may be time consuming with many possible regions of support.

As with Finkelstein's approach, one must be careful with letting the number of step times increase with the sample size as this violates the basic assumption to perform formal (likelihood-ratio) significance tests. As mentioned above, and ad-hoc approach to deal with this problem is to round or group the censoring times.

The third and last approach deals with speeding up the computational procedure to compute the semiparametric solution, ignoring the possible problem of asymptotics.

5.3.3 Iterative convex minorant algorithm

Pan (1999) proposed to speed up and improve the computation of Finkelstein's estimator to the case when there are many interval-censored observations. To this end, he suggested to apply an extension of the iterative convex minorant (ICM) algorithm to fit the PH model for interval-censored data. The original ICM algorithm of Groeneboom and Wellner (1992) does not allow for covariates. Pan reformulated the ICM as a generalized gradient projection (GGP) method such that it can be used to find the NPMLE of the Cox model for interval-censored observations. The GGP method is, as its name suggests, a more general maximization technique that encompasses Newton-type maximization but now allowing also for constraints. The constraint option is especially useful for estimating survival or cumulative distributions since they enjoy the monotonicity constraint when applied to the ordered set of censoring times. However, in contrast to the standard Newton-Raphson method, the method uses only the diagonal elements of the Hessian simplifying calculations which speeds up the computations with a large number of unknown parameters where the standard Newton-Raphson method will result in computational problems. Further, technical details of the algorithm are described in Appendix F.1.

5.3.4 Grouped proportional hazards model

Grouped failure time data are a type of interval-censored data that occur when each subject corresponds to a member of a set of nonoverlapping intervals. The grouped proportional hazards model (Pierce et al., 1979; Prentice and Gloeckler, 1978) is similar to the PH model described in Section 2.3.1 but now with a discrete survival variable with a finite set of values $0 = s_0 < s_1 < \ldots < s_{K+1} < \infty$. The grouped PH model assumes relation (2.13) at times s_j, which implies that for the ith individual ($i = 1, \ldots, n$)

$$P(T_i > s_j \mid \boldsymbol{X}_i) \; = \; S(s_j \mid \boldsymbol{X}_i) = S_0(s_j)^{\exp(\boldsymbol{X}_i'\boldsymbol{\beta})}.$$

Thus with $\theta_j = \log\{S_0(s_{j-1})/S_0(s_j)\}$, we obtain Expression (5.11) showing that this model becomes a special case of Finkelstein's model. Hence, the software used before can be used also in this case. However, we can also see that, with R_j the set of indices of subjects who have an observation at s_j and E_j the set of indices of subjects who fail at s_j ($j = 1, \ldots, K$) the likelihood is proportional to

$$L(\boldsymbol{\beta}, S_0) \; = \; \prod_{j=1}^{K} \prod_{i \in E_j} \{1 - S(s_j \mid \boldsymbol{X}_i)\} \prod_{i \in R_j - E_j} S(s_j \mid \boldsymbol{X}_i).$$

Note that this likelihood is the same as the likelihood of a binary response model with event probabilities $1 - S(s_j \mid \boldsymbol{X}_i)$. The grouped survival model is therefore equivalent to a binary response model with a complementary log-log link function. Thus, standard statistical software can be used to fit this model.

5.3.5 Practical applications

As confirmed by many other simulation studies, appropriately taking into account the interval-censored character of the data gives improved estimates of the model parameters. This was the conclusion taken by Sun and Chen (2010) who compared Finkelstein's method with simple imputation methods (right- or mid-point imputation) followed by an analysis of right-censored data. They concluded that Finkelstein's method is preferable but mid-point imputation may be considered when the intervals are narrow and assessment imbalance is less an issue. Right-point imputation should be avoided.

5.3.5.1 Two examples

We now analyze two data sets: the breast cancer data set and the Signal Tandmobiel data set. The first data set is small, while the second data set is large. For both data sets we compare the performance of statistical software based on Finkelstein's approach, with an algorithm based on Farrington's approach and software based on the extension of the ICM algorithm to Cox PH model with interval-censored data. We compare the estimated regression parameters and the NPMLE of the baseline hazard function. The grouped proportional hazards model is illustrated on the Signal Tandmobiel data set.

Example 5.5 Breast cancer study

Fitting Finkelstein's PH model on the early breast cancer data yields an estimate of 0.7974 for the treatment effect with a hazard ratio of 2.22 (95% CI: 1.11 – 4.43). We again conclude that radiotherapy-only therapy significantly ($P = 0.0010$, score test) increases the time to breast retraction in comparison to primary radiation therapy and adjuvant chemotherapy.

With Farrington's approach the minimal set of step times chosen from 40 time points is equal to $\{4, 8, 12, 17, 23, 30, 34, 39, 48\}$. Adding time point 20 implied a significant model improvement ($-2 \log L$ reduced from 277.0 to 271.3), with a significant treatment effect ($P = 0.0068$) and hazard ratio of 2.21 (95% CI: 1.25 – 3.90). Adding an additional step time did not further significantly (at 0.05) increase the likelihood.

□

Example 5.6 Signal Tandmobiel study

We compare the emergence distributions of the mandibular right first premolar (tooth 44) between boys and girls. Fitting Finkelstein's PH model is quite challenging because there are 645 different time points at which the tooth was examined. The SAS macro %Intev_Cens1 was stopped after 14 hours computations without a solution. With the enhanced ICM algorithm, the ic_sp function in the R package icenReg produced a solution within a few minutes. With a hazard ratio (girls over boys) of 1.20 (95% CI: 0.95 – 1.52) it appears that tooth 44 emerges earlier in girls than boys but not significantly (P = 0.1196). However, it is questionable that the asymptotics hold here given the large number of time points.

Farrington's approach selected 7 time points as minimal set of step times. Five more points were added and reduced $-2\log L$ from 1 304.3 to 1 276.8. This model produced essentially the same result as with Finkelstein's approach: HR=1.21 (95% CI: 0.96 – 1.51), $P = 0.1017$. Although 3 813 models were fitted in the process of finding the best fitting model, the analysis took only 40 minutes.

□

Example 5.7 Signal Tandmobiel study

To illustrate the grouped proportional hazards model, we verified whether a significant gender difference in emergence distributions can be shown of the maxillary right first premolar (tooth 14). For illustrative purposes, we grouped the survival times further into intervals of 1 year ranging from 7 to 11 years, i.e., intervals $(0,7], (7,8], (8,9], (9,10], (10,11], (11,\infty)$. Again no significant impact of gender on the emergence distribution of tooth 14 could be shown ($P = 0.76$). The hazard ratio (girls over boys) is equal to 0.94.

□

5.3.5.2 R solution

Finkelstein's score test is implemented in the package glrt (Zhao and Sun, 2015), unfortunately without the estimation of the hazard ratio.

Inspired by Pan (1999), Anderson-Bergman (2017) implemented in the R package icenReg an algorithm that uses a conditional Newton-Raphson step for the regression parameters and an ICM step for the baseline survival parameters, rather than one single ICM step for both parameter sets.

Since the grouped proportional hazards model is equivalent to a binary response model with a complementary log-log link, one can use the glm function to estimate the model parameters. To this end, each discrete time unit for each subject must be treated as a separate observation. For each of these observations, the response is dichotomous, corresponding to whether or not the subject has the event of interest in the time unit.

Example 5.5 in R (glrt **package**)
The package expects a $n \times 3$ data matrix with the lower and upper bounds of the intervals in the first two columns and a treatment indicator ranging from 0 to k in the third column. The lower bound of left-censored observations is set to zero and the upper bound of right-censored data to 99. An indicator variable for treatment is created.

The score test can be applied by using either of the following two commands.

```
> gLRT(breast, method = "score", M = 50, inf = 99)
```

```
> ScoreTest(breast)
```

A P-value of 0.0023 is obtained.

Alternatively, one can use the test of Sun et al. (2005) with gamma and rho equal to zero, see Section 4.1.7, which is then equivalent to the score test of Finkelstein.

```
> gLRT(breast, method = "glrt2", M = 50,inf = 99)
```

Now $P = 0.0070$ is obtained. Note that the score vector is the same for both implementations but due to a different calculation of the covariance matrix, a slightly different P-value is obtained.

□

Example 5.5 in R (icenReg **package**)
The function ic_sp *from the* R *package* icenReg *fits a semiparametric PH model to interval-censored data. In Section 5.1.2 the requirements for the data are described. The covariance matrix for the regression coefficients is estimated via bootstrapping. The number of samples is determined by the option* bs_samples. *For large data sets, this can become slow so parallel processing can be used to take advantage of multiple cores via the* R *package* doParallel. *This functionality is illustrated in the corresponding program in the supplementary materials. The model can be fitted using*

```
> spmodel <- ic_sp(Surv(low, upp, type = "interval2") ~ treat,
+                  data = breastCancer, model = "ph", bs_samples = 1000)
```

The output of the model is shown below.

```
Model:   Cox PH
Baseline:  semi-parametric
Call: ic_sp(formula = Surv(low, upp, type = "interval2") ~ treat,
  data = breastCancer, model = "ph", bs_samples = 1000, B = c(0, 1))

                    Estimate Exp(Est) Std.Error z-value       p
treatradio+chemo     0.7974     2.22     0.324   2.461 0.01385

final llk =  -133.0342
```

(Cont.)

```
Iterations =  56
Bootstrap Samples =  1000
```

□

Example 5.7 in R

For this example, we grouped the emergence times into intervals of 1 year, i.e., $(0, 7], \ldots, (11, \infty)$. The variable cs_grouped_age *contains the grouped emergence time and ranges from 7 to 12 (interval $(11, \infty)$). An indicator (censoring) variable* y *for the interval was also added.*

```
> data("tandmob", package = "icensBKL")
> T14 <- subset(tandmob, select = c("IDNR","CS_age","CS_14","GENDER"))
> T14$cs_grouped_age <- pmin(ceiling(T14$CS_age), 12)
>
> # create extended data set T14e
> T14e <- NULL
> for (i in 1:nrow(T14)){
+   for (age in 7:T14$cs_grouped_age[i]){
+       y <- T14$CS_14[i] * (abs(age - T14$cs_grouped_age[i]) < 0.01)
+       T14e <- rbind(T14e, cbind(T14[i,], age, y))
+ }}
```

Below we show the effect of this coding for the child with IDNR*=78. At 10.4 years, tooth 14 of the child had already emerged (*cs_14*=1). In this way, the 500 original observations are converted into a data frame with 1 703 records. However, since for none of the children's tooth 14 emerged before the age of 7, we omitted this interval from the analysis. This reduces the data frame to 1 203 records.*

```
> subset(T14e, IDNR == 78,
+       select = c("IDNR", "cs_grouped_age", "CS_14", "age", "y"))
```

```
   IDNR cs_grouped_age CS_14 age y
7    78             11     1   7 0
71   78             11     1   8 0
72   78             11     1   9 0
73   78             11     1  10 0
74   78             11     1  11 1
```

The next statements fit a complementary log-log model to the binary response y *with explanatory variables* age *and* GENDER *(boys = 0, girls = 1). The variable* age *is treated categorically. The function* glm *uses iteratively reweighted least squares as the default method to estimate the parameters.*

```
> # fit complementary log-log model
> fit.cloglog <- glm(y ~ factor(age) + GENDER - 1, T14e2,
+              family = binomial(link = cloglog))
> summary(fit.cloglog)
```

```
Coefficients:
               Estimate Std. Error z value Pr(>|z|)
factor(age)8   -4.44109    0.45632  -9.732  < 2e-16 ***
factor(age)9   -3.93491    0.41873  -9.397  < 2e-16 ***
factor(age)10  -2.75436    0.29399  -9.369  < 2e-16 ***
factor(age)11  -1.41056    0.20420  -6.908 4.93e-12 ***
factor(age)12   0.63666    0.17180   3.706 0.000211 ***
GENDER         -0.05934    0.19380  -0.306 0.759444
---
Signif. codes:  0 *** 0.001 ** 0.01 * 0.05 . 0.1   1
```

The hazard ratio (girls over boys) is equal to exp(−0.0593) = 0.94. This can be similarly done for the age classes.

□

5.3.5.3 SAS solution

The macro %Intev_Cens1 is based on Finkelstein's approach, see Sun and Chen (2010). The macro can be found in the supplementary materials. For Finkelstein's score test, the macro %ICSTEST and the procedure ICLIFETEST can also be used. Farrington's approach has been implemented with self-written macros that make use of PROC NLMIXED.

The grouped proportional hazards model is equivalent to a binary response model with a complementary log-log link, hence the LOGISTIC or GENMOD procedures can be used to fit the model.

Example 5.5 in SAS (macro %Intev_Cens1)

Right-censored observations must have a missing value as right endpoint. The censoring variable (0=right-censored, 1=not right-censored) is specified in the argument Cnsr. *The maximum duration (by default 1 000) is specified in the argument* RgtCnV. *The data sets for storing the parameter estimates and standard errors and for the different tests are given in the arguments* TrtEst *and* TrtTest, *respectively. Since the class option is not operational, categorical variables must be specified as dummy variables. For more information about the other options, we refer to the documentation of the macro.*

First an indicator variable treat *for treatment and the right censoring indicator* censor *are created. The code below provides the macro call.*

```
%Intev_Cens1(
 Dsin =  breast /*Input SAS data set;*/
,LeftT = lower  /*The left bound of the observed time interval*/
,RightT = upper /*The right bound of the observed time interval*/
,Cnsr=censor    /*Right censoring variable*/
,RgtCnV=100     /*Maximum Duration*/
,TrtGrp = treat /*Treatment variable, named other than trt.*/
,Covar =        /*Cont. covariates, >1 cov. separated by spaces*/
,Mclass =       /*Cat. covariates, >1 cov. separated by spaces,
                            so far, it is not used.*/
,OptFunc = 1    /*Options for optimization functions;
             Default, 1 - NLPNRA; 2 - NLPNRR; 3 - NLPQN; 4 - NLPCG;*/
,TrtEst = EstB  /*output data set: estimates and SEs*/
,TrtTest = Test /*output data set: test statistics and p-values*/
,OutDrv =       /*The directory keeping the output data sets */
,Alf=0          /*Whether to use modified method*/
,SmPtb=0        /*small value add to exact survival time
                  when original Finkelstein is used*/
,SmLen=0        /*small interval Length*/
,Rdn=0          /*Rounding or not*/);
```

No direct output is provided but the results are stored in data sets EstB and Test. The estimated hazard ratio is exp(0.79743) = 2.22 (95% CI: 1.11 *–* 4.43*). The three reported tests, the Wald test (*$P = 0.0236$*), the score test (*$P = 0.0010$*) and the likelihood ratio test (*$P = 0.0051$*) confirm the significant treatment effect. However, the score test differs from the result obtained with the R-package* glrt *(*$P = 0.0023$*, see Section 5.3.5.2) and from that reported by Finkelstein (1986) (*$P = 0.004$*).*

□

Example 5.5 in SAS (macro %ICSTEST)
Alternatively, one can use the ICSTEST macro which reports the test of Sun et al. (2005) with gamma and rho equal to zero and is equivalent to the score test of Finkelstein (1986). After loading the macro, one can use the following command.

```
%ICSTEST(data=breast, left=lower, right=upper, group=treat);
```

Now, a P-value of 0.0070 is obtained.

□

Example 5.5 in SAS (PROC ICLIFETEST)
The ICLIFETEST procedure can also be used to get the score test.

```
PROC ICLIFETEST DATA=breast;
TIME (lower,upper);
TEST treat/WEIGHT=FINKELSTEIN;
RUN;
```

We obtain a P-value of 0.0069.

□

Example 5.5 in SAS (PROC NLMIXED)
First, the lower limit of the left-censored observations is changed into a missing value (SAS requirement), and the variables idnr, identifying the individuals, and treat, a treatment indicator, are defined.

The minimal (n=9) and full set (n=40) of time points are determined by the macro %timepoints_PHmixed and stored here in the data sets tmintimepoints and tmaxtimepoints, respectively. Next, the macro %add_dummy_PHmixed augments the input data set with the genuine interval-censored observations. Hereby, 51 additional records are created, the response variable y and the dummy variables D1 to D9 represent the minimal set. For both macros, the left and right endpoints of the intervals must be defined in the arguments lower= (by default lower) and upper= (by default upper), respectively.

```
%timepoints_PHmixed(data=breast,
                    minimal=tmintimepoints, maximal=tmaxtimepoints);
%add_dummy_PHmixed(data=breast,out=breast2,timepoints=Tmintimepoints);
```

The model based on the minimal set of time points in the baseline hazard and treatment as covariate is fitted with the PROC NLMIXED code below.

```
PROC NLMIXED DATA=breast2;
PARMS theta1=0.1 theta2=0.1 theta3=0.1 theta4=0.1 theta5=0.1
      theta6=0.1 theta7=0.1 theta8=0.1 theta9=0.1
```

```
      beta=0;
BOUNDS theta1>=0, theta2>=0, theta3>=0, theta4>=0, theta5>=0,
       theta6>=0, theta7>=0, theta8>=0, theta9>=0;
dsum=theta1*d1 + theta2*d2 + theta3*d3 + theta4*d4 + theta5*d5 +
     theta6*d6 + theta7*d7 + theta8*d8 + theta9*d9;
bsum=beta*treat;
p=1-(exp(-dsum))**(exp(bsum));
MODEL y ~ binary(p);
ESTIMATE "treatment"  beta;
RUN;
```

The PARMS *statement specifies the initial values for the nine θ-parameters in the baseline hazard and the regression parameter β. The* BOUNDS *ensures that $\theta_j \geq 0 \quad (j = 1, \ldots, 9)$. The next three commands define the systematic part of the model, while the* MODEL *command defines the distribution of the response variable* y*. With the* ESTIMATE *command the estimated regression coefficient of treatment is given and the hazard ratio is computed together with its confidence interval. The $-2 \log L$ value of the model equals 277.0. The treatment effect is significant ($P = 0.005$) with hazard ratio equal to 2.27 (95% CI: 1.28 – 4.02).*

The macro %add_timepoint_PHmixed *automates the procedure of adding one additional step time to the current model to select the model with the largest significant increase in the log-likelihood. It repeatedly calls the macro* %add_dummy_PHmixed *and from there the* PROC NLMIXED *code. The macro is called as follows:*

```
%add_timepoint_PHmixed(data=breast,
                       baset=Tmintimepoints, allt=Tmaxtimepoints,
                       parms=%str(beta=0), model=%str(beta*treat),
                       fit=fitstat1, out=finaltimepoints,
                       step=0, printinterim=0);
```

The above macro makes use of the data set breast *and the data sets with the minimal and full set of time points. The* parms= *argument contains the initial value of the treatment regression parameter (default=1), but initial values for the theta parameters are specified by the* init= *argument (default=0.1). In the* model= *argument, the model is specified. The* parms= *and* model= *argument are put within a %str command. The log-likelihood values of all models fitted are stored in a database. The first database is called* fitstat1_1. *When a better fitting model (based on another step time) is found, the new model is the base model and the process repeats. For the second step, all models are stored in* fitstat1_2 *and so on. By default only the final model is shown, except when* printinterim=1 *is specified. The* step= *argument is by default zero indicating that the procedure will be repeated until no significant* *reduction in log-likelihood is found anymore or all time points are added to the model. When the argument is equal to m (> 0), the procedure stops after m steps. The final set of time points in the model is saved in a data set specified by the* out= *argument.*

The final model has ten θ parameters, with $t = 20$ added to the base model. The addition yielded a significant reduction in $-2 \log L$ (from 277.0 to 271.3, $P = 0.017$). All θ-parameters are nonnegative with $\widehat{\theta}_6$ set to zero. Farrington (1996) and Collett (2003) did not include this additional point because no significant reduction in log-likelihood could be obtained without at least one of the θ-parameters becoming negative. Our macro-constraints θ parameters to be positive, without this constraint $\widehat{\theta}_6$ becomes negative. Compared to the treatment effect in the base model, the log hazard ratio slightly reduced to 0.7921 but remained significant ($P = 0.0068$). The corresponding hazard ratio equals to 2.21 (95% CI: 1.25 – 3.90).

□

Example 5.7 in SAS

As in Section 5.3.5.2, the observed survival times are treated as discrete, with the current status data grouped into intervals of 1 year ranging from 7 up to 11 years. We first created the necessary variables for the analysis. The effect of augmenting the data for the record with IDNR *= 78 is shown in Section 5.3.5.2 for the* R *solution. Since for none of the children tooth 14 emerged before the age of 7, we omitted the interval (0,7] from the analysis.*

The next statements fit a complementary log-log model to the binary response y *with explanatory variables* age *and* gender *(boys = 0, girls = 1). The* EVENT='1' *option ensures that the probability that $y = 1$ is modelled.*

```
PROC LOGISTIC DATA=t14 OUTEST=estimates;
WHERE age>7;
CLASS age / PARAM=GLM;
MODEL y(EVENT='1')=age gender / NOINT LINK=cloglog TECHNIQUE=newton;
RUN;
```

The PARAM=GLM *option in the* CLASS *statement creates an indicator column in the design matrix for each level of* age*. The option* NOINT *is specified to prevent any redundancy in estimating the coefficients of* age*. The* LINK=cloglog *argument asks for a complementary log-log link. With the* TECHNIQUE=newton *argument we ask for the Newton-Raphson algorithm to compute the MLE of the parameters.*

The parameter estimates are:

```
                Analysis of Maximum Likelihood Estimates

                                       Standard          Wald
Parameter        DF    Estimate           Error    Chi-Square    Pr > ChiSq
```

(Cont.)

age	8	1	-4.4407	0.4563	94.7152	<.0001
age	9	1	-3.9349	0.4188	88.2606	<.0001
age	10	1	-2.7544	0.2936	87.9987	<.0001
age	11	1	-1.4106	0.2040	47.7875	<.0001
age	12	1	0.6367	0.1749	13.2558	0.0003
GENDER		1	-0.0593	0.1942	0.0934	0.7599

Gender is not significant ($P = 0.7599$), with hazard ratio equal to 0.94. Note that the standard error is slightly higher (0.1942 vs 0.1938) than obtained using Fisher's scoring method (see Section 5.3.5.2).

With the OUTEST= *argument, the parameter estimates are stored in a data set called* estimates *for further processing.*

Alternatively, the GENMOD *procedure could be used. Example code is available in the corresponding program in the supplementary materials.*

□

5.4 Multiple imputation approach

The multiple imputation (MI) approach has been suggested three decades ago and is an immense popular method to deal with missing data. The EM-algorithm works with the completed log-likelihood and thereby implicitly imputes missing data. On the other hand, the imputation approach imputes explicitly the missing part in the data, and then analyzes the data with a (more) standard statistical technique. Rubin (1987) has shown that, to take into account the statistical uncertainty of the imputed values, multiple imputation is needed leading to multiple imputed data sets and multiple estimators of the parameters of interest. Rubin also showed how to combine the estimators from the imputed data sets into one global estimator with statistical uncertainty obtained from the statistical uncertainty of the estimator obtained within each imputed data set and its variability between the data sets.

Censored observations can be considered as missing data and are therefore suitable for a MI approach. In this section we discuss one MI approach for interval-censored survival times that finds its origin in a sampling approach that can be considered as a Bayesian version of the EM algorithm.

5.4.1 Data augmentation algorithm

Pan (2000a) proposed two multiple imputation schemes for analyzing Cox's PH model for interval-censored data. The basic idea is to impute exact survival times from interval-censored data and to take advantage of a standard method for right-censored data. The two approaches date back to the Data Augmentation (DA) algorithm suggested in Tanner and Wong (1987). The DA algorithm essentially relies on a Bayesian Markov chain Monte Carlo approach to obtain the posterior distribution. As for the EM algorithm, the DA algorithm assumes the existence of latent data such that determining the posterior goes easier. The Bayesian approach and the data augmentation idea will be treated in Chapter 9.

Suppose the parameters of interest are $\boldsymbol{\theta}$ and a data set $\boldsymbol{y} = (y_1, \ldots, y_n)'$ has been observed. Suppose also the existence of latent data $\boldsymbol{z}$, then the posterior density can be written as

$$p(\boldsymbol{\theta} \mid \boldsymbol{y}) = \int p(\boldsymbol{\theta} \mid \boldsymbol{z}, \boldsymbol{y}) p(\boldsymbol{z} \mid \boldsymbol{y}) \mathrm{d}\boldsymbol{z}, \tag{5.13}$$

with $p(\boldsymbol{z} \mid \boldsymbol{y})$ the predictive density of the latent data $\boldsymbol{z}$ given $\boldsymbol{y}$ and $p(\boldsymbol{\theta} \mid \boldsymbol{z}, \boldsymbol{y})$ denotes the conditional density of $\boldsymbol{\theta}$ given the augmented data $\{\boldsymbol{z}, \boldsymbol{y}\}$. It is also assumed that $p(\boldsymbol{\theta} \mid \boldsymbol{z}, \boldsymbol{y})$ is easier to evaluate (by sampling) than the original posterior $p(\boldsymbol{\theta} \mid \boldsymbol{y})$. The predictive density $p(\boldsymbol{z} \mid \boldsymbol{y})$ can be written as

$$p(\boldsymbol{z} \mid \boldsymbol{y}) = \int_{\Theta} p(\boldsymbol{\theta} \mid \boldsymbol{z}, \boldsymbol{y}) p(\boldsymbol{\theta} \mid \boldsymbol{y}) \mathrm{d}\boldsymbol{\theta}. \tag{5.14}$$

Plugging Expression (5.14) into Expression (5.13) yields an integral equation to solve for $p(\boldsymbol{\theta} \mid \boldsymbol{y})$:

$$g(\boldsymbol{\theta}) = \int K(\boldsymbol{\theta}, \boldsymbol{\phi}) g(\boldsymbol{\phi}) \mathrm{d}\boldsymbol{\phi}, \tag{5.15}$$

with $K(\boldsymbol{\theta}, \boldsymbol{\phi}) = \int p(\boldsymbol{\theta} \mid \boldsymbol{z}, \boldsymbol{y}) p(\boldsymbol{z} \mid \boldsymbol{\phi}, \boldsymbol{y}) \mathrm{d}\boldsymbol{z}$, and $\boldsymbol{\phi}$ is a clone of $\boldsymbol{\theta}$ in Expression (5.15). This integral equation leads to an iterative procedure to determine $g(\boldsymbol{\theta})$ starting with $g_0(\boldsymbol{\theta})$, i.e.,

$$g_{j+1}(\boldsymbol{\theta}) = (Tg_j)(\boldsymbol{\theta}) \quad (j = 0, 1, \ldots), \tag{5.16}$$

with $Tf(\boldsymbol{\theta}) = \int K(\boldsymbol{\theta}, \boldsymbol{\phi}) f(\boldsymbol{\phi}) \mathrm{d}\boldsymbol{\phi}$.

In general, Equation (5.16) cannot be solved analytically, and a sampling approach has been therefore suggested, which goes as follows:

1. Generate $\boldsymbol{z}^1, \ldots, \boldsymbol{z}^M$ from the current approximation to $p(\boldsymbol{z} \mid \boldsymbol{y})$
2. Update approximation to $p(\boldsymbol{\theta} \mid \boldsymbol{y})$ as

$$g_{j+1}(\boldsymbol{\theta}) = \frac{1}{M} \sum_{m=1}^{M} p(\boldsymbol{\theta} \mid \boldsymbol{z}^m, \boldsymbol{y}). \tag{5.17}$$

To generate the latent data, Tanner and Wong (1987) suggest

A1. Sample $\boldsymbol{\theta}$ from $g_j(\boldsymbol{\theta})$, which is the current approximation to $p(\boldsymbol{\theta} \mid \boldsymbol{y})$

A2. Sample $\boldsymbol{z}$ from $p(\boldsymbol{z} \mid \boldsymbol{\theta})$, with $\boldsymbol{\theta}$ generated in step A1.

Two approximations to the original DA algorithm were then proposed: the *Poor Man's Data Augmentation (PMDA) algorithm* and the *Asymptotic Normal Data Augmentation (ANDA) algorithm*. The PMDA approach is based on the result that for n large, $p(\boldsymbol{z} \mid \boldsymbol{y}) \approx p(\boldsymbol{z} \mid \boldsymbol{y}, \widehat{\boldsymbol{\theta}})$ with $\widehat{\boldsymbol{\theta}}$ the posterior mode based on the observed data. Sampling from the predictive density is then replaced by sampling from $p(\boldsymbol{z} \mid \boldsymbol{y}, \boldsymbol{\theta}(j))$ with $\boldsymbol{\theta}(j)$ the current approximation to $\widehat{\boldsymbol{\theta}}$. The ANDA algorithm is based on the assumption that for a large sample size $p(\boldsymbol{\theta} \mid \boldsymbol{z}^m, \boldsymbol{y})$ can be well approximated by a normal distribution, so that $g_{j+1}(\boldsymbol{\theta})$ in Expression (5.17) is replaced by

$$g_{j+1}(\boldsymbol{\theta}) = \frac{1}{M} \sum_{m=1}^{M} \mathcal{N}(\boldsymbol{\theta} \mid \widehat{\boldsymbol{\theta}}_m, \widehat{\boldsymbol{\Sigma}}_m), \tag{5.18}$$

with $\widehat{\boldsymbol{\theta}}_m$ the posterior mode of the augmented posterior and $\widehat{\boldsymbol{\Sigma}}_m$ the corresponding covariance matrix. Note that the DA algorithm is based on the posterior density $p(\boldsymbol{\theta} \mid \boldsymbol{y})$ and therefore a prior for $\boldsymbol{\theta}$ needs to be specified.

5.4.2 Multiple imputation for interval-censored survival times

Wei and Tanner (1991) applied the above DA algorithm to analyze right-censored survival times in combination with an AFT model. Their intention was to provide a distribution-free method to fit an AFT model to right-censored data. The PMDA and the ANDA approaches were then combined with the Kaplan-Meier estimator. From simulations they concluded that the ANDA approach is preferable in practice for small to medium-sized studies. Pan (2000a) applied the PMDA and ANDA approach on interval-censored data. Note, however, that none of these two approaches mention the choice of a prior. But from the context it is clear that for all parameters a flat prior is chosen, such that the posterior mode coincides with the MLE. We will now describe in more detail the MI approaches suggested in Pan (2000a).

For interval-censored observations, the intervals $\lfloor l_i, u_i \rfloor$ $(i = 1, \ldots, n)$ constitute the observed data $\boldsymbol{y}$ in the above description of the DA algorithm. While the true but unobserved survival times $\boldsymbol{T} = (T_1, \ldots, T_n)'$ represent the latent data $\boldsymbol{z}$. The vector $\boldsymbol{\theta}$ consists of the regression coefficients $\boldsymbol{\beta}$ and the parameters that determine S_0, the baseline survival distribution. The design matrix with rows $\boldsymbol{X}_i$ $(i = 1, \ldots, n)$ is denoted as $\boldsymbol{X}$. For the PMDA and the ANDA algorithm multiple imputed data sets will be generated in an iterative algorithm. Let $\widehat{\boldsymbol{\theta}}(j)$ represent the estimate of $\boldsymbol{\theta}$ at the jth iteration and superscript m the mth imputed data set.

Pan (2000a) proposed the following multiple imputation scheme to analyze interval-censored survival times making minimal assumptions about the baseline survival distribution:

1. At start of iteration j, we have as estimate $\widehat{\boldsymbol{\theta}}(j) = \{\widehat{\boldsymbol{\beta}}(j), \widehat{S}_0(j)\}$.

2. Generate M sets of possibly right-censored observations $\{\boldsymbol{z}^m, \boldsymbol{X}\} \equiv \{\boldsymbol{T}^m, \boldsymbol{\delta}^m, \boldsymbol{X}\}$ $(m = 1, \ldots, M)$ from $p(\boldsymbol{z} \mid \boldsymbol{y}, \widehat{\boldsymbol{\theta}}(j))$ based on the interval-censored observations. That is, for each observation $(\lfloor l_i, u_i \rfloor, \boldsymbol{X}_i)$ $(i = 1, \ldots, n)$ and for $(m = 1, \ldots, M)$:

   - If $u_i = \infty$ (right censoring): $T_i^m = L_i$ and $\delta_i^m = 0$;
   - If $u_i < \infty$ (interval censoring): sample Y_i from the distribution $\widehat{S}_0(j)^{\exp(\boldsymbol{X}_i'\widehat{\boldsymbol{\beta}}(j))}$ and let $T_i^m = Y_i$ and $\delta_i^m = 1$.

3. For the mth $(m = 1, \ldots, M)$ imputed data set, with standard software:

   - Fit a PH model based on $\{\boldsymbol{T}^m, \boldsymbol{\delta}^m, \boldsymbol{X}\}$, to obtain $\widehat{\boldsymbol{\beta}}^m(j)$ and its covariance estimate $\widehat{\boldsymbol{V}}^m(j)$;
   - Calculate the Breslow estimate $\widehat{S}_0^m(j)$ of the baseline survival based on $\{\boldsymbol{T}^m, \boldsymbol{\delta}^m, \boldsymbol{X}\}$ and $\widehat{\boldsymbol{\beta}}^m(j)$.

4. Then combine the estimates of the M imputed data sets to update the approximation to $p(\boldsymbol{\theta} \mid \boldsymbol{y})$:

$$
\begin{aligned}
\widehat{\boldsymbol{\beta}}(j+1) &= \frac{1}{M}\sum_{m=1}^{M} \widehat{\boldsymbol{\beta}}^m(j), \\
\widehat{S}_0(j+1) &= \frac{1}{M}\sum_{m=1}^{M} \widehat{S}_0^m(j), \\
\widehat{\boldsymbol{V}}(j+1) &= \frac{1}{M}\sum_{m=1}^{M} \widehat{\boldsymbol{V}}^m(j) + \left(1 + \frac{1}{M}\right) \frac{\sum_{m=1}^{M}\left(\widehat{\boldsymbol{\beta}}^m(j) - \widehat{\boldsymbol{\beta}}(j+1)\right)^2}{M-1}.
\end{aligned}
$$

5. Then: $j \leftarrow j + 1$. Go to step 1 until $\widehat{\boldsymbol{\beta}}(j)$ converges.

As starting values $\widehat{\boldsymbol{\beta}}(0) = \boldsymbol{0}$ is taken, while $\widehat{S}_0(0)$ is obtained from the Breslow estimates of the baseline survival of M imputed data sets whereby z_i^m is randomly drawn from the uniform distribution $U(l_i, u_i)$ with $T_i^m = Y_i$ and $\delta_i^m = 1$ for $m = 1, \ldots, M$. Note that this just means that at the start $p(z_i \mid y_i)$ is a uniform on $[l_i, u_i]$ $(m = 1, \ldots, M)$. Pan (2000a) suggests to take M equal to 10.

Wei and Tanner (1991) noticed with simulations that the PMDA may underestimate its true variability when missingness is severe (60% right censoring was used in their simulations). The ANDA approach seems to be better in this respect. As indicated above, the ANDA approach is different in two aspects from the PMDA approach. First, it is assumed that the posterior can be approximated well by a mixture of normals, i.e.,

$$g_{j+1}(\boldsymbol{\beta}) = \frac{1}{M}\sum_{m=1}^{M}\mathcal{N}\left(\boldsymbol{\beta}\,\middle|\,\widehat{\boldsymbol{\beta}}^{m}(j), \widehat{\boldsymbol{V}}^{m}(j)\right).$$

Note that no explicit expression for $g_{j+1}(\boldsymbol{\beta})$ is given for the PMDA approach. In step 2 above, one first samples M times from $g_j(\boldsymbol{\beta})$ to obtain $\widehat{\boldsymbol{\beta}}^{m}(j)$ ($m = 1, \ldots, M$) then the same thing is done as for the PMDA algorithm. The other parts are also the same as those of PDMA.

Example 5.8 Breast cancer study
As an illustration we again compare the two treatments for the early breast cancer study. As suggested by Pan (2000a), we used 10 imputations with the PMDA approach. Fitting a PH model yields an estimate of 0.7141 (R) or 0.8301 (SAS) for the treatment effect which corresponds with a hazard ratio of 2.04 (95% CI: 1.17 – 3.57) and 2.29 (95% CI: 1.30 – 4.03), respectively. We again conclude that radiotherapy-only therapy significantly ($P = 0.0122$ and $P = 0.0039$, respectively) increases the time to breast retraction in comparison to primary radiation therapy and adjuvant chemotherapy. The ANDA approach yields an estimate of 0.7184 (R), which corresponds with a hazard ratio of 2.05 (95% CI: 1.17 – 3.59). Pan (2000a) reported an estimate of 0.90 but our results are closer to the estimate from Finkelstein's model (0.80).

□

5.4.2.1 R solution

The two imputation approaches of Pan (2000a) are implemented in the R package MIICD (Delord, 2017). With the function MIICD.coxph a model using the PMDA or ANDA approach may be fitted.

Example 5.8 in R
The lower and upper boundaries of the intervals must be stored in a data frame using the names "left" and "right". In addition, for right-censored observations the upper boundary must be set to Inf *and the lower boundary for left-censored observations to zero.*

The following commands will estimate the PH model with the PMDA approach. The number of imputations is set with the argument m, *here equal to 10. The number of iterations is set with the argument* k = 700. *The formula does not contain a* Surv *object as the names of the lower and upper bounds of the observed intervals are fixed to "left" and "right".*

```
> library("MIICD")
> breastCancer.PMDA <- MIICD.coxph( formula = ~ treat ,
+   data = breastCancer, k = 700 , m = 10,  method = "PMDA" )
```

Convergence of the parameters can be judged from a time series plot of each parameter. But, since sampling is involved, the estimates show quite some variability. Therefore you should verify that the final estimate is close to, for instance, the mean of last 20 estimates.

```
> plot(breastCancer.PMDA, type = "coef", coef = 1)
> mean(tail(as.vector(breastCancer.PMDA$"Coef_seq"), 20)) -
+   breastCancer.PMDA$"Coef."
> print(breastCancer.PMDA)
```

The output below shows a hazard ratio of 2.04, which is significantly different from 1 ($P = 0.0122$).

```
Coefficients:
                        coef   exp(coef)   se(coef)       z        p
treat: radio+chemo    0.7141       2.042     0.2849   2.506   0.0122
```

By using method = "ANDA" *in the call, the ANDA approach is fitted. This yielded similar results but with a slightly higher standard error.*

□

5.4.2.2 SAS solution

Using the procedures PHREG and MIANALYZE, the PMDA multiple imputation approach of Pan (2000a) is implemented in a SAS macro %CoxPH_MI. Efron's method is used to handle ties.

Example 5.8 in SAS

The %CoxPH_MI *macro expects a unique identification variable (here* IDNR*) for the subjects. Left-censored observations may contain either a zero or a missing value in the lower limit. The following commands fit a PH model using the PMDA multiple imputation approach to the breast cancer example.*

```
%CoxPH_MI(data=breast,id=idnr,
          left=lower,right=upper,
          class=tx, cov=tx,
          nimpute=10, maxiter=200, nlastconv=20);
```

The first 4 arguments are required fields, the remaining arguments are optional. The variables lower en upper *indicating the left and right endpoints of the interval-censored observations should be specified in the macro-variables* LEFT *and* RIGHT*, respectively. The covariates, including potential interactions, should be specified in the macro-variable* COV*. Because* tx *is not coded as a binary variable, it should be specified in the optional macro-variable* CLASS *in order to be treated categorically. The number of imputations (default = 10) is controlled by the argument* nimpute*. The procedure stops when* *the difference between the parameter estimates becomes smaller than the value specified in the* conv= *argument (0.0001 by default) or the maximal number of iterations (*maxiter=100 *by default) is reached. Although Pan (2000a) mentioned good convergence properties, convergence is not guaranteed. For this reason, we suggest to base the convergence criterion on the mean value of a given predetermined number of the last iterations rather than comparing it to previous value as suggested in Pan (2000a). In the current example, we used 20 iterations as set by the argument* nlastconv=. *By default, the standard PHREG output on the different imputed data sets is omitted but the output displays the* PROC MIANALYZE *output for the different iterations. Using the* printinterim=1 *argument, one can request to obtain also the* PHREG *output. At the end, also the data set with the final estimates is printed. Part of the final* PROC MIANALYZE *output is shown below. The first part of the output (not shown) provides the number of imputations and several multiple imputations statistics like the within and between imputed data sets variances for each parameter. The second part displays the combined inference of the different parameter estimates.*

```
              Parameter Estimates (10 Imputations)

Parameter                                   Estimate       Std Error

txRadiotherapy___Chemotherapy               0.830140        0.287763

                Parameter Estimates (10 Imputations)

Parameter                                95% Confidence Limits        DF

txRadiotherapy___Chemotherapy             0.266116      1.394165     36690

              Parameter Estimates (10 Imputations)

Parameter                                    Minimum         Maximum

txRadiotherapy___Chemotherapy               0.749753        0.869668

                Parameter Estimates (10 Imputations)

                                                                t for H0:
Parameter                                      Theta0   Parameter=Theta0

txRadiotherapy___Chemotherapy                       0               2.88

      Parameter Estimates (10 Imputations)

Parameter                              Pr > |t|
```

(Cont.)

```
txRadiotherapy___Chemotherapy        0.0039
```

We observe a significant treatment effect ($P = 0.0039$) with a hazard ratio of 2.29 (= exp(0.830140)). In addition, the results of the final analysis are stored by default in a data set called out_coxPH_MI. *The name can be changed by specifying the* out= *argument. In case one jointly tests variables, e.g., in case of a categorical variable with more than two categories, the* outMI= *argument must be given. With a test statement in* PROC MIANALYZE *and using the* outMI= *data set as input data set, a combined test could be performed with the following commands (assuming* age *was also added to the model and the* outMI=*outMIbreast was specified):*

```
PROC MIANALYZE DATA=outMIbreast;
MODELEFFECTS txRadiotherapy___Chemotherapy age;
test1: TEST txRadiotherapy___Chemotherapy=0, age=0 / MULT;
RUN;
```

□

5.5 Model checking

5.5.1 Checking the PH model

Several tools have been suggested for checking the appropriateness of a survival model in the presence of right-censored data. In Section 2.3.1.2 some methods have been reviewed for checking the PH model for right-censored data. While for right-censored data the baseline hazard does not need to be estimated, such luxury cannot be enjoyed with interval-censored data. Hence all PH diagnostics automatically involve the baseline survival distribution.

In the parametric case, global adequacy of the PH model can be verified by, e.g., embedding the model into a larger class of models and then applying a significance test to evaluate whether the simpler model is adequate enough for the data at hand. Or one could contrast the fit of the parametric PH model (baseline survival distribution and regression coefficients) to that obtained from the semiparametric PH model of Finkelstein or a smooth fit. However, this does not provide a test for the PH assumption. One might install, therefore, a two-stage procedure that first checks the baseline survival distribution by contrasting the parametric estimate with the NPMLE and in case of close agreement a further check is done on the PH assumption. The problem is, however, that violation of the PH assumption can distort also the estimation of the baseline survival distribution.

A second option to validate the PH assumption is a graphical check. Using the $\log\left[-\log\{\widehat{S}(t \mid X)\}\right]$ transformation of the survival curves, the difference between subjects with different covariates should be constant for the PH model.

A third possibility is to evaluate the PH model with residuals. In Section 2.3.1.2 two residuals were discussed: the Lagakos and deviance residuals. These two residuals were chosen because extensions to interval-censored observations exist. In Farrington (2000) these, and two other types of residuals, were suggested for the interval-censored case. But before embarking on these generalizations, we could opt for applying the residuals for right-censored observations on the M imputed data sets generated by the method of Pan (2000a) described in Section 5.4.2. Then for each of the imputed data sets the residuals can be calculated and the M duplicates could then be overlaid in one graph together with, e.g., their median. A classical graphical check for right-censored survival times could then be applied on the residuals. No need to say that this is just an informal check. But, unfortunately, all residual plots for validating the assumed PH model for interval-censored survival times are largely informal. We now discuss the two residuals recommended for their use in Farrington (2000) and which are relatively easy to implement in statistical software.

Lagakos residuals

The construction of the Lagakos residual starts with the Cox-Snell residuals (Cox and Snell, 1968), which are based on the result that when T has a survival distribution $S(t)$, then $-\log S(T)$ has an exponential distribution with mean one. This suggests to look at the intervals $\left[-\log \widehat{S}(l_i), -\log \widehat{S}(u_i)\right]$ $(i = 1, \ldots, n)$ and to compare these with similar intervals from an exp(1)-distribution. Since working with intervals is not easy, adjusted Cox-Snell residuals r_i^C $(i = 1, \ldots, n)$ have been suggested. They are based on the expected value of a unit exponential variable restricted to the interval $\left[-\log \widehat{S}(l_i),\ -\log \widehat{S}(u_i)\right]$, i.e.

$$r_i^C = \frac{S(l_i \mid X_i)\{1 - \log S(l_i \mid X_i)\} - S(u_i \mid X_i)\{1 - \log S(u_i \mid X_i)\}}{S(l_i \mid X_i) - S(u_i \mid X_i)}.$$

The estimated residuals, $\widehat{r}_i^C$, are obtained by replacing the unknown survival distributions by their estimates. These residuals become the classical Cox-Snell residuals when $(u_i - l_i) \to 0\,(i = 1, \ldots, n)$. Further, it can be shown that (conditional on the observed survival times) the expected value of $r_i^C(l_i, u_i)$ is equal to 1. However, Farrington (2000) does not recommend the Cox-Snell residuals but prefers the *Lagakos residuals* (Barlow and Prentice, 1988; Lagakos, 1981), which are in fact the centered adjusted Cox-Snell residuals, also called *Martingale residuals* and given by

$$\widehat{r}_i^L = 1 - \widehat{r}_i^C = \frac{\widehat{S}(l_i \mid X_i)\log\{\widehat{S}(l_i \mid X_i)\} - \widehat{S}(u_i \mid X_i)\log\{\widehat{S}(u_i \mid X_i)\}}{\widehat{S}(l_i \mid X_i) - \widehat{S}(u_i \mid X_i)}.$$

To check the PH model, the residuals can be plotted in different ways as in the case of right-censored data. One can plot them against continuous covariates to examine the functional form of covariates in the PH model. Marked departures from the zero mean may indicate the need for a transformation of the covariate. One can also plot the residuals against covariates not included in the PH model to assess the need for inclusion of this covariate into the model. Finally, one can construct an index plot of the residuals for identifying outlying observations.

Deviance residuals

The deviance residual (Therneau et al., 1990) is essentially a log-transformed version of the Lagakos residual in the case of right-censored data. For interval-censored data, the residual still can be defined but is no longer a simple transformation of the Lagakos residual. The deviance residual is defined as

$$\widehat{r}_i^D = \text{sign}(u_i^L)\left[2\log\left\{\frac{\widehat{S}_0(l_i)^{\eta_i} - \widehat{S}_0(u_i)^{\eta_i}}{\widehat{S}_0(l_i)^{\exp(\boldsymbol{X}_i'\widehat{\boldsymbol{\beta}})} - \widehat{S}_0(u_i)^{\exp(\boldsymbol{X}_i'\widehat{\boldsymbol{\beta}})}}\right\}^{1/2}\right]$$

where $\eta_i = \frac{\log\{\widehat{H}_0(u_i)\} - \log\{\widehat{H}_0(l_i)\}}{\widehat{H}_0(u_i) - \widehat{H}_0(l_i)}$ and where $\eta_i = 0$ if $u_i = \infty$ and $\eta_i = \infty$ if $l_i = 0$. Note that the deviance residual does not generally have zero expectation.

5.5.2 R solution

We are not aware that any of the residuals are implemented in a R function. However, the Lagakos residuals are easily implemented for a given survival curve.

The function diag_baseline from the R package icenReg allows the user to easily use the transformation $\log\left[-\log\{\widehat{S}(t|X)\}\right]$ to diagnose whether the PH assumption holds. In particular, it takes a single covariate and stratifies the data on different levels of that covariate. Then, it fits the semiparametric regression model (adjusting for all other covariates in the data set) on each of these strata and extracts the baseline survival function. If the stratified covariate satisfies the PH assumption, the difference between these transformed baseline survival functions should be approximately constant. By default, the transformed survival functions with the overall means subtracted off are plotted. If the PH assumption holds, the mean centered curves should appear as approximately parallel lines. The result of the code below applied to Example

5.1 is shown in Figure 5.1. Given that the transformed survival curves cross, the PH assumption might be questioned. However, crossings may also occur by chance even when the PH assumption holds.

```
> diag_covar(modelWeibull)
```

5.5.3 SAS solution

The procedure ICPHREG has the option to calculate the Lagakos and deviance residuals and produce corresponding plots. When the following option is added to the ICPHREG statement the deviance (RESDEV) and Lagakos (RESLAG) residuals will be plotted. Note that ODS GRAPHICS must be turned on to produce the plots.

```
PLOTS=RESDEV(INDEX XBETA) RESLAG(INDEX XBETA)
```

The suboptions INDEX and XBETA request that the residuals are plotted versus the observation number or linear predictor, respectively. If omitted, the residuals are plotted versus the observation number. Unfortunately, a bug in

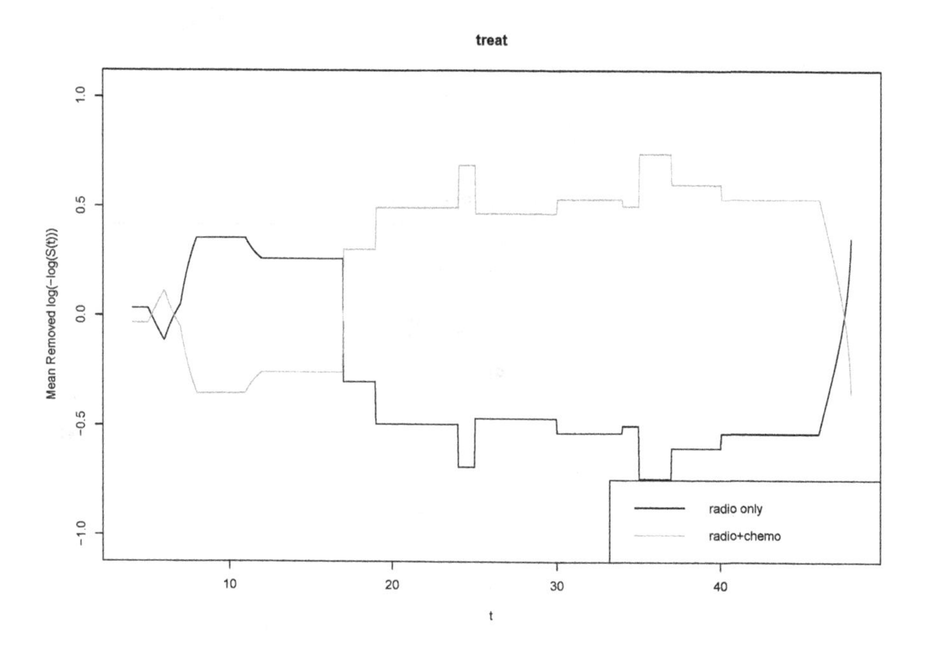

FIGURE 5.1: Breast cancer study. Validation of PH assumption for treatment using the transformed survival function $\log\left[-\log\{\widehat{S}(t \mid X)\}\right]$ (overall means subtracted off), obtained with R package icenReg.

the calculation of the Lagakos and deviance residuals for left-censored observations was observed in SAS/STAT 14.1 (see SAS note 59917). A workaround suggested by the SAS Institute was to replace the lower limit of left-censored observations by a small value (e.g., change $(0, 5]$ into $(10^{-8},\ 5])$ in order to make it a genuine interval-censored observation for which the calculations are performed correctly. This small change will have no impact on the results and provide correct residuals. Please check the support pages of SAS to verify if a hot fix is already available or the problem is corrected in the next SAS/STAT (maintenance) version. Figure 5.2 shows the output for Exam-

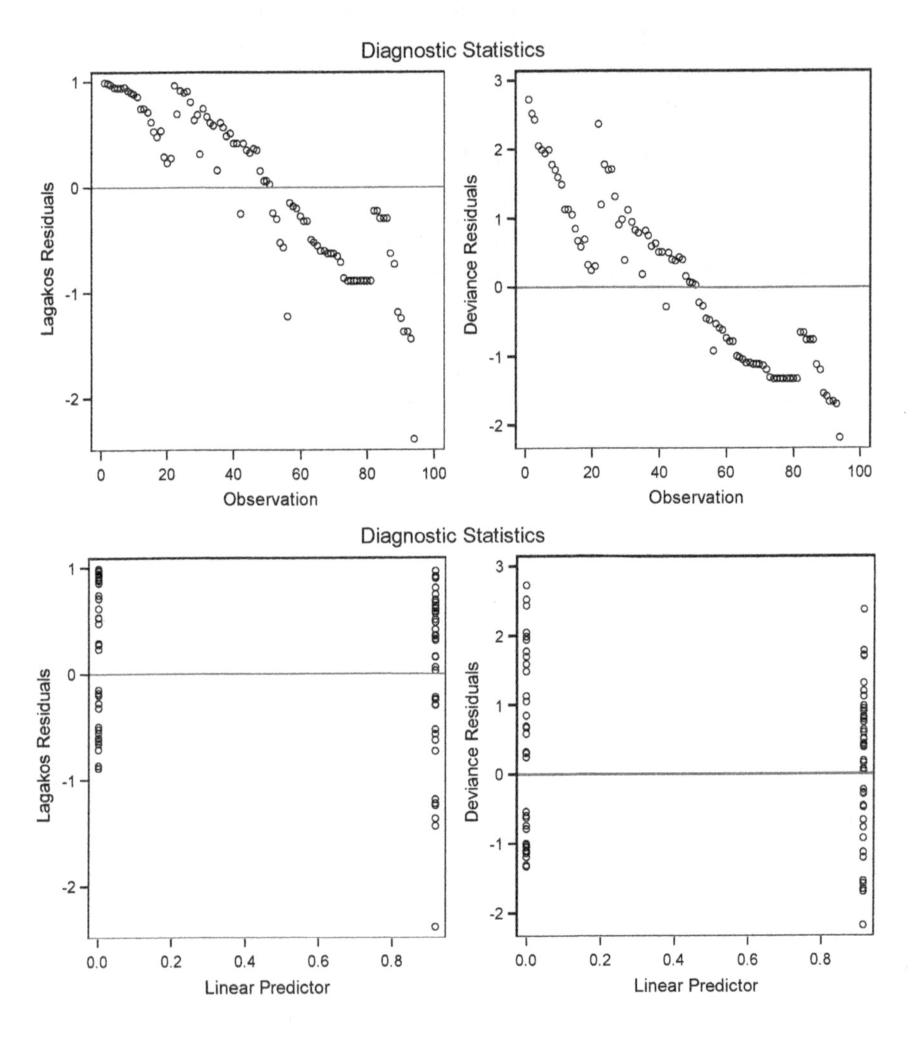

FIGURE 5.2: Breast cancer study. Lagakos and deviance residuals versus observation number (upper panel) or treatment (lower panel), obtained with SAS procedure ICPHREG.

ple 5.2. The downward trend in the plots versus the observation number is due to the specific ordering performed in the data set and not a problem. We do, however, observe one outlying Lagakos residual, namely the last one which corresponds to $(48, \infty)$. It is induced by the fact that the estimated survival probability at 48 months for this individual is reduced from 0.233 to 0.092 when treatment is added to the model. Lagakos residuals take values in $(-\infty, 1)$. This lack of symmetry of the Lagakos residuals can make it difficult to detect outliers, particularly for positive $\widehat{r}_i^L$. Because deviance residuals are more spread out than Lagakos residuals, outliers at the top may be better detected. However, in this example no positive outlying residual could be detected.

In the supplementary materials, the program for analyzing Example 5.8 contains code to create a residual plot on which the residuals calculated on the 10 imputed data sets are plotted and their median values highlighted.

Also with SAS, a figure like Figure 5.1 can be easily created.

5.6 Sample size calculation

It is recommended to take interval censoring into account when planning a new study. One aspect of the planning is the calculation of the required study size. For the PH model, research and corresponding software appear to be lacking. However, Kim et al. (2016) proposed a method for calculating the sample size and power for interval-censored data which assumes that T follows a parametric AFT model (see Chapter 6). Since the method assumes that the log-transformed event times follow a Weibull distribution, it implicitly also models a PH model (see Section 2.3.2). Hence, the available software described in Section 6.5 can be used here too. Alternatively, for other (more complex) baseline hazard functions, a simulation study could be set up.

5.7 Concluding remarks

In this section we have reviewed numerous implementations of the PH model for interval-censored data. The popularity of the PH model has much to do with its popularity for right-censored data. However, the attractive feature of the partial likelihood approach cannot be relied upon in the interval censoring case. For all approaches, from parametric to semiparametric, the baseline survival distribution must be estimated jointly with the regression coefficients.

TABLE 5.2: Sensory shelf life study. Hazard ratio and 95% confidence intervals (CI) for the sensory shelf life of yoghurt of adults versus children obtained from PH models with the baseline survival hazard estimated with different software and two single imputation methods using Efron's method to handle ties. NC=Not calculated due to computational difficulties.

baseline hazard of PH Model	**software**	**HR**	**95% CI**
Finkelstein	SAS macro	3.37	1.96 – 5.82
	R package icenReg	3.37	1.91 – 5.97
Farrington	SAS macro	3.38	1.95 – 5.79
piecewise exponential	PROC ICPHREG	3.38	1.96 – 5.82
	R package frailtypack	NC	NC
cubic splines	PROC ICPHREG	3.38	1.96 – 5.82
penalized splines (5 knots)	R package frailtypack	3.39	1.97 – 5.85
parametric: Weibull	PROC ICPHREG	3.41	1.98 – 5.87
	R package frailtypack	3.41	1.98 – 5.87
parametric: log-normal	R package icenReg	3.35	1.94 – 5.78
multiple imputation (PMDA)	SAS macro	3.34	1.94 – 5.76
	R package MIICD	3.08	1.79 – 5.31
single imputation PH Model			
mid-point imputation	PROC PHREG	3.33	1.94 – 5.73
right-point imputation	PROC PHREG	3.13	1.83 – 5.37

Since most often the parametric distribution of the hazard function is unknown, the practical use of the parametric PH model is rather limited in practice. Nevertheless, the following distributions may be easily examined using existing statistical software: exponential, Weibull, (generalized) gamma, log-normal and log-logistic.

The piecewise exponential baseline survival model is a flexible and easy approach to estimate the baseline survival curve and may be sufficient from a practical point of view since most often the interest lies not in the baseline survival function but in the regression coefficients. This modelling option is supported by the R function frailtyPenal and the ESQP method in PROC ICPHREG. Should one need a more smooth version of the baseline survival distribution, one could make use of the R function frailtyPenal and the SAS procedure ICPHREG. Both are quite user-friendly, which cannot be said of the SAS macro implementation of the method of Zhang and Davidian (2008).

The semiparametric PH model of Finkelstein (1986) is more challenging to fit, especially where there are many different time points at which the subjects are evaluated. The function ic_sp function of R package icenReg provides the easiest and quickest implementation of the method. A score test for $\boldsymbol{\beta} = \mathbf{0}$ has been developed too. However, because the model assumes a discrete baseline survival distribution with the number of parameters that increase with the sample size, the regularity conditions for asymptotic results of the score test might be violated. Grouping of observations may therefore be necessary. Farrington (1996) proposed a way to limit the number of time points taken into account. The method is implemented in a set of SAS macros but it can be quite time consuming for a data set with even a moderate number of observed time points.

Finally, one can use the multiple imputation approach of Pan (2000a) which is implemented in R function MIICD.coxph and SAS macro %CoxPH_MI. However, note that for larger data sets the method may be quite computationally intensive.

In Example 5.9, we analyzed the sensory shelf life data introduced in Section 1.6.4 using most of the techniques described in this chapter and compared the results. In addition, we compared the results with those obtained from a single imputation (middle and right) approach.

Example 5.9 Sensory shelf life study

We compare the sensory shelf life of whole-fat, stirred yoghurt with strawberry pulp between adults and children. Table 5.2 displays the hazard ratio and 95% CI for the approaches considered in this chapter. In addition we have analyzed the data with a single imputation (mid- and right-point) technique. The hazard rates are quite similar for all models taking into account the interval-censored nature of the data with exception of the MI approach using the R package MIICD which yielded a lower estimate. This lower result

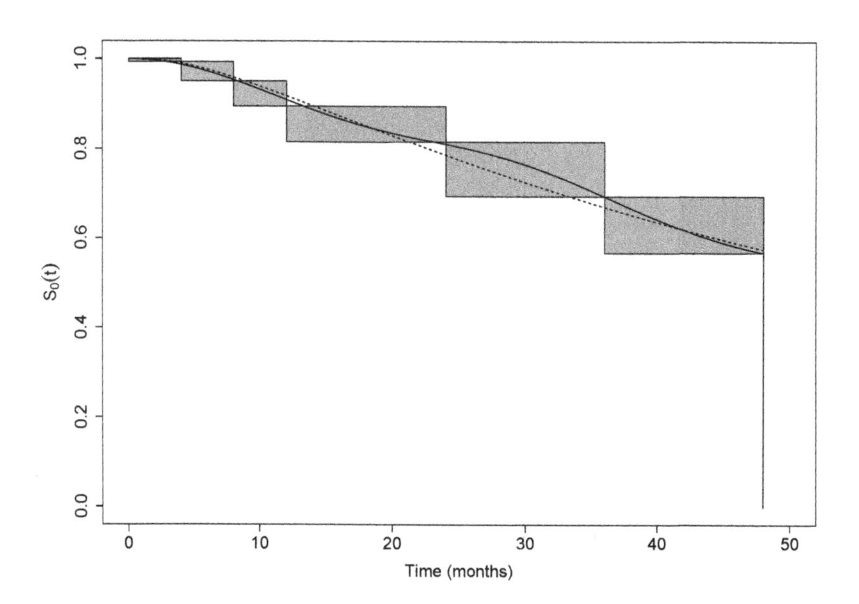

FIGURE 5.3: Sensory shelf life study. Baseline survival function for Finkelstein's model (gray areas), PH model with log-normal (dashed line) or penalized splines (full line) estimated baseline, obtained with R packages icenReg and frailtypack.

remained quite stable in different sensitivity analyses with different seeds, longer runs and increased multiple imputations. Given that also in the breast cancer analysis (see Example 5.8) a lower estimate was observed, we performed a very small simulation study. The mean bias was observed to be both negative and positive in different settings. Hence, that we observed for these two examples a lower estimate is probably a chance finding.

The single mid-point imputation model provided a similar point estimate as most of the interval-censored analyses. On the other hand, the right-point imputation model yielded a lower hazard ratio and about 10% narrower 95% CI because we have artificially reduced the uncertainty in the data.

The baseline survival function for Finkelstein's model and the PH models with a log-normal baseline and a penalized splines estimated baseline are depicted in Figure 5.3. Again, no relevant differences are observed.

□

Chapter 6

The accelerated failure time model

The accelerated failure time (AFT) model is another popular regression model often used to analyze survival data. In this chapter, we show how the AFT model can be used for interval-censored survival times. As before, assume that the survival data are the realization of n random variables $T_1, \ldots, T_n$, with a distribution depending on covariate vectors $\boldsymbol{X}_i = (X_{i1}, \ldots, X_{ip})'$ ($i = 1, \ldots, n$), but only known to lie in an interval $\lfloor l_i,\, u_i \rfloor$. The censoring type is indicated by the variable δ_i ($i = 1, \ldots, n$) (see Table 1.1). Let

$$\boldsymbol{\mathcal{D}} = \bigl\{\lfloor l_i,\, u_i \rfloor,\, \delta_i,\, \boldsymbol{X}_i : \, i = 1, \ldots, n\bigr\}$$

denote the observed data and $Y_i = \log(T_i)$ ($i = 1, \ldots, n$).

From Section 2.3.2 we know that the AFT model relates linearly the logarithm of the survival time to covariates, i.e.,

$$Y = \log(T) = \boldsymbol{X}'\boldsymbol{\beta} + \varepsilon, \tag{6.1}$$

where $\boldsymbol{\beta} = (\beta_1, \ldots, \beta_p)'$ represents unknown regression parameters and ε is an error random variable. Further, if $\hbar_0$ and S_0 denote the baseline hazard and survival function, respectively, of the random variable $T_0 = \exp(\varepsilon)$, we recapitulate that Model (6.1) implies the following hazard and survival function for a subject with covariate vector $\boldsymbol{X}$:

$$\begin{aligned} \hbar(t \mid \boldsymbol{X}) &= \hbar_0\bigl\{\exp(-\boldsymbol{X}'\boldsymbol{\beta})\, t\bigr\}\, \exp(-\boldsymbol{X}'\boldsymbol{\beta}), \\ S(t \mid \boldsymbol{X}) &= S_0\bigl\{\exp(-\boldsymbol{X}'\boldsymbol{\beta})\, t\bigr\}, \end{aligned}$$

and we can also write

$$T = \exp(\boldsymbol{X}'\boldsymbol{\beta})\, T_0 \tag{6.2}$$

showing that the effect of a covariate is acceleration or deceleration of the baseline event time.

Throughout the chapter, we assume that the errors $\varepsilon_1, \ldots, \varepsilon_n$ in

$$Y_i = \log(T_i) = \boldsymbol{X}_i'\boldsymbol{\beta} + \varepsilon_i \qquad (i = 1, \ldots, n), \tag{6.3}$$

are independent and identically distributed with density $g(e)$. In other words, we will assume that given covariates, the response times $T_1, \ldots, T_n$ are independent.

In the remainder of this chapter, we discuss frequentist methods to estimate the parameters of the AFT Model (6.2) for an interval-censored response.

In Section 6.1, we assume a parametric distribution for the model error terms and illustrate the maximum likelihood inference. In Section 6.2, we relax the parametric assumptions by modelling the error distribution using one of the smoothing techniques introduced in Section 3.3. We continue with two short sections. Section 6.3 discusses the SNP approach for an AFT model suggested by Zhang and Davidian (2008) and introduced in Section 5.2.2 for the PH model. Also Section 6.4 is quite short, relying mostly on the discussion in Section 5.5. In Section 6.5, we describe an approach to calculate the necessary study size when planning a new study likely to experience interval censoring. Concluding remarks are given in Section 6.6.

6.1 Parametric model

As seen in Section 2.3.2, semiparametric estimation of the AFT model is not straightforward even with right-censored data. This is not better with interval-censored data. Therefore, we first treat fully parametric AFT models.

From Equations (6.1) and (6.3) we see the AFT model is in fact a standard linear regression model with a logarithmically transformed survival time. With fully observed responses and with a linear model, the most popular estimation method is the Least Squares approach. However, since we have here (interval-) censored responses this approach does not work, and we need to rely on general maximum likelihood theory. In classical linear models the least squares estimate (LSE) coincides with the maximum likelihood estimate (MLE) for a normal distribution of the error term. With survival times, the normal distribution does not take that central position anymore and we will consider instead a variety of densities for the error term. Often, the error density $g(e)$ belongs to a particular location and scale family with an unknown location parameter $\mu \in \mathbb{R}$ and scale parameter $\sigma > 0$, with the most popular choice the family of normal distributions $\mathcal{N}(\mu, \sigma^2)$. That is, the density g of the error term ε can be written as

$$g(e) = \frac{1}{\sigma}\, g^{\star}\Bigl(\frac{e-\mu}{\sigma}\Bigr),$$

where $g^{\star}$ is the standardized density (with a fixed mean and variance, not necessarily being zero and one, respectively) of a particular family of distributions. Then, we can rewrite AFT Model (6.1) as

$$Y = \log(T) = \mu + \boldsymbol{X}'\boldsymbol{\beta} + \sigma\,\varepsilon^{\star}, \qquad (6.4)$$

where $\varepsilon^{\star}$ is the standardized error term with density $g^{\star}(e^{\star})$. We refer to μ as the intercept. Summarized, one has to estimate the intercept μ, the regression coefficients $\boldsymbol{\beta}$ and the scale parameter σ. Popular choices for the distribution of the standardized error term include the standard normal with

$$g^{\star}(e^{\star}) = \frac{1}{\sqrt{2\pi}} \exp\Big\{-\frac{(e^{\star})^2}{2}\Big\}, \tag{6.5}$$

the standard logistic with

$$g^{\star}(e^{\star}) = \frac{\exp(e^{\star})}{\big\{1 + \exp(e^{\star})\big\}^2}, \tag{6.6}$$

and the standard type I least extreme value distribution with

$$g^{\star}(e^{\star}) = \exp\big\{e^{\star} - \exp(e^{\star})\big\}. \tag{6.7}$$

6.1.1 Maximum likelihood estimation

The maximum likelihood procedure for estimating the parameters of a parametric AFT model follows the approach explained in Section 3.2.1. The $(p+2)$ parameters of interest are $\boldsymbol{\theta} = \big(\mu,\, \boldsymbol{\beta}',\, \log(\sigma)\big)'$, whereby the logarithm of the scale parameter σ is estimated to avoid constrained optimization when calculating the MLE. Under the assumption of independent and noninformative censoring, the likelihood of AFT Model (6.3) based on $\mathcal{D}$ is given by $L(\boldsymbol{\theta}) = \prod_{i=1}^{n} L_i(\boldsymbol{\theta})$, where, in accordance with Equation (1.13) in Section 1.5.1,

$$L_i(\boldsymbol{\theta}) = \oint_{l_i}^{u_i} f_i(t;\, \boldsymbol{\theta})\mathrm{d}t, \tag{6.8}$$

with

$$\begin{aligned} f_i(t;\, \boldsymbol{\theta}) &= \frac{1}{\sigma\, t}\, g^{\star}\Big(\frac{\log(t) - \mu - \boldsymbol{X}_i'\boldsymbol{\beta}}{\sigma}\Big) \\ &= \frac{1}{t}\, g\Big(\log(t) - \boldsymbol{X}_i'\boldsymbol{\beta}\Big) \qquad (i = 1, \ldots, n). \end{aligned} \tag{6.9}$$

The MLE $\widehat{\boldsymbol{\theta}}$ is obtained by numerically maximizing $\ell(\boldsymbol{\theta}) = \log\big\{L(\boldsymbol{\theta})\big\}$.

Selection of the error distribution

In practice, the parametric distribution for the error term of the AFT model is rarely known. Rather, its choice is based on a model selection procedure. Since AFT models with different error distributions are generally not nested, information criteria such as AIC and BIC are used to select the best fitting model with the smallest value indicating a better fit.

Predicted survival distribution

The fitted AFT model can be used to establish the (predicted) survival distribution for a specific subject with covariate vector $\boldsymbol{X}_{pred}$. The survival distribution depends on the choice of the error distribution, which we have limited here to: normal (6.5), logistic (6.6), and type I least extreme value distribution (6.7). These choices lead to the log-normal, the log-logistic and the Weibull distribution, respectively, for the survival time T. In Appendix B we have reviewed these two-parameter survival distributions and parametrized them using γ (shape) and α (which scales the survival distribution of T). Given $\boldsymbol{X}_{pred}$ and parameters μ, $\boldsymbol{\beta}$, σ of AFT Model (6.4), the survival parameters γ and $\alpha = \alpha_{pred}$ follow from Equation (B.2) and are equal to

$$\gamma = \frac{1}{\sigma}, \qquad \alpha_{pred} = \exp\bigl(\mu + \boldsymbol{X}'_{pred}\boldsymbol{\beta}\bigr).$$

Model based covariate specific survival and hazard functions can be obtained using formulas (B.3), (B.6) and (B.8) for our three chosen error distributions. Formulas (B.4), (B.7) and (B.9) give the model-based covariate specific mean and median survival times. For other error distributions, the same general principles lead to all required statistical output.

Example 6.1 Signal Tandmobiel study

In Example 3.2, we used parametric models to estimate the distribution of the emergence times of tooth 44 (permanent right mandibular first premolar) of boys. In Example 4.1 we showed that the emergence of tooth 44 is significantly accelerated when primary tooth 84 (primary mandibular right first molar) has exhibited caries in the past. Here, we shall examine whether the emergence distributions of boys and girls differ and whether caries on tooth 84 has impact.

In the following, let T_i $(i = 1, \ldots, n)$ be the emergence time of tooth 44 of the ith child with X_{i1} its gender (0 for boys, 1 for girls), X_{i2} the binary DMF (Decayed, Missing or Filled, see Section 1.6.7) status tooth 84 (DMF = 1, caries present currently or in the past, 0 = otherwise) and $X_{i3} = X_{i1} \cdot X_{i2}$ be the gender by DMF interaction (also denoted as gender:DMF). Since the DMF status of the primary tooth 84 is unknown for four children, the sample size in this example is $n = 496$.

Model

To model the dependence of the emergence time T_i on the covariate vector $\boldsymbol{X}_i = (X_{i1}, X_{i2}, X_{i3})'$ the following AFT model will be considered

$$\log(T_i) = \mu + \boldsymbol{X}'_i\boldsymbol{\beta} + \sigma\,\varepsilon^\star_i \qquad (i = 1, \ldots, n), \tag{6.10}$$

where $\boldsymbol{\beta} = (\beta_1, \beta_2, \beta_3)'$ and $\varepsilon^\star_i$ $(i = 1, \ldots, n)$ are the i.i.d. error terms with a suitable parametric density with a fixed mean and variance. Hence, the vector of unknown parameters to be estimated using the method of maximum likelihood is $\boldsymbol{\theta} = \bigl(\mu, \beta_1, \beta_2, \beta_3, \log(\sigma)\bigr)'$.

TABLE 6.1: Signal Tandmobiel study. Parametric AFT model for emergence of tooth 44. Model fit statistics obtained with R package survival.

Error distribution	$\ell(\widehat{\boldsymbol{\theta}})$	AIC	BIC
Normal	−623.6	1257.2	1278.3
Logistic	−625.8	1261.6	1282.6
Extreme value	−644.6	1299.2	1320.2

TABLE 6.2: Signal Tandmobiel study. AFT model for emergence of tooth 44 with a normal error distribution. Parameter estimates and related quantities of the statistical inference obtained with R package survival.

Parameter	Estimate	SE	95% CI	P-value
Intercept (μ)	2.38	0.01	(2.36, 2.40)	<0.0001
Gender[girl] (β_1)	−0.0456	0.0146	(−0.0741, −0.0170)	0.0018
DMF[1] (β_2)	−0.0751	0.0181	(−0.1105, −0.0397)	<0.0001
Gender:DMF (β_3)	0.0480	0.0261	(−0.0031, 0.0991)	0.0659
log(σ)	−2.15	0.04		
Scale (σ)	0.12			

Choice of the error distribution

In Table 6.1 the maximized log-likelihood, AIC and BIC, are shown for AFT Model (6.10) whereby the error distribution is one of the above three selected distributions. Since the number of unknown parameters is the same (five) for all considered models, the choice of the information criteria does not matter for selecting the error distribution. We note that there is some preference for the normal error distribution, which implies a log-normal distribution for the emergence times. Note also, that AIC and BIC differ less than 5 for the normal and the logistic distribution, so that the preference for the normal errors is not strong. In the following, we shall focus though on the log-normal model.

Model-based inference

Table 6.2 provides parameter estimates, 95% CIs and Wald P-values of the above specified AFT model. The gender by DMF interaction term is borderline non-significant at 0.05 ($P = 0.0659$) indicating that, strictly speaking, DMF does not have a different impact on the emergence times for tooth 44 for boys and girls. This could also be evaluated using the likelihood ratio test (LRT) by comparing the current AFT model to a simpler AFT model involving only covariates X_{i1} and X_{i2} with the same normal error distribution. This leads to the maximized log-likelihood of $\ell\big(\widehat{\boldsymbol{\theta}}_0\big) = -625.3$ and a LRT statistic equal to

3.37 with a P-value of 0.0663. Although the interaction term is not significant, we wish to include it here for illustrative purposes.

We now look at the acceleration factors of emergence due to decayed primary predecessors, i.e., due to DMF = 1. Since the gender by DMF interaction term is included in the AFT model, the acceleration factor must be different for boys and girls.

For boys ($X_{i1} = 0$) the acceleration factor $AF_{boys}(\text{DMF})$ due to DMF = 1 is equal to $\exp(\beta_2)$, with estimated value and 95% CI obtained by exponentiating the corresponding values from the DMF[1] (β_2) row of Table 6.2. That is, $\widehat{AF}_{boys}(\text{DMF}) = \exp(-0.0751) = 0.928$ which means that the age of emergence of tooth 44 following a decayed predecessor 84 is on average equal to 92.8% of the age of emergence for a sound predecessor. The corresponding 95% CI is (0.895, 0.961). Finally, the effect of DMF for boys is expressed by the Wald P-value for the coefficient β_2, which was highly significant ($P < 0.0001$).

The calculation of the acceleration factor due to DMF = 1 for girls and its confidence interval are slightly more complicated since the DMF effect for girls is given by $\omega = \beta_2 + \beta_3 = \boldsymbol{c}'\boldsymbol{\theta}$, where $\boldsymbol{c} = (0, 0, 1, 1, 0)'$. A test of the

FIGURE 6.1: Signal Tandmobiel study. Survival functions for emergence of permanent tooth 44 in gender by DMF groups based on a parametric AFT model with the normal error distribution, obtained with R package survival.

null hypothesis $\text{H}_0 : \omega = 0$ evaluates whether for girls a decayed primary predecessor significantly affects the emergence distribution of tooth 44. To estimate ω one computes $\widehat{\omega} = \widehat{\beta}_2 + \widehat{\beta}_3 = -0.0272$ leading to $\widehat{AF}_{girls}(\text{DMF}) = \exp(\widehat{\omega}) = 0.973$. That is, for girls, the effect of a decayed primary predecessor on the emergence of the permanent successor is less profound than for boys. The age of emergence of a permanent tooth following a decayed predecessor is on average equal to 97.3% of the age of emergence when the predecessor was sound.

To perform the Wald test for $\text{H}_0 : \omega = 0$ and to calculate the confidence intervals for ω and $AF_{girls}(\text{DMF})$, we need the standard error of $\widehat{\omega}$. This is obtained from the estimated covariance matrix $\widehat{\boldsymbol{V}}$ of $\widehat{\boldsymbol{\theta}}$ (see Equation (1.15)) as

$$\widehat{\text{SE}}(\widehat{\omega}) = \sqrt{\boldsymbol{c}'\widehat{\boldsymbol{V}}\boldsymbol{c}},$$

where for our model,

$$\widehat{\boldsymbol{V}} = \widehat{\mathsf{var}}(\widehat{\boldsymbol{\theta}})$$
$$= \begin{pmatrix} 0.000114 & -0.000113 & -0.000113 & 0.000113 & 0.0000501 \\ -0.000113 & 0.000212 & 0.000113 & -0.000212 & -0.0000121 \\ -0.000113 & 0.000113 & 0.000326 & -0.000326 & -0.0000203 \\ 0.000113 & -0.000212 & -0.000326 & 0.000680 & 0.0000128 \\ 0.0000501 & -0.0000121 & -0.0000203 & 0.0000128 & 0.00179 \end{pmatrix}.$$

Hence $\widehat{\text{SE}}(\widehat{\omega}) = 0.0188$. The Wald test statistic is then $Z = \widehat{\omega}/\widehat{\text{SE}}(\widehat{\omega}) = -1.444$ corresponding to $P = 0.149$. That is, now the impact of the decayed primary predecessor on the emergence of its permanent successor is not significant. The 95% CI for ω calculated with (1.18) is then (−0.0640, 0.0097). The 95% CI for the acceleration factor due to DMF = 1 of girls is obtained by exponentiating the bounds of the CI for ω, and it is equal to (0.938, 1.010).

Model-based estimated survival distributions

We now calculate model-based estimated survival functions for selected combinations of covariates. Namely, we wish to estimate the emergence distribution of boys with DMF = 0, boys with DMF = 1, girls with DMF = 0, and girls with DMF = 1 which corresponds to covariate combinations: $\boldsymbol{X}_{pred,1} = (0,\, 0,\, 0)'$, $\boldsymbol{X}_{pred,2} = (0,\, 1,\, 0)'$, $\boldsymbol{X}_{pred,3} = (1,\, 0,\, 0)'$, and $\boldsymbol{X}_{pred,4} = (1,\, 1,\, 1)'$, respectively. Model-based estimated survival functions for these four combinations of covariates are shown in Figure 6.1. A quick visual appreciation of the model fit can be done by contrasting the model-based survival distributions with the corresponding NPMLE of the survival function (see Section 3.1). In Figure 6.2, we can observe a rather good agreement between the NPMLE and the model-based survival function except for

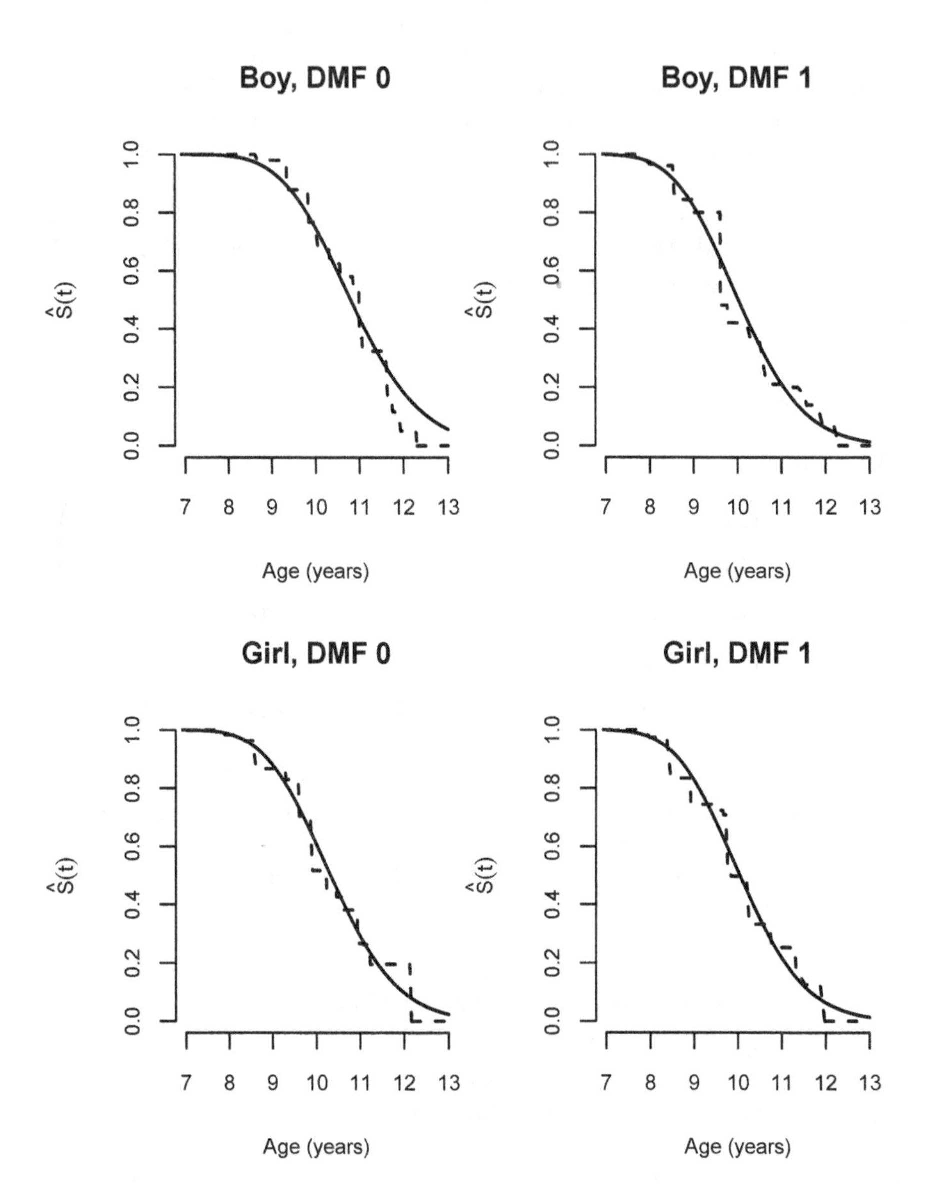

FIGURE 6.2: Signal Tandmobiel study. Survival functions for emergence of permanent tooth 44 in gender by DMF groups based on a parametric AFT model with the normal error distribution (solid lines) compared to the NPMLE (dashed lines), obtained with R packages survival and Icens, respectively.

boys with DMF = 0 where some disagreement is seen in the right tail of the distribution. This may be due to a nonnormal error distribution or that the scale parameter σ in AFT Model (6.4) differs across the four groups. How to remedy the described disagreement by extending the basic AFT model is discussed in Section 6.2.

□

6.1.2 R solution

Maximum likelihood estimation for the parametric AFT model is provided by the function survreg of the package survival. More details on the function are given in Appendix D.4.

Example 6.1 in R

Upon loading the data, we create a data frame tandDMF84 *containing only those children with nonmissing covariate values.*

Choice of the error distribution

AFT Model (6.10) with the three considered error distributions (normal, logistic, type I extreme value) corresponding to the log-normal, log-logistic, and Weibull distribution of the emergence times are fitted and stored as components of the list fit1.

```
> fit1 <- list()
> fit1[["lognormal"]] <- survreg(Surv(L44, R44, type = "interval2")
+            ~ fGENDER + fDMF_84 + fGENDER:fDMF_84,
+            data = tandDMF84, dist = "lognormal")
> fit1[["loglogistic"]] <- survreg(Surv(L44, R44, type = "interval2")
+           ~ fGENDER + fDMF_84 + fGENDER:fDMF_84,
+           data = tandDMF84, dist = "loglogistic")
> fit1[["weibull"]] <- survreg(Surv(L44, R44, type = "interval2")
+            ~ fGENDER + fDMF_84 + fGENDER:fDMF_84,
+            data = tandDMF84, dist = "weibull")
```

The maximized log-likelihood values of the fitted models shown in Table 6.1 are obtained with:

```
> loglik <- sapply(fit1, "[[", "loglik")[2,]
> print(loglik)
```

```
lognormal loglogistic     weibull
-623.6146   -625.8079   -644.5844
```

The information criteria for the models in Table 6.1 are calculated and printed with

```
> n <- nrow(tandDMF84)        ## sample size
> nu0 <- 5                    ## number of parameters
>
> ModelFit1 <- data.frame(LogLik = loglik,
+         AIC = -2 * loglik + 2 * nu0,
+         BIC = -2 * loglik + log(n) * nu0)
> print(ModelFit1)
```

```
               LogLik      AIC      BIC
lognormal   -623.6146 1257.229 1278.262
loglogistic -625.8079 1261.616 1282.649
weibull     -644.5844 1299.169 1320.202
```

Model-based inference

Further, we create an object Fit1 containing the fitted AFT model with the normal error distribution used for later inference. The summary method is used to print the most important results, many of which are shown in Table 6.2.

```
> Fit1 <- fit1[["lognormal"]]
> summary(Fit1)
```

```
Call:
survreg(formula = Surv(L44, R44, type = "interval2") ~ fGENDER +
    fDMF_84 + fGENDER:fDMF_84, data = tandDMF84, dist = "lognormal")
                            Value Std. Error      z        p
(Intercept)                2.3795     0.0107 222.56 0.00e+00
fGENDERgirl               -0.0456     0.0146  -3.13 1.77e-03
fDMF_84caries             -0.0751     0.0181  -4.16 3.22e-05
fGENDERgirl:fDMF_84caries  0.0480     0.0261   1.84 6.59e-02
Log(scale)                -2.1493     0.0423 -50.83 0.00e+00

Scale= 0.117

Log Normal distribution
Loglik(model)= -623.6   Loglik(intercept only)= -635.6
        Chisq= 23.9 on 3 degrees of freedom, p= 2.6e-05
Number of Newton-Raphson Iterations: 4
n= 496
```

The output shows the MLEs of the intercept μ, the regression coefficient vector $\boldsymbol{\beta}$, and the log-scale parameter together with their (asymptotic) standard errors, Wald test statistics and P-values (see Section 1.5.2). In addition are reported: the estimated scale parameter σ, the maximized value of the model log-likelihood, and also the maximized value of the log-likelihood of the model where the regression coefficient vector $\boldsymbol{\beta}$ is equal to a vector of zeros. The Chisq *value of* 23.9 *and the P-value of* 2.6e-05 *is the likelihood ratio test statistic and the related P-value evaluating whether the current model is significantly better than the model containing only the intercept. At the bottom, the number of Newton-Raphson iterations to find the MLE and the sample size are printed.*

The Wald confidence intervals shown in Table 6.2 are obtained using the confint *method:*

```
> confint(Fit1, level = 0.95)
```

```
                                    2.5 %      97.5 %
(Intercept)                    2.358591822  2.40050298
fGENDERgirl                   -0.074142577 -0.01700256
fDMF_84caries                 -0.110529835 -0.03970119
fGENDERgirl:fDMF_84caries -0.003146241  0.09907176
```

The LRT that compares two nested models is obtained with the anova *function applied to the objects holding the two model fits. The significance of the gender:DMF interaction by the means of the LRT is achieved using:*

```
> Fit2 <- survreg(Surv(L44, R44, type = "interval2")
+              ~ fGENDER + fDMF_84,
+            data = tandDMF84, dist = "lognormal")
> anova(Fit2, Fit1)
```

```
                              Terms Resid. Df    -2*LL
1                  fGENDER + fDMF_84       492 1250.602
2 fGENDER + fDMF_84 + fGENDER:fDMF_84       491 1247.229
              Test Df Deviance   Pr(>Chi)
1                  NA       NA         NA
2 +fGENDER:fDMF_84  1 3.372424 0.06629616
```

The value of the LRT statistic is found on the second row of the Deviance *column with the related P-value in column* Pr(>Chi).

The acceleration factor of emergence due to DMF = 1 for boys including the 95% CI is calculated using the following commands:

```
> beta2 <- coef(Fit1)["fDMF_84caries"]
> ci.beta2 <- confint(Fit1, level = 0.95)["fDMF_84caries",]
> AFboys <- c(exp(beta2), exp(ci.beta2))
> names(AFboys) <- c("Estimate", "Lower", "Upper")
> print(AFboys)
```

```
 Estimate     Lower     Upper
0.9276363 0.8953596 0.9610766
```

The estimated covariance matrix $\widehat{V}$ of the MLE $\widehat{\theta}$ is obtained from the object Fit1 using the function vcov as (output not shown):

```
> vcov(Fit1)
```

This estimated covariance can then be used to calculate the standard errors for the estimated linear combinations of the model parameters, needed for the Wald tests and confidence intervals. This is illustrated for the log-acceleration factor $\omega = \beta_2 + \beta_3$ due to DMF for girls, its standard error, and the 95% confidence interval.

```
> cc <- c(0, 0, 1, 1, 0)                         ## coefficients of the
>                                                ## linear combination
> theta <- summary(Fit1)$table[, "Value"]        ## MLE
>
> omega <- as.numeric(crossprod(cc, theta))
> se.omega <- as.numeric(sqrt(t(cc) %*% vcov(Fit1) %*% cc))
> ci.omega <- omega + c(-1, 1) * se.omega * qnorm(0.975)
>
> Omega <- c(omega, se.omega, ci.omega)
> names(Omega) <- c("Estimate", "Std. Error", "Lower", "Upper")
> print(Omega)
```

```
    Estimate   Std. Error        Lower        Upper
-0.027152749  0.018806122 -0.064012072  0.009706573
```

Further, we calculate the P-value of the Wald test for the null hypothesis $H_0 : \omega = 0$.

```
> Z.omega <- omega / se.omega
> P.omega <- 2 * pnorm(-abs(Z.omega))
> print(P.omega)
```

```
[1] 0.1487882
```

Finally, we calculate the acceleration factor of emergence due to DMF = 1 for girls and the 95% CI with

```
> AFgirls <- c(exp(omega), exp(ci.omega))
> names(AFgirls) <- c("Estimate", "Lower", "Upper")
> print(AFgirls)
```

```
 Estimate     Lower     Upper
0.9732126 0.9379937 1.0097538
```

Model-based estimated survival distributions

To produce Figure 6.2, we first create a dense grid of t-values in which we evaluate the survival functions. Secondly, we create a vector of labels for the considered covariate combinations.

```
> ltgrid <- 200
> tgrid <- seq(7, 13, length = ltgrid)
> n.combinations <- 4
> x.label <- paste(rep(c("Boy", "Girl"), each = 2), ", ",
+                  rep(c("DMF 0", "DMF 1"), 2), sep = "")
> print(x.label)
```

```
[1] "Boy, DMF 0"  "Boy, DMF 1"  "Girl, DMF 0" "Girl, DMF 1"
```

Thirdly, we create a data frame replicating four times the time grid in the first column and replicating 200-times each covariate combination.

```
> pdata <- data.frame(t      = rep(tgrid, n.combinations),
+                     gender = rep(c(0, 1), each = 2*ltgrid),
+                     dmf    = rep(c(0, 1, 0, 1), each = ltgrid))
> pdata <- transform(pdata, gender.dmf = gender*dmf)
> print(pdata[c(1:2, 201:202, 401:402, 601:602),])
```

```
           t gender dmf gender.dmf
1   7.000000      0   0          0
2   7.030151      0   0          0
...
201 7.000000      0   1          0
202 7.030151      0   1          0
...
401 7.000000      1   0          0
```

(Cont.)

```
402 7.030151        1   0         0
...
601 7.000000        1   1         1
602 7.030151        1   1         1
```

In the next step, we calculate the values of the linear predictor $\boldsymbol{X}'_{pred}\widehat{\boldsymbol{\beta}}$ *corresponding to each row of the data frame* pdata.

```
> Eta1 <- coef(Fit1)["(Intercept)"] + as.numeric(
+             as.matrix(pdata[, c("gender", "dmf", "gender.dmf")])
+             %*% coef(Fit1)[-1])
```

The values of the survival function of the AFT model with a normal error distribution are calculated using the plnorm *function where the linear predictor is used as its* meanlog *argument, and the estimated scale* $\widehat{\sigma}$ *as its* sdlog *argument:*

```
> S1all <- plnorm(pdata$t, meanlog = Eta1, sdlog = Fit1$scale,
+                 lower.tail = FALSE)
```

For the logistic error distribution (log-logistic distribution of T *obtained with* dist = "loglogistic" *in the* survreg *call), the following call to the* pllogis *function is needed to get the estimated survival function (inverted scale of the AFT model is the shape of the log-logistic distribution, exponentiated linear predictor is the scale of the log-logistic distribution):*

```
> S1all <- pllogis(pdata$t, shape = 1 / Fit1$scale, scale = exp(Eta1),
+                  lower.tail = FALSE)
```

Analogously, a call to function pweibull *is needed to get the estimated survival function for the extreme value error distribution (Weibull distribution of* T *obtained with* dist = "weibull" *in the* survreg *call):*

```
> S1all <- pweibull(pdata$t,
+                   shape = 1 / Fit1$scale, scale = exp(Eta1),
+                   lower.tail = FALSE)
```

We then create a list (one list component for each covariate combination) of data frames containing the estimated values of the survival function (the first five rows of this data frame for boys with DMF = 0 are printed).

```
> S1 <- list()
> for (i in 1:n.combinations){
+   S1[[i]] <- data.frame(t = tgrid,
+                         S = S1all[((i - 1)*ltgrid + 1):(i*ltgrid)])
+ }
```

```
> names(S1) <- x.label
> print(S1[[1]])
```

```
         t         S
1 7.000000 0.9999005
2 7.030151 0.9998849
3 7.060302 0.9998672
4 7.090452 0.9998470
5 7.120603 0.9998240
...
```

Finally, we use the calculated values to obtain Figure 6.1.

The data frame S1 *was further used to draw a graph on Figure 6.2. The code is available in the supplementary materials.*

□

6.1.3 SAS solution

The procedures of Section 3.2.5 can be used for the AFT model, since it is simply a parametric model depending now on covariates. Hence, an AFT model can be fitted using the SAS/STAT procedure LIFEREG or the SAS/QC procedure RELIABILITY. The LIFEREG procedure produces P-values and information criteria, which are not available in the RELIABILITY procedure. On the other hand, in the RELIABILITY procedure the scale parameter is allowed to depend on covariates. We start with an analysis using the SAS procedure LIFEREG.

Example 6.1 in SAS (PROC LIFEREG)

We first create a data set with the children having nonmissing covariates.

The log-normal *model of Table 6.2 can be fitted with the following* SAS *commands:*

```
PROC LIFEREG DATA=tandmob;
*CLASS gender;
MODEL (L44 R44) = gender dmf_84 gender*dmf_84 / D=LNORMAL;
RUN;
```

The option D=LNORMAL *requests the log-normal AFT model to be fitted. For the log-logistic or Weibull AFT models, we ask for* LLOGISTIC *or* WEIBULL, *respectively. The* LIFEREG *procedure can also fit models to untransformed data by specifying the* NOLOG *option, but then we are not dealing anymore with AFT models. In Appendix E.1, distributions that can be fitted with the* LIFEREG *procedure are listed.*

The LIFEREG *procedure has no* ESTIMATE *statement. Hence, to calculate the effect of a linear combination of parameters, the model must be reparametrized to provide the requested parameter immediately from the output or by postprocessing the parameter estimates and their corresponding covariance matrix as illustrated above (see Section 6.1.2). Reparametrization can often be done by choosing an appropriate reference category for a categorical variable. For instance, uncommenting the CLASS statement (delete the star) will make the females the reference category for gender. Adding the optional argument* ORDER= *after the* LIFEREG *statement determines how* SAS *chooses the reference category of categorical variables, see the* SAS *manual for possibilities. To postprocess the parameter estimates and covariance matrix, adding the options* COVOUT *and* OUTEST= *to the* LIFEREG *statement writes the necessary estimates into a data set. In the supplementary materials, an example program illustrating the postprocessing with* PROC IML *can be found.*

The following parameter estimates are produced.

Analysis of Maximum Likelihood Parameter Estimates

Parameter	DF	Estimate	Standard Error	95% Confidence Limits		Chi-Square	Pr > ChiSq
Intercept	1	2.3795	0.0107	2.3586	2.4005	49532.0	<.0001
gender	1	-0.0456	0.0146	-0.0741	-0.0170	9.77	0.0018
dmf_84	1	-0.0751	0.0181	-0.1105	-0.0397	17.28	<.0001
gender*dmf_84	1	0.0480	0.0261	-0.0031	0.0991	3.38	0.0659
Scale	1	0.1166	0.0049	0.1073	0.1266		

Note that the LIFEREG *procedure reports the estimate of* σ *and not of* $\log(\sigma)$.

□

Example 6.1 in SAS (PROC RELIABILITY)

Alternatively, the SAS/QC procedure RELIABILITY *may be used. Below we used this procedure to fit the log-normal model to the dental data.*

```
PROC RELIABILITY DATA=tandmob;
DISTRIBUTION LOGNORMAL;
MODEL (L44 R44) = gender dmf_84 gender*dmf_84;
run;
```

The output (not shown) is similar to the output of the LIFEREG *procedure with the exception that no P-values nor fit statistics (e.g., AIC) are provided. However, with the* RELIABILITY *procedure one can let the scale parameter depend on covariates. For instance, adding the next line of code fits a log-normal model with a different scale parameter for each gender.*

```
LOGSCALE gender;
```

□

6.2 Penalized Gaussian mixture model

In the previous section, we have assumed a specific parametric error distribution for the AFT model and hence a specific survival distribution. Note that the semiparametric approach whereby the error density g is left unspecified in the AFT model is difficult to handle with interval-censored data. Nevertheless, the smoothing techniques introduced in Section 3.3 can be invoked to relax the strict parametric assumptions on g or its standardized counterpart $g^{\star}$. In that section no covariates were involved, but it is obvious that the same ideas can be used in a regression context. As a result, we argue that the resulting AFT model will be more robust against misspecification of the error distribution, or equivalently, of the baseline survival distribution. A possible approach, based on the concept of *a finite mixture* distribution (see Section 3.3.3), is introduced in this section.

We again assume the following AFT model:

$$Y = \log(T) = \boldsymbol{X}'\boldsymbol{\beta} + \varepsilon, \tag{6.11}$$

but now combined with a penalized Gaussian mixture for the error distribution, which was introduced in Section 3.3.3. Now the density $g(e)$ is given by

$$g(e) = \frac{1}{\tau} \sum_{k=-K}^{K} w_k \varphi_{\mu_k, \sigma_0^2}\Big(\frac{e - \alpha}{\tau}\Big). \tag{6.12}$$

As explained in Section 3.3.3, the number of components $(2K + 1)$ and the means $\boldsymbol{\mu} = (\mu_{-K}, \ldots, \mu_K)'$ are chosen in advance. The means are fixed at equidistant knots centered around zero, i.e., $\mu_k = k\delta$, $k = -K, \ldots, K$, for some $\delta > 0$. The boundary knots (μ_{-K} and μ_K) are chosen such that the interval (μ_{-K}, μ_K) covers a high probability region of the zero-mean, unit-variance distribution. Typically, a relatively high number of knots $(2K + 1)$ are chosen (≈ 30). Possible choices for the fixed parameters following from recommendations given in Komárek et al. (2005a) are: $K = 15$, $\delta = 0.3$, $\sigma_0 = (2/3)\delta = 0.2$. Finally, also the standard deviation σ_0 is fixed in advance. The unknown parameters of the error Model (6.12) are the vector of weights $\boldsymbol{w} = (w_{-K}, \ldots, w_K)'$, the shift parameter α and the scale parameter $\tau > 0$.

The standardized error term ε^* is now defined as

$$\varepsilon^* = \frac{\varepsilon - \alpha}{\tau},$$

its density follows from (6.12) and is given as

$$g^*(e^*) = \tau\, g(\alpha + \tau\, e^*) = \sum_{k=-K}^{K} w_k \varphi_{\mu_k, \sigma_0^2}(e^*).$$

To estimate the weights $\boldsymbol{w}$, we have to take into account the following constraints:

$$0 < w_k < 1 \ (k = -K, \ldots, K), \qquad \sum_{k=-K}^{K} w_k = 1, \tag{6.13}$$

$$\mathsf{E}(\varepsilon^*) = \sum_{k=-K}^{K} w_k \mu_k = 0, \tag{6.14}$$

$$\mathsf{var}(\varepsilon^*) = \sum_{k=-K}^{K} w_k (\mu_k^2 + \sigma_0^2) = 1. \tag{6.15}$$

Constraints (6.13) ensure that $g(e)$ is a density, constraints (6.14) and (6.15) are needed to make the shift and scale parameters α and τ identifiable. At the same time, they ensure that the standardized error term ε^* has zero mean and unit variance. Consequently, when we write the AFT model as

$$Y = \log(T) = \mu + \boldsymbol{X}'\boldsymbol{\beta} + \sigma\,\varepsilon^\star,$$

the constraints (6.14) and (6.15) provide direct equalities $\mu = \alpha$ and $\sigma = \tau$.

To avoid estimation under the constraints (6.13), the computations are based on the transformed weights $\boldsymbol{a} = (a_{-K}, \ldots, a_K)'$, with $a_0 = 0$ and

$$w_k = w_k(\boldsymbol{a}) = \frac{\exp(a_k)}{\sum_{j=-K}^{K} \exp(a_j)} \qquad (k = -K, \ldots, K).$$

To avoid the remaining constraints (6.14) and (6.15), two coefficients, say, a_{-1} and a_1, can be expressed as functions of $\boldsymbol{a}_{-(-1,0,1)} = (a_{-K}, \ldots, a_{-2}, a_2, \ldots, a_K)'$. The estimated parameters of the error model are then: the shift α, the scale τ, the reduced vector of the transformed mixture weight $\boldsymbol{a}_{-(-1,0,1)}$. The complete vector of unknown parameters of the AFT Model (6.11) with the error Density (6.12) is then

$$\boldsymbol{\theta} = (\boldsymbol{a}'_{-(-1,0,1)},\ \alpha,\ \gamma_0,\ \boldsymbol{\beta}')',$$

with $\gamma_0 = \log(\tau)$ to deal with the constraint $\tau > 0$.

6.2.1 Penalized maximum likelihood estimation

Mean-scale model

We will see in Example 6.2 that flexibly modelling the error density of the AFT model not necessarily guarantees that the survival distribution fits well for all possible covariate values. A possible reason can be that not only the location of the survival distribution depends on covariates but also its scale. Hence a possible improvement of the model is to allow also the scale parameter τ to depend on a vector of covariates, let say $\boldsymbol{Z}$. This leads to an extension of the AFT Model (6.11) whereby both the mean and the scale parameters depend on covariates, the so-called *mean-scale* AFT model proposed in Lesaffre et al. (2005). More specifically, the dependence of the scale parameter τ on covariates is modelled as

$$\log(\tau) = \gamma_0 + \boldsymbol{Z}'\boldsymbol{\gamma}, \tag{6.16}$$

where γ_0 and $\boldsymbol{\gamma}$ are unknown parameters. That is,

$$\tau = \tau(\gamma_0, \boldsymbol{\gamma}) = \exp\bigl(\gamma_0 + \boldsymbol{Z}'\boldsymbol{\gamma}\bigr).$$

The complete vector of unknown parameters is then

$$\boldsymbol{\theta} = (\boldsymbol{a}'_{-(-1,0,1)}, \alpha, \gamma_0, \boldsymbol{\gamma}', \boldsymbol{\beta}')'.$$

In the following, a model where no covariates are assumed to influence the scale parameter τ, i.e., where $\tau = \log(\gamma_0)$ and the complete vector of unknown parameters is given by (6.2), will be called the *mean* AFT model.

Penalized maximum likelihood

For both the mean and the mean-scale model, we shall use the method of penalized maximum likelihood (PMLE) introduced in Section 3.3.3 as a principal tool for estimation and inference. That is, the PMLE of $\boldsymbol{\theta}$ is obtained by maximizing the penalized log-likelihood

$$\ell_P(\boldsymbol{\theta}; \lambda) = \ell(\boldsymbol{\theta}) \; - \; q(\boldsymbol{a}; \lambda), \tag{6.17}$$

where $q(\boldsymbol{a}; \lambda)$ is the penalty term, see Equation (3.21), and $\lambda > 0$ is a fixed tuning parameter controlling the smoothness of the fitted error distribution. The log-likelihood $\ell(\boldsymbol{\theta})$ is, as before,

$$\ell(\boldsymbol{\theta}) = \sum_{i=1}^{n} \log\bigl\{L_i(\boldsymbol{\theta})\bigr\}, \qquad L_i(\boldsymbol{\theta}) = \oint_{l_i}^{u_i} f_i(t; \boldsymbol{\theta})\mathrm{d}t \quad (i = 1, \ldots, n),$$

where now the survival density f_i of the ith event time T_i is given by

$$f_i(t; \boldsymbol{\theta}) = \frac{1}{t}\, g\bigl(\log(t) - \boldsymbol{X}'_i\boldsymbol{\beta}\bigr) = \frac{1}{\tau\, t} \sum_{k=-K}^{K} w_k\, \varphi_{\mu_k, \sigma_0^2}\biggl(\frac{\log(t) - \alpha - \boldsymbol{X}'_i\boldsymbol{\beta}}{\tau}\biggr) \qquad (i = 1, \ldots, n).$$

To select the optimal value of the smoothing parameter λ, the procedure described in Section 3.3.3 is followed. Let $\widehat{\boldsymbol{\theta}}(\lambda)$ maximize the penalized log-likelihood (6.17) for given $\lambda > 0$ and let

$$\widehat{\boldsymbol{H}} = -\frac{\partial^2 \ell_P\bigl\{\widehat{\boldsymbol{\theta}}(\lambda)\bigr\}}{\partial\boldsymbol{\theta}\partial\boldsymbol{\theta}'}, \qquad \widehat{\boldsymbol{I}} = -\frac{\partial^2 \ell\bigl\{\widehat{\boldsymbol{\theta}}(\lambda)\bigr\}}{\partial\boldsymbol{\theta}\partial\boldsymbol{\theta}'}.$$

To find the optimal value of λ, a grid search is performed. That is, we look for $\widehat{\lambda}$ which minimizes Akaike's information criterion

$$\mathsf{AIC}(\lambda) = -2\ell\{\widehat{\boldsymbol{\theta}}(\lambda)\} + 2\nu(\lambda), \tag{6.18}$$

where

$$\nu(\lambda) = \mathrm{trace}\big(\widehat{\boldsymbol{H}}^{-1}\widehat{\boldsymbol{I}}\big)$$

is the effective number of parameters.

The final parameter estimate is then

$$\widehat{\boldsymbol{\theta}} = \widehat{\boldsymbol{\theta}}\big(\widehat{\lambda}\big).$$

As described in Section 3.3.3, asymptotic inference of the elements of the parameter vector $\boldsymbol{\theta}$ can either be based on the *pseudo-covariance* matrix

$$\widehat{\mathsf{var}}_P\big(\widehat{\boldsymbol{\theta}}\big) = \widehat{\boldsymbol{H}}^{-1} \tag{6.19}$$

or the *asymptotic covariance* matrix

$$\widehat{\mathsf{var}}_A\big(\widehat{\boldsymbol{\theta}}\big) = \widehat{\boldsymbol{H}}^{-1}\,\widehat{\boldsymbol{I}}\,\widehat{\boldsymbol{H}}^{-1}. \tag{6.20}$$

It is also explained in Section 3.3.3 that the pseudo-covariance matrix is usually preferred.

Survival distribution for a new subject

Also now we can estimate the survival distribution for a new subject characterized by specific values for the covariates. For example, the expectation of the survival time T_{pred} generated by the assumed model with the covariates $\boldsymbol{X}_{pred}$ in (6.11) and $\boldsymbol{Z}_{pred}$ in (6.16) is given by

$$\mathsf{E}\big(T_{pred}\big) = \sum_{k=-K}^{K} w_k \exp\Big(\alpha + \boldsymbol{X}'_{pred}\boldsymbol{\beta} + \tau_{pred}\mu_k + \frac{\tau_{pred}\,\sigma_0^2}{2}\Big), \tag{6.21}$$

where $\tau_{pred} = \exp\big(\gamma_0 + \boldsymbol{Z}'_{pred}\boldsymbol{\gamma}\big)$. The survival function, density and the hazard function are given by

$$S_{pred}(t;\,\boldsymbol{\theta}) = \sum_{k=1}^{K} w_k \left[1 - \Phi_{\mu_k,\,\sigma_0^2}\left\{\frac{\log(t) - \alpha - \boldsymbol{X}'_{pred}\boldsymbol{\beta}}{\tau_{pred}}\right\}\right],$$

$$f_{pred}(t;\,\boldsymbol{\theta}) = \frac{1}{\tau_{pred}\,t}\sum_{k=-K}^{K} w_k\,\varphi_{\mu_k,\,\sigma_0^2}\left\{\frac{\log(t) - \alpha - \boldsymbol{X}'_{pred}\boldsymbol{\beta}}{\tau_{pred}}\right\},$$

$$\hbar_{pred}(t;\,\boldsymbol{\theta}) = \frac{f_{pred}(t;\,\boldsymbol{\theta})}{S_{pred}(t;\,\boldsymbol{\theta})}.$$

Estimated values of above quantities are obtained by replacing the model parameters by their estimated values.

Example 6.2 Signal Tandmobiel study
We revisit Example 6.1 and fit again the AFT model to evaluate the dependence of the emergence time T of tooth 44 on gender of a child and the caries status of tooth 84 expressed by its dichotomized DMF score. Recall that the covariate vector $\boldsymbol{X}$ of the AFT Model (6.11) is $\boldsymbol{X} = (X_1, X_2, X_3)'$, where X_1 equals one for girls and zero for boys, X_2 is a dichotomized DMF score where a value of one means that the primary predecessor was decayed, missing due to caries or filled and $X_3 = X_1 \cdot X_2$ is the gender by DMF interaction covariate. Two models will now be considered:

Mean model where the scale parameter does not depend on covariates, i.e., $\log(\tau) = \gamma_0$.

Mean-scale model where the scale parameter is allowed to depend on the same vector of covariates as the mean, i.e., $\log(\tau) = \gamma_0 + \boldsymbol{Z}'\boldsymbol{\gamma}$, where $\boldsymbol{Z} = \boldsymbol{X}$.

Note that the mean model in this example is the same model as in Example 6.1 except for the distribution of the error term. Remember that the AFT model with a normal distribution of the error term was found as the best fitting in Example 6.1. Nevertheless, as seen on Figure 6.2, the fit is not fully satisfactory for boys with DMF = 0. We now show that the situation will not change much if we use a flexible model for the distribution of the error terms but let only the location parameter depend on the covariates (the mean model). On the other hand, the mean-scale model provides more flexibility in fitting the survival distribution in the gender by DMF groups and leads in this particular situation to clearly a better fit. While the common shape of the survival distributions in the four gender by DMF groups is still assumed, the variability is now allowed to be different.

The penalized mixture used in this example was the same as in Example 3.4. That is, we used a basis standard deviation $\sigma_0 = 0.2$ and an equidistant set of 41 knots ranging from -6 to 6 yielding a distance of $\delta = 0.3$ between the two consecutive knots. The third order penalty ($s = 3$ in Equation (3.21)) was considered to calculate the penalized maximum-likelihood estimates of model parameters.

Degree of smoothing and the error distribution

For both the mean and the mean-scale model, the estimated smoothing parameter λ was relatively high, see Table 6.3. This results in smooth fitted error distributions as shown in Figure 6.3. Low values for the effective number of parameters were found, i.e., 5.94 and 8.61 for the mean and mean-scale model, respectively. Note that the numbers of "standard" parameters in the

TABLE 6.3: Signal Tandmobiel study. Mean and mean-scale model for emergence of tooth 44 with a PGM error distribution. Model fit statistics. Obtained with R package smoothSurv.

Model	λ/n	Number of param.	$\nu(\lambda)$	$\ell(\widehat{\boldsymbol{\theta}})$	AIC	BIC
Mean	2.7	43	5.94	−622.4	1 256.7	1 281.7
Mean-scale	7.4	46	8.61	−618.8	1 254.8	1 291.0

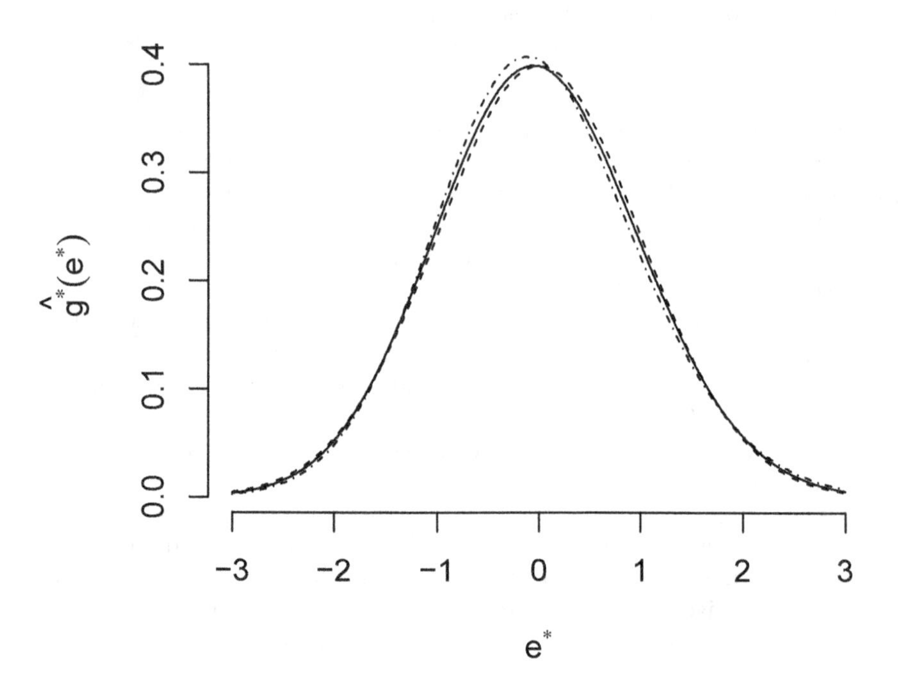

FIGURE 6.3: Signal Tandmobiel study. Estimated densities of the standardized error term in the AFT models for emergence of permanent tooth 44 based on mean model (dotted-dashed line), mean-scale model (solid line) compared to a standard normal density (dashed line), obtained with R function smoothSurvReg.

two models is 5 and 8, respectively, so that for both models less than one degree of freedom was consumed to estimate the distributional shape.

Comparison of models

Table 6.3 shows the mean-scale model is chosen by AIC, while BIC selects the mean model. Nevertheless, since the mean model is a submodel of the mean-scale model, it is also possible to perform a formal comparison of the two models by testing the null hypothesis $\gamma_1 = \gamma_2 = \gamma_3 = 0$ in the mean-scale model. To this end, we can use the Wald test based on the penalized ML estimates shown in Table 6.5 and the corresponding submatrix of the pseudo-covariance matrix (6.19). The Wald test statistic equals 8.71 which leads to the P-value of 0.0334, when comparing to a χ^2_3 distribution and thus the mean-scale model is preferred over the mean model.

Estimates and inference

Penalized maximum likelihood estimates of the most important model parameters and the acceleration factors due to DMF = 1 are shown in Tables 6.4 and 6.5 together with the quantities needed for statistical inference based on the fitted models. The acceleration factors are defined in the same way as in Example 6.1, that is,

$$AF_{boys}(\text{DMF}) = \exp(\beta_2), \qquad AF_{girls}(\text{DMF}) = \exp(\beta_2 + \beta_3).$$

Standard errors are derived from the pseudo-covariance matrix (6.19), the confidence intervals and the P-values are Wald-based.

It appears that the covariates DMF and Gender:DMF are not important in the scale part of the mean-scale model. This can be jointly tested by a Wald test of the null hypothesis $\gamma_2 = \gamma_3 = 0$. Here we obtained a test statistic equal to 2.78 corresponding to a P-value of 0.2488. We can therefore conclude that only gender influences the variability of the emergence times. Nevertheless, to keep this illustrative example simple, we shall continue with the description of the results based on the originally fitted mean and mean-scale models with estimates given in Tables 6.4 and 6.5.

Model-based survival distributions

Model-based estimates of the survival functions are compared to their NPMLE for the two covariate combinations (boys with DMF = 0, boys with DMF = 1) in Figure 6.4. Results for girls (not shown) are similar. It is seen that the mean-scale model provides only for boys with DMF = 0 a different fitted survival function than the mean model. Now the parametric curve and the NPMLE agree better than for the mean model. Clearly, modelling also the scale parameter as a function of covariates helped.

Further, we used Equation (6.21) and the delta method to calculate model-based estimates of the mean emergence times together with SEs and CIs for

TABLE 6.4: Signal Tandmobiel study. Mean AFT model for emergence of tooth 44 with a PGM error distribution. Parameter estimates and related quantities of the statistical inference. Obtained with R package smoothSurv.

Parameter	Estimate	SE	95% CI	P-value
Mean parameters				
Intercept (μ)	2.38	0.01	(2.36, 2.41)	<0.0001
Gender[girl] (β_1)	−0.0496	0.0150	(−0.0789, −0.0203)	0.0009
DMF[1] (β_2)	−0.0787	0.0183	(−0.1146, −0.0427)	<0.0001
Gender:DMF (β_3)	0.0508	0.0264	(−0.0009, 0.1025)	0.0541
Scale parameters				
Intercept ($\gamma_0 = \log(\sigma)$)	−2.12	0.05	(−2.22, −2.02)	
Scale (σ)	0.12		(0.11, 0.13)	
Acceleration factors due to DMF				
AF_{boys}(DMF)	0.924	0.017	(0.892, 0.958)	<0.0001
AF_{girls}(DMF)	0.973	0.018	(0.937, 1.009)	0.1429

TABLE 6.5: Signal Tandmobiel study. Mean-scale AFT model for emergence of tooth 44 with a PGM error distribution. Parameter estimates and related quantities of the statistical inference. Obtained with R package smoothSurv.

Parameter	Estimate	SE	95% CI	P-value
Mean parameters				
Intercept (μ)	2.38	0.01	(2.36, 2.39)	<0.0001
Gender[girl] (β_1)	−0.0385	0.0144	(−0.0668, −0.0103)	0.0075
DMF[1] (β_2)	−0.0703	0.0177	(−0.1049, −0.0356)	<0.0001
Gender:DMF (β_3)	0.0423	0.0274	(−0.0113, 0.0960)	0.1220
Scale parameters				
Intercept (γ_0)	−2.34	0.08	(−2.49, −2.18)	<0.0001
Gender[girl] (γ_1)	0.286	0.105	(0.081, 0.491)	0.0062
DMF[1] (γ_2)	0.211	0.127	(−0.037, 0.459)	0.0954
Gender:DMF (γ_3)	−0.214	0.181	(−0.568, 0.140)	0.2358
Acceleration factors due to DMF				
AF_{boys}(DMF)	0.932	0.016	(0.900, 0.965)	<0.0001
AF_{girls}(DMF)	0.972	0.020	(0.933, 1.013)	0.1823

TABLE 6.6: Signal Tandmobiel study. Estimated mean emergence times based on the AFT models with the PGM error distribution. Obtained with R package smoothSurv.

Group	Estimate	SE	95% CI
Mean model			
Boys, DMF = 0	10.92	0.13	(10.67, 11.18)
DMF = 1	10.10	0.15	(9.80, 10.40)
Girls, DMF = 0	10.40	0.11	(10.18, 10.61)
DMF = 1	10.11	0.17	(9.78, 10.43)
Mean-scale model			
Boys, DMF = 0	10.81	0.10	(10.60, 11.01)
DMF = 1	10.10	0.16	(9.79, 10.41)
Girls, DMF = 0	10.43	0.12	(10.19, 10.68)
DMF = 1	10.15	0.19	(9.78, 10.52)

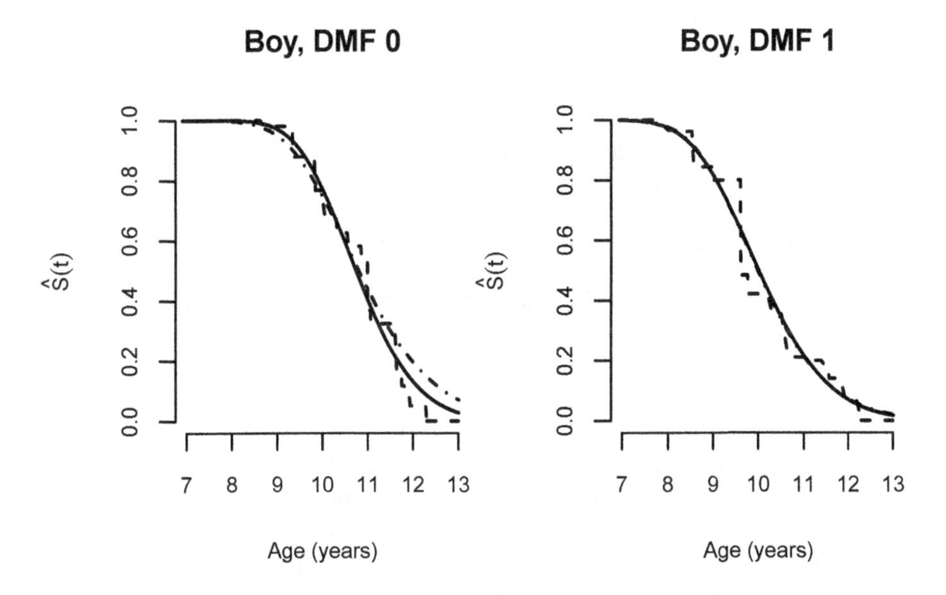

FIGURE 6.4: Signal Tandmobiel study. Estimated survival functions for emergence of permanent tooth 44 in DMF groups for boys based on a AFT model with the PGM error distribution, the mean model (dotted-dashed lines), the mean-scale model (solid lines) compared to NPMLE (dashed lines), obtained with R packages smoothSurv and Icens.

the four groups, see Table 6.6. While the two fitted survival functions obtained using the mean and the mean-scale model differ the most for boys with DMF = 0, the difference between the estimated emergence times is rather low, being only 0.11 years, i.e., about 40 days.

□

6.2.2 R solution

The penalized maximum likelihood estimation of the AFT model is implemented in the R package smoothSurv, see also Appendix D.6 for a comprehensive description of the main routines of this package.

Example 6.2 in R

We first load the data and create a data frame tandDMF84 *containing only those children with nonmissing values for the covariates included in the model.*

Model fitting

The mean model is fitted using:

```
> library("smoothSurv")
> pfit1 <- smoothSurvReg(Surv(L44, R44, type = "interval2")
+           ~ fGENDER + fDMF_84 + fGENDER:fDMF_84,
+           data = tandDMF84, info = FALSE)
```

To fit the considered mean-scale model, we run:

```
> pfit2 <- smoothSurvReg(Surv(L44, R44, type = "interval2")
+           ~ fGENDER + fDMF_84 + fGENDER:fDMF_84,
+           logscale = ~ fGENDER + fDMF_84 + fGENDER:fDMF_84,
+           data = tandDMF84, info = FALSE)
```

As with the smoothSurvReg *function (see Section 3.3.3), arguments* sdspline, by.knots, dist.range *and* difforder *may be used to choose the different values of the basis standard deviation σ_0, the distance δ between the two consecutive knots (mixture means), the boundary knots μ_{-K} and μ_K and the order of the penalty s than those used by default ($\sigma_0 = 0.2$, $\delta = 0.3$, $\mu_{-K} = -6$, $\mu_K = 6$, $s = 3$). In the remainder of this example, we limit ourselves to the mean-scale model. The results are stored in object* pfit2.

Summary of the model and basic inference

A summary of the fitted model is obtained using:

```
> print(pfit2)
```

```
Call:
smoothSurvReg(formula = Surv(L44, R44, type = "interval2") ~
    fGENDER + fDMF_84 + fGENDER:fDMF_84, logscale = ~fGENDER +
    fDMF_84 + fGENDER:fDMF_84, data = tandDMF84, info = FALSE)

Estimated Regression Coefficients:
                                    Value Std.Error
(Intercept)                       2.37535  0.009155
fGENDERgirl                      -0.03854  0.014416
fDMF_84caries                    -0.07026  0.017672
fGENDERgirl:fDMF_84caries         0.04235  0.027382
LScale.(Intercept)               -2.33566  0.079738
LScale.fGENDERgirl                0.28617  0.104523
LScale.fDMF_84caries              0.21096  0.126503
LScale.fGENDERgirl:fDMF_84caries -0.21421  0.180688
                                 Std.Error2       Z      Z2
(Intercept)                        0.009092 259.449 261.255
fGENDERgirl                        0.014405  -2.673  -2.675
fDMF_84caries                      0.017637  -3.975  -3.983
fGENDERgirl:fDMF_84caries          0.027361   1.547   1.548
LScale.(Intercept)                 0.076536 -29.292 -30.517
LScale.fGENDERgirl                 0.104405   2.738   2.741
LScale.fDMF_84caries               0.125729   1.668   1.678
LScale.fGENDERgirl:fDMF_84caries   0.180372  -1.186  -1.188
                                          p         p2
(Intercept)                       0.000e+00  0.000e+00
fGENDERgirl                       7.509e-03  7.465e-03
fDMF_84caries                     7.023e-05  6.793e-05
fGENDERgirl:fDMF_84caries         1.220e-01  1.217e-01
LScale.(Intercept)               1.322e-188 1.554e-204
LScale.fGENDERgirl                6.184e-03  6.126e-03
LScale.fDMF_84caries              9.539e-02  9.337e-02
LScale.fGENDERgirl:fDMF_84caries  2.358e-01  2.350e-01

Penalized Loglikelihood and Its Components:
     Log-likelihood: -618.7988
            Penalty: -0.183553
   Penalized Log-likelihood: -618.9824
```

(Cont.)

```
Degree of smoothing:
   Number of parameters: 46
                   Mean parameters: 4
                  Scale parameters: 4
                 Spline parameters: 38

                   Lambda: 7.389056
              Log(Lambda): 2
                       df: 8.60792

AIC (higher is better):  -627.4067

Number of Newton-Raphson Iterations:  5
n = 496
```

The section "Estimated Regression Coefficients" shows in column Value *estimates of the intercept μ, the regression coefficients $\boldsymbol{\beta}$, and the coefficients γ from the Model (6.16) for the scale parameter $\tau = \sigma$ (rows introduced by* LScale*). Further, columns* Std.Error *and* Std.Error2 *show standard errors derived from the pseudo-covariance (6.19) and the asymptotic covariance (6.20), respectively. Columns* Z, Z2, p, p2 *are the corresponding Wald statistics and P-values. Columns* Value, Std.Error, p *also provided entries corresponding columns of sections* "Mean parameters" *and* "Scale parameters" *of Table 6.5.*

The "Penalized Loglikelihood and Its Components" and "Degree of smoothing" sections show results from penalized maximum likelihood estimation and the selected smoothing parameter λ. We point out that the value of Lambda *in the output is the optimal value of λ/n and similarly,* Log(Lambda) *is the optimal value of $\log(\lambda/n)$. The row* df *then provides the effective number of parameters $\nu(\lambda)$. Finally, we note that now the AIC reported in the row* AIC (higher is better) *is based on an alternative definition, namely,*

$$\mathsf{AIC}_{alter}(\lambda) = \ell\{\widehat{\boldsymbol{\theta}}(\lambda)\} - \nu(\lambda).$$

That is, $\mathsf{AIC}(\lambda)$ defined in (6.18) is equal to $-2\,\mathsf{AIC}_{alter}(\lambda)$. To obtain the AIC (and BIC) according to definition (6.18) we may use the following code:

```
> n <- nrow(pfit2$y)
> AIC2 <- -2 * pfit2$aic
> BIC2 <- -2 * pfit2$loglik[1, "Log Likelihood"]
+         + log(n) * pfit2$degree.smooth[1, "df"]
> print(c(AIC = AIC2, BIC = BIC2))
```

```
     AIC      BIC
1254.813 1291.023
```

The values of $\mathsf{AIC}_{alter}(\lambda)$ *together with the corresponding values of the effective degrees of freedom* $\nu(\lambda)$ *and the maximized penalized log-likelihood* $\ell_P\{\widehat{\boldsymbol{\theta}}(\lambda)\}$ *and log-likelihood* $\ell\{\widehat{\boldsymbol{\theta}}(\lambda)\}$ *are shown for all values of* λ *in the grid search using (only a part of the output shown):*

```
> pfit2$searched
```

```
     Lambda Log(Lambda)       AIC       df PenalLogLik
1 7.3890561           2 -627.4067 8.607920   -618.9824
2 2.7182818           1 -627.6135 9.004831   -618.7929
3 1.0000000           0 -627.9611 9.488698   -618.6250
4 0.3678794          -1 -628.2247 9.914535   -618.4733
...
     LogLik nOfParm fail
1 -618.7988      46    0
2 -618.6087      46    0
3 -618.4724      46    0
4 -618.3102      46    0
...
```

Note that the columns Lambda *and* Log(Lambda) *provide the values of* λ/n *and* $\log(\lambda/n)$*, respectively.*

Estimates of the weights $\boldsymbol{w}$ *together with their standard errors can be obtained with the following code (output not shown):*

```
> print(pfit2, spline = TRUE)
```

The Wald confidence intervals for the model parameters shown in Table 6.5 are calculated using:

```
> confint(pfit2, level = 0.95)
```

```
                                   2.5 %      97.5 %
(Intercept)                   2.35740987  2.39329829
fGENDERgirl                  -0.06679304 -0.01028464
fDMF_84caries                -0.10489213 -0.03561875
fGENDERgirl:fDMF_84caries    -0.01132003  0.09601361
LScale.(Intercept)           -2.49194112 -2.17937456
LScale.fGENDERgirl            0.08131014  0.49103233
```

(Cont.)

```
LScale.fDMF_84caries              -0.03698448  0.45889708
LScale.fGENDERgirl:fDMF_84caries -0.56835152  0.13993130
```

The error distribution

The fitted error density can be produced with the following command (see Figure 6.5):

```
> plot(pfit2, resid = FALSE)
```

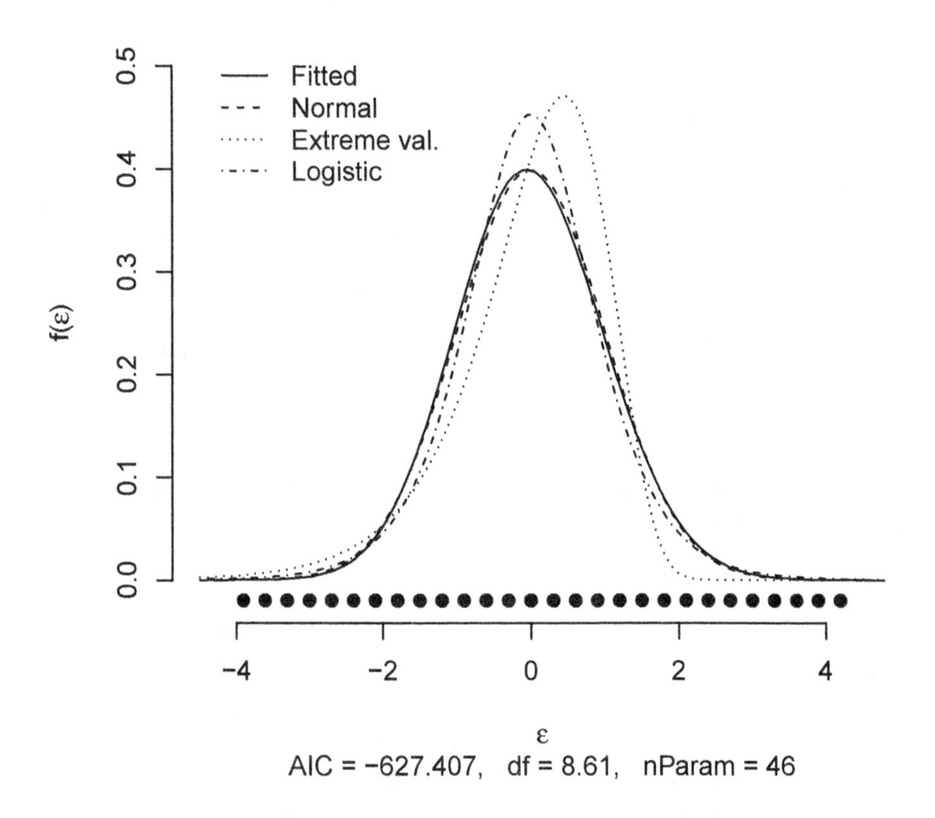

FIGURE 6.5: Signal Tandmobiel study. Mean-scale AFT model for emergence of tooth 44 with a PGM error distribution. Estimated density of the standardized error term compared to parametric densities, obtained with R function smoothSurvReg.

The figure shows the estimated density of the standardized error term $\varepsilon^\star$ of the mean-scale model and compares its fit to that of three popular densities for AFT models. The code for a similar plot (Figure 6.3) comparing the fitted densities of the mean model, the mean-scale model and the density of a standard normal distribution is also available in the supplementary materials.

Comparison of models

The Wald test of the null hypothesis $\gamma_1 = \gamma_2 = \gamma_3 = 0$ to compare the improvement in fit of the mean-scale model over the mean model can be performed using the code below. It takes the estimated values of coefficients γ_1, γ_2, γ_3 stored in the regres *component of the* pfit2 *object, a proper submatrix of the pseudo-covariance matrix (6.19) stored in the* var *component of the* pfit2 *object and calculates in a standard way the Wald test statistic (1.17) and the corresponding P-value.*

```
> coefName <- paste("LScale.",
+       c("fGENDERgirl", "fDMF_84caries", "fGENDERgirl:fDMF_84caries"),
+       sep = "")
> gamma <- pfit2$regres[coefName, "Value"]
> Vgamma <- pfit2$var[coefName, coefName]
> X2 <- as.numeric(t(gamma) %*% solve(Vgamma, gamma))
> pval <- pchisq(X2, df = 3, lower.tail = FALSE)
> WT1 <- c(Statistic = X2, Pvalue = pval)
> print(WT1)
```

```
 Statistic      Pvalue
8.71261073 0.03336628
```

Additional inference

The Wald test to possibly simplify the mean-scale model by dropping the DMF and Gender:DMF covariates from the scale part of the model is obtained analogously using the following code:

```
> coefName <- paste("LScale.",
+       c("fDMF_84caries", "fGENDERgirl:fDMF_84caries"), sep = "")
> gamma <- pfit2$regres[coefName, "Value"]
> Vgamma <- pfit2$var[coefName, coefName]
> X2 <- as.numeric(t(gamma) %*% solve(Vgamma, gamma))
> pval <- pchisq(X2, df = 2, lower.tail = FALSE)
> WT2 <- c(Statistic = X2, Pvalue = pval)
> print(WT2)
```

```
Statistic    Pvalue
 2.781898  0.248839
```

The estimated acceleration factors due to decayed primary predecessors reported in Table 6.5, SEs, 95% CIs and P-values are calculated as follows. First, we extract from the object pfit2 a vector of estimated values of parameters μ, β_1, β_2, β_3 and the proper submatrix of the pseudo-covariance matrix (6.19):

```
> coefName <- c("(Intercept)", "fGENDERgirl", "fDMF_84caries",
+               "fGENDERgirl:fDMF_84caries")
> beta2 <- pfit2$regres[coefName, "Value"]
> names(beta2) <- coefName
> V2 <- pfit2$var[coefName, coefName]
```

Secondly, we define a matrix with coefficients of the linear combinations leading to the calculated acceleration factors for boys and girls, respectively.

```
> L <- matrix(c(0, 0, 1, 0,  0, 0, 1, 1), nrow = 2, byrow = TRUE)
> rownames(L) <- c("Boys", "Girls")
> colnames(L) <- coefName
> print(L)
```

```
      (Intercept) fGENDERgirl fDMF_84caries fGENDERgirl:fDMF_84caries
Boys            0           0             1                         0
Girls           0           0             1                         1
```

Thirdly, we calculate the logarithms of the estimated acceleration factors which are the linear combinations of the estimated regression coefficients, their standard errors and related P-values.

```
> logAF2 <- as.numeric(L %*% beta2)
> names(logAF2) <- rownames(L)
> print(logAF2)
```

```
       Boys       Girls
-0.07025544 -0.02790866
```

```
> print(SE.logAF2 <- sqrt(diag(L %*% V2 %*% t(L))))
```

```
      Boys      Girls
0.01767211 0.02092351
```

```
> P.AF2 <- 2 * pnorm(-abs(logAF2 / SE.logAF2))
```

Finally, we compute the estimates of the acceleration factors, SEs using the delta method and 95% CIs.

```
> SE.AF2 <- exp(logAF2) * SE.logAF2
> qq <- qnorm(0.975)
> AF2 <- data.frame(Value = exp(logAF2),
+                   Std.Error = SE.AF2,
+                   Lower = exp(logAF2 - qq * SE.logAF2),
+                   Upper = exp(logAF2 + qq * SE.logAF2),
+                   P.value = P.AF2)
> print(AF2)
```

```
          Value  Std.Error     Lower     Upper      P.value
Boys  0.9321557 0.01647315 0.9004216 0.9650081 7.023174e-05
Girls 0.9724772 0.02034763 0.9334032 1.0131869 1.822556e-01
```

Model-based survival distributions

To calculate the model-based estimates of the survival distributions for specific combinations of covariates, we first specify a matrix Covars *containing the requested covariate combinations in its rows.*

```
> x.label <- paste(rep(c("Boy", "Girl"), each = 2), ", ",
+                  rep(c("DMF 0", "DMF 1"), 2), sep = "")
> Covars <- data.frame(GENDER = c(0, 0, 1, 1), DMF_84 = c(0, 1, 0, 1))
> Covars <- transform(Covars, GENDER.DMF = GENDER * DMF_84)
> Covars <- as.matrix(Covars)
> rownames(Covars) <- x.label
> print(Covars)
```

```
            GENDER DMF_84 GENDER.DMF
Boy, DMF 0       0      0          0
Boy, DMF 1       0      1          0
Girl, DMF 0      1      0          0
Girl, DMF 1      1      1          1
```

Estimated survival functions (all covariate combinations in one plot, cfr. upper panel of Figure 6.6) are obtained using the following code.

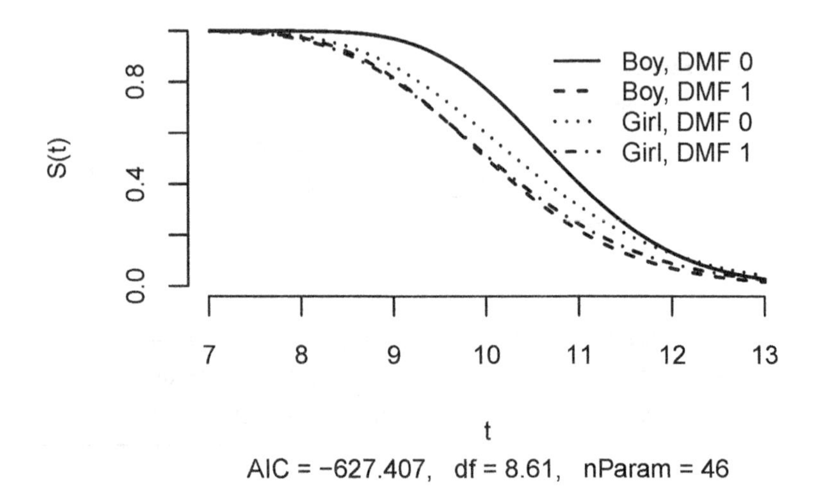

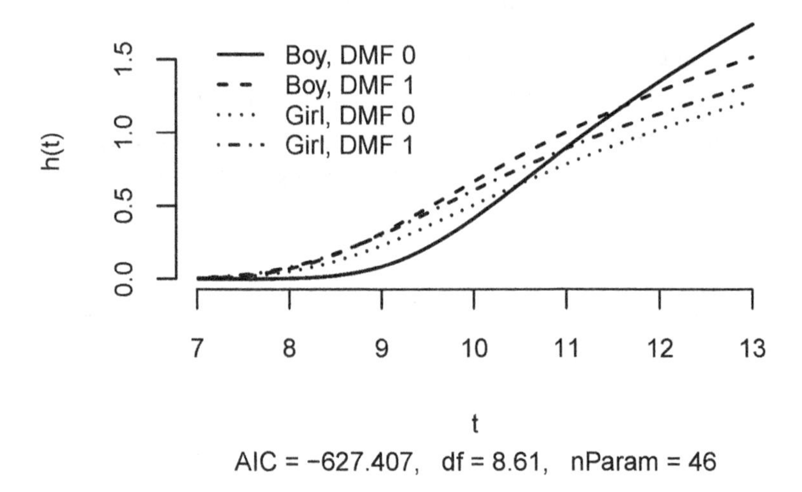

FIGURE 6.6: Signal Tandmobiel study. Estimated survival (upper panel) and hazard (lower panel) functions based on the mean-scale PGM AFT model for emergence of permanent tooth 44, obtained with R package smoothSurv.

```
> S2 <- survfit(pfit2, cov = Covars, logscale.cov = Covars,
+        plot = TRUE, xlim = c(7, 13), by = 0.1, legend = x.label)
```

Analogously, the estimated hazard functions are calculated and plotted (cfr. lower panel of Figure 6.6) using:

```
> h2 <- hazard(pfit2, cov = Covars, logscale.cov = Covars,
+        plot = TRUE, xlim = c(7, 13), by = 0.1, legend = x.label)
```

Also the survival densities can be obtained (plot not shown) using:

```
> f2 <- fdensity(pfit2, cov = Covars, logscale.cov = Covars,
+            plot = TRUE, xlim = c(7, 13), by = 0.1, legend = x.label)
```

Coordinates of the calculated curves stored in objects S2, h2 *and* f2 *can also be used to draw the plots in other formats. For example, the solid curves in Figure 6.4 (estimated survival functions based on the mean-scale model) were drawn using the following code:*

```
> par(mfrow = c(2, 2))
> for (i in 1:4){
+   plot(S2[, "x"], S2[, paste("y", i, sep = "")],
+          type = "l", col = "black",
+          xlab = "Age (years)", ylab = expression(hat(S)(t)),
+          xlim = c(7, 13), ylim = c(0, 1))
+   title(main = x.label[i])
+ }
```

Finally, we show how to calculate the estimated mean emergence times for specific covariate combinations including their standard errors and CIs reported in Table 6.6. To this end, we may use the estimTdiff *function which serves two purposes: (i) inference on the differences in the mean survival times, (ii) inference on the mean survival times themselves which is here of interest. First, we define a matrix with the coefficients of the linear combinations of the parameter vectors* $(\mu, \beta_1, \beta_2, \beta_3)'$ *and* $(\gamma_0, \gamma_1, \gamma_2, \gamma_3)'$ *in the considered AFT model that leads to the expression of the emergence times of boys and girls with DMF* $= 0$*:*

```
> Cov1 <- as.matrix(data.frame(GENDER = c(0, 1),
+              DMF = c(0, 0), GENDER.DMF = c(0, 0)))
> rownames(Cov1) <- c("Boy:DMF=0", "Girl:DMF=0")
> print(Cov1)
```

```
            GENDER DMF GENDER.DMF
Boy:DMF=0        0   0          0
Girl:DMF=0       1   0          0
```

Analogously we define a matrix with the linear combination coefficients leading to the emergence times of boys and girls with DMF = 1:

```
> Cov2 <- as.matrix(data.frame(GENDER = c(0, 1),
+               DMF = c(1, 1), GENDER.DMF = c(0, 1)))
> rownames(Cov2) <- c("Boy:DMF=1", "Girl:DMF=1")
> print(Cov2)
```

```
            GENDER DMF GENDER.DMF
Boy:DMF=1        0   1          0
Girl:DMF=1       1   1          1
```

Finally, we apply the estimTdiff *function where we use the matrix* Cov1 *as its arguments* cov1 *and* logscale.cov1 *and the matrix* Cov2 *as its arguments* cov2 *and* logscale.cov2.

```
> estimTdiff(pfit2, cov1 = Cov1, cov2 = Cov2,
+                  logscale.cov1 = Cov1, logscale.cov2 = Cov2)
```

```
...
Estimates of Expectations
and Their 95% Confidence Intervals
             ET1 Std.Error Lower Upper      Z p
Boy:DMF=0  10.81    0.1032 10.60 11.01 104.67 0
Girl:DMF=0 10.43    0.1246 10.19 10.68  83.78 0

             ET2 Std.Error Lower Upper     Z p
Boy:DMF=1  10.10    0.1592 9.785 10.41 63.42 0
Girl:DMF=1 10.15    0.1886 9.777 10.52 53.80 0

                        E(T1 - T2) Std.Error   Lower  Upper     Z
Boy:DMF=0 - Boy:DMF=1       0.7082    0.1894  0.3369 1.0795 3.738
Girl:DMF=0 - Girl:DMF=1     0.2877    0.2252 -0.1537 0.7292 1.278
                                p
Boy:DMF=0 - Boy:DMF=1   0.0001851
Girl:DMF=0 - Girl:DMF=1 0.2014205
```

Estimated mean emergence times corresponding to the specified covariates combinations are printed in the section "Estimates of Expectations and Their 95% Confidence Intervals." The final part of the output shows the inference on the differences between the mean emergence times corresponding to the covariates combinations specified by matrices Cov1 *and* Cov2.

□

6.3 SemiNonParametric approach

In Section 5.2.2, a PH model with a smooth baseline hazard proposed by Zhang and Davidian (2008) was introduced. The same methodology may also be used to fit an AFT model with a "SemiNonParametric" (SNP) smooth error distribution.

Using the SNP approach, Model (6.3) translates into

$$\log(T_i) = \boldsymbol{X}_i'\boldsymbol{\beta} + \varepsilon_i = \boldsymbol{X}_i'\boldsymbol{\beta} + \mu + \sigma Z_i$$

where ε_i and Z_i are i.i.d., and the density of Z_i may be well-approximated by the normal or exponential SNP formulation, see Section 5.2.2. We assume that $\varepsilon_i = \log(T_{0i})$, where T_0 has survival function $S_0(t)$ and "smooth" density $f_0(t)$ that may be approximated by (5.4) or (5.5). For fixed K, this leads to approximations to the conditional survival and density functions of T given $\boldsymbol{X}$, $S(t \mid \boldsymbol{X};\boldsymbol{\beta})$ and $f(t \mid \boldsymbol{X};\boldsymbol{\beta})$, given by

$$\begin{aligned} S_K(t \mid \boldsymbol{X};\boldsymbol{\beta},\boldsymbol{\theta}) &= S_{0,K}\big(t\mathrm{e}^{\boldsymbol{X}'\boldsymbol{\beta}};\, \boldsymbol{\theta}\big), \\ f_K(t \mid \boldsymbol{X};\boldsymbol{\beta},\boldsymbol{\theta}) &= \mathrm{e}^{\boldsymbol{X}'\boldsymbol{\beta}} f_{0,K}\big(t\mathrm{e}^{\boldsymbol{X}'\boldsymbol{\beta}};\, \boldsymbol{\theta}\big). \end{aligned}$$

Zhang and Davidian (2008) propose to select adaptively the K-base density combination based on the HQ criterion (see Section 1.5.2) over all combinations of base density (normal, exponential) and $K = 0, 1, \ldots, K_{max}$.

Combining the SNP approach of Zhang and Davidian (2008) and the order selection test of Aerts et al. (1999), Nysen et al. (2012) proposed a goodness of fit test for (interval) censored data. The underlying idea is to accept a given parametric model if and only if it is chosen by a model selection criterion, namely a modified AIC. Nysen et al. (2012) adopted the available SAS macros from Zhang and Davidian (2008) to test a given model. Unfortunately, covariates are not yet allowed in the model and also the log-normal model is the only available distribution that may be tested at this moment. Therefore, we do not discuss the macro but provide it in the supplementary materials.

Example 6.3 Breast cancer study
Compare again the two treatments for the early breast cancer study, but

now assume that the error of the AFT model can be well approximated by either (5.4) or (5.5) for a fixed K. The best fitting model according to the HQ criterion (HQ = 295.72) is the model with an exponential base function and $K = 0$. The estimate of 0.9164 for the treatment effect (hazard ratio of 2.50, 95% CI: 1.44 – 4.35) is the same as in Example 5.3 because this AFT model is also a PH model.

□

6.3.1 SAS solution

The model can be fitted using the SAS macros kindly provided by Zhang and Davidian. The program is similar to the solution of Example 5.3 with now the option AFT set to 1 in the main macro %SNPmacroint. The full program is available in the supplementary materials.

6.4 Model checking

Similar checks as for the PH model (see Section 5.5) may be performed for the AFT model. A parametric AFT model without covariates or a single categorical covariate can be compared with the NPMLE, as illustrated in Figure 6.2. A parametric AFT model may be compared with a smooth AFT model, as illustrated in Figure 6.5.

The (adjusted) Cox-Snell, Lagakos and deviance residuals defined in Section 5.5 may be displayed in an index plot or versus a covariate included in the model in order to detect outliers and questionable patterns.

6.5 Sample size calculation

6.5.1 Computational approach

The interval-censored character of the data needs to be taken into account when designing a study. Kim et al. (2016) proposed a method for calculating the sample size and power for interval-censored data which assumes that T follows a parametric AFT model. In their approach, the authors assume that the log-transformed event times follow a Weibull distribution. This approach and software implementation can also be used with PH regression models since at the design stage of a study often quite simple parametric assumptions (such as exponential distributions of the survival times) are made. Because the Weibull distribution is an extension of the exponential distribution and a Weibull PH model is equivalent to a Weibull AFT model, the current approach also applies to PH regression models assuming Weibull distributions with interval-censored data.

In the two-group setting, let an indicator variable X define the two groups. The hypothesis of interest is then $\text{H}_0 : \beta = 0$ vs $\text{H}_1 : \beta \neq 0$ and $\exp(\beta)$ represents the hazard ratio between the groups with $X = 1$ and $X = 0$. Further it is assumed that both groups have the same censoring/dropout distribution. For this two-group comparison, the Wald statistic is given by

$$X_W^2 = \frac{\widehat{\beta}^2}{\widehat{\mathsf{var}}(\widehat{\beta})}, \tag{6.22}$$

where $\widehat{\beta}$ is the MLE of β. Under H_0, X_W^2 has asymptotically a (central) χ_1^2-distribution. Under the alternative H_1, X_W^2 is asymptotically distributed as a noncentral $\chi_{1,(\omega)}^2$-distribution with noncentrality parameter ω equal to the RHS of Expression (6.22) with $\widehat{\beta}$ and $\widehat{\mathsf{var}}(\widehat{\beta})$ replaced by the true β and $\mathsf{var}\,(\widehat{\beta})$. Let α denote the specified type I error rate and $\chi_{1,1-\alpha}^2$ represent the $100(1-\alpha)\%$ critical value from a central χ_1^2 distribution, then the power for rejecting H_0 with the Wald test under H_1 is

$$\mathsf{P}\big(X_W^2 > \chi_{1,1-\alpha}^2\big). \tag{6.23}$$

Kim et al. (2016) assumed that there are K equidistant visits in the new study and a uniformly distributed dropout rate. Following Lyles et al. (2007), they created an extended data set which contains for each of the n patients all potential event/censoring interval outcomes ($2 * K + 1$ possibilities) with weights from the prespecified Weibull distribution and censoring distribution in the design of the study. For each individual the weights sum up to 1. Fitting a Weibull model to this extended data set using standard software yields an estimate of $\mathsf{var}\,(\widehat{\beta})$. The power may be calculated using (6.23) for the given sample size n.

Kim et al. (2016) demonstrated in their simulations based on various Weibull models that their method works well assuming the correct distribution and performing the generalized log-rank test of Sun et al. (2005).

Example 6.4 A new breast cancer study

Suppose that we wish to design a new breast cancer trial in which patients will be examined every 3 months for one year and every 6 months thereafter until 48 months. Hence, there are 10 planned visits in case no breast retraction occurs or the patient does not drop out. Patients will be allowed to schedule their visit up to 14 days before or after a planned visit. Assume that in the radiotherapy only group the event of interest follows an exponential distribution with at 48 months 60% of the patients having reached the event. A hazard ratio of 2 (primary radiation therapy and adjuvant chemotherapy over radiotherapy only) is found to be clinically relevant. In addition, it is expected that yearly 5% of the patients will drop out of the study.

With 2×52 patients, the study will have 80% power to show a significant treatment effect at 0.05. A power of 90% is obtained when the study is based on 2×69 patients.

□

6.5.2 SAS solution

The method of Kim et al. (2016) is implemented in a SAS macro %SampleSizeIC which can be found in the supplementary materials. In contrast to Kim et al. (2016), it does not allow for uniformly distributed deviations for the first visit only, but rather allows for triangularly distributed deviations for all visits such that more weight is put on visits closely to the planned visits. In addition, different distances between the visits are allowed. A detailed description of all options is given in the header of the macro.

Example 6.4 in SAS
The macro calculates the power for a given sample size which is provided in the argument n. *By default, it assumes that both groups have the same size. For trials with different group sizes, the group sizes may alternatively be entered in arguments* n1 *and* n2. *In search of the required sample size with a given power, the macro allows to provide a minimal total sample size to be verified in the argument* nmin. *By an even number of steps determined by the argument* nstep *(=2, by default), the macro will examine all sample sizes from* n *to* nmin. *The results are written to the log and stored in a data set determined by the argument out. The total number of planned visits should be specified in the argument* nvisits. *The distance between visits is given in the argument* vspan. *If not all distances are equal, a list must be provided. If deviations from the planned visits are allowed, the argument* vdeviation *determines the total width. As we allowed 14 days or 0.5 month before or after the planned visit, the total width is 1 month. By default, the same amount is allowed before and after the visit. However, specifying argument* devratio *(=0.5, by default) allows to change this. As we assumed an exponentially distributed survival curve for one group, the shape argument of the Weibull distribution is set to 1. The specific distribution is determined by specifying the percentage of survivors in argument* psurvt *at the time point specified in argument* t. *Here, 40% survivors at 48 months. The hazard ratio is specified in the argument* hr. *The percentage dropout (uniformly distributed) at the end of the study, not necessarily equal to the argument* t, *is specified in the argument* dropout. *With a 4-year study and a 5% yearly dropout, 20% of the patient will be dropped out at 48 months. It is possible to provide a probability that a specific visit is missed (equal for all visits) in the argument* pmissing *but we assumed that patients did not miss a visit. The code below shows the macro call.*

```
%SampleSizeIC(n=110, nmin=100, nstep=2,
```

```
nvisits=10, vspan=3 3 3 3 6 6 6 6 6 6, vdeviation=1,
pmissing=0, dropout=0.2, hr=2, shape=1, psurvt=0.4, t=48, seed=123);
```

In the log, we observed the following results. The study will have at least 80% power if 104 patients are randomized in a 1:1 ratio to both groups. A similar call would show that if the number of patients randomized increases to 138, 90% power would have been obtained.

```
Total sample size=110 (n1=55, n2=55)  power=82.63%
Total sample size=108 (n1=54, n2=54)  power=81.94%
Total sample size=106 (n1=53, n2=53)  power=81.23%
Total sample size=104 (n1=52, n2=52)  power=80.49%
Total sample size=102 (n1=51, n2=51)  power=79.73%
Total sample size=100 (n1=50, n2=50)  power=78.95%
```

□

6.6 Concluding remarks

Much less attention has been devoted in the statistical literature to the development of flexible AFT regression models for interval-censored data. The reason is clear. Namely the PH regression model still takes a central position which it inherited from its major role for right-censored data. This section, has focused therefore only on two AFT regression models: a parametric model and a flexible model based on a mixture of Gaussian densities. Software is available for both approaches to derive all required summary measures and predictions. If the parametric distribution for the error term of the AFT model is known, the model may be easily fitted with the R function survreg or SAS procedures LIFEREG or RELIABILITY. Otherwise, one can fit an AFT model with a penalized Gaussian mixture as error distribution using the R function smoothSurvReg.

We ended this chapter with the illustration of an approach to compute the necessary sample size for a newly designed study involving interval-censored observations. Although developed for an AFT model, we argued that it can also be used for studies using the PH regression model.

Using the same small simulation study introduced in Section 4.3, we also fitted a Weibull AFT model in all settings. For the equal and unequal censoring scheme, the type I error rate was again under control. In case of a hazard ratio of 1.05 and under the equal censoring scheme, there was up to 12% drop in power for interval-censored responses compared to the analysis with the true event times. However, for the unequal censoring scheme the mid-point imputation had 0% power, while a 24% drop in power was seen with the proper interval-censored analysis in comparison to the analysis making use of the true event times.

Chapter 7

Bivariate survival times

In some applications one is interested in the association of two or more interval-censored responses. In Chapter 1 we have seen that the association is biasedly estimated if the interval-censored character of the responses is ignored. This case must be distinguished from doubly interval-censored survival times where the onset as well as the end of a time period are interval-censored. Such data will be treated in Chapter 8.

Bivariate interval-censored data consist of rectangles $R_i = \lfloor l_{1i}, u_{1i} \rfloor \times \lfloor l_{2i}, u_{2i} \rfloor$ $(i = 1, \ldots, n)$ which contain the unobserved times to the events of interest T_{1i} and T_{2i}, respectively. A graphical representation of such observations is given in Figure 7.1. Note that the rectangles can be half-planes, half-lines or points when there are two right-censored observations, one right-censored and one exact observation or two exact observations, respectively.

In this chapter we start with nonparametric frequentist approaches. As will be seen, these approaches are extensions of Peto's and Turnbull's solutions, but are computationally considerably more complex. The pure parametric models are again relatively straightforward to apply, but are often not entirely

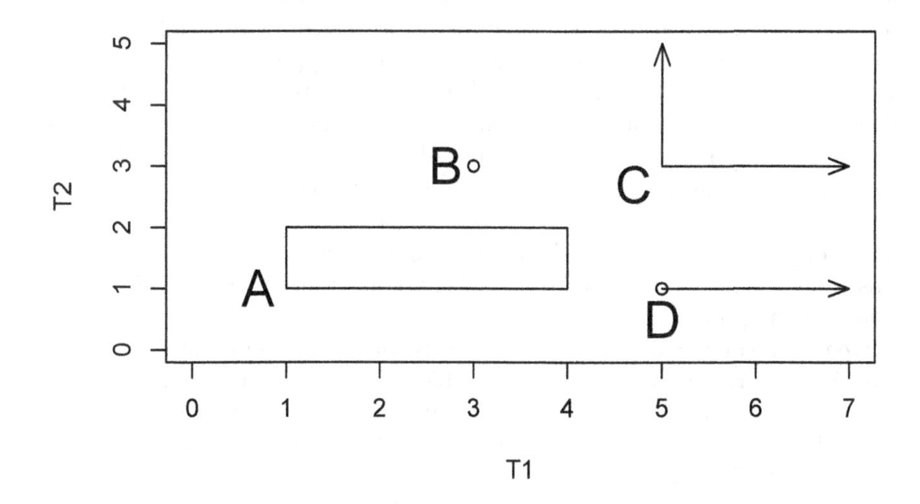

FIGURE 7.1: Graphical representation of 4 bivariate interval-censored observations. A: $(1, 4] \times (1, 2]$, B: 3×3, C: $(5, \infty) \times (3, \infty)$, D: $5 \times (1, \infty)$.

satisfactory since they lack flexibility. The flexible bivariate survival models are a compromise between the nonparametric and parametric approaches and are extensions of the smoothing methods seen in Section 6.2. Yet another approach is to define the bivariate survival model via a copula function which describes the association between the two survival times and to combine it with (flexible) marginal survival distributions. At the end of the chapter, we also look at methods to estimate bivariate association measures which can be used to check the goodness of fit of the proposed copula models.

7.1 Nonparametric estimation of the bivariate survival function

As in the univariate case (see Section 3.1), the estimation of the nonparametric maximum likelihood estimator (NPMLE) for bivariate interval-censored survival data can be split up in two steps: finding the regions of possible support and maximizing the likelihood.

7.1.1 The NPMLE of a bivariate survival function

Finding the regions of possible support

Betensky and Finkelstein (1999b) generalized Peto's and Turnbull's argument to bivariate interval-censored data. That is, information on the bivariate nonparametric survival function is limited to a number of rectangles bearing (possibly) non-zero mass. The regions of possible support are the rectangles that are the nonempty intersections of the observed rectangles such that no other intersection is contained within, or the observed rectangle itself if it has no intersection with any of the other observed rectangles. Formally, B is a region of possible support if $B = \bigcap_{\forall R_j : R_j \cap B \neq \emptyset} R_j$. Note that the number of regions of possible support can be larger than the original number of observations. This is illustrated in Figure 7.2 where 5 observations result in 6 regions of possible support.

We use the artificial example in Betensky and Finkelstein (1999b) to illustrate the construction of regions of possible support in Figure 7.3. Note that observation 6 reduces to a one-dimensional interval because the event in the second dimension is exactly observed ($[7, 8] \times [3, 3]$). The 4 regions of possible support are indicated in gray.

Betensky and Finkelstein (1999b) proposed a simple algorithm to calculate the regions of possible support. The first step in the search process consists in making pairwise intersections of all observed rectangles and keeping the (nonempty) intersections or the rectangle itself (if there is no intersection

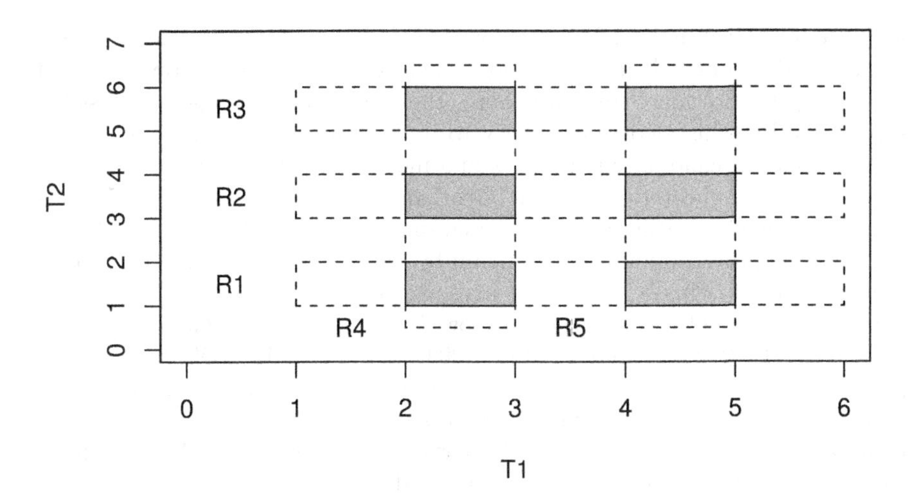

FIGURE 7.2: Artificial example with 6 regions of possible support indicated in gray from 5 observations ($R_i, i = 1, \ldots, 5$).

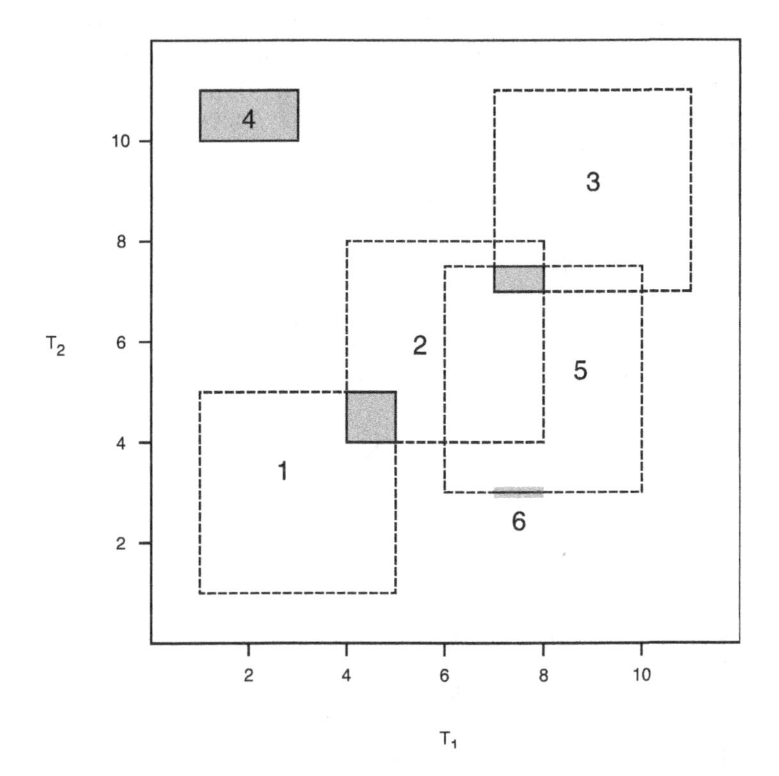

FIGURE 7.3: An artificial data set with 6 observed rectangles and their corresponding 4 regions of support indicated in gray.

with the other rectangles). This step produces a new set of rectangles, and one can again determine their intersections. The procedure is then iterated until the rectangles do not change anymore, these are the rectangles of possible support. While this algorithm is easy to implement, there may be an overly large number of candidate rectangles in a moderately sized data set.

Using graph theoretic concepts, Gentleman and Vandal (2001) proposed a more efficient algorithm based on a marginal approach. Bogaerts and Lesaffre (2004) improved upon this algorithm by directly searching for the regions of possible mass. Finally, the height map algorithm given by Maathuis (2005) is the most efficient algorithm up to now. The algorithms of Gentleman and Vandal (2001), Bogaerts and Lesaffre (2004) and Maathuis (2005) have a complexity of at most $O(n^5)$, $O(n^3)$ and $O(n^2)$, respectively. The algorithm of Maathuis (2005) has memory requirements of $O(n^2)$, compared to a requirement of $O(n^3)$ for the algorithm of Bogaerts and Lesaffre (2004). Details about the three algorithms can be found in Appendix F.2.

Estimating the NPMLE of a bivariate survival function

Finding the regions of possible support is only the first part in determining the NPMLE of a bivariate survival function. Once the regions of possible support are determined, the NPMLE can be estimated by constrained maximization of the likelihood. Let $\mathcal{B} = \{B_j = \lfloor p_{1j}, q_{1j} \rfloor \times \lfloor p_{2j}, q_{2j} \rfloor, j = 1, \ldots, m\}$ denote the disjoint rectangles that constitute the regions of possible support of the NPMLE of F. Define

$$\alpha_{ij} = I\big(B_j \subset \lfloor l_{1i}, u_{1i} \rfloor \times \lfloor l_{2i}, u_{2i} \rfloor\big),$$

with $I(\cdot)$ the indicator function and

$$s_j = F(B_j) = F(q_{1j}, q_{2j}) - F(p_{1j}, q_{2j}) - F(q_{1j}, p_{2j}) + F(p_{1j}, p_{2j}),$$

$(i = 1, \ldots, n; j = 1, \ldots, m)$. Then the likelihood function $L(F)$ can be written as

$$L(\boldsymbol{s}) = \prod_{i=1}^{n} \sum_{j=1}^{m} \alpha_{ij} s_j$$

with $\boldsymbol{s} = (s_1, \ldots, s_m)'$, and the NPMLE of F is determined by maximizing $L(\boldsymbol{s})$ over $\boldsymbol{s}$ subject to $s_j \geq 0$ $(j = 1, \ldots, m)$ and $\sum_{j=1}^{m} s_j = 1$.

Gentleman and Vandal (2001) discuss various algorithms, using either convex optimization or viewing it as mixture problem. An EM-algorithm was suggested by Dempster et al. (1977), while Böhning et al. (1996) suggested the vertex exchange method, which is an example of mixture estimation.

Properties of the bivariate NPMLE

In addition to representational non-uniqueness, see Section 3.1, there may be now also *mixture non-uniqueness*, pointed out by Gentleman and Vandal

(2002). This non-identifiability problem occurs only in the bivariate case and happens when there exists at least 2 different candidate solutions $\widehat{\boldsymbol{s}}_1$ and $\widehat{\boldsymbol{s}}_2$ which yield the same maximum likelihood value. Gentleman and Geyer (1994) suggested a sufficient condition for the uniqueness of the solution. Namely, the NPMLE is unique if the rank of the $(n \times m)$ *cliques matrix* $\boldsymbol{A} = (\alpha_{ij})$ is equal to m. This guarantees that the log-likelihood function is strictly concave and thus has a unique maximum. This is, however, not a necessary condition since the NPMLE may still be unique even when $\boldsymbol{A}$ has not full rank, which in practice often occurs. In addition, Gentleman and Geyer (1994) proved that a simpler sufficient condition for the uniqueness of the NPMLE may be based on $\boldsymbol{A}^*$, the submatrix of $\boldsymbol{A}$ that consists of all columns of $\boldsymbol{A}$ with $s_j > 0$. More specifically, uniqueness occurs when the rank of $\boldsymbol{A}^*$ is equal to its number of columns.

For the bivariate case, several consistency results have been shown under different settings (e.g., van der Vaart and Wellner (2000), Wong and Yu (1999), Schick and Yu (2000)), with the most general result developed by Yu et al. (2006).

With a unique NPMLE a goodness of fit test can be developed for a parametric model for the bivariate failure times by comparing the parametric with the nonparametric estimate.

Example 7.1 AIDS clinical trial

As an illustration, we take the AIDS Clinical Trial ACTG 181 where interest lies in examining the bivariate distribution of shedding of CMV in the urine and blood, and colonization of MAC in the sputum or stool. The 204 bivariate censored observations lead to 32 regions of possible support and to 13 regions of positive mass. Table 7.1 displays all regions of possible support with the assigned mass. Because the rank of the cliques submatrix $\boldsymbol{A}^*$ has maximal rank 13, the NPMLE is unique.

□

7.1.2 R solution

The bivariate NPMLE for a given bivariate distribution may be estimated using the function computeMLE from the package MLEcens (Maathuis, 2013). The package uses the height map algorithm of Maathuis (2005) and a combination of sequential quadratic programming and the support reduction algorithm of Groeneboom et al. (2008) to maximize the likelihood.

Example 7.1 in R

First, we load the data and create a matrix mat *containing the left and right endpoints of shedding of CMV and colonization of MAC. Missing values are not allowed. Hence, we replace the left endpoint by zero for left-censored observations and the right endpoint by a sufficiently high (say 99) value for*

TABLE 7.1: AIDS clinical trial. Thirty-two possible regions of support and the assigned mass s_j ($j = 1, \ldots, 32$) for the NPMLE of the bivariate distribution of shedding of CMV and colonization of MAC, obtained with R package MLEcens.

j	$[p_{1j}, q_{1j}] \times [p_{2j}, q_{2j}]$	s_j	j	$[p_{1j}, q_{1j}] \times [p_{2j}, q_{2j}]$	s_j
1	$[0,0]\times[0,0]$	0.014	17	$[6,6]\times[6,6]$	0.015
2	$[3,3]\times[0,0]$	0	18	$[9,9]\times[6,6]$	0
3	$[6,6]\times[0,0]$	0	19	$[12,12]\times[6,6]$	0
4	$[9,9]\times[0,0]$	0	20	$[15,15]\times[6,6]$	0
5	$[12,12]\times[0,0]$	0.005	21	$[18,\infty]\times[6,6]$	0
6	$[15,15]\times[0,0]$	0.042	22	$[9,9]\times[9,9]$	0.010
7	$[18,\infty]\times[0,0]$	0	23	$[3,3]\times[12,12]$	0
8	$[0,0]\times[3,3]$	0	24	$[6,6]\times[12,12]$	0
9	$[3,3]\times[3,3]$	0	25	$[21,\infty]\times[15,15]$	0.044
10	$[6,6]\times[3,3]$	0	26	$[6,6]\times[18,\infty]$	0.063
11	$[9,9]\times[3,3]$	0	27	$[21,\infty]\times[18,\infty]$	0.267
12	$[12,12]\times[3,3]$	0	28	$[0,0]\times[21,\infty]$	0.308
13	$[15,15]\times[3,3]$	0	29	$[3,3]\times[21,\infty]$	0.087
14	$[18,\infty]\times[3,3]$	0	30	$[15,15]\times[21,\infty]$	0.022
15	$[0,0]\times[6,6]$	0	31	$[12,12]\times[24,\infty]$	0.053
16	$[3,3]\times[6,6]$	0	32	$[9,9]\times[27,\infty]$	0.071

right-censored observations. The bivariate cdf is estimated using the following code. The argument B = c(1, 1, 1, 1) *indicates that all the boundaries are closed. Half-open intervals (*B = c(0, 1, 0, 1)*) are the default.*

```
> library("MLEcens")
> npmle <- computeMLE(R = mat, B = c(1, 1, 1, 1))
```

The argument npmle$p *contains the mass assigned to each region of support and* npmle$rects *contains the regions of support. Make sure that* npmle$conv *= 1 (not shown) to assure that the procedure has converged.*

```
> npmle
$p
 [1] 0.013676984 0.307533525 0.087051863 0.014940282 0.062521573
 [6] 0.010009349 0.071073995 0.004836043 0.053334241 0.042456241
[11] 0.021573343 0.044427509 0.266565054

$rects
      [,1] [,2] [,3] [,4]
 [1,]    0    0    0    0
 [2,]    0    0   21   99
 [3,]    3    3   21   99
```

(Cont.)

```
 [4,]   6   6   6   6
 [5,]   6   6  18  99
 [6,]   9   9   9   9
 [7,]   9   9  27  99
 [8,]  12  12   0   0
 [9,]  12  12  24  99
[10,]  15  15   0   0
[11,]  15  15  21  99
[12,]  21  99  15  15
[13,]  21  99  18  99
```

Using the function reduc *we calculate the height map (*hm = TRUE*) and cliques matrix (*cm = TRUE*). The height map can be visualized using the* plotHM *function (result not shown).*

```
> res <- reduc(mat, B = c(1, 1, 1, 1), hm = TRUE, cm = TRUE)
> plotHM(res$hm, mat, main = "Height map")
```

From the cliques matrix, we determine the columns which correspond to the regions of possible support with mass > 0. Because this submatrix has a maximal rank of 13, the solution is unique. The setkey *function from the package* data.table *assists in selecting the right columns of the clique matrix. The function* rankMatrix *from the package* Matrix *calculates the rank of the matrix.*

```
> library("Matrix")
> library("data.table")
> M1 <- setkey(data.table(res$rects))
> M2 <- setkey(data.table(npmle$rects))
> index <- na.omit(M1[M2, which = TRUE])
> subcm <- res$cm[index,]
> rankMatrix(subcm)
> nrow(npmle$rects)
```

The package MLEcens *contains also functions to plot the marginal cdf or survival functions (*plotCDF1*), the bivariate cdf or survival functions (*plotCDF2*), the univariate (*plotDens1*) or bivariate* plotDens2 *density plots (all not illustrated).*

□

7.1.3 SAS solution

The SAS macro %bivnpmle is based on the algorithm of Bogaerts and Lesaffre (2004) to determine the regions of possible support. Additionally, the macro estimates the NPMLE making use of the SAS/OR® procedure NLP.

Example 7.1 in SAS

The macro %bivnpmle assumes standard SAS coding of interval-censored data (see Table 1.7). To internally process the data, the missing values of left- and right-censored observations are by default replaced by −9 999 and 9 999, respectively. The user should verify that these boundaries are appropriate for the data at hand and if not, adapt the arguments min= and max=, respectively. In addition, a default shift=0.000001 value is used to solve the problem of potential ties (see Section 3.1). In our example, no adaptations were needed. Therefore, the NPMLE can be estimated using the following call:

```
%bivnpmle(data=icdata.ACTG181,
          l1=l1, r1=r1, l2=l2, r2=r2,
          l1open=n, r1open=n, l2open=n, r2open=n,
          outsup=support, outnpmle=npmle, noprint=1);
```

The left and right endpoints of the two interval-censored times must be given in the l1=, r1= and l2=, r2= arguments, respectively. By default, the macro assumes that the intervals are half-open. However, the intervals are closed in our example. To indicate whether a boundary is open (l1open=, r1open=, l2open= and r2open=) must be set to n (=no), while for an open boundary y (=yes) must be specified. The regions of possible support and the final NPMLE are saved in data sets defined by the outsup= and the outnpmle= arguments, respectively. Finally, the option noprint=1 suppresses the output of the NLP procedure. To process larger data sets more efficiently, a split= option is available. By default, no splitting is applied. This option allows to split the observations into parts. We recommend to take splitting smaller than the number of regions of possible support for the univariate NPMLE for one of the two dimensions, but it would be too technical to go deeper into the details of this option. The macro %bivnpmle also saves the full cliques matrix and the submatrix of the cliques matrix consisting of all columns corresponding to those possible regions of support with positive mass. These matrices are called by default cliques and cliquesp, respectively. Using PROC IML, the rank of the submatrix can be calculated using the commands below. Because the matrix has a maximal rank of 13, the found solution is unique.

```
PROC IML;
USE cliquesp;
READ ALL INTO a;
CLOSE cliquesp;
```

```
nsup=NCOL(a);
r=ROUND(TRACE(GINV(a)*a));
PRINT nsup r;
QUIT;
```

□

7.2 Parametric models

7.2.1 Model description and estimation

We now assume a parametric model for the data. We denote the bivariate cdf evaluated in (t_1, t_2) as $F_{1,2}(t_1, t_2) \equiv F(t_1, t_2)$ with $S(t_1, t_2)$ the corresponding bivariate survival function and $F_1(t_1)$ and $F_2(t_2)$ the two marginal cdfs. We further assume that the cdf F (and S) is continuous with density f and is known up to a finite dimensional parameter vector $\boldsymbol{\theta} = (\theta_1, \ldots, \theta_p)'$. To indicate this, we write $F(t) = F(t; \boldsymbol{\theta})$, $S(t) = S(t; \boldsymbol{\theta})$, and $f(t) = f(t; \boldsymbol{\theta})$.

Maximum likelihood will provide us the estimates of the model parameters. These estimates are obtained immediately when the chosen bivariate distribution is supported by the software, otherwise some extra programming is required.

The likelihood for n independent interval-censored event times is given by $L(\boldsymbol{\theta}) = \prod_{i=1}^{n} L_i(\boldsymbol{\theta})$, where $L_i(\boldsymbol{\theta})$ is defined as

$$L_i(\boldsymbol{\theta}) = \oint_{l_{1i}}^{u_{1i}} \oint_{l_{2i}}^{u_{2i}} f(t_1, t_2; \boldsymbol{\theta})\, \mathrm{d}t_1 \mathrm{d}t_2.$$

Another popular approach to analyze bivariate survival models is to make use of the fact that the two survival times are obtained on the same individual and therefore must be correlated. This correlation can be modelled using a frailty term and leads to frailty models, which are in fact random effects models for survival data. A general overview of such models can be found in, e.g., Hougaard (2000), Duchateau and Janssen (2008) and Wienke (2010). In Chapter 8 we cover parametric frailty models in the general multivariate setting. Hence, we defer further discussion of this important class of models to the chapter.

Example 7.2 Signal Tandmobiel study
As an illustration, we determine the bivariate distribution of the emergence times of teeth 34 and 44 (permanent left and right mandibular first premolar, respectively) of boys. We assume that the emergence times are bivariate log-normally distributed. Under these model assumptions, tooth 34 has a geometric mean emergence time of 10.58 year (= exp(2.36)) (95% CI: 10.40 – 10.76).

Tooth 44 has a geometric mean emergence time of 10.51 year (= exp(2.35)) (95% CI: 10.33 – 10.68). The emergence times are highly correlated with an estimated Pearson correlation of 0.75 on the log scale.

□

7.2.2 R solution

To our knowledge, no built-in function is available in R to fit a parametric bivariate model to interval-censored data. One has to specify the likelihood and maximize it via an optimizing function like optim from the R package stats.

Example 7.2 in R

We load the data and select the left and right endpoints of teeth 34 and 44 for the boys. The function bivlognorm *(available in the supplementary materials) calculates the log-likelihood of a bivariate log-normal model using the functions* pnorm *and* pbivnorm *(from the package* pbivnorm*) for the cdf of an univariate normal distribution and bivariate normal distribution, respectively.*

```
> data(tandmob, package = "icensBKL")
> boys <- subset(tandmob, fGENDER == "boy",
+               select = c("L34", "R34", "L44", "R44"))
> library("pbivnorm")
```

Using the optim *function, the log-likelihood (calculated in* fn=bivlognorm*) is maximized via box-constrained maximization (*method="L-BFGS-B"*) constraining the standard deviation to be greater than zero and the correlation between* -1 *and* 1*. The lower and upper boundaries are given in the arguments* lower *and* upper*, respectively. A marginal fit using the* survreg *function (see Section 3.2) provides the starting values for the means and standard deviations, while the starting value for the correlation is based on the mid-points of the interval-censored observations. The Hessian matrix is stored to determine the standard errors afterwards.*

```
> result <- optim(c(2.3577, 2.3519, 0.1140, 0.1136, 0.68972),
+    method = "L-BFGS-B", fn = bivlognorm, hessian = TRUE,
+    L1 = boys$L34, R1 = boys$R34, L2 = boys$L44, R2 = boys$R44,
+    lower = c(-Inf, -Inf, 0, 0, -1), upper = c(Inf, Inf, Inf, Inf, 1))
```

The solution is given below. Exponentiating the first two estimates of result$par *provides the estimated geometric mean emergence times.*

```
> print(result$par)
```

```
[1] 2.3561177 2.3507655 0.1164397 0.1127800 0.7577435
```

```
> invH <- solve(result$hessian)
> se <- sqrt(diag(invH))
> print(se)
```

```
[1] 0.008642565 0.008378444 0.006778075 0.006593319 0.035608868
```

□

7.2.3 SAS solution

For a standard parametric model, i.e., covered by a SAS built-in base function, the procedure NLMIXED can be used to obtain the MLE. PROC IML is also an option making use of the QUAD function (performs integration of a function over an interval) in combination with maximization procedures (not shown).

Example 7.2 in SAS
For a bivariate log-normal distribution the PROBBNRM *and* PROBNORM *functions together with the* NLMIXED *procedure provide the MLE.*

```
PROC NLMIXED DATA=icdata.tandmob ;
WHERE gender=0;                    /* select boys */
PARMS mu1=2.3578 mu2=2.3519 sigma1=0.1140 sigma2=0.1136 rho=0.68972;
/* right and interval censored */
     IF (L34^=. and R34=. and L44^=. and R44^=.) THEN
   like =  PROBNORM((log(R44)-mu2)/sigma2)
          -PROBBNRM((log(L34)-mu1)/sigma1,(log(R44)-mu2)/sigma2,rho)
          -PROBNORM((log(L44)-mu2)/sigma2)
          +PROBBNRM((log(L34)-mu1)/sigma1,(log(L44)-mu2)/sigma2,rho);
/* right and left censored */
...
llike=LOG(like);
MODEL IDNR~GENERAL(llike);
RUN;
```

The PARMS *statement defines the unknown parameters and sets the initial values. The starting values for μ_d and σ_d $(d = 1, 2)$ are obtained from the marginal log-normal survival models (using* PROC LIFEREG*, see Section*

3.2.5). The starting value for ρ is based on the Pearson correlation (using PROC CORR*) on the mid-point imputed interval-censored observations.*

The next lines specify the log-likelihood function by calculating the contribution to the likelihood for each possible combination of censoring type (left, right or interval), see supplementary materials for details. At the end the logarithm is taken. The model statement uses the general likelihood function as distribution (see Section 5.1.3). Recall that it does not matter which dependent variable (here IDNR*) you use in the model statement, only it cannot contain missing values.*

Below part of the output is shown.

Parameter Estimates

Parameter	Estimate	Standard Error	DF	t Value	Pr > |t|	Alpha
mu1	2.3587	0.008551	256	275.85	<.0001	0.05
mu2	2.3516	0.008344	256	281.82	<.0001	0.05
sigma1	0.1150	0.007009	256	16.41	<.0001	0.05
sigma2	0.1131	0.006809	256	16.61	<.0001	0.05
rho	0.7575	0.03602	256	21.03	<.0001	0.05

□

7.3 Copula models

7.3.1 Background

The term "copula" originates from Latin, and refers to "connecting" or "joining together." In a statistical context, "copulas" refer to the way in which random variables relate to each other. A copula is a function that joins a multivariate probability distribution to a collection of univariate marginal probability functions. The marginal distribution describes the way in which a random variable acts "on its own," while the copula function describes how they "come together" to determine the multivariate distribution. Copulas extract the dependence structure from the joint distribution function, and so "separate out" the dependence structure from the marginal distribution functions.

According to Nelsen (1998), copulas can be seen from two angles: "From one point of view, copulas are functions that join or couple multivariate distribution functions to their one-dimensional marginal distribution functions. Alternatively, copulas are multivariate distribution functions whose one-dimensional margins are uniform on the interval (0,1)." In general, a copula can be $d\,(\geq 2)$-dimensional but we will limit ourselves here to the 2-dimensional case.

Wang and Ding (2000) use copulas to fit case I bivariate interval-censored data. Sun et al. (2006) and Bogaerts and Lesaffre (2008b) treat case II and arbitrary bivariate interval-censored data (see Section 1.2.3), respectively.

Let the range of a function F be denoted as $RanF$. Formally, a 2-dimensional *copula function* (or briefly a copula) is defined as a function C with $[0,1] \times [0,1]$ as domain and $[0,1]$ as range with the following properties:

1. $C(a,0) = 0 = C(0,b)$ for every $a,b \in [0,1]$;
2. $C(a,1) = a$ and $C(1,b) = b$ for every $a,b \in [0,1]$;
3. For all $(a_1,b_1),(a_2,b_2) \in [0,1] \times [0,1]$ with $a_1 \leq a_2$ and $b_1 \leq b_2$, we have:
$$C(a_2,b_2) - C(a_1,b_2) - C(a_2,b_1) + C(a_1,b_1) \geq 0.$$

Hence any bivariate distribution function with standard uniform distributions is a copula.

The relation between the copula and the bivariate distribution function is shown by Sklar (1959). The well-known *Sklar theorem* states that if F is a bivariate joint distribution function with marginal distributions F_1 and F_2, then there exists a copula C such that for all $x, y \in \mathbb{R}$,

$$F(x,y) = C\big(F_1(x),\, F_2(y)\big). \tag{7.1}$$

If F_1 and F_2 are continuous, then C is unique; otherwise, C is uniquely determined on $RanF_1 \times RanF_2$. Conversely, if C is a copula and F_1 and F_2 are distribution functions, then the function F defined by (7.1) is a joint distribution function with marginal distributions F_1 and F_2.

The copula can also be defined in terms of a survival function S (Nelsen, 1998), then we obtain a *survival copula*. Let T_1 and T_2 have marginal survival functions $S_1(t)$ and $S_2(t)$, respectively, and joint survival function $S(t_1,t_2) = P(T_1 > t_1,\, T_2 > t_2)$. The survival copula is given by

$$S(t_1,t_2) = \breve{C}_\alpha\big(S_1(t_1),\, S_2(t_2)\big), \tag{7.2}$$

where $\breve{C}_\alpha$ is a specific copula with parameter (vector) α which regulates the association between T_1 and T_2. The copula C of F and its associated survival copula $\breve{C}$ are related according to $\breve{C}(a,b) = a + b + C(1-a,\, 1-b) - 1$.

Some well-known copulas that are used further in this book are the *Clayton copula*, the *normal copula* and the *Plackett copula*.

The *Clayton model* (Clayton, 1978) assumes a constant local cross-ratio (see Section 7.5.1), evaluating the degree of dependence at a single time point, defined in general as

$$\theta_L(t_1,t_2) \;=\; S(t_1,t_2) \cdot \frac{\partial^2 S(t_1,t_2)}{\partial t_1 \partial t_2} \Big/ \left\{ \frac{\partial S(t_1,t_2)}{\partial t_1} \cdot \frac{\partial S(t_1,t_2)}{\partial t_2} \right\}. \tag{7.3}$$

Independence corresponds with $\theta_L = 1$, positive dependence with $\theta_L > 1$ and negative dependence with $\theta_L < 1$. The local cross-ratio has a natural interpretation in conditional hazard rates.

For $\theta_L > 0$ but $\theta_L \neq 1$, the Clayton copula is given by

$$\breve{C}^C_{\theta_L}(a,b) = (a^{1-\theta_L} + b^{1-\theta_L} - 1)^{\frac{1}{1-\theta_L}}.$$

Another parametrization of the Clayton copula is given by

$$\breve{C}^C_{\vartheta}(a,b) = (a^{-\vartheta} + b^{-\vartheta} - 1)^{-\frac{1}{\vartheta}},$$

with $\vartheta = \theta_L - 1$ whereby $\vartheta \in [-1,\infty) \setminus \{0\}$. The Clayton copula is also known as the Kimeldorf-Sampson copula (Kimeldorf and Sampson, 1975) or the Pareto family of copulas (Nelsen, 1998).

Bivariate normally distributed data yield the *normal or Gaussian copula*, given by

$$\breve{C}^G_{\rho}(a,b) = \Phi_\rho\big(\Phi^{-1}(a), \Phi^{-1}(b)\big),$$

where Φ_ρ denotes the standard bivariate normal distribution function with correlation ρ. The normal copula does not have a simple closed form, but can be expressed as an integral. In two dimensions for $|\rho| < 1$ we have that

$$\breve{C}^G_{\rho}(a,b) = \int_{-\infty}^{\Phi^{-1}(a)} \int_{-\infty}^{\Phi^{-1}(b)} \frac{1}{2\pi(1-\rho^2)^{1/2}} \exp\left\{\frac{-(x^2 - 2\rho xy + y^2)}{2(1-\rho^2)}\right\} dxdy.$$

The *Plackett model* (Plackett, 1965) assumes a constant global cross-ratio function (see also Section 7.5.1), defined in general as

$$\theta_G(t_1,t_2) = \frac{S(t_1,\,t_2)\{1 - S_1(t_1) - S_2(t_2) + S(t_1,\,t_2)\}}{\{S_1(t_1) - S(t_1,\,t_2)\}\{S_2(t_2) - S(t_1,\,t_2)\}}. \tag{7.4}$$

Thus, for the global cross-ratio the bivariate space is divided at each location (t_1, t_2) into 4 quadrants producing four quadrant probabilities and $\theta_G(t_1, t_2)$. The Plackett family of copulas is defined for $\theta_G > 0$, $\theta_G \neq 1$ as

$$\breve{C}^P_{\theta_G}(a,b) = \frac{\{1 + (\theta_G - 1)(a+b)\} - \sqrt{\{1 + (\theta_G - 1)(a+b)\}^2 - 4ab\theta_G(\theta_G - 1)}}{2(\theta_G - 1)}$$

and for $\theta_G = 1$,

$$\breve{C}^P_1(a,b) = \lim_{\theta_G \to 1} \breve{C}_{\theta_G}(a,b) = ab.$$

In Figure 7.4 the densities of the three copulas are presented for selected parameter values.

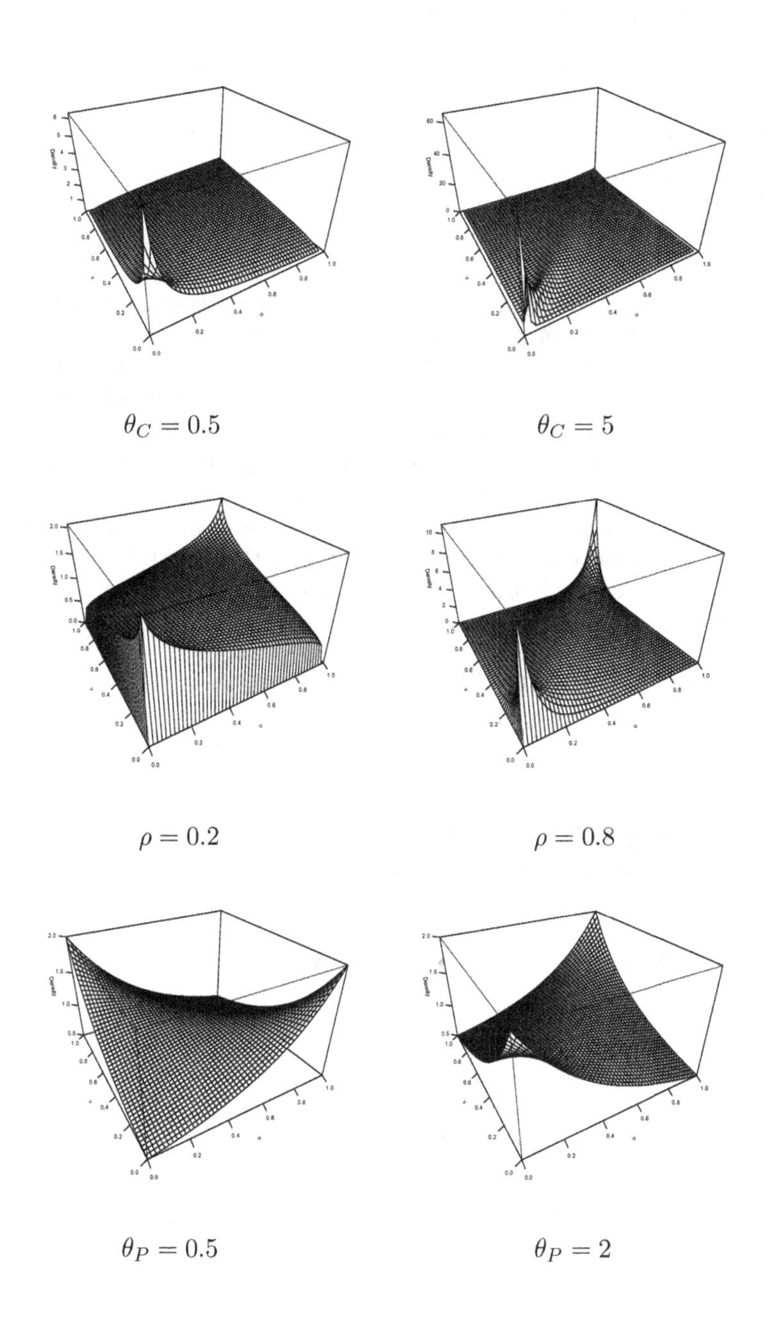

FIGURE 7.4: Density plots for (top) Clayton copula ($\theta_C = 0.5, 5$), (middle) normal copula ($\rho = 0.2, 0.8$) and (bottom) Plackett copula ($\theta_P = 0.5, 2$).

7.3.2 Estimation procedures

As before, the pair of survival times (T_1, T_2) is observed to lie in the rectangle $\lfloor l_1, u_1 \rfloor \times \lfloor l_2, u_2 \rfloor$ with $0 \leq l_j < u_j \leq \infty$ $(j = 1, 2)$. The censoring indicators for a left- and interval-censored T_j $(j = 1, 2)$ are $\delta_j^{(1)}$ and $\delta_j^{(2)}$, producing the vector $\boldsymbol{\delta} = (\delta_1^{(1)}, \delta_1^{(2)}, \delta_2^{(1)}, \delta_2^{(2)})'$. Again, we assume that (T_1, T_2) are independent of (L_1, U_1, L_2, U_2), but of course (L_1, U_1) and (L_2, U_2) may be dependent. The sample of n i.i.d. bivariate intervals $\lfloor l_{1i}, u_{1i} \rfloor \times \lfloor l_{2i}, u_{2i} \rfloor$ $(i = 1, \ldots, n)$ is denoted below as $\boldsymbol{\mathcal{D}}$.

Wang and Ding (2000) and Sun et al. (2006) investigated parameter estimation for the Clayton copula when the bivariate survival model is expressed as (7.2). The three above-described copulas were investigated by Bogaerts and Lesaffre (2008b). We now describe their approach which allows the dependence parameter α to depend on covariates. For the Clayton and Plackett copula, the dependence parameter (θ_L and θ_G, respectively) is modelled on the log-scale, i.e., $\log(\alpha_i) = \boldsymbol{\gamma}' \boldsymbol{Z}_i$ where $\boldsymbol{\gamma} = (\gamma_0, \ldots, \gamma_{p-1})'$ is a p-dimensional vector of unknown regression parameters and $\boldsymbol{Z}_i = (z_{0i}, \ldots, z_{p-1,i})'$ is a p-dimensional vector of covariates measured on the i-th subject. For the normal copula, the dependence will be modelled modulo a Fisher transformation, i.e., $\frac{1}{2} \log \dfrac{1 + \alpha_i}{1 - \alpha_i} = \boldsymbol{\gamma}' \boldsymbol{Z}_i$. Then under Model (7.2) and for n i.i.d. observations, the log-likelihood function is given by

$$\log L(\boldsymbol{\gamma}, S_1, S_2 \mid \boldsymbol{\mathcal{D}}) = \sum_{i=1}^{n} \ell(\boldsymbol{\gamma}, S_1, S_2 \mid \boldsymbol{X}_i, \boldsymbol{\delta}_i).$$

Each individual likelihood contribution can be written as a sum of 9 different terms depending on whether the observation is left-, interval- or right-censored in both dimensions, i.e.,

$$\begin{aligned}
\ell(\boldsymbol{\gamma}, S_1, S_2 \mid \boldsymbol{X}, \delta) \;=\;& \delta_1^{(1)} \delta_2^{(1)} \log S_{11}(\boldsymbol{\gamma} \mid \boldsymbol{X}) \\
&+ \delta_1^{(1)} \delta_2^{(2)} \log S_{12}(\boldsymbol{\gamma} \mid \boldsymbol{X}) \\
&+ \delta_1^{(1)} (1 - \delta_2^{(1)} - \delta_2^{(2)}) \log S_{13}(\boldsymbol{\gamma} \mid \boldsymbol{X}) \\
&+ \delta_1^{(2)} \delta_2^{(1)} \log S_{21}(\boldsymbol{\gamma} \mid \boldsymbol{X}) \\
&+ \delta_1^{(2)} \delta_2^{(2)} \log S_{22}(\boldsymbol{\gamma} \mid \boldsymbol{X}) \\
&+ \delta_1^{(2)} (1 - \delta_2^{(1)} - \delta_2^{(2)}) \log S_{23}(\boldsymbol{\gamma} \mid \boldsymbol{X}) \\
&+ (1 - \delta_1^{(1)} - \delta_1^{(2)}) \delta_2^{(1)} \log S_{31}(\boldsymbol{\gamma} \mid \boldsymbol{X}) \\
&+ (1 - \delta_1^{(1)} - \delta_1^{(2)}) \delta_2^{(2)} \log S_{32}(\boldsymbol{\gamma} \mid \boldsymbol{X}) \\
&+ (1 - \delta_1^{(1)} - \delta_1^{(2)}) (1 - \delta_2^{(1)} - \delta_2^{(2)}) \log S_{33}(\boldsymbol{\gamma} \mid \boldsymbol{X})
\end{aligned}$$

omitting the subindex i for clarity and

$$
\begin{aligned}
S_{11}(\gamma \mid \boldsymbol{X}) &= P(T_1 \leq l_1, T_2 \leq l_2) \\
&= 1 - S_1(l_1) - S_2(l_2) + \breve{C}_\alpha\big(S_1(l_1), S_2(l_2)\big), \\
S_{12}(\gamma \mid \boldsymbol{X}) &= P(T_1 \leq l_1, l_2 < T_2 \leq u_2) \\
&= S_2(l_2) - S_2(u_2) \\
&\quad + \breve{C}_\alpha\big(S_1(l_1), S_2(u_2)\big) - \breve{C}_\alpha\big(S_1(l_1), S_2(l_2)\big), \\
S_{13}(\gamma \mid \boldsymbol{X}) &= P(T_1 \leq l_1, T_2 > u_2) \\
&= S_2(u_2) - \breve{C}_\alpha\big(S_1(l_1), S_2(u_2)\big), \\
S_{21}(\gamma \mid \boldsymbol{X}) &= P(l_1 < T_1 \leq u_1, T_2 \leq l_2) \\
&= S_1(l_1) - S_1(u_1) \\
&\quad + \breve{C}_\alpha\big(S_1(u_1), S_2(l_2)\big) - \breve{C}_\alpha\big(S_1(l_1), S_2(l_2)\big), \\
S_{22}(\gamma \mid \boldsymbol{X}) &= P(l_1 < T_1 \leq u_1, l_2 < T_2 \leq u_2) \\
&= \breve{C}_\alpha(S_1(l_1), S_2(l_2)) - \breve{C}_\alpha\big(S_1(l_1), S_2(u_2)\big) \\
&\quad - \breve{C}_\alpha\big(S_1(u_1), S_2(l_2)\big) + \breve{C}_\alpha\big(S_1(u_1), S_2(u_2)\big), \\
S_{23}(\gamma \mid \boldsymbol{X}) &= P(l_1 < T_1 \leq u_1, T_2 > u_2) \\
&= \breve{C}_\alpha(S_1(l_1), S_2(u_2)) - \breve{C}_\alpha\big(S_1(u_1), S_2(u_2)\big), \\
S_{31}(\gamma \mid \boldsymbol{X}) &= P(T_1 > u_1, T_2 \leq l_2) \\
&= S_1(u_1) - \breve{C}_\alpha\big(S_1(u_1), S_2(l_2)\big), \\
S_{32}(\gamma \mid \boldsymbol{X}) &= P(T_1 > u_1, l_2 < T_2 \leq u_2) \\
&= \breve{C}_\alpha\big(S_1(u_1), S_2(l_2)\big) - \breve{C}_\alpha\big(S_1(u_1), S_2(u_2)\big), \\
S_{33}(\gamma \mid \boldsymbol{X}) &= P(T_1 > u_1, T_2 > u_2) \\
&= \breve{C}_\alpha\big(S_1(u_1), S_2(u_2)\big).
\end{aligned}
$$

A full maximum likelihood procedure requires to estimate the marginal and the copula parameters jointly. This may be computationally cumbersome, and therefore a two-stage procedure based on a pseudo-likelihood has been proposed. In the two-stage procedure first the model parameters of S_1 and S_2 are estimated, yielding the marginal estimates $\widehat{S}_1$ and $\widehat{S}_2$. The pseudo-log-likelihood $\ell(\gamma, \widehat{S}_1, \widehat{S}_2)$ is then maximized with respect to γ. The two-stage procedure has been proposed by Sun et al. (2006), Shih and Louis (1995) and Wang and Ding (2000).

Wang and Ding (2000) and Sun et al. (2006) propose to model the marginal distributions nonparametrically. Bogaerts and Lesaffre (2008b) propose to model the marginal distributions with an AFT model with a flexible error term. Their approach allows also for incorporation of covariates in the marginal distributions.

Variance estimation for the copula parameters remains complicated. Godambe (1991) derived an asymptotic covariance matrix of the two-stage parameter estimates by making use of inference functions. Since the calculations require complicated second-order derivatives, Sun et al. (2006) rather proposed a bootstrap procedure. Bootstrap samples of size n with replacement are drawn independently M (= fixed) times from the observed data $\mathcal{D}$. This yields M estimators $\{\widetilde{\gamma}_m; m = 1, \ldots, M\}$ of γ allowing to compute the variance of all parameters. The validity of this bootstrap procedure is still a topic of research, but simulations suggest that it works well in practical settings (Bogaerts and Lesaffre, 2008b; Sun et al., 2006).

Finally, there is the question which copula model to choose for the data at hand. One possibility is to fit several copulas to the data and pick the one with the smallest, say, AIC, but now based on the pseudo-likelihood. However, this does not mean that the chosen model fits the data well. Little research has been done to develop goodness of fit procedures for copulas with censored data and there is basically nothing on interval-censored data. In the absence of covariates, the penalized Gaussian mixture (PGM) model (see Section 7.4.1) provides a goodness of fit for the chosen copula. For instance, a graphical comparison of the local cross-ratio function (Section 7.5) derived from the PGM model and the parametric copula model for different cross-sections may help in evaluating a chosen copula.

Example 7.3 Signal Tandmobiel study

As an illustration, we analyzed the emergence times of the maxillar first premolars (contralateral teeth 14 and 24) for both boys and girls. The marginal distributions were assumed to follow an AFT model with gender as a covariate and with a flexible error term as suggested by Komárek et al. (2005a). We used 40 knots to model the distribution of the error term. The dependence between the emergence times was modelled by the three discussed copula functions whereby each time the association parameter was allowed to depend on gender.

As shown earlier (Leroy et al., 2003), a significant gender effect was found for the emergence times of the contralateral teeth 14 and 24. In girls the maxillar first premolars emerge earlier than in boys ($P < 0.001$).

Based on AIC, the Plackett copula was fitting best with pseudo-likelihood equal to $-1\,120$ (as compared to $-1\,148$ for the normal copula and $-1\,155$ for the Clayton copula). The estimated parameters (on the log-scale) were 3.23 for the intercept and 0.50 for gender. On the original scale, this results in a global odds ratio of 23.8 and 36.8 for boys and girls, respectively. Gender, however, does not significantly impact the association parameter ($P = 0.24$).

□

7.3.3 R solution

The approach of Bogaerts and Lesaffre (2008b) is implemented in the function fit.copula available in the package icensBKL. The function makes use of the package smoothSurv to fit the marginal distributions.

Example 7.3 in R

The R code for the Plackett copula applied to the Signal Tandmobiel data is shown below. Time to emergence of tooth 14 and 24 is modelled using AFT models with gender impacting the mean and scale parameter (see logscale *argument). A grid of candidate* λ *values is provided in the argument* lambda*. The correlation between the emergence times of teeth 14 and 24 is also allowed to depend on gender (argument* cov *contains the covariate gender). For the normal and Clayton model one changes the* copula *argument to "normal" or "clayton," respectively.*

```
> T1424.plackett <- fit.copula(tandmob,
+       copula = "plackett", init.param = NULL, cov = ~GENDER,
+       marginal1 = Surv(L14, R14, type = 'interval2') ~ GENDER,
+       logscale1 = ~GENDER, lambda1 = exp(3:(-3)),
+       marginal2 = Surv(L24, R24,type = 'interval2') ~ GENDER,
+       logscale2 = ~ GENDER, lambda2 = exp(3:(-3)),
+       bootstrap = FALSE)
```

The output (not shown) displays the search for the optimal λ *value in both marginals and the minimization procedure of minus the pseudo-log-likelihood. Here, log(lambda)=1 and 0 yielded the smallest AIC value for the first and second marginal, respectively. The maximum pseudo-likelihood value was obtained for the Plackett copula model* ($\ell = -1\,119.7$).

Once the best fitting copula model is chosen, a refit of the model with the bootstrap *argument set to TRUE is applied in order to determine the variance of the estimated parameters. We used 200 bootstrap samples as suggested by Sun et al. (2006). The smoothing parameters* λ_1 *and* λ_2 *are now restricted to the optimal choice obtained in the first fit.*

```
> T1424.plackett.bs <- fit.copula(tandmob,
+       copula = "plackett", init.param = NULL, cov = ~GENDER,
+       marginal1 = Surv(L14, R14, type = 'interval2') ~ GENDER,
+       logscale1 = ~GENDER, lambda1 = exp(1),
+       marginal2 = Surv(L24, R24, type = 'interval2') ~ GENDER,
+       logscale2 = ~GENDER, lambda2 = exp(0),
+       bootstrap = TRUE, nboot = 200)
```

By exponentiating the (on the log-scale) estimated copula parameter, one obtains the global odds ratio for both boys and girls.

The gender effect is not statistically significant, as seen below.

```
> 2*(1-pnorm(abs(T1424.plackett.bs$fit["Copula.param2"]/
+                T1424.plackett.bs$fit["sdBS.Copula.param2"])))
```

```
Copula.param2
    0.2370915
```

□

7.4 Flexible survival models

7.4.1 The penalized Gaussian mixture model

The univariate penalized Gaussian mixture (PGM) approach described in Sections 3.3.3 and 6.2 can also be applied in the bivariate setting.

Let $g(y_1, y_2)$ represent the joint density of $(Y_1,\, Y_2)'$, $Y_d = \log(T_d)$ $(d = 1, 2)$ with marginal densities $g_1(y_1)$ and $g_2(y_2)$. A smooth approximation of this density based on a sample of size n can be obtained from a penalized weighted sum of bivariate uncorrelated normal densities located on a predefined grid. The method is based on the penalized smoothing procedure of Eilers and Marx (1996). With the prespecified grid points $\boldsymbol{\mu}_{k_1,k_2} = (\mu_{1,k_1}, \mu_{2,k_2})'$ $(k_1 = 1, \ldots, K_1;\ k_2 = 1, \ldots, K_2)$, we assume that

$$\begin{pmatrix} Y_1 \\ Y_2 \end{pmatrix} \sim \sum_{k_1=1}^{K_1} \sum_{k_2=1}^{K_2} w_{k_1,k_2}\, \mathcal{N}_2(\boldsymbol{\mu}_{k_1,k_2}, \boldsymbol{\Sigma}),$$

where $\mathcal{N}_2(\cdot)$ indicates a bivariate normal distribution, where

$$\boldsymbol{\Sigma} = \begin{pmatrix} \sigma_1^2 & 0 \\ 0 & \sigma_2^2 \end{pmatrix}$$

is a diagonal covariance matrix with prespecified values of σ_1^2 and σ_2^2. Further, $w_{k_1,k_2} > 0, \forall k_1,\, k_2$ and $\sum_{k_1=1}^{K_1} \sum_{k_2=1}^{K_2} w_{k_1,k_2} = 1$. The aim is to estimate the weights $w_{k_1,k_2} (k_1 = 1, \ldots, K_1;\ k_2 = 1, \ldots, K_2)$ by maximizing the likelihood thereby keeping the grid points fixed. Unconstrained MLEs are obtained by expressing the log-likelihood as a function of $\boldsymbol{a} = \big(a_{k_1,k_2} :\ k_1 = 1, \ldots, K_1;\ k_2 =$

$1, \ldots, K_2)'$ using

$$w_{k_1,k_2} = w_{k_1,k_2}(\boldsymbol{a}) = \frac{\exp(a_{k_1,k_2})}{\sum_{j_1=1}^{K_1}\sum_{j_2=1}^{K_2}\exp(a_{j_1,j_2})}.$$

To ensure identifiability, we may take one of the a_{k_1,k_2}'s equal to zero.

A penalty term involving differences of the a-coefficients (Eilers and Marx, 1996), given by

$$q(\boldsymbol{a};\boldsymbol{\lambda}) = \frac{\lambda_1}{2}\sum_{k_1=1}^{K_1}\sum_{k_2=1+s}^{K_2}\left(\Delta_1^s a_{k_1,k_2}\right)^2 + \frac{\lambda_2}{2}\sum_{k_2=1}^{K_2}\sum_{k_1=1+s}^{K_1}\left(\Delta_2^s a_{k_1,k_2}\right)^2$$

prevents overfitting. In the above expression, $\boldsymbol{\lambda} = (\lambda_1,\, \lambda_2)'$, $\lambda_1 > 0$ and $\lambda_2 > 0$ are penalty or "smoothing" parameters. Furthermore, Δ_d^s is the s-th order backward difference operator in the d-th dimension ($d = 1, 2$), iteratively defined (for the first dimension) as $\Delta_1^s a_{k_1,k_2} = \Delta_1^{s-1} a_{k_1,k_2} - \Delta_1^{s-1} a_{k_1,k_2-1}$ for $s > 0$ and $\Delta_1^0 a_{k_1,k_2} = a_{k_1,k_2}$.

Given λ_1, λ_2, the penalized log-likelihood $\ell_P(\boldsymbol{a};\boldsymbol{\lambda}) = \ell(\boldsymbol{a}) - q(\boldsymbol{a};\boldsymbol{\lambda})$ is maximized with respect to $\boldsymbol{a}$, yielding estimates $\widehat{a}_{k_1,k_2}(k_1 = 1, \ldots, K_1;\ k_2 = 1, \ldots, K_2)$. The optimum λ_1 and λ_2 may be obtained by minimizing AIC. The "effective number of parameters" can be written as a function of the penalized and unpenalized likelihood (Gray, 1992).

The above procedure provides a parametric approach yielding a smooth solution (Ghidey et al., 2004). In the absence of censoring, it resembles a histogram estimator with a penalty term for roughness. For a small sample size, the penalty term avoids spiky solutions. For large sample sizes, the smoothing is not really necessary but provides more numerical stability. In case of censoring, Bogaerts and Lesaffre (2003) and Bogaerts (2007) showed in a simulation study good results for this approach, except for heavy (75%) left or right censoring. In that case, the w's corresponding to regions with limited information show high uncertainty reflected in their 95% confidence intervals. In a Bayesian context, Komárek and Lesaffre (2006) used this approach for survival problems involving interval-censored data, see also Chapter 13.

Example 7.4 Signal Tandmobiel study

We examine the bivariate distribution of the emergence times of the contralateral (left and right) maxillary first premolars (teeth 14 and 24) for boys. Based on a SAS macro `%smooth`, we produced Figure 7.5 which depicts the smooth approximation of the bivariate distribution. One can observe that the two emergence times are highly correlated. This is also confirmed in Section 7.5 where we compute the association between these emergence times.

□

7.4.2 SAS solution

The SAS macro %smooth, making use of SAS IML, fits a penalized Gaussian mixture model to bivariate interval-censored data, with C++. In Windows, the dynamic link library *kentau.dll* is required and must have access to a dynamic link version of the GNU Scientific Library (GSL) (Galassi et al., 2009). The SAS macro, a compiled 32-bit dll-file of the C++ code and the GSL (version 1.15) link libraries are available in the supplementary materials.

Example 7.4 in SAS
The macro %smooth can be called using the following code.

```
%smooth(data=icdata.tandmob ,
        left1=l34, right1=r34, left2=l44, right2=r44,
        equallambda=1, gridsearch=1, k=3, id=idnr,
        gridpoints1=20, gridpoints2=20,
        lambda1=%str(0.1,1,10,100,500,1000), lambda2=,
        max_grid1=5, max_grid2=5,
        outres=outtandmob, outcoef=outcoeftandmob);
```

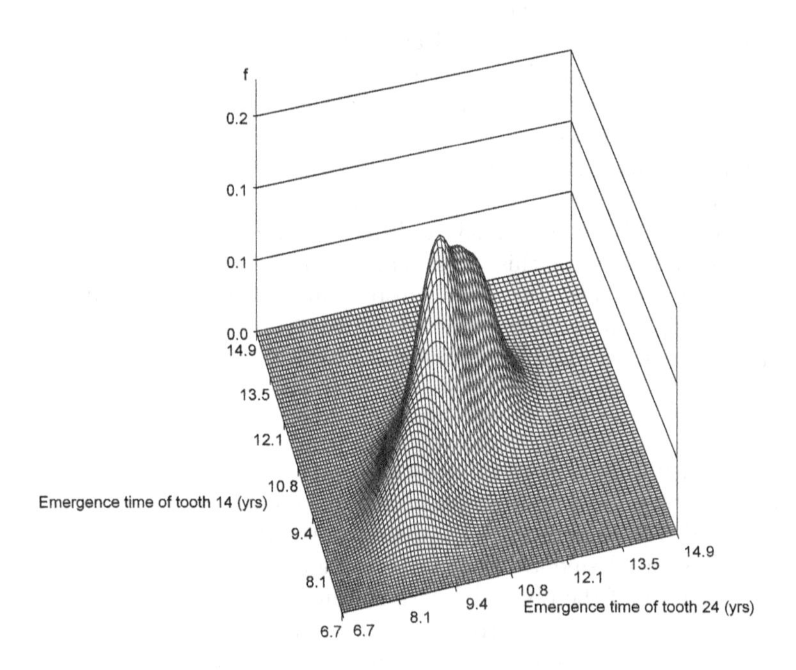

FIGURE 7.5: Signal Tandmobiel study. Density of penalized Gaussian mixture model for emergence of permanent teeth 14 and 24 obtained from SAS using macro %smooth.

The left and right boundaries of the first variable are specified in the arguments left1= *and* right1=, *respectively.* Equallambda=1 *requests that* λ_1 *and* λ_2 *are equal, with zero the default value (not necessary equal). A grid search is performed when* gridsearch=1 *using the values specified in the argument* lambda1 *(and* lambda2, *not applicable in this example because we chose* equallambda=1*). Also a quadratic search (*gridsearch=0*) for the optimal smoothing parameter is possible but can be quite time consuming. A third order difference penalty (*k=3*) and a 20 by 20 grid (*gridpoints1=20 *and* gridpoints2=20*) were taken here. The mean and standard deviation from a standard log-normal fit on the first variable determines the grid points in the first dimension.* Max_grid1=5 *specifies an equally distributed grid around the mean up to 5 standard deviations wide in both directions, and similarly for the second dimension. The search history is saved in the data set specified in the argument* outres=. *The output data set contains the smoothing parameters* λ_d*, the AIC value, the actual number of parameters (ADF), the convergence status of the NLPNRR call in IML and the coefficients* w_{k_1,k_2} *and* a_{k_1,k_2}. *Negative values for the convergence status indicate a problem. The best fitting model is stored in the data set specified in* outcoef=, *which contains two records. For each of two dimensions the following is reported: the fixed mean and standard deviation* σ_d *of the normal components, the location of the grid points (*$\boldsymbol{\mu}_{k_1,k_2}$*), the coefficients* w_{k_1,k_2} *and* a_{k_1,k_2}*, the smoothing parameters* λ_d*, the AIC value, ADF, the convergence status. In addition, several global and local association measures, which are discussed in Section 7.5, are given. The summary of the search procedure for the optimal lambdas is also displayed in the log. Here, the best fitting model was chosen with* $\lambda_1 = \lambda_2 = 10$ *and AIC = 2 222.4. A graphical display of the density function is shown in Figure 7.5, exhibiting a high correlation between the two emergence times. In Section 7.5 the reported association measures are defined. The code to produce the figure is provided in the supplementary materials.*

□

7.5 Estimation of the association parameter

Measures of association have been extensively studied for completely observed data. Some measures have been extended to right-censored data, but less work has been devoted to interval-censored data. For bivariate right-censored data, Oakes (1982) suggested a method to estimate τ, which is the difference between the probabilities of concordance and discordance of bivariate data. A pair of bivariate observations (U_1, V_1) and (U_2, V_2) are said to be concordant if $U_1 < U_2$ and $V_1 < V_2$ or if $U_1 > U_2$ and $V_1 > V_2$ and they are discordant if $U_1 < U_2$ and $V_1 > V_2$ or if $U_1 > U_2$ and $V_1 < V_2$.

For interval-censored data, Betensky and Finkelstein (1999a) suggested to combine Oakes's approach with multiple imputation which "completes" as much as possible the interval-censored data based on an estimated bivariate survival distribution. Bogaerts and Lesaffre (2008a) make use of the PGM method to compute several global and local measures of association. Also based on copula models, measures of association can be determined (Bogaerts and Lesaffre, 2008b). In this section we first introduce some measures of association that might be useful to characterize the dependence of two survival outcomes. Some of these measures were introduced in the context of copulas. Here we introduce them in general.

7.5.1 Measures of association

Let f and g represent the bivariate densities of (T_1, T_2) and $(Y_1, Y_2) \equiv (\log(T_1), \log(T_2))$, respectively. and $F(t_1, t_2)$, $G(y_1, y_2)$ the corresponding cdfs with survival functions $S_F(t_1, t_2)$ and $S_G(y_1, y_2)$. The corresponding univariate marginal distributions are denoted as $F_1(\cdot)$ $(G_1(\cdot))$ and $F_2(\cdot)$ $(G_2(\cdot))$.

Association measures characterize the association's strength between T_1 and T_2. A desirable property of an association measure is that it is invariant with respect to strictly increasing transformations and varies between -1 and 1. We distinguish between global and local measures. A global measure is based on all values in the support of the bivariate distribution, in contrast to a local measure which measures the association at a particular location. A specific area of interest of local dependence is the tails of the bivariate distribution. Tail dependence refers to the degree of dependence in the corner of the lower-left quadrant or upper-right quadrant of a bivariate distribution. More formally, lower tail dependence is defined as $\lim_{q \to 0} \mathsf{P}\big(T_2 \leq F_2^{-1}(q) \mid T_1 \leq F_1^{-1}(q)\big)$ and upper tail dependence as $\lim_{q \to 1} \mathsf{P}\big(T_2 > F_2^{-1}(q) \mid T_1 > F_1^{-1}(q)\big)$ (McNeil et al., 2005). We show below that it is insightful to characterize the bivariate distributions by these association measures.

Well-known global measures of association are the Pearson correlation coefficient, Kendall's tau and the Spearman rank correlation coefficient. Pearson correlation has been originally defined for bivariate normally distributed random variables, but Embrechts et al. (1999) showed that this is also a useful measure for elliptical distributions. For most survival distributions, the Pearson correlation may be less attractive. For instance, the Pearson correlation is constrained in $[-0.09; 0.67]$ for T_1 and T_2 log-normal distributed random variables both with mean 0 and standard deviation equal to 1 and 2, respectively. This interval becomes arbitrarily small for increasing standard deviation of T_2 so that its interpretation becomes difficult.

Kendall's tau (Hougaard, 2000) is more appropriate for survival distributions and is equal to

$$\tau = 4 \cdot \int S_F(t_1, t_2) \mathrm{d}F(t_1, t_2) - 1. \tag{7.5}$$

τ estimates the difference between the probabilities of concordance and discordance and can also be written as $2\pi - 1$ where π represents the probability of concordance. This is best understood by the interpretation provided by Hougaard (2000), which we translated here to the dental example. Namely, consider teeth 14 and 24 for two children. Given that tooth 14 of the first child emerges before tooth 14 of the second child, π equals the probability that also tooth 24 of the first child emerges before tooth 24 of the second child.

Spearman's correlation is a nonparametric global association measure, which is invariant to marginal monotonic transformations. The population version of Spearman's correlation (Joe, 1997) is given by

$$\rho_S = 12 \cdot \int_0^1 \int_0^1 S(S_1^{-1}(r_1), S_2^{-1}(r_2)) \mathrm{d}r_1 \mathrm{d}r_2 - 3.$$

The cross-ratio function suggested by Clayton (1978) and Oakes (1989) and introduced in Section 7.3.1 is a local measure of association. It evaluates how locally the joint density of the survival times behaves in relation to an independent survival function with the same marginal densities. The cross-ratio function has also an interpretation in conditional hazard rates (Oakes, 1989), namely

$$\theta_L(t_1, t_2) = \frac{\hbar_1(t_1 \mid T_2 = t_2)}{\hbar_1(t_1 \mid T_2 \geq t_2)} = \frac{\hbar_2(t_2 \mid T_1 = t_1)}{\hbar_2(t_2 \mid T_1 \geq t_1)},$$

where $\hbar_1$ and $\hbar_2$ are the hazard functions for T_1 and T_2, respectively.

For a better understanding, we interpret the measure in the context of our dental example. Let T_1 and T_2 represent the time to emergence of teeth 14 and 24, respectively, then $\theta_L(t_1, t_2)$ is the ratio of the risk for emergence for tooth 14 in children of age t_1 whose tooth 24 emerged at the age of t_2, compared to the risk of emergence for tooth 14 in children of age t_1 whose tooth 24 was not yet emerged at the age t_2.

The global cross-ratio $\theta_G(t_1, t_2)$ is, in contrast to its name, to some extent also locally defined since it is computed at each (t_1, t_2). See Equation (7.4) for an expression of $\theta_G(t_1, t_2)$. But in contrast to the local cross-ratio, the whole support of the bivariate survival function is involved in the computation.

In Section 7.3.1 we have seen that the Clayton model is characterized to have a constant cross-ratio function, while for the Plackett model the global cross-ratio is constant.

7.5.2 Estimating measures of association

When a parametric model for the bivariate interval-censored data can be assumed, the association measure could be estimated by plugging-in the parameter estimates in the population expression of the association. For instance, fitting a bivariate normal model provides naturally a Pearson correlation, while a gamma frailty model (see Chapter 8) provides Kendall's τ. The parametric approach has the obvious drawback that it may be hard to choose the correct distribution, especially with bivariate interval-censored observations.

Using a copula can be a way-out to a too strictly specified parametric model. We now show that Spearman's rho and Kendall's tau can be expressed as a function of the copula. Spearman's rho is defined as

$$\begin{aligned}
\rho_S &= 12\int_0^\infty\int_0^\infty F_1(t_1).F_2(t_2)\mathrm{d}F(t_1,t_2) - 3 \\
&= 12\int_0^1\int_0^1 r_1.r_2\mathrm{d}C(r_1,r_2) - 3 \\
&= 12\int_0^1\int_0^1 C(r_1,r_2)\mathrm{d}r_1\mathrm{d}r_2 - 3.
\end{aligned}$$

Kendall's tau is defined as

$$\begin{aligned}
\tau &= 4\int_0^\infty\int_0^\infty F(t_1,t_2)\mathrm{d}F(t_1,t_2) - 1 \\
&= 4\int_0^1\int_0^1 C(r_1,r_2)\mathrm{d}C(r_1,r_2) - 1.
\end{aligned}$$

For several copulas, an explicit relation between the copula parameter and the association measure can be established. Namely, for the Clayton copula Kendall's tau and the parameter θ are related as $\tau = \frac{\theta}{\theta+2}$. Note that only positive correlations can be modelled with the Clayton copula.

For a Plackett copula with parameter $\theta > 0$, Spearman's rho is given by

$$\rho_S(\theta) = \frac{\theta+1}{\theta-1} - \frac{2\theta}{(\theta-1)^2}\log(\theta). \tag{7.6}$$

Kendall's τ cannot be expressed in a closed form.

For a Gaussian copula, the Pearson correlation ρ is a parameter of the model. In addition, Spearman's correlation and Kendall's tau for the Gaussian copula are: $\rho_S = \frac{6}{\pi}\arcsin\big(\frac{1}{2}\rho\big)$ and $\tau = \frac{2}{\pi}\arcsin(\rho)$. A proof of these relations can be found in McNeil et al. (2005).

Note that the independence ($\rho = 0$), comonotonicity ($\rho = 1$) and countermonotonicity ($\rho = -1$) copulas are special cases of the Gaussian copula. Therefore, the Gaussian copula can be thought of as a dependence structure that interpolates between perfect positive and negative dependence, where the parameter ρ represents the strength of dependence.

Finally, of the three considered copulas, tail dependence is only allowed for the Clayton copula.

Bogaerts and Lesaffre (2008a) suggested to estimate the association measure based on the PGM model introduced in Section 7.4.1. Their approach consists in replacing the functionals by their estimated counterparts determined from the bivariate smoothed function in the population version of the association measure. The method can be applied to global as well as local association measures. The details of the calculations can be found in Bogaerts and Lesaffre (2008a).

For Kendall's tau, this approach consists in replacing G by the cumulative distribution of the bivariate smoothed function in Expression (7.5) expressed on the log scale, leading to the estimate of τ:

$$\widehat{\tau} = 4 \cdot \sum_{k_1=1}^{K_1} \sum_{k_2=1}^{K_2} \sum_{l_1=1}^{K_1} \sum_{l_2=1}^{K_2} \widehat{w}_{k_1,k_2} \widehat{w}_{l_1,l_2} \Phi\left(\frac{\mu_{1,k_1} - \mu_{1,l_1}}{\sqrt{2}\sigma_1}\right) \Phi\left(\frac{\mu_{2,k_2} - \mu_{2,l_2}}{\sqrt{2}\sigma_2}\right) - 1$$

where $\widehat{w}_{k_1,k_2}$ and $\widehat{w}_{l_1,l_2}$ are the estimated coefficients and Φ denotes the univariate standard normal cdf. Given $\widehat{w}_{k_1,k_2}$, the calculation of $\widehat{\tau}$ is readily done.

Analogously, given the $\widehat{w}_{k_1,k_2}$, Spearman's rho is readily estimated by

$$\widehat{\rho}_s = 12 \cdot \sum_{k_1} \sum_{k_2} \sum_{l_1} \sum_{l_2} \sum_{j_1} \sum_{j_2} \widehat{w}_{k_1,k_2} \widehat{w}_{l_1,l_2} \widehat{w}_{j_1,j_2} \Phi\left(\frac{\mu_{1,k_1} - \mu_{1,j_1}}{\sqrt{2}\sigma_1}\right) \Phi\left(\frac{\mu_{2,l_2} - \mu_{2,j_2}}{\sqrt{2}\sigma_2}\right) - 3,$$

but the computations are somewhat harder because of the extra summation.

Also local measures of association can be determined in this way. For the cross-ratio function the approach consists in replacing $S(t_1, t_2)$ by the estimated survival function of the bivariate smoothed function in Expression (7.3) and leads to the following expression for the estimate of $\theta_L(t_1, t_2)$:

$$\widehat{\theta}_L(t_1, t_2) = \frac{\sum_{k_1} \sum_{k_2} \widehat{w}_{k_1,k_2} \Phi\left(-\frac{t_1 - \mu_{1,k_1}}{\sigma_1}\right) \Phi\left(-\frac{t_2 - \mu_{2,k_2}}{\sigma_2}\right)}{\sum_{k_1} \sum_{k_2} \widehat{w}_{k_1,k_2} \Phi\left(-\frac{t_1 - \mu_{1,k_1}}{\sigma_1}\right) \varphi\left(\frac{t_2 - \mu_{2,k_2}}{\sigma_2}\right)} \times \frac{\sum_{k_1} \sum_{k_2} \widehat{w}_{k_1,k_2} \varphi\left(\frac{t_1 - \mu_{1,k_1}}{\sigma_1}\right) \varphi\left(\frac{t_2 - \mu_{2,k_2}}{\sigma_2}\right)}{\sum_{k_1} \sum_{k_2} \widehat{w}_{k_1,k_2} \varphi\left(\frac{t_1 - \mu_{1,k_1}}{\sigma_1}\right) \Phi\left(-\frac{t_2 - \mu_{2,k_2}}{\sigma_2}\right)}, \quad (7.7)$$

where φ denotes the univariate standard normal density. When (7.7) is evaluated at the values of the grid, the evaluation of $\widehat{\theta}_L(t_1, t_2)$ can be done computationally efficient because most components are already available from the calculation of the bivariate density.

Using the delta method, one can easily derive the asymptotic variance and a (95%) confidence interval for any association measure. Further, for association measures estimated from two (or more) independent groups of subjects a two-sample Z-test (or ANOVA test) can be derived to test their equality.

For the calculation of τ in the presence of interval-censored data, Betensky and Finkelstein (1999a) extended the method of Oakes (1982) who calculated τ in the presence of right-censored data. The method, however, is too dependent on the percentage right censoring, see Bogaerts (2007).

Example 7.5 Signal Tandmobiel study
As an illustration, we reanalyzed Examples 7.3 and 7.4 where the emergence times of teeth 14 and 24 for both boys and girls were fitted.

The PGM model applied to boys yields a Kendall's tau of 0.54 (95% CI: 0.48 – 0.60) and a Spearman's rho of 0.72 (95% CI: 0.64 – 0.80). These values confirm that the emergence times of teeth 14 and 24 are highly correlated, as seen in Figure 7.5.

Based on AIC, the Plackett copula was best fitting with a global odds ratio obtained from Expression (7.4) of 23.8 and 36.8 for boys and girls, respectively. The corresponding Spearman's rho's (see Expression (7.6)) are 0.80 and 0.85, respectively. Kendall's τ was approximated (see below) to yield a value of 0.61 and 0.68 for boys and girls, respectively which difference is not significant ($P = 0.23$).

By plotting the cross-ratio function determined from the PGM model together with that of the different copula models for different levels of one margin, one can compare in a graphical manner the goodness of fit of the best fitting copula. Although best fitting, one can observe in Figure 7.6 that the Plackett copula does not seem to capture the dependence adequately when conditioning on tooth 14. A similar figure can be obtained while conditioning on the tooth 24.

□

7.5.3 R solution

Example 7.5 in R (Association measures)
In Section 7.3.3, we analyzed teeth 14 and 24 for both boys and girls with the Clayton, normal and Plackett copula. The Plackett copula was chosen as best fitting. Using Expression (7.6), one can estimate Spearman's correlation from the Plackett copula. Kendall's tau is approximated numerically using, e.g., the function adaptIntegrate *from the package* cubature *(Narasimhan and Johnson, 2017).*

```
> library("cubature")
> f <- function(x){
+    u <- x[1]
+    v <- x[2]
+    ((a*(1 + (-1 + a)*v + u*(-1 + a + 2*v - 2*a*v))*
+    (1 + (-1 + a)*(u + v) - sqrt(-4*(-1 + a)*a*u*v +
+    (1 + (-1 + a)*(u + v))^2)))/
+    (2*(-1 + a)*(-4*(-1 + a)*a*u*v + (1 + (-1 + a)*(u + v))^2)^(3/2)))
+ }
> a <- 23.8
```

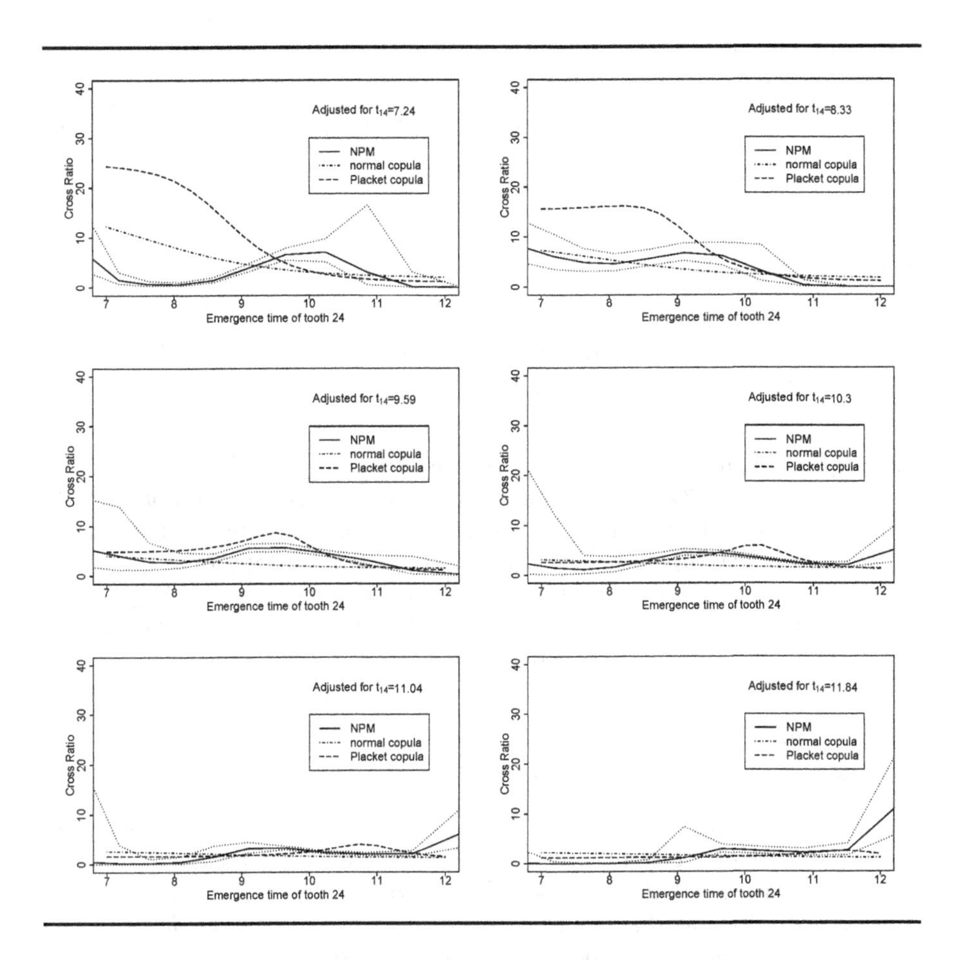

FIGURE 7.6: Signal Tandmobiel study. The estimated cross-ratio function $\theta(t_{14}, t_{24})$ at different values of t_{14} for the maxillar first premolars for boys derived from the penalized Gaussian mixture (PGM) approach, the normal copula and the Plackett copula for fitting the emergence times of contralateral teeth 14 and 24. A pointwise 95% confidence interval for the cross-ratio derived from the PGM is represented by dots. The computations are based on a R program available in the supplementary materials.

```
> intb <- adaptIntegrate(f, c(0, 0), c(1, 1), tol = 1e-4)
> print(4*intb$integral - 1)
```

```
[1] 0.6175957
```

The program used to create the graphs in Figure 7.6 is available in the supplementary materials.

□

7.5.4 SAS solution

Using the %smooth macro that fits a PGM model (see Section 7.4.1), the global association measures Kendall's tau, Spearman's rho and the local cross-ratio function can be determined.

Example 7.5 in SAS (Association measures)
In Example 7.4 a PGM model was fitted to the emergence times of teeth 14 and 24 for boys. In the data set Outcoeftandmob *an estimate of Kendall's tau (variable* Esttau*) and Spearman's rho (variable* EstSpearman*) is given. Using the corresponding variances for these estimators (variables* VarTau *and* VarSpearman*, respectively), 95% confidence intervals can be constructed. The following commands can be used in a data step to calculate a 95% CI for Kendall's tau:*

```
LL=EstTau - PROBIT(1-.05/2)*SQRT(VarTau);
UL=EstTau + PROBIT(1-.05/2)*SQRT(VarTau);
```

In addition, the local cross-ratio function evaluated at the grid points are included (variables LD1-LD400*) together with the corresponding variances (variables* VarLogLD1-VarLogLD400*) of the log-transformed local cross-ratio function. As illustrated above they can be useful in evaluating the goodness of fit of copula (or other) models when no covariates are involved.*

□

7.6 Concluding remarks

As for univariate problems, in the bivariate case a good strategy is to compare parametric solutions to the NPMLE. For an R-user the function computeMLE from the package MLEcens (Maathuis, 2013) is currently the best available. In SAS, the macro %bivnpmle can perform the calculations.

However, for large sample sizes it is probably best to switch to the more efficient R function. No doubt, most interesting research questions always involve covariates requiring parametric or flexible models. Except for the bivariate log-normal model, fitting a parametric bivariate model is less trivial due to the lack of pre-built functions for bivariate distributions. A copula model with flexible marginal distributions may be fitted in R using the function fit.copula available in the package icensBKL. However, the choice of the copula is limited to the Clayton, Plackett or Gaussian copula. Alternatively, the user can fit a penalized Gaussian mixture model with the SAS macro %smooth.

Chapter 8

Additional topics

In Chapter 7 we have discussed various methods to analyze bivariate interval-censored survival times. We have, however, omitted the treatment of doubly interval-censored observations. The reason is that, although the data are bivariate, they show a specific characteristic. Namely, for doubly interval-censored observations the first event must always happen before the second event and therefore the first true survival time must always be shorter than the second. In addition, we are not interested now in the correlation of the true survival times, but rather in the true time between the first and second event, i.e., we wish to estimate the distribution of the gap time and examine the factors that determine that distribution. In Section 8.1, frequentist methods to analyze doubly interval-censored observations data will be reviewed and illustrated. In Section 8.2, we look at frailty regression models, which are called upon when data exhibit an hierarchical structure. While both shared as well as correlated frailty models will be considered, only the first type will be treated here. We will look at parametric frailty models for interval-censored survival times. In the Bayesian part of the book, we will relax the parametric assumptions. Frailty models are quite popular in statistics, but they have the drawback that the regression coefficients have a subject-specific interpretation and are therefore more difficult to interpret. In the same section, the Generalized Estimating Equations (GEE) approach will be discussed for interval-censored survival times. GEE methods constitute another popular class of models that can deal with correlated data. The GEE approach has an attractive property that the regression coefficients have a population-averaged interpretation and are therefore easier to interpret. In addition, the GEE approach does not expect the association structure to be correctly specified. In fact the correlation structure is considered to be a nuisance for GEE models. Note that the two approaches may also be useful to analyze higher (than 2) dimensional multivariate interval-censored survival times. In Section 8.3, we discuss the extension of the biplot to interval-censored observations. The biplot is a popular graphical exploratory technique to obtain insight into the relationship of multivariate data and the variables in a 2-dimensional figure. The extension of the biplot to multivariate interval-censored observations could be useful to graphically represent the ordering of events in a 2-dimensional plot, which was of interest in the Signal Tandmobiel study.

8.1 Doubly interval-censored data

8.1.1 Background

In Section 1.2.4 we introduced doubly interval-censored observations. Such survival times occur when the onset of the period at risk and the time of event are both interval-censored. In clinical trials, the origin of survival is often the time of randomization (and the start of treatment) but it could also be the time that a patient enters a particular state. An example of this is found in HIV research where the onset of HIV can only be established at a doctor's visit and therefore the time to HIV is interval censored. When the event of interest is AIDS, then for the same reason the end of the period at risk is right censored or interval censored. We then say that the time to AIDS from HIV is doubly right censored (DR) or doubly interval (DI) censored. Another example of doubly interval-censored observations can be found in the Signal Tandmobiel study when time to caries of a permanent tooth was of interest. A proper analysis involves the time at risk of the tooth, i.e., the time from emergence of that tooth to caries. But, since emergence and caries of a tooth in an epidemiological study can only be established at (planned) visits to the dentist, both the onset and the end of the time span are interval censored. A graphical representation of a DI-censored survival time has been given in Figure 1.2.

A naive approach to the analysis of DI-censored survival times is to reduce them to interval-censored (or even right-censored) times. Suppose T_i^O denotes the true onset of the period at risk and T_i^E the true time when the event of interest occurs for the ith subject ($i = 1, \dots, n$). However, what is observed in the data is $\lfloor l_i^O, u_i^O \rfloor$ for T_i^O and $\lfloor l_i^E, u_i^E \rfloor$ for T_i^E. We are interested in $T_i^E - T_i^O = T_i$, which is called the *gap time*. To explore the distribution of T_i or the effect that covariates have on that distribution, one might assume that T_i is interval-censored observed in the interval $\lfloor l_i^E - u_i^O, u_i^E - l_i^O \rfloor$. Proceeding in this way, however, is not appropriate as first pointed out by De Gruttola and Lagakos (1989). They showed that the distribution of T_i^O as well as the distribution of T_i^E needs to be modelled to obtain an unbiased estimator of the distribution of the gap time. This theoretical result implies that dedicated software must be used to deal with DI-censored survival times. It therefore often happens that one applies the reduced likelihood approach, whereby (1) only the gap time is modelled as above, or (2) the interval for the onset is replaced by the mid-point or (3) both the begin and the end interval are replaced by mid-points. Such ad hoc approaches allow to use standard software to analyze DI-censored survival times. Dejardin and Lesaffre (2013) show that only when the intervals are small the reduced likelihood approach may give valid estimates of the gap time distribution. In all other cases one must model both the distribution of T_i^O and T_i^E.

Thus, one needs to maximize the full likelihood of the data to estimate the parameters of the gap distribution. De Gruttola and Lagakos (1989) introduced a method for doing this in the presence of DI-censored data. Their method is based on discretizing the distribution based on prespecified mass points. Kim et al. (1993) used the same idea to propose an extension of the Cox PH model to DI-censored data. These authors pointed out that identifiability problems may occur, especially when two points are jointly included or excluded from all intervals or when some observed intervals do not contain any mass points. So, it is difficult to specify the mass points in advance without reference to the actual observed data. Sun et al. (1999) and Sun et al. (2004) proposed to estimate the regression coefficients in the gap distribution by integrating out the distribution of the onset. Goggins et al. (1999) described a Monte-Carlo EM algorithm for a PH model for T_i. Pan (2001) proposed a multiple imputation approach. However, the aforementioned approaches are restricted to doubly right-censored data and are therefore not suitable in general. Bayesian approaches have also been proposed in the literature (see Chapter 13), which are especially useful for exploratory purposes. Dejardin and Lesaffre (2013) have suggested a stochastic EM algorithm to analyze semiparametrically the impact of covariates on the distribution of T_i in the presence of DI-censored survival times. In contrast to the above described procedures, the method does not rely on prespecified mass points and accounts for the impact of covariates on T_i^O.

In the following, interest lies in estimating the impact of covariates $\boldsymbol{X}$ on the distribution of the gap time $T = T^E - T^O$, denoted as $S_T(t \mid \boldsymbol{X}) = 1 - F_T(t \mid \boldsymbol{X})$. The PH assumption will be exploited for the effect of the covariates. That is, it is assumed:

$$S_T(t \mid \boldsymbol{X}) = S_{0,T}(t)^{\exp(X'\boldsymbol{\beta}_T)}.$$

We refer to $S_{0,T}(t) = S_T(t \mid \boldsymbol{X} = \boldsymbol{0})$ as baseline distribution, and $\lambda_0(t)$ denotes the corresponding baseline hazard with $\Lambda_0(t)$ the cumulative hazard. Further, the density of T is denoted as f_T. The regression parameters influencing the distribution of the gap time are denoted as $\boldsymbol{\beta}_T$, abbreviated as $\boldsymbol{\beta}$. Similarly, $S_{T^O}(t^O \mid \boldsymbol{X}) = 1 - F_{T^O}(t^O \mid \boldsymbol{X})$ denotes the distribution of T^O and f_{T^O} its density, also depending on covariates in a PH manner. While the focus is on estimating $\boldsymbol{\beta}$ in the presence of DI-censored survival times, right-censored T^E data (DR survival times) are allowed by setting the upper bound of the interval for T^E to ∞. Further, independent censoring is assumed, see Section 1.4.2, and also independence between T^O and T^E. These are classical assumptions for the treatment of DI-censored survival times.

The likelihood for DI-censored data can be written as

$$\begin{aligned} &L(\boldsymbol{\psi}, \boldsymbol{\theta} \mid t_i^O \in \lfloor l_i^O, u_i^O \rfloor, t_i^E \in \lfloor l_i^E, u_i^E \rfloor, i = 1, \ldots, n) \\ &\quad = \prod_{i=1}^n \int_{l_i^O}^{u_i^O} \int_{l_i^E}^{u_i^E} f_{T^O}(t^O \mid \boldsymbol{\psi}, \boldsymbol{X}) f_T(t^E - t^O \mid \boldsymbol{\theta}, \boldsymbol{X}) \mathrm{d}t^O \, \mathrm{d}t^E, \quad (8.1) \end{aligned}$$

where $\boldsymbol{\psi}$ and $\boldsymbol{\theta}$ are the parameters of the unknown densities f_{T^O} and f_T, respectively. We are interested in the estimation of $\boldsymbol{\theta}$ with minimal assumptions ($\boldsymbol{\psi}$ are treated as nuisance parameters).

Note that integration in Equation (8.1) must be done over admissible values of T^O and T. In Figure 8.1 (inspired by Figure 1 in De Gruttola and Lagakos (1989), but adapted to the continuous case), we show three admissible cases defined by the following values for $\lfloor l_i^O, u_i^O \rfloor$ and $\lfloor l_i^E, u_i^E \rfloor$.

Case 1: $\lfloor 2.3, 4.1 \rfloor$, $\lfloor 6.5, 6.5 \rfloor$, i.e., T^E is known.

Case 2: $\lfloor 1.2, 3.4 \rfloor$, $\lfloor 4.7, 5.9 \rfloor$, i.e., both T^O and T^E are interval-censored.

Case 3: $\lfloor 4.5, 5.6 \rfloor$, $\lfloor 7.9, \infty)$, i.e., T^O is interval-censored and T^E is right censored.

The admissible values for case 2 are obtained from the following reasoning. The possible (min, max) value for T when (1) $T^O = 1.2$ is $(4.7-1.2,\, 5.9-1.2)$ and when (2) $T^O = 3.4$ is $(5.8-3.4,\, 4.7-3.4)$.

Without assumptions on f_T, the integral in Equation (8.1) cannot be computed. When T^O is observed and T right-censored, the semiparametric full likelihood underlying the Cox PH model can be written as (see Klein and Moeschberger, 2003):

$$L(\boldsymbol{\theta} \mid t_i, \boldsymbol{\delta}_i, i = 1, \ldots, n) = \prod_{i=1}^{n} \Big\{ \exp(\boldsymbol{X}_i'\beta)\lambda(t_i) \Big\}^{\delta_i} \exp\Big\{ -\Lambda(t_i) \exp(\boldsymbol{X}_i'\beta) \Big\}$$

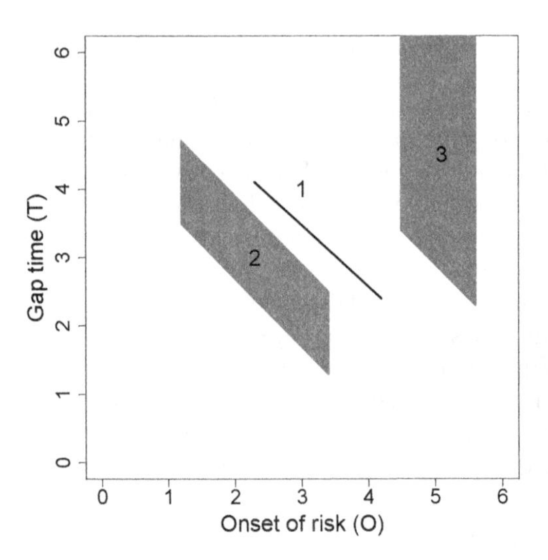

FIGURE 8.1: Three cases of doubly interval-censored survival times, the cases are defined in the text.

where t_i is the right-censored datum and $\delta_i = 1$ is the event indicator ($u_i^E < \infty$) and 0 otherwise. No assumptions on f_T are required when using the usual semiparametric Cox PH model. Here $\boldsymbol{\theta} = (\boldsymbol{\beta}', \lambda_1, \ldots, \lambda_d)'$, with $\lambda_1, \ldots, \lambda_d$ hazard parameters, d the number of events and $\lambda(t_i) = \lambda_i$ when t_i is the event time and 0 otherwise, $\Lambda(t_i) = \sum_{j:t_j \leq t_i} \lambda_j$. The $\boldsymbol{\beta}$ parameters are estimated by the partial likelihood technique and the hazard parameters by the Breslow estimator. However, in case of DI-censored survival times t_i is not exactly observed but is either interval-censored (i.e., $t_i \in [l_i^E - t_i^O, u_i^E - t_i^O]$) when $\delta_i = 1$ or right-censored (i.e., $t_i \in [l_i^E - t_i^O, \infty)$) when $\delta_i = 0$. But, we must be reminded that t_i^O is also not exactly observed but lies in $\lfloor l_i^O, u_i^O \rfloor$.

The proposed method consists in assuming that T^O and T are unobserved (but known to lie in observed intervals), and uses the EM algorithm (Dempster et al., 1977) to derive the parameters of interest based on the right-censored data likelihood. In the EM algorithm, the E-step computes at iteration $k+1$ the expectation of the log-likelihood with respect to the missing values (here t_i^O and t_i), given the observed data and the parameters at iteration k and is defined as $Q_{k+1}(\boldsymbol{\theta} \mid \boldsymbol{\theta}^k)$. The M-step subsequently maximizes Q_{k+1} with respect to $\boldsymbol{\theta}$. However, no closed form for Q_{k+1} can be derived and thus one cannot maximize Q_{k+1} easily in the M-step. Therefore, the Stochastic EM algorithm (StEM) introduced by Celeux and Diebolt (1985) was used. The details of the algorithm can be found in Dejardin and Lesaffre (2013). The StEM algorithm does not suffer from identifiability problems as with the aforementioned algorithms, but individuals with overlapping intervals must also be excluded. Extensive simulations showed that the StEM algorithm provides less biased results than the reduced likelihood approaches and previously suggested approaches in the literature. There was also a large bias in the estimated standard deviation for the mid-point approach, while unbiased for the StEM algorithm. The simulations also revealed that the coverage probability of the 95% CI for the model parameters obtained with the StEM algorithm is close to 0.95.

Example 8.1 Survey on mobile phone purchases

The data set on mobile phone purchases is introduced in Section 1.6.5. The data set is based on a survey with the aim to find the factors that stimulate the mobile phone users to exchange their mobile phone for a new one. In particular, we are interested in establishing whether gender, the size of the household and age have an effect on the lifetime of a mobile phone (time span until user buys a new phone). The analysis is based on 478 subjects for whom data were available to answer the question.

The analysis is based on several R functions (developed by David Dejardin) using the aforementioned stochastic EM approach. Currently, the functions only handle single covariates. We evaluated here the effect of gender, household size with 5 levels (1, ..., 4, 5 (or more) people in family) and age group with 6 levels. Our analysis showed that gender does not significantly impact the time to change phone but the household size and age did. Namely, the Wald P-value for gender is 0.29, for the household size $P < 0.001$ and for age groups $P < 0.001$. Figure 8.2 provides a forest plot of the log(hazard) estimates together with 95% CIs.

For illustration purposes, we also analyzed the data with two ad hoc approaches: (1) an approach that treats the gap time as an interval-censored covariate and (2) an approach based on the mid-points of the onset and end of the gap times. The results for these two approaches were obtained with the Finkelstein's PH model and the standard Cox PH model for right-censored data, respectively. The results are shown in Table 8.1. This comparison shows that, for this example, approach (1) still yielded similar results, but for approach (2) quite different estimates were obtained.

□

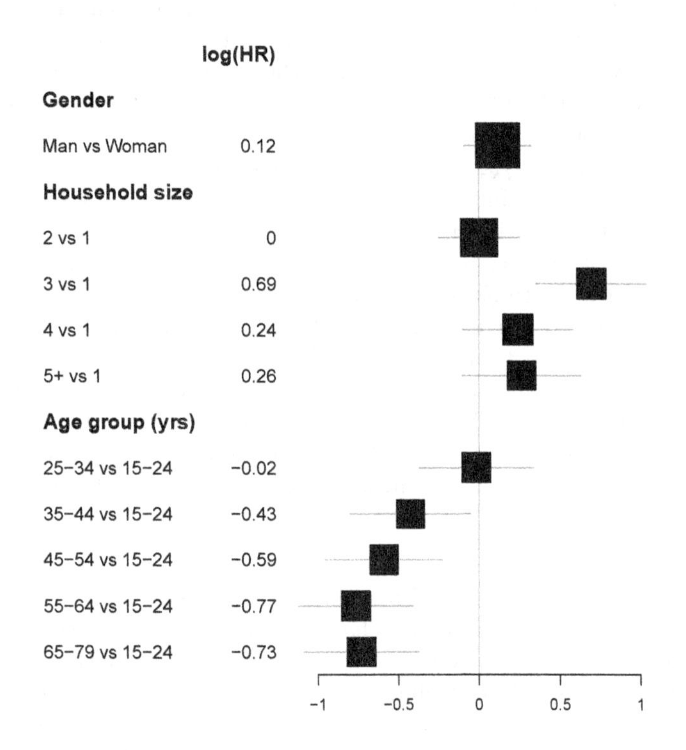

FIGURE 8.2: Mobile study. Forest plot of log(hazard) estimates together with 95% CIs showing the impact of gender, household size and age on the time to change phone, obtained from R.

TABLE 8.1: Mobile study. Log-hazard estimates with 95% CIs showing the impact of gender, household size and age on the time to change phone obtained from two ad hoc approaches treating the gap time as an interval-censored covariate (Method 1) and using the mid-points of the onset and end of the gap times (Method 2), obtained with function ic_sp of R package icenReg and function coxph of R package survival, respectively.

Parameter	log(HR)	95% CI	P-value
	Method 1		
Gender			0.2268
Man vs Woman	0.13	(−0.08, 0.35)	
Household size			0.0011
2 vs 1	−0.09	(−0.34, 0.17)	
3 vs 1	0.71	(0.32, 1.09)	
4 vs 1	0.23	(−0.13, 0.60)	
5+ vs 1	0.22	(−0.23, 0.67)	
Age group (yrs)			< 0.0001
25–34 vs 15–24	0.06	(−0.37, 0.48)	
35–44 vs 15–24	−0.38	(−0.84, 0.09)	
45–54 vs 15–24	−0.57	(−1.01, −0.13)	
55–64 vs 15–24	−0.79	(−1.22, −0.36)	
65–79 vs 15–24	−0.75	(−1.16, −0.35)	
	Method 2		
Gender			0.9560
Man vs Woman	0.01	(−0.18, 0.19)	
Household size			0.0005
2 vs 1	−0.24	(−0.48, 0)	
3 vs 1	0.42	(0.10, 0.75)	
4 vs 1	0.10	(−0.23, 0.42)	
5+ vs 1	−0.06	(−0.40, 0.29)	
Age group (yrs)			< 0.0001
25–34 vs 15–24	0.48	(0.14, 0.82)	
35–44 vs 15–24	0.11	(−0.24, 0.47)	
45–54 vs 15–24	−0.08	(−0.42, 0.26)	
55–64 vs 15–24	−0.27	(−0.61, 0.06)	
65–79 vs 15–24	−0.27	(−0.61, 0.05)	

8.1.2 R solution

Example 8.1 in R

The R *functions require specific names for the interval boundaries. Namely,* L *and* R *for the lower and upper interval boundary of the previous phone purchase date, and* SL *and* SR *for the subsequent lower and upper interval boundary of the current phone purchase date, respectively. First, the mobile data are loaded (available in the package* icensBKL*). The boundary values are extracted from the mobile data set and stored in a data frame. The covariate* gender *is copied into the covariate* cov*, which is the covariate the* R *function expects as single covariate. Further, the subjects with overlapping intervals must be excluded (not present in our data).*

```
> data("mobile", package="icensBKL")
> dataan <- data.frame(L = mobile$PLL, R = mobile$PUL,
+                      SL = mobile$CLL, SR = mobile$CUL,
+                      cov = mobile$gender)
```

Some further data preparation and function loading is performed as shown below. The original scale of days is transformed into months starting from the first recorded time. In fact, the commands below change the scale of the boundaries of the intervals. We then load the functions, which is done in the wrapper file load_StEM_binary.R *which calls the* survival *package and source files as well as defines some background variables.*

```
> dataan[, -5] <- (dataan[, -5] - min(dataan[, -5]))/30.4375
> ## load functions
> source("load_StEM_binary.R")
```

The StEM *estimation for doubly interval-censored data happens in 3 steps. In the first step, the distribution of the origin is estimated via the StEM method of Pan (2001). This is performed with the following code:*

```
> ## Computes the distribution of T^O
> ppdistU <- Udistfun(data_ = dataan)
```

The output of the procedure is the distribution of the origin, here the time to previous purchase. Also, the estimated regression parameters of the covariates included in the model are given. By default, the covariate impacts both T^O *and* T^E *and thereby possibly impacting the time between the current and previous purchase. Using this distribution and parameter value, the impact of the covariate on the gap time (i.e., the actual* StEM *algorithm) is performed using the first command below. The* sun.est *function has as arguments:* data*: data set with appropriate variables,* nmcint*: number of data sets at each* StEM *iteration (100) and maximum number of* StEM *iterations (500),* betaU*: estimated covariate parameter on the distribution of* T^O, S0U*: estimated baseline distribution of* T^O*. The second command deals with the estimation of the variance of the parameter.*

```
> sui_gender <- sun.est(data_ = dataan,  nmcint = c(rep(100, 500)),
+                  betaU = ppdistU$betaU, S0U = ppdistU$S0U)
> resvar_gender <- varcalc(sui_gender, dataan)
```

The object resvar_gender *contains the estimated parameter using the* StEM *algorithm along with its estimated variance, with* meanbeta *giving the estimated parameter and* corrvar *its 'corrected' variance obtained by a print statement. This is the variance corrected for the randomness of the* StEM*, see Goggins et al. (1998).*

```
> print(resvar_gender[,c(3,5)])
```

```
   meanbeta       corrvar
1 0.1294863    0.009477124
```

For variables with multiple levels, an adapted code is used. We give here the example for the household size variable with 5 levels:

```
> dataan <- data.frame(L = mobile$PLL, R = mobile$PUL, SL = mobile$CLL,
+                          SR = mobile$CUL, cov = mobile$hhsize)
+      [!(mobile$CLL == mobile$PLL & mobile$CUL - mobile$PUL <= 1),]
> dataan[,-5] <- (dataan[,-5] - min(dataan[,-5]))/30.4375
> ## load functions
> source("load_StEM_mlevel.R")
> ppdistU_hh <- Udistfun(data_ = dataan)
> sui_hh <- sun.est.ml(data_ = dataan, nmcint = c(rep(100,500)),
+             crit = 0.0001, betaU = t(ppdistU_hh$betaU),
+             S0U = ppdistU_hh$S0U)
> resvar_hh <- varcalc(sui_hh,dataan)
```

The output obtained from the above R *commands is again obtained in* resvar_hh*:*

```
> print(resvar_hh)
```

```
$betaest
  data_$cov2     data_$cov3     data_$cov4    data_$cov5+
9.385657e-05   6.917359e-01   2.383096e-01   2.615357e-01

$varest
```

(Cont.)

```
              datacv$cov2  datacv$cov3  datacv$cov4  datacv$cov5+
datacv$cov2    0.01625571   0.01023115   0.01054588   0.01025562
datacv$cov3    0.01023115   0.03076856   0.01125531   0.01066406
datacv$cov4    0.01054588   0.01125531   0.03004570   0.01008442
datacv$cov5+   0.01025562   0.01066406   0.01008442   0.03502375

$naivevar
            [,1]        [,2]        [,3]        [,4]
[1,] 0.014548712 0.009667942 0.009737084 0.009784857
[2,] 0.009667942 0.027468839 0.009983854 0.009874542
[3,] 0.009737084 0.009983854 0.027502393 0.009877443
[4,] 0.009784857 0.009874542 0.009877443 0.031219754
```

The element varest *contains the 'corrected' variance estimator. Note that the naive variance estimator (element* naivevar*) is also provided (which corresponds to the average of the covariance matrices on the last iteration of the StEM). The naive variance estimator is provided for information but underestimates the true variance, as stated earlier, because it does not account for the randomness of the StEM, see Goggins et al. (1998).*

The R *code to calculate the P-values and to create Figure 8.2 can be found in the supplementary materials. Also the* R *code to perform the ad hoc analyses can be found there in a separate program.*

□

8.2 Regression models for clustered data

In this section we consider clustered survival times, which show an hierarchical data structure. Examples of such data can be found in randomized clinical trials where patients are recruited from different centers and followed up in time until an event happens. Patients are then clustered in centers. Another example is the emergence time of teeth in a mouth, or the time until a cavity on teeth occurs. Data from one cluster look more similar than data from different clusters and the same is true for survival times. The hierarchical structure of the data therefore induces a correlation in the survival times from the same cluster and this correlation needs to be taken into account in the analysis.

We review in this section two approaches that can deal with correlated and interval-censored survival times: frailty models and the Generalized Estimating Equations approach.

8.2.1 Frailty models

Frailty models are random effects models for hierarchical survival data. They are commonly used in the analysis of survival data whenever there is an hierarchical structure in the data. There are two types of frailty models: *shared frailty models* and *correlated frailty models*. We will start and focus here on the first type of frailty models.

Shared frailty models

Shared frailty models are commonly used in the analysis of multivariate survival data. The dependence between the members of a cluster is modelled by means of a shared random effect or *frailty term*. Such a random effect can be incorporated in both PH (Chapter 5) and AFT (Chapter 6) regression models. In both cases, it is assumed that, given the frailty variable(s), the event times in a cluster are independent. We start with the parametric PH frailty model.

The parametric PH shared frailty model is simply Model (2.11) in which a frailty term is incorporated as follows:

$$\hbar_{ij}(t \mid \boldsymbol{X}_{ij}, Z_i) = Z_i \, \hbar_0(t) \exp(\boldsymbol{X}_{ij}'\boldsymbol{\beta}) \quad (i = 1, \ldots, n, j = 1, \ldots, m_i), \quad (8.2)$$

where $\hbar_0(t)$ denotes the baseline hazard, Z_i stands for the frailty term which follows a distribution g_Z with mean 1, $\boldsymbol{X}_{ij}$ represents the covariate values of the jth subject in the ith cluster and $\boldsymbol{\beta}$ is the vector of (fixed effects) regression coefficients. A popular choice for g_Z is the gamma distribution $\text{Gamma}(\alpha, \alpha)$, which has mean 1 and variance $1/\alpha$. When a parametric baseline hazard is chosen, we deal with a *parametric shared frailty model*. Note that alternatively one could assume that $b_i = \log(Z_i)$ has a normal distribution. In that case, we can write Expression (8.2) as:

$$\hbar_{ij}(t \mid \boldsymbol{X}_{ij}, b_i) = \hbar_0(t) \exp(\boldsymbol{X}_{ij}'\boldsymbol{\beta} + b_i).$$

Further, in a parametric shared frailty (PH) model also the baseline hazard needs to be specified explicitly. Often a Gompertz or Weibull hazard for $\hbar_0(t)$ is assumed.

To obtain the MLEs, the marginal log-likelihood, obtained after integrating out the frailties, is maximized. Assume interval-censored survival times $\boldsymbol{\mathcal{D}}_i \equiv \{\lfloor l_{ij}, u_{ij} \rfloor \ (j = 1, \ldots, m_i)\} = \{\boldsymbol{\mathcal{D}}_{i1}, \ldots, \boldsymbol{\mathcal{D}}_{im_i}\}$ in the ith cluster with censoring indicators $\delta_{i1}, \ldots, \delta_{im_i}$, then the conditional likelihood of the ith cluster given the frailty term is given by $L_i(\boldsymbol{\theta} \mid b_i, \boldsymbol{X}_i, \boldsymbol{\mathcal{D}}_i) = \prod_{j=1}^{m_i} L_{ij}(\boldsymbol{\theta} \mid b_i, \boldsymbol{X}_{ij}, \boldsymbol{\mathcal{D}}_{ij})$ with $\boldsymbol{X}_i = \{\boldsymbol{X}_{i1}, \ldots, \boldsymbol{X}_{im_i}\}$ representing the covariates in the ith cluster and $L_{ij}(\boldsymbol{\theta} \mid b_i, \boldsymbol{X}_{ij}, \boldsymbol{\mathcal{D}}_{ij})$ the likelihood contribution of the jth measurement in the ith cluster as in Expression (1.13). The total marginal likelihood over all clusters is then computed as

$$L(\boldsymbol{\theta} \mid \boldsymbol{X}, \boldsymbol{\mathcal{D}}) = \prod_{i=1}^{n} \int L_i(\boldsymbol{\theta} \mid b_i, \boldsymbol{X}_i, \boldsymbol{\mathcal{D}}_i) f(b_i; \boldsymbol{\psi}) \mathrm{d}b_i,$$

with $f(b_i;\boldsymbol{\psi})$ the density of the random effect depending on parameters $\boldsymbol{\psi}$ and $\boldsymbol{X} = \{\boldsymbol{X}_1,\ldots,\boldsymbol{X}_n\}$.

For most frailties, no closed form of the marginal frailty distribution is available. An exception is the gamma frailty model with a Weibull baseline hazard (Duchateau and Janssen, 2008). For this reason it is also a frequently used frailty model. For the other frailties, numerical approximations must be applied. SAS and R can deal with log-normally distributed frailties, i.e., when $b_i = \log(Z_i)$ follows a normal distribution.

The parametric AFT shared frailty model is similar to Model (6.4) to which a random intercept is added:

$$\log(T_{ij}) \;=\; \mu + b_i + \boldsymbol{X}'_{ij}\boldsymbol{\beta} + \sigma\,\varepsilon^*_{ij} \quad (i = 1,\ldots,n, j = 1,\ldots,m_i),$$

where the ingredients in the above model have the same meaning as for the PH model, but of course enjoy a different interpretation because they are part of a different model. Further, b_i is the random intercept for cluster i, which is assumed to follow a density g_b with mean 0 and ε^* is the standardized error term with a density $g^*_\varepsilon(e)$. For $g^*_\varepsilon(e)$ often a standard normal or logistic density is taken. For g_b, a normal distribution is a popular choice. Similar as for the parametric PH shared frailty model, we can write down the marginal likelihood.

Note that, while the marginal likelihood is computed to estimate the model parameters, the regression coefficients $\boldsymbol{\beta}$ have a *subject-specific interpretation.* This means that their interpretation is conditional on the random effects. In general, the subject-specific interpretation of a regression coefficient is more complicated than a marginal interpretation. For example, the *marginal effect* or *population-averaged effect* of a regression coefficient is similar to that in a classical regression model. We refer to Molenberghs and Verbeke (2005) for more details on the difference between random effects and marginal models and the interpretation of regression coefficients. Examples of a marginal model are the parametric models seen in previous chapters, the copula models and the marginal model treated in Section 8.2.2.

Further, we note that the frailty model can handle also genuine multivariate survival times of a general dimension. In that case, $m_i = m$ for $i = 1,\ldots,n$. Of course, this is a useful method only when the conditional independence assumption is realistic for the data at hand. In Section 8.2.2 and in the Bayesian part of the book, we will treat models for multivariate interval-censored data with a general correlation structure.

The above models are rather restrictive since the frailty distribution and the baseline survival distribution need to be specified explicitly. There are two ways to relax the assumptions. First, with a computational trick also other frailty distributions may be fitted with standard software. Namely, Nelson et al. (2006) suggested to use the probability integral transformation (PIT) method. Here we describe the approach suggested by Liu and Yu (2008). The trick consists of a likelihood reformulation method for non-normal random effects models such that these models can also be easily estimated using

Gaussian quadrature tools. Let the random effect have a density $g_b(b_i; \boldsymbol{\psi})$. The marginal likelihood for the ith subject can be expressed as

$$L_i = \int \exp(\ell_i^b) g_b(b_i; \boldsymbol{\psi}) \mathrm{d}b_i,$$

where $\ell_i^b = \log L_i(\boldsymbol{\theta} \mid b_i, \boldsymbol{X}_i, \boldsymbol{\mathcal{D}}_i)$ is the conditional log-likelihood for the i-th subject given the frailty ($\equiv$ given the random effect b_i) obtained from the PH or AFT frailty model. Reformulating the above integrand with respect to $g_b(b_i; \boldsymbol{\psi})$ to that with respect to a standard normal density $\varphi(b_i)$ yields

$$\begin{aligned} L_i &= \int \exp(\ell_i^a) \frac{g_b(b_i; \boldsymbol{\psi})}{\varphi(b_i)} \varphi(b_i) \mathrm{d}b_i \\ &= \int \exp(\ell_i^a + \ell_i^b - \ell_i^c) \varphi(b_i) \mathrm{d}b_i, \end{aligned} \tag{8.3}$$

where $\ell_i^b = \log\{g_b(b_i; \boldsymbol{\psi})\}$ and $\ell_i^c = \log\{\varphi(b_i)\}$. To implement this trick in software, the density function $\varphi(\cdot)$ must have a closed form or be available by a function call.

The second way to relax parametric assumptions, is to make use of a smoothing technique for the baseline hazard and/or the frailty distribution. Similar to the methodology described in Section 5.2.3, Rondeau et al. (2003) adapted the use of penalized likelihood estimation to a gamma frailty model. However, the majority of smoothing approaches for frailty models make use of the Bayesian framework. We will return to these implementations in Section 13.1.

We now illustrate the use of a PH frailty model on the mastitis data.

Example 8.2 Mastitis study

To illustrate the PH shared frailty model, the effect of parity (1, 2 to 4 and > 4 calvings) and the position of the udder quarter (front or rear) on the time to inflammation of the udder. Four models will be considered varying in the frailty distribution: gamma or log-normal and the baseline hazard: Weibull or piece-wise constant. Also the semiparametric penalized likelihood approach of Rondeau et al. (2003), which fits the baseline hazard smoothly, was applied. Table 8.2 shows the parameter estimates for the different models. For all four models, it can be concluded that cows with 5 or 6 offsprings develop significantly quicker inflammation of the udder, but whether the udder is located at the front or the rear does not play an important role. Based on AIC, the best fitting model is the gamma frailty model with a piece-wise constant estimated hazard.

□

8.2.1.1 R solution

The package frailtypack allows to fit a PH model with a shared gamma frailty making use of semiparametric penalized likelihood estimation (using

TABLE 8.2: Mastitis study. Parameter estimates and model fit statistics of shared frailty models obtained from SAS procedure NLMIXED (first 2 models) and the R package frailtypack (last 2 models). γ and α denote the shape and scale parameter of the Weibull baseline hazard, respectively. θ denotes the variance of the frailty term. N.A.= Not Applicable.

Frailty	Log-normal		Gamma		Gamma		Gamma	
Baseline hazard	Weibull		Weibull		Piece-wise constant		Spline	
	Est.	SE	Est.	SE	Est.	SE	Est.	SE
β_{rear}	0.174	0.123	0.180	0.122	0.190	0.125	0.183	0.120
β_{par24}	0.037	0.373	−0.201	0.336	−0.288	0.330	−0.090	0.297
β_{par56}	1.878	0.540	1.400	0.486	1.618	0.499	1.340	0.441
γ	2.015	0.115	1.936	0.109	N.A.		N.A.	
α	2.015	0.277	1.184	0.159	N.A.		N.A.	
θ	2.518	0.534	1.599	0.279	1.325	0.241	1.283	0.241
−2 LogL	1 456.0		1 460.1		1 399.3		1413.1	
df	6.0		6.0		14.0		13.9	
AIC	1 468.0		1 472.1		1 427.3		1 440.8	

splines or a piecewise constant hazard function with equidistant intervals) or parametric (Weibull) estimation of the hazard function.

Example 8.2 in R

In order to use the function frailtyPenal *from the package* frailtypack, *the data must be organized in a particular way. The LHS of the formula must be a* SurvIC(lower, upper, event) *object. For right-censored data, the lower and upper bound must be equal. The variable* event *should be a censoring indicator with 0 for right-censored data and 1 for interval-censored data. Left-censored data are considered to be interval-censored data with a 0 in the lower bound. The next two lines of code create new variables* upper *and* lower *based on the existing variables* ll *and* ul *to comply with these requirements and to express them in quarters of a year rather than days. The latter is done for computational reasons. The variable* censor *is already specified.*

```
> mastitis$lower <- ifelse(is.na(mastitis$ll), 0, mastitis$ll/91.31)
> mastitis$upper <- ifelse(is.na(mastitis$ul),
+                          mastitis$ll/91.31, mastitis$ul/91.31)
```

We first fit a gamma frailty model using semiparametric penalized likelihood estimation.

```
> GammaSp001 <- frailtyPenal(SurvIC(lower, upper, censor) ~
+                          cluster(cow) + rear + par24 + par56,
```

```
+                              data = mastitis,
+                              n.knots = 9, kappa = 0.01)
```

The shared frailty term is added to the model by adding a grouping variable cluster(cow) *in the RHS of the model, indicating which udder quarters share the same frailty. The number of knots specified in* n.knots *argument corresponds to the (n.knots + 2) splines functions for the approximation of the baseline hazard function. The allowed range is between 4 and 20. With the smoothing parameter (*kappa*) fixed at 100, 9 knots produced the smallest likelihood cross-validation criterion (LCV) value. Varying the smoothing parameter on the grid (0.0001, 0.001, 0.01, 0.1, 1, 10, 100, 1000) selected* $\kappa = 0.01$ *with a LCV of 1.7951. If necessary, the maximum amount of iterations may be augmented using the argument* maxit.

```
  Shared Gamma Frailty model parameter estimates
  using a Penalized Likelihood on the hazard function

            coef exp(coef) SE coef (H) SE coef (HIH)         z         p
rear   0.1827036  1.200459    0.120685      0.120437  1.513884 0.1300600
par24 -0.0901748  0.913771    0.296717      0.293026 -0.303908 0.7612000
par56  1.3969721  4.042940    0.441419      0.438067  3.164730 0.0015523

 Frailty parameter, Theta: 1.28256 (SE (H): 0.241396 ) p = 5.3884e-08

      penalized marginal log-likelihood = -706.57
      Convergence criteria:
      parameters = 5.44e-07 likelihood = 2.34e-06 gradient = 4.34e-10

      LCV = the approximate likelihood cross-validation criterion
            in the semi parametrical case     = 1.79509

      n= 400
      n events= 317  n groups= 100
      number of iterations:  8

      Exact number of knots used:  9
      Value of the smoothing parameter:  0.01, DoF:  13.86
```

Two standard errors are reported as output: SE (H) *which is estimated by inverting the Hessian matrix of the penalized likelihood and* SE (HIH) *a sandwich estimator equal to* $\boldsymbol{H}^{-1}\boldsymbol{I}\boldsymbol{H}^{-1}$ *where* $\boldsymbol{H}$ *and* $\boldsymbol{I}$ *are the Hessian of the penalized and unpenalized likelihood, respectively. The sandwich estimator provides generally more robust standard errors. The baseline hazard (*type.plot = "Hazard"*) or survival (*type.plot = "Survival"*) function can be plotted together with 95% confidence intervals (output not shown) as follows.*

```
> plot(GammaSp001, type.plot = "Hazard", conf.bands = TRUE)
> plot(GammaSp001, type.plot = "Survival", conf.bands = TRUE)
```

Replacing the arguments n.knots *and* kappa *by* hazard = "Weibull" *in the above call will fit a parametric frailty model with a Weibull baseline hazard. However, note that this analysis yielded different results from those reported in Table 8.2 because it was impossible to let the model converge to the result obtained in Table 8.2, even after providing initial values for the β-parameters and the variance of the frailty. A frailty model with a piece-wise constant hazard can be fitted using the arguments* hazard = "Piecewise-equi" *and* nb.int = 10, *indicating the number of pieces to be used (between 1 and 20, here 10).* □

8.2.1.2 SAS solution

Shared frailty models may be implemented in SAS using the procedure NLMIXED. By default, the random effects added in the NLMIXED procedure are assumed to follow a normal distribution. However, with the above explained computational trick suggested by Liu and Yu (2008) other distributions may be fitted.

First some data management is needed. In order to add only one contribution to the likelihood per cow, an indicator variable lastid is created. Also, a non-missing outcome variable dummy is created together with the variable event indicating the different types of censoring (interval, left and right).

```
DATA Mastitis;
 SET icdata.Mastitis;
 BY cow;
IF last.cow THEN lastid=1; ELSE lastid=0;
dummy=1;
     IF censor=1 and ll>0 THEN event=3; /* IC */
ELSE IF censor=1 and ll=. THEN event=2; /* LC */
ELSE IF censor=0          THEN event=1; /* RC */
RUN;
```

The NLMIXED procedure allows for only normally distributed random effects, which corresponds to a log-normal frailty distribution. The RANDOM statement specifies that the frailty b has a normal distribution with variance theta. By default adaptive Gaussian quadrature (Pinheiro and Bates, 1995) is used to maximize the marginal likelihood, with the number of quadrature points specified with the option QPOINTS. With the option COV the covariance matrix of the parameters is asked for. We assumed here a Weibull hazard function. With the BOUNDS statement, we constrain the parameters gamma, theta and alpha to be positive. The next statements define the likelihood. Based on the event variable, we calculate the contribution to the likelihood for an interval-, left- and right-censored observation.

```
PROC NLMIXED DATA=Mastitis QPOINTS=20 COV;
BOUNDS gamma>0, theta>0, alpha>0;
IF ll>0 THEN
S1 = exp(-exp(b)*(1/alpha*(ll/91.31))**gamma*
        exp(beta1*rear+beta2*par24+beta3*par56));
S2 = exp(-exp(b)*(1/alpha*(ul/91.31))**gamma*
        exp(beta1*rear+beta2*par24+beta3*par56));
     IF event=3 THEN lik=S1-S2;
else IF event=2 THEN lik=1-S2;
else IF event=1 THEN lik=S1;
llik=LOG(lik);
MODEL dummy~GENERAL(llik);
RANDOM b~NORMAL(0,theta) SUBJECT=cow;
RUN;
```

We applied the likelihood reformulation method of Liu and Yu (2008) to fit a gamma frailty model. The following statements calculate the extra log-likelihood contributions ℓ_i^b and ℓ_i^c (see Equation (8.3)). These commands should be inserted after the BOUNDS statement in the program above.

```
expb=EXP(b);
loggammaden =b/theta - 1/theta*expb
             -1/theta*LOG(theta)-LGAMMA(1/theta);
lognormalden=-b*b/2 - LOG(SQRT(2*CONSTANT('PI')));
```

After the calculation of the log-likelihood contribution llik of each cow, the extra terms per cluster are added to the contribution of the last cow in the cluster.

```
IF lastid=1 THEN llik=llik + loggammaden - lognormalden;
```

The output from both models is omitted, but the results can be consulted in Table 8.2.

Correlated frailty models

The correlated frailty model was introduced by Yashin et al. (1995) and is an extension of the shared frailty model. The extension consists in assuming a multivariate frailty term which itself has a correlation structure. In the case of bivariate survival times, this model would assume a bivariate random effects vector say with a bivariate normal distribution. The shared frailty model is a special case of the correlated frailty model with correlations between the frailties equal to one. Hens et al. (2009) fitted a correlated frailty model to serological bivariate current status data. For this, they constructed a bivariate gamma distribution for the frailty term. In the appendix to their paper, Hens et al. (2009) provide SAS code to fit the correlated gamma frailty model for current status data using the NLMIXED procedure.

8.2.2 A marginal approach to correlated survival times

The bivariate parametric regression models of Chapter 7 allow for a marginal interpretation of the regression coefficients. However, these models allow only bivariate survival responses. In addition, even when interest lies only in the marginal regression coefficients, all previously discussed models must pay attention to the correct specification of the correlation structure. In this section we discuss a marginal model where interest lies only in the marginal regression coefficients, hence treating the correlation as a nuisance. The analysis will be based on the *Generalized Estimating Equations (GEE)* (Liang and Zeger, 1986) approach.

8.2.2.1 Independence working model

For modelling paired (right-censored) survival data with covariates, Huster et al. (1989) derived a GEE approach allowing to draw inferences about the marginal distributions while treating the dependence between the members of the cluster as nuisance. More specifically, the method allows to specify the marginal distributions independently of the association structure and in addition leaves the nature of the dependence between the event times of the cluster members unspecified. The parameters in the marginal model are estimated by maximum likelihood under the independence assumption. This model was called the *Independence Working Model (IWM)* by Huster et al. (1989). In a second step, the standard errors of the estimated parameters are adjusted with a sandwich correction.

Bogaerts et al. (2002) extended the method to interval-censored clustered data with cluster size ≥ 2. Assume we have m event times per cluster. Let T_{ij} be the random variable representing the exact time until the jth member of cluster i ($i = 1, \ldots, n; j = 1, \ldots, m$) has experienced the event. Assume also that the marginal survival functions S_j and marginal densities f_j ($j = 1, \ldots, m$) are both parametrized with the vector $\boldsymbol{\theta}$ containing all parameters of all member specific parameters. Let $\lfloor l_{ij}, u_{ij} \rfloor$ be the interval which contains T_{ij} in the ith cluster. Further, let δ_{ijk} ($k = 1, \ldots, 4$) be the indicator variables for an exact, right-, left- or interval-censored observation, respectively, for the jth member of the ith cluster. The likelihood contribution for cluster i can then be written as

$$L_{IWM,i}(\boldsymbol{\theta}) = \prod_{j=1}^{m} \Big[f_j(l_{ij};\boldsymbol{\theta})^{\delta_{ij1}} \, S_j(l_{ij};\boldsymbol{\theta})^{\delta_{ij2}} \, \big\{1 - S_j(u_{ij};\boldsymbol{\theta})\big\}^{\delta_{ij3}} \times \big\{S_j(l_{ij};\boldsymbol{\theta}) - S_j(u_{ij};\boldsymbol{\theta})\big\}^{\delta_{ij4}} \Big].$$

The IWM-likelihood is then equal to $\prod_i L_{IWM,i}(\boldsymbol{\theta})$. Under the correct specification of the marginal distributions, the IWM-estimates $\widehat{\boldsymbol{\theta}}$, which maximize the IWM-likelihood, can be shown to be consistent.

Further, in the independence case and following Royall's (1986) approach, Bogaerts et al. (2002) derived consistent estimators for the variance under model misspecification. One can show that under mild regularity conditions (Royall, 1986) for a consistent estimator $\widehat{\boldsymbol{\theta}}$, $\sqrt{n}(\widehat{\boldsymbol{\theta}}-\boldsymbol{\theta})$ converges in distribution to $\mathcal{N}\big(\mathbf{0},\ \boldsymbol{\Lambda}(\boldsymbol{\theta})\big)$ with

$$\boldsymbol{\Lambda}(\boldsymbol{\theta}) = n\,\boldsymbol{\Upsilon}(\boldsymbol{\theta})^{-1}\,\mathsf{E}\{\boldsymbol{U}(\boldsymbol{\theta})\boldsymbol{U}(\boldsymbol{\theta})'\}\boldsymbol{\Upsilon}(\boldsymbol{\theta})^{-1},$$

where $\boldsymbol{U}$ is the score vector and $\boldsymbol{\Upsilon}(\boldsymbol{\theta}) = \mathsf{E}\Big[-\partial^2 \log\{L_{IWM}(\boldsymbol{\theta})\}/\partial\boldsymbol{\theta}\partial\boldsymbol{\theta}'\Big]$ is the expected information matrix, respectively. In practice, $\boldsymbol{\Lambda}(\boldsymbol{\theta})$ can be estimated by

$$\widehat{\boldsymbol{\Lambda}}(\widehat{\boldsymbol{\theta}}) = n\,\boldsymbol{H}(\widehat{\boldsymbol{\theta}})^{-1}\left\{\sum_i \boldsymbol{U}_i(\widehat{\boldsymbol{\theta}})\boldsymbol{U}_i(\widehat{\boldsymbol{\theta}})'\right\}\boldsymbol{H}(\widehat{\boldsymbol{\theta}})^{-1},$$

where $\boldsymbol{H}$ is the matrix of second derivatives of $\log(L_{IWM})$.

The multivariate analysis allows us to calculate an overall P-value for intra-cluster questions. In case of a significant result, pairwise tests and corresponding confidence intervals can be obtained as well.

Example 8.3 Signal Tandmobiel study

Research questions such as "Is there a left-right symmetry with respect to the mean (median) emergence times?" are of interest to a dentist. This intra-subject question involves the comparison of the correlated emergence times of teeth from the same child. As an illustration, we verify whether there is a left-right symmetry in the emergence times of the first premolars in the four quadrants after a correction for gender, i.e., teeth 14 (right) vs 24 (left) in the upper jaw and teeth 34 (left) vs 44 (right) in the lower jaw. Overall no significant difference in emergence times between the first premolars could be shown ($P = 0.1465$).

□

8.2.2.2 SAS solution

Bogaerts et al. (2002) provided the SAS macro %IWM to fit marginal log-logistic distributions. Marginal distributions other than the log-logistic distribution can be fitted by adapting the code in the macro. Note that it is not sufficient to adapt only the PROC LIFEREG parts in the beginning. Also the first and second derivatives of the likelihood must be adapted.

Example 8.3 in SAS

First we create a data set which has a record for each tooth. The left and right bounds are stored in ll *and* ul*, respectively. Note that left-censored observations must have a missing value in the left bound and right-censored observations a missing value in the right bound. Both conditions are already satisfied with our data. In addition, we create indicator variables for the four different teeth.*

```
DATA tandmob;
 SET icdata.tandmob;
ll=L14; ul=R14; tooth14=1; tooth24=0; tooth34=0; tooth44=0; OUTPUT;
ll=L24; ul=R24; tooth14=0; tooth24=1; tooth34=0; tooth44=0; OUTPUT;
ll=L34; ul=R34; tooth14=0; tooth24=0; tooth34=1; tooth44=0; OUTPUT;
ll=L44; ul=R44; tooth14=0; tooth24=0; tooth34=0; tooth44=1; OUTPUT;
KEEP idnr LL UL tooth14 tooth24 tooth34 tooth44 gender;
RUN;
```

The following call fits a model for the 4 premolars. The variable identifying the cluster must be specified in the argument id*. The model to be fitted is specified in argument* x*. Note that class variables are not allowed and must be replaced by dummy variables. A common scale factor was assumed but the macro has the ability to let the scale depend on covariates by specifying the argument* z=*. Finally, there is the possibility to estimate the emergence time, the difference in emergence times and to perform global testing. As an illustration, we estimated the median tooth emergence time of tooth 14 for boys, compared the median tooth emergence time of teeth 14 and 24, and tested if there is an overall left-right difference, as well as for the two pairs of teeth (14 vs 24 and 34 vs 44). Note that the difference should be specified as two entries separated with a comma and both included in a %*str*-command.*

```
%iwm(data=tandmob,
     left=ll, right=ul ,
     id=idnr, x=gender tooth14 tooth24 tooth34 tooth44,
     int=n,
     outcon=outcon,
     est1=%str(0 1 0 0 0),
     estlab1=%str(median emergence time for boys of T14),
           dif1=%str(0 1 0 0 0,
               0 0 1 0 0),
     diflab1=%str(diff. emergence times between T14 and T24 for boys),
     con1=%str(0 1 -1 0  0,
               0 0  0 1 -1),
     conlab1=%str(join test left-right),
     con2=%str(0 1 -1 0 0),
     conlab2=%str(left-right T14-T24),
```

```
con3=%str(0 0 0 1 -1),
conlab3=%str(left-right T34-T44));
```

The macro outputs the standard LIFEREG *output that does not correct for clustering. The following estimates are also given: the log-scale parameters (*zeta*), the scale parameters (*sigma_z*), the inverse scale parameters (*gamma*), the iteration history of the log-likelihood (*iter_ll*), the final parameter estimates (*beta *and* scale*), the maximum change in parameter estimates at convergence (*crit*), the change in parameter estimates (*delta*), the score vector (*U*), the naive and robust covariance and correlation matrix, the naive and robust standard errors, test statistics and P-values, the robust covariance matrix after transformation of the scale factor and finally the requested estimates, differences and contrasts. Only part of the output is shown below.*

```
labelx        beta      scale       crit
GENDER   -0.036365 0.0704214 2.069E-10
TOOTH14  2.3671596
TOOTH24  2.3593517
TOOTH34  2.3615912
TOOTH44  2.3574042

        Naive Standard Errors Robust Standard Errors
GENDER              0.0062040917            0.0105454356
TOOTH14              0.006993907            0.0077194293
TOOTH24             0.0068629032            0.0072861438
TOOTH34             0.0069096458             0.007524689
TOOTH44             0.0069699158             0.007590467

        Naive p-value Robust p-value
GENDER          0.0000         0.0006
TOOTH14         0.0000         0.0000
TOOTH24         0.0000         0.0000
TOOTH34         0.0000         0.0000
TOOTH44         0.0000         0.0000

median emergence time for boys of T14
Estimate Standard Error
  10.667          0.082
95% Confidence Interval
10.505657    10.828444

diff. emergence times between T14 and T24 for boys
Estimate Standard Error
   0.083          0.048
95% Confidence Interval
```

(Cont.)

```
-0.010595      0.1765218

join test left-right
Test statistic      Degrees of Freedom    P-value
          3.84                       2     0.1465
```

□

8.3 A biplot for interval-censored data

8.3.1 Classical biplot

The biplot (Gabriel, 1971) is a graph that displays information on both individuals and variables of a data matrix in a low-dimensional space, often of dimension $m = 2$. Various types of biplots have been suggested. When the focus of analysis is on the relationship between variables, the individuals in the biplot are displayed as points while variables are displayed as vectors. Alternatively, individuals can also be represented as vectors and the variables as points.

The construction of the biplot goes as follows. The two-way data matrix $\boldsymbol{Y}$ of n subjects and m variables with $n > m$ can be factorized according to the singular value decomposition (SVD):

$$\boldsymbol{Y} = \boldsymbol{U}\boldsymbol{\Lambda}\boldsymbol{V}', \tag{8.4}$$

with $\boldsymbol{U}$ a $n \times m$ matrix that satisfies $\boldsymbol{U}'\boldsymbol{U} = \mathbb{I}_m$ and $\boldsymbol{V}$ a $m \times m$ matrix that satisfies $\boldsymbol{V}\boldsymbol{V}' = \mathbb{I}_m$. The columns of $\boldsymbol{U}$ are the 'individual' vectors and are called the left singular vectors. The rows of $\boldsymbol{V}$ are the 'variable' vectors and are called the right singular vectors. In both sets, the vectors are orthonormal. The matrix $\boldsymbol{\Lambda}$ is the $m \times m$ diagonal matrix of singular values arranged in decreasing order of magnitude, i.e., $\mathsf{diag}(\lambda_1, \ldots, \lambda_m)$ with $\lambda_1 \geq \lambda_2 \geq \ldots \geq \lambda_m \geq 0$. For the (i, j)th element of $\boldsymbol{Y}$, Equation (8.4) implies

$$y_{ij} = \sum_{s=1}^{m} \lambda_s u_{is} v_{js} = \boldsymbol{u}_i'\boldsymbol{v}_j, \qquad (i = 1, \ldots, n; j = 1, \ldots, m),$$

where $\boldsymbol{u}_i = (\lambda_1^{\alpha} u_{i1}, \ldots, \lambda_m^{\alpha} u_{im})'$ and $\boldsymbol{v}_j = (\lambda_1^{1-\alpha} v_{j1}, \ldots, \lambda_m^{1-\alpha} v_{jm})'$, with $0 \leq \alpha \leq 1$. Hence, the (i, j)th element of $\boldsymbol{Y}$ can be obtained as the scalar product of the vectors $\boldsymbol{u}_i$ and $\boldsymbol{v}_j$. If $\lambda_1 \geq \ldots \geq \lambda_d \gg \lambda_{d+1}$, then

$$y_{ij} \approx \widetilde{y}_{ij} = \sum_{s=1}^{d} \lambda_s u_{is} v_{js} = \widetilde{\boldsymbol{u}}_i'\widetilde{\boldsymbol{v}}_j, \tag{8.5}$$

with $\widetilde{\boldsymbol{u}}_i = (\lambda_1^\alpha u_{i1}, \ldots, \lambda_d^\alpha u_{id})'$ and $\widetilde{\boldsymbol{v}}_j = (\lambda_1^{1-\alpha} v_{j1}, \ldots, \lambda_d^{1-\alpha} v_{jd})'$. Another way to present the approximation is

$$y_{ij} \approx \|\widetilde{\boldsymbol{u}}_i\| \|\widetilde{\boldsymbol{v}}_j\| \cos\theta_{ij},$$

where $\|\cdot\|$ denotes the Euclidean length and θ_{ij} is the angle between $\widetilde{\boldsymbol{u}}_i$ and $\widetilde{\boldsymbol{v}}_j$. The part $\|\widetilde{\boldsymbol{v}}_j\| \cos\theta_{ij}$ is equal to the projection of variable $\widetilde{\boldsymbol{v}}_j$ onto the direction of $\widetilde{\boldsymbol{u}}_i$. Hence, y_{ij} is reconstructed by the projection multiplied by the length of $\widetilde{\boldsymbol{u}}_i$.

Thus, if there are only d important singular values, then d dimensions are enough to represent (approximately) the original data values. Eckart and Young (1936) showed that the best d-dimensional least squares approximation of the matrix $\boldsymbol{Y}$ can be obtained from the SVD. That is, $\sqrt{\sum_{i=1}^n \sum_{j=1}^m (y_{ij} - \widetilde{y}_{ij})^2}$ is minimized by taking $\widetilde{y}_{ij}$ from Equation (8.5). Typically the first d values of $u_{is}, (s = 1, \ldots, d)$ and $v_{js}, (s = 1, \ldots, d)$ are plotted in an d-dimensional plot to get insight into the relationships between variables and the interplay between individuals and variables. Most often $d = 2$, which then provides an extension of a classical scatterplot. Further, when the columns in $\boldsymbol{Y}$ are mean-centered, the SVD solution corresponds to the PCA solution. There are three popular choices of α (0, 0.5, 1), whereby the choice depends on what aspect one wishes to present best, see Gower and Hand (1996) for more details.

8.3.2 Extension to interval-censored observations

Cecere et al. (2013) extended the biplot to interval-censored data with the emergence times of permanent teeth in the Signal Tandmobiel study as an application. Now the jth variable for subject i is only known to lie in an interval $\lfloor l_{ij}, u_{ij} \rfloor$, i.e., the data are intervals and the interest goes to obtain insight in the structure of the true values y_{ij} $(i = 1, \ldots, n; j = 1, \ldots, m)$. For our application, y_{ij} represents (the logarithm of) the emergence time of the jth permanent tooth (i.e., teeth 14, 15 and 16) of the ith child. The extension of the biplot must satisfy the requirement that $\widetilde{y}_{ij} \in \lfloor l_{ij}, u_{ij} \rfloor$. Since only finite intervals can be handled, right-censored observations are changed into interval-censored observations with a sufficiently large upper bound.

The algorithm searches for a configuration of n vectors $\widetilde{\boldsymbol{u}}_i$ representing the individuals and a configuration of d vectors $\widetilde{\boldsymbol{v}}_j$ in dimension d $(d = 2)$ such that the projections of the points representing the variables $\widetilde{\boldsymbol{v}}_j$ onto each vector $\widetilde{\boldsymbol{u}}_i$ correspond as closely as possible to the unknown optimally transformed data $\widetilde{y}_{ij}$, which satisfy the interval constraints. In other words, the vectors $\widetilde{\boldsymbol{u}}_i$ and $\widetilde{\boldsymbol{v}}_j$ should satisfy that $\widetilde{\boldsymbol{u}}_i'\widetilde{\boldsymbol{v}}_j$ is as close as possible to $\widetilde{y}_{ij} \in \lfloor l_{ij}, u_{ij} \rfloor$ across all i and j. Note that now $\widetilde{y}_{ij}$, as well as $\widetilde{\boldsymbol{u}}_i$, as $\widetilde{\boldsymbol{v}}_j$ $(i = 1, \ldots, n; j = 1, \ldots, m)$ must be determined iteratively. Formally and in matrix notation, a solution to the problem is obtained by minimizing the loss function

$$L(\widetilde{\boldsymbol{U}}, \widetilde{\boldsymbol{V}}, \widetilde{\boldsymbol{Y}}) = \|\widetilde{\boldsymbol{Y}} - \widetilde{\boldsymbol{U}}\widetilde{\boldsymbol{V}}'\|^2 = \sum_{i=1}^{n}\sum_{j=1}^{m}(\widetilde{y}_{ij} - \widetilde{\boldsymbol{u}}_i'\widetilde{\boldsymbol{v}}_j)^2 \tag{8.6}$$

over all configurations $\widetilde{\boldsymbol{U}}$, $\widetilde{\boldsymbol{V}}$ and $\widetilde{\boldsymbol{Y}}$ subject to $l_{ij} \leq \widetilde{y}_{ij} \leq u_{ij}$. The $n \times d$ matrix $\widetilde{\boldsymbol{U}}$ has columns $\widetilde{\boldsymbol{u}}_i$, the $d \times d$ matrix $\widetilde{\boldsymbol{V}}$ has columns $\widetilde{\boldsymbol{v}}_j$ and the $n \times d$ matrix $\widetilde{\boldsymbol{Y}}$ consists of optimally transformed data constructed from the original intervals. Although common in PCA, no standardization of the variables is performed here because the aim is to reconstruct the data values and visually represent them in a biplot. The algorithm that minimizes (8.6) is an iterative two-step minimization procedure:

1. Fix the optimally transformed data $\widetilde{\boldsymbol{Y}}$ and find configurations $\widetilde{\boldsymbol{U}}$ and $\widetilde{\boldsymbol{V}}$ such that L is minimized over all matrices $\widetilde{\boldsymbol{U}}$ and $\widetilde{\boldsymbol{V}}$.

2. Fix the configurations $\widetilde{\boldsymbol{U}}$ and $\widetilde{\boldsymbol{V}}$ and estimate the optimally transformed data $\widetilde{\boldsymbol{Y}}$ subject to its constraints.

A solution to step 1 is a well-known result in classical PCA and it is given by a singular value decomposition of $\widetilde{\boldsymbol{Y}}$. In step 2, $\widetilde{y}_{ij}$ is set to $\max(l_{ij}, \min(u_{ij}, \sum_{s=1}^{d} u_{is}v_{js}))$ to guarantee that $\widetilde{y}_{ij} \in \lfloor l_{ij}, u_{ij} \rfloor$. Both update steps are least-squares optimal over one of the sets of parameters given the other is kept fixed. Alternating these two steps guarantee that the least squares loss function L, which is bounded from below, descends in each step. Hence, the algorithm will convergence.

Cecere et al. (2013) also proposed two types of diagnostic measures. The first one is the DAF (dispersion accounted for) of the reconstructed data and is similar to the variance accounted for in the case $\widetilde{\boldsymbol{Y}}$ would be column centered and have unit column variance. DAF takes values between 0 and 1, with the value 1 in case of perfect reconstruction. The second type of diagnostic measure can be derived when the rank order of the data is of interest, like in Example 8.4 below. One can compare how many times the rank order in $\widetilde{\boldsymbol{Y}}$ matches the reconstructed rank order in the biplot and how often it is wrong by only 2, 3 or more interchanges. The greater the number of matches, the better the model is doing with respect to preserving the ranking of the variables per sample. We refer to Cecere et al. (2013) for more technical details.

Example 8.4 Signal Tandmobiel study

We want to examine the emergence pattern of the two permanent premolars (teeth 14 and 15) and first permanent molar (tooth 16) in the upper right quadrant of the mouth for boys. Figure 8.3 displays the biplot. The dots represent teeth 14, 15 and 16. The crosses represent the 256 children. The arrow points to one randomly chosen individual. The projections of the teeth onto the arrow provides the sequence of emergence of the teeth. For this individual, tooth 16 emerged first, followed by teeth 14 and 15. Children in the gray area all have the same emergence sequence. It is a feature of the biplot that the space divides into sectors in which all children have the same emergence sequence. The DAF was equal to 0.975 indicating a good fit. In $\widetilde{\boldsymbol{Y}}$, ties were observed in y_{ij} for 72 children. Assuming these ties follow the same sequence as the reconstructed emergence times $\widetilde{\boldsymbol{u}}_i'\widetilde{\boldsymbol{v}}_j$, all sequences were similar as those of the estimated optimally transformed data $\widetilde{y}_{ij}$. The majority of children (73%) had tooth emergence sequence 16-15-14. The sequence 16-

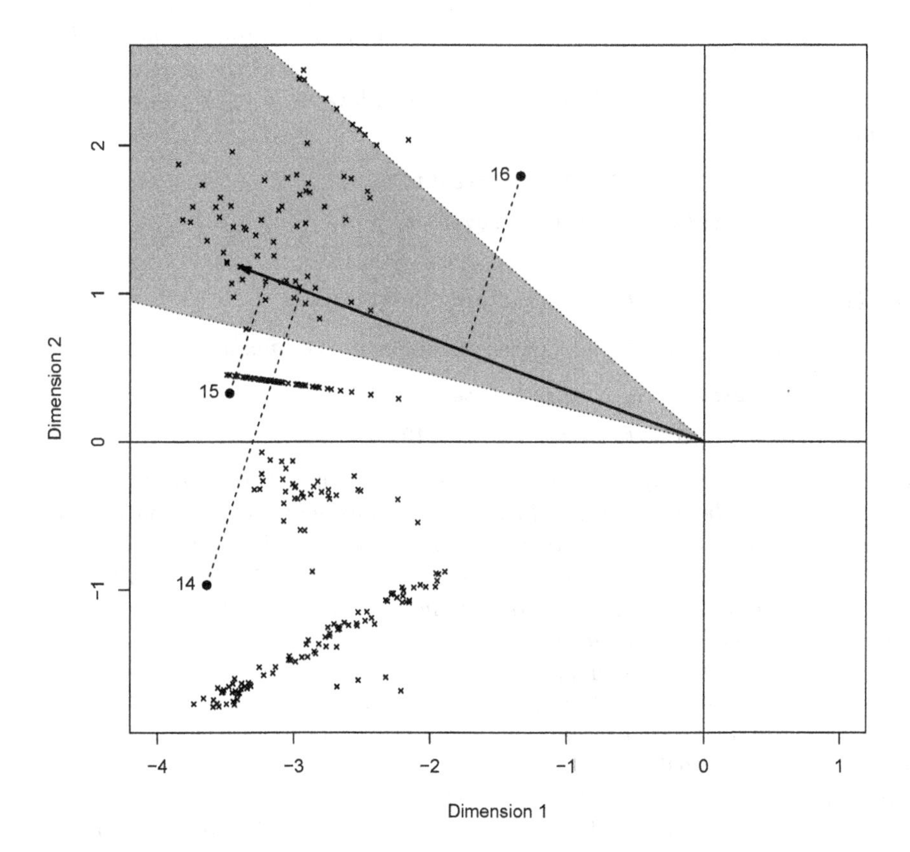

FIGURE 8.3: Signal Tandmobiel study. Biplot for 256 boys of teeth 14, 15 and 16, obtained with R package icensBKL. Points are teeth, each cross represents a child. Projection of the teeth onto the arrow of a particular child indicates the sequence of emergence. All children in the gray area have the same sequence of emergence.

14-15 was seen in 24% of the children and the remaining 3% had the sequence 14-16-15.

□

8.3.3 R solution

The biplot for interval-censored data can be prepared by the function icbiplot included in the R package icensBKL which accompanies this book.

Example 8.4 in R

First the Tandmobiel data are loaded and the boys are selected. The left and right endpoints of the 3 teeth are combined into the vectors L *and* R. *The lower limit of left-censored observations is set to zero and the upper limit of right-censored observations is set to a high enough value (e.g., 20) representing infinity.*

```
> L <- cbind(Boys$L14, Boys$L15, Boys$L16)
> R <- cbind(Boys$R14, Boys$R15, Boys$R16)
> L[is.na(L)] = 0
> R[is.na(R)] = 20      ## 20 = infinity in this case
```

The IC-Biplot is then calculated with the following command:

```
> icb <- icbiplot(L, R, p = 2, MaxIter = 10000, tol = 1e-6,
+                   plotit = TRUE, seed = 12345)
```

L *and* R *are the only required arguments. Argument* p $= 2$ *states that the original data should be approximated by* $d = 2$ *dimensional vectors. Convergence is obtained when two successive iterations yield a relative difference in the loss function* L *smaller than the value specified in* tol *(*10^{-6} *by default). The maximum number of iterations is determined by* MaxIter *(10 000 by default). Note that one needs to verify that the final number of iterations performed (*icb$iter*) is lower than the maximal number specified. No warning message is provided when the maximal number of iterations is reached but convergence is not yet obtained. A basic plot (not shown) is provided when the* plotit *argument is set to* TRUE. *The* seed = *argument is used to set the seed for the random number generator. By default, the current* R *system seed is used.*

The program to create the enhanced Figure 8.3 is available in the supplementary materials.

We also performed the two diagnostic measures for the IC-biplot. The DAF value equals to 0.975, which indicates a good fit, and can be obtained using

```
> print(icb$DAF)
```

Given that the ordering of the data is important in our example, we also verified how often the rank in $\widetilde{\boldsymbol{Y}}$ *matches the reconstructed ranks in the biplot. The ranks for both can be found as follows:*

```
> orderXY <- t(apply(icb$X %*% t(icb$Y), 1, rank))
> orderH <- t(apply(icb$H, 1, rank))
```

Note that in the above code, $\boldsymbol{H}$ *plays the role of* $\widetilde{\boldsymbol{Y}}$, $\boldsymbol{X}$ *plays the role of* $\widetilde{\boldsymbol{U}}$, $\boldsymbol{Y}$ *plays the role of* $\widetilde{\boldsymbol{V}}$. *Note also that there are ties in the* $\widetilde{y}_{ij}$. *The maximum difference in rank is 0.5 indicating that, assuming ranks are in the same order for the ties, a perfect reconstruction of the ranks is present. Ties occurred for 72 children.*

```
> max(orderXY - orderH)
```

```
[1] 0.5
```

```
> sum(apply(orderH, 1, function(x) any((trunc(x) - x) < 0)))
```

```
[1] 72
```

The frequencies of the specific sequences can be obtained using

```
> table(apply(orderXY, 1, paste, collapse = ","))
```

```
1,3,2  2,3,1  3,2,1
    8     61    187
```

The sequence 1, 3, 2 corresponds then to the sequence of emerging of teeth 14-16-15 and occurs 8 times. The other sequences are obtained analogously.

□

8.4 Concluding remarks

What is reviewed in this section is just the tip of the iceberg. We have taken here a personal choice of additional topics where extensions were developed to deal with interval-censored survival times. But no doubt, interval censoring can be included into each existing statistical approach and possibly turned into a relatively complex one. In the next part of the book, we will treat complex problems involving interval-censored survival times analyzed with the Bayesian approach.

Part III

Bayesian methods for interval-censored data

Chapter 9

Bayesian concepts

We now review the basic concepts of Bayesian inference with a focus on statistical modelling. We start in Section 9.1 with parametric Bayesian approaches, which are by far the most popular. This popularity has much to do with the widespread use of WinBUGS and related software. We review in this section also the choice of a Bayesian summary measure to express the importance of model parameters. We then introduce in that section the Data Augmentation (DA) approach, which assumes that additional data are available such that the posterior computations become easier. The DA approach is a popular tool in hierarchical models whereby the random effects are treated as parameters, but assumed to be known when sampling the full conditional distributions for the estimation process. In combination with Markov chain Monte Carlo (MCMC) algorithms, the model parameters are then relatively easy to estimate. Also, in Section 9.1 we describe the principles that are the basis for the MCMC algorithms. No doubt, these algorithms have drastically changed the Bayesian landscape and greatly contributed to its popularity nowadays. In the same section, we review how to select and check assumed models in a Bayesian context. The Bayesian nonparametric (BNP) and semiparametric (BSP) approach is of a more advanced theoretical level. We cover in Section 9.2 and in the remainder of the book only the basic principles of two classes of BNP and BSP approaches.

At the end of this chapter we discuss the practical implementation of the Bayesian methods for statistical modelling. First, in Section 9.3 we briefly review the available Bayesian software. Up until recently, this was dominated by BUGS-like software such as WinBUGS, OpenBUGS and JAGS. We highlight how these packages differ in the way interval-censored data should be handled. Recently, some SAS survival procedures allow for a Bayesian approach, and a dedicated SAS procedure, MCMC, has been developed that can deal with basically any parametric Bayesian analysis. Software for BNP and BSP models was lacking until very recently. This changed with the introduction of the R package DPpackage, see Jara et al. (2011). Some functions from this package allow for interval-censored data. This and other R packages are introduced and exemplified in Chapters 10 to 13. In Section 9.4, the use of some parametric Bayesian software, i.e., BUGS and SAS procedure LIFEREG, is illustrated but only for right-censored data. Some elementary BNP and BSP analyses for right-censored data are illustrated in the same section with self-written R programs, one combined with BUGS.

9.1 Bayesian inference

In the Bayesian paradigm, both the data and the parameter vector $\boldsymbol{\theta}$ are treated as random variables. Besides the probabilistic model to generate the data, *a prior distribution* $p(\boldsymbol{\theta})$ expressing the prior belief about which values of $\boldsymbol{\theta}$ are plausible, must be specified. This prior belief can range from basically "ignorance" to a strong subjective belief. In the first case, a vague (also coined as noninformative) prior distribution is used, while in the second case the prior distribution is called subjective. The prior distribution is combined with the likelihood $L(\boldsymbol{\theta}) \equiv L(\boldsymbol{\theta} \mid \boldsymbol{\mathcal{D}})$, where $\boldsymbol{\mathcal{D}}$ stands for the collected data. Inference is then based on the *posterior distribution* $p(\boldsymbol{\theta} \mid \boldsymbol{\mathcal{D}})$ of the parameters given the data which is calculated using *Bayes*' rule:

$$p(\boldsymbol{\theta} \mid \boldsymbol{\mathcal{D}}) = \frac{L(\boldsymbol{\theta})\, p(\boldsymbol{\theta})}{\int_{\Theta} L(\boldsymbol{\theta})\, p(\boldsymbol{\theta})\, \mathrm{d}\boldsymbol{\theta}} \propto L(\boldsymbol{\theta})\, p(\boldsymbol{\theta}). \tag{9.1}$$

In the Bayesian paradigm, all information about a model parameter is contained in the posterior distribution. Thus, inference in a Bayesian context is done conditional on only the observed data. One says that Bayesian inference is based on the parameter space. This is different from the frequentist approach where inference is based on the sample space. This means that the observed result is contrasted to what would have been obtained if the study had been repeated. An example is the classical P-value where repeated sampling is done under a null hypothesis.

Suppose that $\boldsymbol{\theta} = (\theta_1, \ldots, \theta_q)'$ and $p(\theta_j \,|\, \boldsymbol{\mathcal{D}})$ $(j = 1, \ldots, q)$ are the marginal posterior distributions of the elements of the parameter vector $\boldsymbol{\theta}$ derived from the joint posterior distribution $p(\boldsymbol{\theta} \,|\, \boldsymbol{\mathcal{D}})$. Classically three estimators are considered in a Bayesian setting as point estimates of the elements of $\boldsymbol{\theta}$. The posterior mean equal to $\overline{\theta}_j = \int \theta_j\, p(\theta_j \mid \boldsymbol{\mathcal{D}})\, \mathrm{d}\theta_j$ is preferable to characterize symmetric posterior distributions, while the posterior median $\overline{\theta}_{M,j}$ defined by $0.5 = \int_{-\infty}^{\overline{\theta}_{M,j}} p(\theta_j \mid \boldsymbol{\mathcal{D}})\, \mathrm{d}\theta_j$ is used for asymmetric posterior distributions. These two posterior summary measures are most popular in practice since they can easily be obtained from MCMC output. The third location summary measure is the posterior mode, defined as $\widehat{\boldsymbol{\theta}}_M = \arg\max_{\boldsymbol{\theta}} p(\boldsymbol{\theta} \mid \boldsymbol{\mathcal{D}})$, and is less used as it is not reported by classical Bayesian software. Uncertainty about the model parameters can be expressed by the posterior standard deviations $\overline{\sigma}_j$, equal to the square root of $\overline{\sigma}_j^2 = \int (\theta_j - \overline{\theta}_j)^2 p(\theta_j \mid \boldsymbol{\mathcal{D}}) \mathrm{d}\theta_j$. Another possibility is to use a Bayesian version of the confidence interval, now called a credible interval constructed from the posterior distribution. Two types of credible intervals are in use, as explained in Section 9.1.4.

For an extensive introduction into the area of Bayesian statistics, see, e.g., Robert (2007), Carlin and Louis (2008), Lesaffre and Lawson (2012), and Gelman et al. (2013).

9.1.1 Parametric versus nonparametric Bayesian approaches

Bayes' rule combines the prior distribution with the likelihood function. The latter describes the probabilistic mechanism that generates the data. A statistical method that assumes a specific class of distributions (with a finite number of parameters) for the data is called *parametric*. For instance, many classical statistical tests assume that the data have a Gaussian distribution $\mathcal{N}(\mu, \sigma^2)$ whereby μ and σ^2 are estimated from the data. In a Bayesian approach, we put a prior $p(\mu, \sigma^2)$ on these parameters to specify our prior belief which Gaussian distributions are to be preferred. But often, one is not sure about the choice of the distribution. In that case, one might wish to relax the distributional assumption, e.g., by assuming a larger family of distributions or not to rely on any distributional assumption at all. The latter is characteristic for the classical nonparametric methods such as the Wilcoxon-Mann-Whitney test, which is based on the overall ranks for the data to compare two distributions. In a BNP approach the class of possible distributions is much larger justifying the adjective "nonparametric". However, the statistical basis for these methods is quite advanced. In Section 9.2 we provide a brief overview of some popular BNP and BSP approaches. In Section 9.4 we provide some elementary BNP and BSP analyses for right-censored survival times using currently available software.

9.1.2 Bayesian data augmentation

The computation of the likelihood for interval-censored data is more involved. Indeed, the maximum likelihood method involves integration as seen in Equation (1.13). Combined with the optimization of the likelihood this could become prohibitively complex even for simple models.

In the Bayesian paradigm, the parameters $\boldsymbol{\theta}$ are assumed to be random and the posterior distribution $p(\boldsymbol{\theta} \mid \boldsymbol{\mathcal{D}})$ is used for inference. It could be useful to augment the vector of unknowns by specifically chosen *auxiliary variables*, say $\boldsymbol{\psi}$. For example, when for some subjects their response is missing, we could assign to the unobserved response a parameter in addition to the model parameters. Inference can then be based on the joint posterior distribution $p(\boldsymbol{\theta}, \boldsymbol{\psi} \mid \boldsymbol{\mathcal{D}})$. Indeed, all (marginal) posterior characteristics of $\boldsymbol{\theta}$ (mean, median, credible intervals) are the same regardless of whether they are computed from $p(\boldsymbol{\theta} \mid \boldsymbol{\mathcal{D}})$ or $p(\boldsymbol{\theta}, \boldsymbol{\psi} \mid \boldsymbol{\mathcal{D}})$ since

$$p(\boldsymbol{\theta} \mid \boldsymbol{\mathcal{D}}) = \int p(\boldsymbol{\theta}, \boldsymbol{\psi} \mid \boldsymbol{\mathcal{D}}) \, \mathrm{d}\boldsymbol{\psi}. \tag{9.2}$$

Since the above procedure augments the original set of data, it is called *(Bayesian) Data Augmentation.* Note that in Section 5.4 we have introduced the Data Augmentation algorithm which is based on the same principle but uses a specific numerical technique to obtain an expression of the posterior distribution.

For censored data, matters simplify considerably if the unknown true event times t_i are explicitly considered to make part of the vector of unknowns, i.e., $\boldsymbol{\psi} = (t_i \,:\, i = 1, \ldots, n,$ for which T_i is censored$)'$. Then Bayes' rule leads to

$$p(\boldsymbol{\theta},\, \boldsymbol{\psi} \mid \mathcal{D}) \propto L^*(\boldsymbol{\theta},\, \boldsymbol{\psi})\, p(\boldsymbol{\theta},\, \boldsymbol{\psi}), \tag{9.3}$$

where $L^*(\boldsymbol{\theta},\, \boldsymbol{\psi}) = p(\mathcal{D} \mid \boldsymbol{\theta},\, \boldsymbol{\psi})$ is the likelihood given the augmented parameter vector. The term $p(\boldsymbol{\theta},\, \boldsymbol{\psi})$ can further be decomposed as

$$p(\boldsymbol{\theta},\, \boldsymbol{\psi}) = p(\boldsymbol{\psi} \mid \boldsymbol{\theta})\, p(\boldsymbol{\theta}), \tag{9.4}$$

where $p(\boldsymbol{\theta})$ is the (usually tractable) prior distribution of the original parameter vector. The conditional distribution $p(\boldsymbol{\psi} \mid \boldsymbol{\theta})$ is in fact the likelihood would there be no censoring, i.e., $p(\boldsymbol{\psi} \mid \boldsymbol{\theta})$ is free of integrals of the type shown in Expression (1.13). Finally, also the augmented likelihood $L^*(\boldsymbol{\theta},\, \boldsymbol{\psi})$ has typically a tractable form being proportional to a product of zeros and ones as will be illustrated in the following example.

Example 9.1 Log-normal model
Let T_i $(i = 1, \ldots, n)$ be a set of i.i.d. event times following a log-normal distribution. That is, $Y_i = \log(T_i)$ $(i = 1, \ldots, n)$ are i.i.d. with a normal distribution $\mathcal{N}(\mu,\, \sigma^2)$. The parameter vector is $\boldsymbol{\theta} = (\mu,\, \sigma^2)'$. As prior distribution, we can take the conjugate normal-inverse gamma distribution, i.e.,

$$p(\boldsymbol{\theta}) = p(\mu,\, \sigma^2) = p(\mu \mid \sigma^2)\, p(\sigma^2),$$

where $p(\mu \mid \sigma^2)$ is a density of the normal distribution $\mathcal{N}(m,\, s\,\sigma^2)$ and $p(\sigma^2)$ is an inverse gamma density, which means σ^{-2} has a gamma distribution $\mathcal{G}(h_1,\, h_2)$. Parameters $m \in \mathbb{R}$, s, h_1, $h_2 > 0$ are prior hyperparameters which must be chosen by the analyst.

Suppose that all observations are interval-censored and the i-th observed interval $\lfloor l_i,\, u_i \rfloor$ results from checking the survival status at random or fixed times $\boldsymbol{C}_i = \{c_{i0}, \ldots, c_{i,k_i+1}\}$ where $0 = c_{i0} < c_{i1} < \cdots < c_{i,k_i} < c_{i,k_i+1} = \infty$. Hence

$$\mathcal{D} = \big\{\lfloor l_i,\, u_i \rfloor : i = 1, \ldots, n\big\},$$

and $\lfloor l_i,\, u_i \rfloor = \lfloor c_{i,j_i},\, c_{i,j_i+1} \rfloor$ if $T_i \in \lfloor c_{i,j_i},\, c_{i,j_i+1} \rfloor$. The likelihood is then given by

$$L(\boldsymbol{\theta}) = \prod_{i=1}^{n} \int_{l_i}^{u_i} (\sigma\, t)^{-1} \varphi\Big\{\frac{\log(t) - \mu}{\sigma}\Big\} \mathrm{d}t, \tag{9.5}$$

where φ is the density of a standard normal distribution $\mathcal{N}(0,\, 1)$. Let $\boldsymbol{\psi} = \{t_i : i = 1, \ldots, n\}$ be the set of (unobserved) exact event times, the term $p(\boldsymbol{\psi} \mid \boldsymbol{\theta})$ from Expression (9.4) then becomes simply the likelihood of the uncensored sample from a log-normal distribution, i.e.,

$$p(\boldsymbol{\psi} \mid \boldsymbol{\theta}) = \prod_{i=1}^{n} \Bigg[(\sigma\, t_i)^{-1} \varphi\Big\{\frac{\log(t_i) - \mu}{\sigma}\Big\}\Bigg].$$

The augmented likelihood $L^*(\boldsymbol{\theta}, \boldsymbol{\psi})$ has, compared to (9.5), a very simple form, i.e.,

$$\begin{aligned} L^*(\boldsymbol{\theta}, \boldsymbol{\psi}) &= p(\boldsymbol{\mathcal{D}} \mid \boldsymbol{\theta}, \boldsymbol{\psi}) \\ &= \prod_{i=1}^{n} \mathsf{P}\big(T_i \in \lfloor l_i, u_i \rfloor \,\big|\, \boldsymbol{\theta}, T_i = t_i\big) = \prod_{i=1}^{n} \mathsf{P}\big(T_i \in \lfloor l_i, u_i \rfloor \,\big|\, T_i = t_i\big) \\ &= \begin{cases} 1, & \text{if } t_i \in \lfloor l_i, u_i \rfloor \text{ for all } i = 1, \ldots, n, \\ 0, & \text{otherwise.} \end{cases} \end{aligned}$$

□

The idea of *Data Augmentation* was first introduced in the context of the Expectation-Maximization (EM) algorithm (Dempster et al., 1977) and formalized in the context of Bayesian computation by Tanner and Wong (1987). For more complex models with censored data, DA provides a highly appealing alternative to difficult maximum likelihood estimation. Moreover, it is quite natural to include the true event times or the values of latent random effects in the set of unknowns. Finally, the distribution of the true values T_i is provided. For notational convenience, in most Bayesian models, no explicit distinction between genuine model parameters $\boldsymbol{\theta}$ and augmented parameters $\boldsymbol{\psi}$ will be made and all parameters will be simply included in a vector $\boldsymbol{\theta}$.

9.1.3 Markov chain Monte Carlo

The Bayesian approach is based on the posterior distribution $p(\boldsymbol{\theta} \mid \boldsymbol{\mathcal{D}})$ which is obtained using Bayes' formula (9.1) and is proportional to the product of the likelihood and the prior distribution. We have seen that for the analysis of interval-censored data, difficult likelihood evaluations can be avoided by the introduction of a set of suitable auxiliary variables (augmented data). What still needs to be discussed is how the posterior distribution can be computed and how posterior summaries about $\boldsymbol{\theta}$ are determined. Most quantities related to posterior summarization (posterior moments, quantiles, credible intervals/regions, etc.) involve computation of the posterior expectation of some function $Q(\boldsymbol{\theta})$, i.e., computation of

$$\overline{Q(\boldsymbol{\theta})} = \mathsf{E}_{\boldsymbol{\theta}|\boldsymbol{\mathcal{D}}}\big\{Q(\boldsymbol{\theta})\big\} = \int_{\Theta} Q(\boldsymbol{\theta})\, p(\boldsymbol{\theta} \mid \boldsymbol{\mathcal{D}})\, \mathrm{d}\boldsymbol{\theta} = \frac{\int_{\Theta} Q(\boldsymbol{\theta})\, L(\boldsymbol{\theta})\, p(\boldsymbol{\theta})\, \mathrm{d}\boldsymbol{\theta}}{\int_{\Theta} L(\boldsymbol{\theta})\, p(\boldsymbol{\theta})\, \mathrm{d}\boldsymbol{\theta}}. \quad (9.6)$$

The integration in Expression (9.6) is usually high-dimensional and only rarely analytically tractable in practical situations.

MCMC methods avoid the explicit evaluations of integrals. To this end, one constructs a Markov chain with state space Θ whose stationary distribution is equal to $p(\boldsymbol{\theta} \mid \boldsymbol{\mathcal{D}})$. After a sufficient number of burn-in iterations, the current draws follow the stationary distribution, i.e., the posterior distribution of interest. We keep a sample of $\boldsymbol{\theta}$ values, let us say $\boldsymbol{\theta}^{(1)}, \ldots, \boldsymbol{\theta}^{(M)}$ and approximate the posterior expectation (9.6) by

$$\widehat{\overline{Q}} = \frac{1}{M}\sum_{m=1}^{M} Q(\boldsymbol{\theta}^{(m)}). \tag{9.7}$$

The ergodic theorem implies that, under mild conditions, $\widehat{\overline{Q}}$ converges almost surely to $\mathsf{E}_{\boldsymbol{\theta}|\boldsymbol{\mathcal{D}}}\{Q(\boldsymbol{\theta})\}$ as $M \to \infty$ (see, e.g., Billingsley, 1995, Section 24).

Many methods are available to construct the Markov chains with desired properties. Most often used are the Metropolis-Hastings algorithm (Hastings, 1970; Metropolis et al., 1953) and the Gibbs algorithm (Gelfand and Smith, 1990; Geman and Geman, 1984). In the Metropolis-Hastings algorithm, a *proposal density* $q(\cdot \mid \boldsymbol{\theta}^{(m)})$ centered at $\boldsymbol{\theta}^{(m)}$, the current position of the chain, suggests a candidate position $\widetilde{\boldsymbol{\theta}}$ as a next position. This position will be accepted with probability $\min\left(\frac{p(\widetilde{\boldsymbol{\theta}}|\boldsymbol{\mathcal{D}})\, q(\boldsymbol{\theta}^{(m)}|\widetilde{\boldsymbol{\theta}})}{p(\boldsymbol{\theta}^{(m)}|\boldsymbol{\mathcal{D}})\, q(\widetilde{\boldsymbol{\theta}}|\boldsymbol{\theta}^{(m)})},\ 1\right)$. Hence the new position will be accepted or rejected. After a burn-in sample, the Metropolis-Hastings algorithm provides under mild regularity conditions a sample from the posterior. With Gibbs sampling, the *full conditional distributions* or *full conditionals* $p(\theta_j \mid \theta_1, \ldots, \theta_{j-1}, \theta_{j+1}, \ldots, \theta_q, \boldsymbol{\mathcal{D}})$ $(j = 1, \ldots, q)$ need to be determined, or at least must be sampled from. Likewise, it can be shown that under mild regularity conditions sampling "long enough" from all full conditionals will result in a sample from the posterior.

"Long enough" means in practice that there is convergence of the Markov chains and that the posterior summary measures are determined with enough accuracy. Popular convergence diagnostics are the Geweke diagnostic (Geweke, 1992) and the Brooks-Gelman-Rubin (BGR) diagnostic (Gelman and Rubin, 1992; Brooks and Gelman, 1998). The first convergence test contrasts the mean of an early part of the chain to the mean of a late part of the chain. Convergence is then concluded when there is no statistically significant difference between the two means. The BGR diagnostic is based on multiple chains and verifies whether they mix well, i.e., they sample the same (posterior) distribution. Upon convergence, the summary measures should show little variation when further sampling. This is measured by the Monte Carlo Error (MCE). A rule of thumb is to take the MCE smaller than 5% of the posterior standard deviation. The convergence of a chain very much relies on how dependent the elements in the Markov chain are. While there is conditional independence of $\boldsymbol{\theta}^{(m+1)}$ with $\boldsymbol{\theta}^{(m-1)}, \boldsymbol{\theta}^{(m-2)} \ldots$ given $\boldsymbol{\theta}^{(m)}$, the Markov chain elements are in general dependent. The higher this dependence, the slower the convergence. This dependence is measured with the *autocorrelation* of different lags based on a converged Markov chain. The autocorrelation of lag k for parameter θ is defined as the (marginal) correlation between $\theta^{(m)}$ with $\theta^{(m+k)}$ irrespective of m. The autocorrelations are often estimated by a time series approach. A direct measure of the impact of the dependence in the Markov chain is the *Carlin's effective sample size*. The effective sample size n_{eff} for parameter θ is defined as the size of an independent sample $\theta_1, \ldots, \theta_{n_{eff}}$ with the same

Monte Carlo error for its mean as the sampled Markov chain $\theta^{(1)}, \ldots, \theta^{(M)}$. Hence, n_{eff} is the solution of the equation $\overline{\sigma}_\theta / \sqrt{n_{eff}} = \mathrm{MCE}_\theta$, with $\overline{\sigma}_\theta$ the posterior standard deviation of θ and MCE_θ the Monte Carlo error obtained for θ.

A comprehensive introduction into the area of the MCMC can be found, e.g., in Geyer (1992), Tierney (1994) and Besag et al. (1995). More details can be obtained from several books, e.g., Gilks et al. (1996), Chen et al. (2000), Robert and Casella (2004), and Gamerman and Lopes (2006). Further, for more details on the convergence diagnostics of the MCMC, the reader is advised to consult specialized books like Gelman (1996), Carlin and Louis (2008, Section 3.4), Lesaffre and Lawson (2012, Chapter 7), and Gelman et al. (2013, Section 11.4). In the remainder of the book we assume that the reader is familiar with basic tools to monitor convergence from the MCMC output.

9.1.4 Credible regions and contour probabilities

With a frequentist approach, confidence intervals or regions are used to express the uncertainty with which the estimates of $\boldsymbol{\theta}$ were determined. Confidence intervals are nowadays the preferred way to base inference on, but the P-value is still quite common to use. In Bayesian statistics, the role of the confidence regions is played by the *credible regions* and P-values can be replaced by *contour probabilities*. In this section, we briefly discuss their meaning and construction.

Credible regions

For a given $\alpha \in (0,\,1)$, the $100(1-\alpha)\%$ credible region Θ_α for a parameter of interest $\boldsymbol{\theta}$ is defined as

$$\mathsf{P}\big(\boldsymbol{\theta} \in \Theta_\alpha \bigm| \boldsymbol{\mathcal{D}}\big) = 1 - \alpha. \tag{9.8}$$

Equal-tail credible interval

For a univariate parameter of interest θ, the credible region Θ_α can be constructed as a credible interval. This interval can be obtained by setting $\Theta_\alpha = (\theta_\alpha^L,\, \theta_\alpha^U)$, such that

$$\mathsf{P}\big(\theta \leq \theta_\alpha^L \bigm| \boldsymbol{\mathcal{D}}\big) = \mathsf{P}\big(\theta \geq \theta_\alpha^U \bigm| \boldsymbol{\mathcal{D}}\big) = \alpha/2. \tag{9.9}$$

Such an interval is called an *equal-tail credible interval* and is easily constructed when a sample from the posterior distribution of θ (obtained, e.g., using the MCMC technique) is available. Indeed, θ_α^L and θ_α^U are the $100(\alpha/2)\%$ and $100(1-\alpha/2)\%$, respectively, quantiles of the posterior distribution and from the MCMC output they can be estimated using the sample quantiles.

Simultaneous credible bands

For a multivariate parameter of interest, $\boldsymbol{\theta} = (\theta_1, \ldots, \theta_q)'$ we may wish to calculate *simultaneous* probability statements. Besag et al. (1995, p. 30) suggest to compute *simultaneous credible bands.* In that case, Θ_α equals

$$\Theta_\alpha = (\theta^L_{1,\alpha_{uni}}, \theta^U_{1,\alpha_{uni}}) \times \cdots \times (\theta^L_{q,\alpha_{uni}}, \theta^U_{q,\alpha_{uni}}). \tag{9.10}$$

That is, Θ_α is given as a product of univariate equal-tail credible intervals of the same univariate level α_{uni} (typically $\alpha_{uni} \leq \alpha$). As shown by Besag et al. (1995), the simultaneous credible bands can easily be computed when the sample from the posterior distribution is available as only order statistics for each univariate sample are needed. The most intensive part in computation of the simultaneous credible band is to sort the univariate samples. However, when simultaneous credible bands for different values of α are required, this must be done only once. This property is advantageously used when computing the Bayesian counterparts of P-values (see Section 9.1.4). However, note that Held (2004) pointed out that such a hyperrectangular simultaneous credible band may actually cover a large area not supported by the posterior distribution, especially when there is a high posterior correlation in $\boldsymbol{\theta}$.

Highest posterior density region

An alternative credible interval to the equal-tailed CI defined above is the *highest posterior density (HPD) interval.* The generalization of the HPD interval to higher dimensions is called the *highest posterior density (HPD) region.* In both cases, it consists of the parameter values that are most plausible a posteriori. The set Θ_α is then obtained by requiring (9.8) and additionally

$$p(\boldsymbol{\theta}_1 \mid \boldsymbol{\mathcal{D}}) > p(\boldsymbol{\theta}_2 \mid \boldsymbol{\mathcal{D}}) \quad \text{for all } \boldsymbol{\theta}_1 \in \Theta_\alpha,\ \boldsymbol{\theta}_2 \notin \Theta_\alpha. \tag{9.11}$$

For unimodal posterior densities, the HPD region is connected and univariately consists of an interval. In case the posterior distribution is known, the HPD region must be determined with a numerical optimization algorithm. When only a sample is available, then one might use the algorithm of Held (2004).

Contour probabilities

The Bayesian counterpart of the P-value for the hypothesis $\mathrm{H}_0 : \boldsymbol{\theta} = \boldsymbol{\theta}_0$ (typically $\boldsymbol{\theta}_0$ is a vector of zeros) – *the contour probability* P – can be defined as one minus the content of the credible region which just covers $\boldsymbol{\theta}_0$, i.e.,

$$P = 1 - \inf\big\{\alpha : \boldsymbol{\theta}_0 \in \Theta_\alpha\big\}. \tag{9.12}$$

In the Bayesian literature, different definitions of the *contour probability* can be found. Definition (9.12) stems from analogies between Bayesian and frequentist inferential concepts, and was used, e.g., by Held (2004). He suggests to base calculation of (9.12) on the HPD credible region. Nevertheless, from a probabilistic point of view, other than HPD credible regions can be used as well. Finally, note that the term *Bayesian P-value* is reserved to a different concept that is discussed in Section 9.1.5.

In the univariate case, a two-sided contour probability based on the *equal-tail credible interval* is computed quite easily once the sample from the posterior distribution is available since (9.12) can be expressed as

$$P = 2\min\Big\{\mathsf{P}(\theta \le \theta_0 \mid \boldsymbol{\mathcal{D}}),\ \mathsf{P}(\theta \ge \theta_0 \mid \boldsymbol{\mathcal{D}})\Big\}, \tag{9.13}$$

and $\mathsf{P}(\theta \le \theta_0 \mid \boldsymbol{\mathcal{D}})$, $\mathsf{P}(\theta \ge \theta_0 \mid \boldsymbol{\mathcal{D}})$ can be estimated as a proportion of the sample being lower or higher, respectively, than the point of interest θ_0.

In the multivariate case, a two-sided pseudo-contour probability based on the *simultaneous credible band* can be obtained by calculating the simultaneous credible bands Θ_α on various levels α and determining the smallest level, such that $\boldsymbol{\theta}_0 \in \Theta_\alpha$, i.e., by direct usage of Expression (9.12).

9.1.5 Selecting and checking the model

In this section we first discuss two popular criteria to select the best model from a set of models, namely the *Deviance Information Criterion (DIC)* and the *Pseudo-Bayes factor (PSBF)* with the related Logarithm of Pseudo-Marginal Likelihood (LPML). The PSBF is a version of the Bayes factor and increasingly used in practice. Further, many techniques are available to check the appropriateness of the selected model. Here we discuss the posterior predictive check, which is most popular nowadays.

Deviance Information Criterion

The DIC, proposed by Spiegelhalter et al. (2002), is a generalization of AIC to Bayesian models. See Spiegelhalter et al. (2014) for a review of the advantages and disadvantages of using DIC for model selection. A key question is now how to define the number of free parameters in a Bayesian context. Spiegelhalter et al. (2002) suggested p_D, as a Bayesian measure of complexity. More specifically,

$$p_D = \mathsf{E}_{\boldsymbol{\theta}|\boldsymbol{\mathcal{D}}}\big\{-2\ell(\boldsymbol{\theta}) + 2\ell(\overline{\boldsymbol{\theta}})\big\}, \tag{9.14}$$

with $\overline{\boldsymbol{\theta}}$ the posterior mean. p_D can also be written as

$$p_D = \overline{D(\boldsymbol{\theta})} - D(\overline{\boldsymbol{\theta}}), \tag{9.15}$$

whereby $D(\boldsymbol{\theta}) = -2\,\ell(\boldsymbol{\theta}) + 2\log$-likelihood(saturated model) is the *Bayesian deviance*. $\overline{D(\boldsymbol{\theta})}$ is the posterior mean of $D(\boldsymbol{\theta})$, and $D(\overline{\boldsymbol{\theta}})$ is equal to $D(\boldsymbol{\theta})$ evaluated in the posterior mean of the parameters. However, the saturated model term cancels out from the expression of p_D. Note that in BUGS this term is not computed. Much of p_D's credibility comes from the fact that it coincides with frequentist measures of complexity in simple models. Based on p_D, Spiegelhalter et al. (2002) proposed the DIC as a Bayesian model selection criterion, defined as

$$DIC = D(\overline{\boldsymbol{\theta}}) + 2p_D = \overline{D(\boldsymbol{\theta})} + p_D, \tag{9.16}$$

and hence uses a similar penalty term as AIC. As for AIC, DIC aims to evaluate the predictive performance of the chosen model on a future but similar data set.

An attractive aspect of p_D and DIC is that both can readily be calculated from an MCMC run by monitoring $\boldsymbol{\theta}$ and $D(\boldsymbol{\theta})$. Let $\boldsymbol{\theta}^{(1)}, \ldots, \boldsymbol{\theta}^{(M)}$ represent a converged Markov chain, then $\overline{D(\boldsymbol{\theta})}$ can be approximated by $\frac{1}{M}\sum_{m=1}^{M} D(\boldsymbol{\theta}^{(m)})$ and $D(\overline{\boldsymbol{\theta}})$ by $D\big(\frac{1}{M}\sum_{m=1}^{M} \boldsymbol{\theta}^{(m)}\big)$. The rule of thumb for using DIC in model selection is roughly the same as for AIC and BIC, namely a difference in DIC of more than 10 rules out the model with the higher DIC while with a difference of less than 5 there is no clear winner. Note that DIC is subject to sampling variability since it is based on the output of an MCMC procedure.

The implementation of p_D and DIC in BUGS has encouraged greatly their use. But at the same time considerable criticism has been expressed. A major criticism is that DIC lacks a sound theoretical foundation, but there are also practical problems with the use of DIC and p_D. For instance, in some cases p_D is estimated to be negative, which of course does not make sense. Another problem has to do with the focus of the statistical analysis. Namely, the data augmentation algorithm makes it attractive to fit in a Bayesian context hierarchical models whereby all parameters of the models, also the so-called random effects, are jointly estimated. DIC applied to such fitted model estimates the predictive ability of the model on a future sample based on the same clusters as in the original data set, called the conditional DIC. In this case the focus of the analysis rests on the current clusters. This is in contrast to the DIC obtained from fitting a marginal model where the random effects have been integrated out of the likelihood. Then a marginal DIC is obtained where the focus of the analysis lies in the future clusters. Note that a conditional and a marginal DIC are not comparable. Now we argue that in practice one is most often interested in the future clusters, yet that implies the calculation of the marginal likelihood within the MCMC algorithm, which is often quite difficult to achieve. Nowadays, the attitude is to choose the conditional DIC even when the focus of the analysis is on future clusters, simply because it is much easier to compute, but simulation studies appear to indicate that model selection based on the conditional DIC tends to choose too complex models. In Quintero and Lesaffre (2017) sampling methods and an approximation method are explored to compute the marginal DIC for random effects models.

Note that there are different versions of DIC and p_D implemented in Bayesian software, see Lesaffre and Lawson (2012) for an overview. Finally, recently Watanabe (2010) has suggested the Watanabe-Akaike's information criterion (WAIC). This appears to be a promising successor of DIC implemented in the recently developed package Stan (`http://mc-stan.org/`).

Pseudo-Bayes factor

Another criterion is the pseudo-Bayes factor. The PSBF is becoming popular in practice as an alternative to DIC, especially when the latter cannot be computed or when there are doubts about its validity. The PSBF is a version of the Bayes factor. The Bayes factor is a Bayesian tool for hypothesis testing and can be used to compare nested and nonnested models. Suppose that there are two candidate (classes of) models for the data $\boldsymbol{y} = \{y_1, \ldots, y_n\}$: M_1 with parameters $\boldsymbol{\theta}_1$, likelihood $p(\boldsymbol{y} \mid \boldsymbol{\theta}_1, M_1)$ and prior $p(\boldsymbol{\theta}_1 \mid M_1)$ and M_2 with parameters $\boldsymbol{\theta}_2$, likelihood $p(\boldsymbol{y} \mid \boldsymbol{\theta}_2, M_2)$ and prior $p(\boldsymbol{\theta}_2 \mid M_2)$. To choose the most appropriate model for the data at hand, one uses the marginal (averaged) likelihood under each model since the actual values of $\boldsymbol{\theta}_1$ and $\boldsymbol{\theta}_2$ are never known. Hence, one computes for model M_m

$$p(\boldsymbol{y} \mid M_m) = \int p(\boldsymbol{y} \mid \boldsymbol{\theta}_m, M_m)\, p(\boldsymbol{\theta}_m \mid M_m)\, \mathrm{d}\boldsymbol{\theta}_m \quad (m = 1, 2),$$

which is also called the prior predictive distribution for model M_m. Suppose that $p(M_m)$ $(m = 1, 2)$ are the prior probabilities of the two models, e.g., $p(M_1) = p(M_2) = \frac{1}{2}$. The posterior probability for model M_m

$$p(M_m \mid \boldsymbol{y}) = \frac{p(\boldsymbol{y} \mid M_m)\, p(M_m)}{p(\boldsymbol{y} \mid M_1)\, p(M_1) + p(\boldsymbol{y} \mid M_2)\, p(M_2)} \quad (m = 1, 2),$$

tells us which of the two models is most supported by the data. The Bayes factor is then the ratio $BF_{12}(\boldsymbol{y}) = p(\boldsymbol{y} \mid M_1)/p(\boldsymbol{y} \mid M_2)$ and summarizes the evidence in the data for model M_1 compared to model M_2 irrespective of the prior probabilities $p(M_1)$ and $p(M_2)$. When $BF_{12}(\boldsymbol{y})$ is greater than one, then the posterior odds for M_1 is increased compared to the prior odds. For equal prior probabilities, model choice based on the Bayes factor is the same as model choice based on the above model posterior probabilities.

A major difficulty with the Bayes factor is that it poses computational difficulties with vague (improper) priors. Various suggestions were made in the literature to bypass this obstacle. One approach is to use part of the data to update the (improper) prior to a proper posterior and then base the Bayes factor on this posterior. This is the basis for the *Logarithm of Pseudo-Marginal Likelihood* and the PSBF. Suppose that the data are split up into a learning part $\boldsymbol{y}_L$ and a nonoverlapping testing part $\boldsymbol{y}_T$. When $p(\boldsymbol{y} \mid M_m)$ is replaced by $p(\boldsymbol{y}_T \mid \boldsymbol{y}_L, M_m)$ a variant of the Bayes factor is obtained:

$$BF_{12}^{*}(\boldsymbol{y}) = \frac{p(\boldsymbol{y}_T \mid \boldsymbol{y}_L, M_1)}{p(\boldsymbol{y}_T \mid \boldsymbol{y}_L, M_2)}.$$

Suppose now that $\boldsymbol{y}_L = \{y_i\}$ and $\boldsymbol{y}_T = \{\boldsymbol{y}_{(i)}\}$, whereby $\boldsymbol{y}_{(i)}$ is the part of the data without the ith subject. Then $p(y_i \mid \boldsymbol{y}_{(i)}, M_m)$ measures the plausibility of the ith observation when model M_m is estimated from the remaining data. This is called the *Conditional Predictive Ordinate (CPO)*. Hence, CPO_i measures the outlying aspect of the ith observation under model M_m. The pseudo-marginal likelihood for model M_m is defined as $\prod_i p(y_i \mid \boldsymbol{y}_{(i)}, M_m)$ and measures the plausibility of the observed data under model M_m. The logarithm of this pseudo-marginal likelihood is LPML. The ratio of the two pseudo-marginal likelihoods is called the pseudo-Bayes factor. More explicitly, the PSBF for comparing model M_1 with model M_2 is

$$\text{PSBF}_{12} = \frac{\prod_i p(y_i \mid \boldsymbol{y}_{(i)}, M_1)}{\prod_i p(y_i \mid \boldsymbol{y}_{(i)}, M_2)} = \frac{\prod_i \text{CPO}_i(M_1)}{\prod_i \text{CPO}_i(M_2)}.$$

When PSBF_{12} is greater than 1, or equivalently, $\text{LPML}_1 - \text{LPML}_2 > 0$, model M_1 is preferred over model M_2. Of course, as for DIC this difference must be large enough for a stable decision, but there appears no rule of thumb about how much this difference should be.

The pseudo-Bayes factor implicitly corrects for complexity of the model since it is a version of the Bayes factor. Indeed, for a more elaborate model the marginal likelihood is averaged over a larger parameter space thereby correcting for complexity in the computation of the marginal likelihood to obtain the Bayes factor. Finally, note that PSBF suffers from the same problem as DIC for hierarchical models. Namely that there are two versions, a conditional and a marginal, and they are not comparable.

Posterior predictive checks

Suppose that under the hypothesis H_0 we postulate model M_0: $p(\boldsymbol{y} \mid \boldsymbol{\theta})$ for a sample $\boldsymbol{y} = \{y_1, \ldots, y_n\}$ whereby the parameters $\boldsymbol{\theta} \in \Theta_0$ must be estimated from the data. In a frequentist context, H_0 is tested with a goodness of fit test (GOF) statistic $T(\boldsymbol{y})$. The model assumption is then rejected if $T(\boldsymbol{y})$ is too small (or too large) which then indicates a poor model fit. "Too small" is then defined as when $T(\boldsymbol{y})$ falls in the, say, 5% extreme region of possible values of $T(\boldsymbol{y})$ under H_0. In the Bayesian paradigm, one conditions on the observed data, and hence one needs a different technique to check the GOF of the model. In fact, the Bayesian approach makes use of the *posterior predictive distribution (PPD)* $p(\widetilde{\boldsymbol{y}} \mid \boldsymbol{y})$, which represents our knowledge of the distribution of future observations that are supposed to come from the same model as the sample $\boldsymbol{y}$ (but independent of that sample) taking our uncertainty of the model parameters into account via the posterior distribution. The PPD is formally defined as

$$p(\widetilde{\boldsymbol{y}} \mid \boldsymbol{\mathcal{D}}) = \int_{\Theta} p(\widetilde{\boldsymbol{y}} \mid \boldsymbol{\theta})\, p(\boldsymbol{\theta} \mid \boldsymbol{\mathcal{D}})\, \mathrm{d}\boldsymbol{\theta}, \tag{9.17}$$

whereby $\boldsymbol{\mathcal{D}}$ represents the collected sample $\boldsymbol{y}$. Thus in establishing the possible distribution of the future data, the uncertainty of the parameter vector $\boldsymbol{\theta}$

is taken into account. For a *posterior predictive check (PPC)* one samples from the PPD thereby computing each time $T(\widetilde{\boldsymbol{y}})$. For each sampled $T(\widetilde{\boldsymbol{y}})$ one verifies whether it is more extreme than the observed $T(\boldsymbol{y})$. When the proportion of generated values $T(\widetilde{\boldsymbol{y}})$ more extreme than the observed $T(\boldsymbol{y})$ is too small, say, less than 5%, one rejects H_0 in a Bayesian way. This proportion is called the *posterior predictive P-value (PPP-value)*, and is also referred to as the *Bayesian P-value*. Rubin (1984) was the first to formalize the concept of the posterior predictive check. Meng (1994) and Gelman et al. (1996) further developed this proposal into a practical procedure.

Note that the GOF test can be any measure of discrepancy that might indicate deviation of the data from the assumed model. Furthermore, the discrepancy measure may also depend on the parameter vector $\boldsymbol{\theta}$. In that case one works with $D(\boldsymbol{y}, \boldsymbol{\theta})$. In practice, the PPP-value is approximated by $\overline{P}_D$ on a converged Markov chain $\boldsymbol{\theta}^{(1)}, \ldots, \boldsymbol{\theta}^{(M)}$ as follows:

$$\overline{P}_D = \frac{1}{M} \sum_{m=1}^{M} I\left[D\left(\widetilde{\boldsymbol{y}}^{(m)}, \boldsymbol{\theta}^{(m)}\right) \geq D\left(\boldsymbol{y}, \boldsymbol{\theta}^{(m)}\right)\right],$$

with $\widetilde{\boldsymbol{y}}^{(m)}$ is a generated sample of size n from the distribution $p(\boldsymbol{y} \mid \boldsymbol{\theta}^{(m)})$ and I is the indicator function equal 1 if the expression is true, and 0 otherwise.

9.1.6 Sensitivity analysis

A Bayesian analysis depends on the assumed model for the data and the chosen priors. The choice of the model is often speculative, but also the choice of priors is not necessarily straightforward. Prior information can come from expert knowledge or from historical studies. The problem is then to transform this information into a prior distribution. This is not a trivial exercise, especially when the prior information has a qualitative nature. When there is no prior information, we wish to express this ignorance in a vague prior. In a complex model, it might be difficult to ensure that the chosen vague priors indeed bring in almost no (subjective) information into the calculation. For all of these reasons, it is classical to complement the original analysis with additional analyses where the assumed model differs from the original model and/or the priors are varied. This is called a sensitivity analysis. It is common that in each Bayesian analysis a sensitivity analysis is included, to demonstrate the stability of the conclusions under varying assumptions.

9.2 Nonparametric Bayesian inference

Parametric inference is still very popular in both the frequentist and the Bayesian paradigm. In a Bayesian context it is relatively easy to relax parametric assumptions, partly due to the MCMC software. However, as in the frequentist paradigm, there could be the wish or the need to refrain from making distributional assumptions leading then to a nonparametric approach. When also covariates are involved, a nonparametric model is turned into a semiparametric model. In classical survival analysis, the semiparametric approach has been most popular since the introduction of Cox proportional hazards model, briefly reviewed in Section 2.3.1. In the present section, we introduce two related BNP and BSP approaches for survival analysis. In both cases, it is assumed that the survival function $S(t)$ is a realization of a stochastic process. That is, an infinite dimensional prior will be put on the functional space of S. The first BNP approach has been developed for survival applications with a prior defined on the (cumulative) hazard. The second BNP approach inspired many Bayesians to develop a powerful class of BNP approaches specifying an infinite dimensional prior on the (cumulative) distribution function. The first approach handles easily (right) censored observations and can naturally be extended to a BSP approach leading to a proportional hazards model, with the Cox PH model as a limiting case. The second BNP approach originally does not easily allow for censoring, especially when covariates play a role. The early BNP approaches aimed for analytical expressions of the posterior. However, this quickly appeared horrendously complicated especially with censored observations. Breakthroughs were realized when involving MCMC techniques. There the only viable BSP option appears to involve the AFT model. In the two BSP approaches, most often the baseline hazard is viewed as the nuisance part of the model and interest lies in the importance of the covariates.

In Kalbfleisch (1978) and Kalbfleisch and Prentice (2002, Section 11.8) the link between the two BNP approaches for (primarily) uncensored survival times is explained. While several variants and extensions of both approaches have been suggested in the literature, in this section we limit ourselves to describe the ideas behind the original BNP approaches. The technical details are left behind since they are beyond the scope of this book. In Section 9.4 we illustrate the use of two BNP approaches to analyze right-censored survival times. In the next chapters, some of the variations and extensions will be illustrated on interval-censored observations. For a comprehensive treatment on Bayesian survival analysis techniques developed up to about 2000, we refer to Ibrahim et al. (2001). For a comprehensive and up-to-date overview of BNP and BSP approaches, we refer to Müller et al. (2015) and Mitra and Müller (2015). In Kalbfleisch and Prentice (2002) the Bayesian approach to survival analysis is contrasted to the frequentist approach. For a recent and practical review of Bayesian nonparametric regression models, see Hanson and Jara (2013).

Until recently, nonparametric Bayesian software was largely lacking. With the introduction of the R package DPpackage, a great variety of BNP and BSP analyses based on the Dirichlet process and its variants are possible now, see Jara et al. (2011). In Section 9.3 Bayesian software to perform such analyses will be briefly reviewed, but the versatility of the DPpackage will only become clear in the subsequent chapters.

9.2.1 Bayesian nonparametric modelling of the hazard function

Kalbfleisch (1978) suggested to put a prior distribution on the parameters of the cumulative hazard function. Let, as before, T be the random survival time with survival function $S(t)$ for t a particular chosen fixed time point with corresponding cumulative hazard function $H(t) = -\log\{S(t)\}$. Further, assume a partition of $[0, \infty)$ into a finite number k of disjoint intervals $[s_0 = 0, s_1), [s_1, s_2), \ldots, [s_{k-1}, s_k = \infty)$. Note that the grid can be defined by the ordered observed survival times $t_{(1)} \leq t_{(2)} \cdots \leq t_{(n)}$. Let the conditional probability to belong to interval $[s_{j-1}, s_j)$ (i.e., the discrete hazard) be $q_j = \mathsf{P}\big(T \in [s_{j-1}, s_j) \,\big|\, T \geq s_{j-1}, H\big)$, then

$$H(s_j) = \sum_{l=1}^{j} -\log(1 - q_l) = \sum_{l=1}^{j} r_l \quad (j = 1, \ldots, k).$$

Clearly, $H(s_0) = 0$, $H(s_j) > 0$ $(j = 1, \ldots, k)$ and $r_j = -\log(1 - q_j) \geq 0$.

Kalbfleisch (1978) suggested to put a prior distribution on the space of cumulative hazard contributions H by specifying the finite dimensional distributions of $q_1, q_2, \ldots, q_k$ for each possible partition $[s_{j-1}, s_j)$ $(j = 1, \ldots, k)$. Independent prior densities are specified for r_j (equivalent to q_j) defined in such a way that there is consistency between all possible partitions. Namely, the prior for $r_j + r_{j+1}$ should be the same as obtained directly from $[s_{j-1}, s_{j+1})$ (called Kolmogorov consistency conditions). By construction $H(t)$ then becomes a nondecreasing process with independent increments, also called *tail-free* or *neutral to the right.* A possible way to specify this prior is to define the stochastic behavior of r_j $(j = 1, \ldots, k)$, namely:

- Take $c > 0$ a real value, and $S^*(t) = \exp\{-H^*(t)\}$ a particular right-continuous (parametric) survival function;
- Assume that r_j $(j = 1, \ldots, k)$ have independent gamma distributions defined on the grid $[s_0 = 0, s_1), [s_1, s_2), \ldots, [s_{k-1}, s_k = \infty)$, i.e., $r_j \sim \mathcal{G}(\alpha_j - \alpha_{j-1},\ c)$ $(j = 1, \ldots, k)$, with $\alpha_j = cH^*(s_j)$. Note that $\mathcal{G}(0,\ c)$ is defined as the distribution with unit mass at 0, while $\mathcal{G}(\infty - \infty,\ c)$ and $\mathcal{G}(\infty, c)$ have unit mass at ∞.

The above implies that H is a gamma process denoted by $H \sim \mathcal{G}(cH^*, c)$. The interpretation of the parameters H^* and c follows from the partition $[0, t), [t, \infty)$. Indeed taking $s_1 \equiv t$, the assumption of the gamma distribution implies

$$\mathsf{E}\big\{H(t)\big\} = H^*(t) \quad \text{and} \quad \mathsf{var}\big\{H(t)\big\} = H^*(t)/c,$$

hence $H^*(t)$ can be viewed as an initial guess of the mean function around which $H(t)$ varies with precision equal to c. Any realization $H(t)$ from the gamma process is discrete with probability one. In Figure 9.1 we have sampled from the above-defined gamma process with $H^*(t)$ the cumulative hazard function from a Weibull $\mathcal{W}(1.5,\ 7)$ distribution for two values of c and based on a grid of time values shown on the X-axis. One can observe that all realizations $H(t)$ make a jump at the grid points and the variability around the template $H^*(t)$ increases with c decreasing. Further, H^* has only an impact on the estimation process via its value at the grid points.

To deal with covariates, Kalbfleisch (1978) embedded the gamma process prior into a proportional hazards model $\mathsf{P}\big(T_i \geq t_i \,\big|\, \boldsymbol{X}_i,\ H\big) = \exp\big\{-H(t_i)\exp(\boldsymbol{X}_i'\boldsymbol{\beta})\big\}$ with observed data $(t_1, \boldsymbol{X}_1), \ldots, (t_n, \boldsymbol{X}_n)$. Let us assume that the observed survival or censoring times are ordered and distinct, i.e., $t_1 < t_2 < \ldots < t_n$. Define $t_0 = 0$, $t_{n+1} = \infty$ and let now $r_i = H(t_i) - H(t_{i-1})$ and $r_i \sim \mathcal{G}(cH^*(t_i) - cH^*(t_{i-1}), c)$, $i = 1, \ldots, n+1$. Then, assuming no ties in the survival times Kalbfleisch (1978) derived analytically the marginal posterior of $\boldsymbol{\beta}$ under the gamma process prior and with flat improper priors on $(-\infty, \infty)$ for elements of $\boldsymbol{\beta}$. He also showed that the partial Cox likelihood for $\boldsymbol{\beta}$ can be obtained from the marginal posterior of $\boldsymbol{\beta}$ when $c \to 0$. For $c \to \infty$, one obtains the joint likelihood function for $\boldsymbol{\beta}$ and $h^*(t_1), \ldots, h^*(t_n)$, which are the jumps made by $H^*(t)$ at the observed survival times. This result was formally proved in Sinha et al. (2003). The case of ties was also considered, but found to be too complex.

Kalbfleisch (1978) also derived analytically the posterior distribution of $H(t)$. In the absence of ties, Clayton (1991) showed that under a quadratic

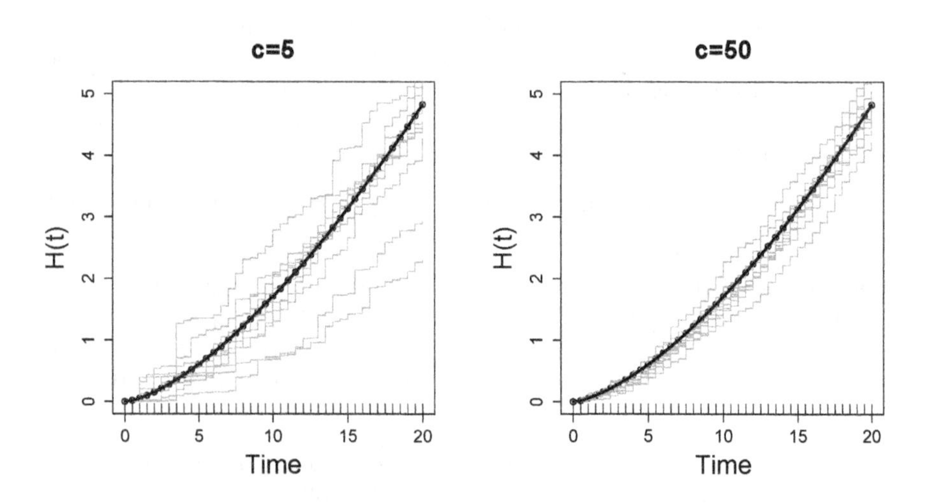

FIGURE 9.1: Illustration of gamma process: Ten realizations of $\mathcal{G}(cH^*, c)$ with $H^*(t)$ the cumulative hazard function of a Weibull$(1.5, 7)$ distribution and for $c = 5,\ 50$. The grid $[s_{j-1}, s_j)$, $(j = 1, \ldots, k)$ is indicated on the X-axis.

loss function the Bayes estimator of $H(t)$ (given $\boldsymbol{\beta}$) is a compromise between the Nelson-Aalen estimate (introduced in Section 2.1.1) and the prior mean $H^*(t)$. When c is large, the prior belief in $H^*(t)$ is poor and the estimator approaches the Nelson-Aalen estimator, while if c is small the estimator is close to the a priori chosen template $H^*(t)$. For this, he combined the counting process formalism with the Bayesian approach. The above analytical derivations quickly become intractable, as seen. It is then advantageous to make use of the MCMC machinery, as is done in Section 9.4 where we illustrate Clayton's approach with a BUGS implementation. In Ibrahim et al. (2001) an MCMC algorithm is provided for grouped survival times where the prior of the cumulative hazard function is again given by a gamma process prior. In the same chapter of the book, the authors discuss alternative semiparametric techniques.

9.2.2 Bayesian nonparametric modelling of the distribution function

The above BNP (and BSP) approach has been developed specifically for survival models. Another BNP approach has been suggested that is more general and specifies the prior on the functional space of cumulative distribution functions or equivalently of survival functions. The paper of Ferguson (1973) introducing the *Dirichlet process (DP) prior* implied the start of an impressive amount of new statistical developments in BNP analyses. The technical level of these developments is, however, quite advanced and beyond the scope of this book. We rather limit ourselves to introducing the concepts of BNP inference and as such will largely follow Müller et al. (2015). Further details and references can be found therein.

The Dirichlet process is defined via the Dirichlet distribution, see Appendix C for more details on the Dirichlet distribution and its relation to the Dirichlet process. The formal definition of the Dirichlet process (adapted to survival context) is:

Let $c > 0$ and $S^*(t)$ be a right-continuous survival function defined on $[0, \infty)$ ($\mathbb{R}^+$, with Borel σ-field). A Dirichlet Process (DP) with parameters $(c,\ S^*(t))$ is a random survival function S, which for each (measurable) finite partition $\{B_1, B_2, \ldots, B_k\}$ of $\mathbb{R}^+$ assigns probabilities $S(B_j)$ $(j = 1, \ldots, k)$ such that, the joint distribution of the vector $(S(B_1), S(B_2), \ldots, S(B_k))$ is the Dirichlet distribution $\mathcal{D}ir(cS^*(B_1),\ cS^*(B_2),\ \ldots,\ cS^*(B_k))$.

Using Kolmogorov's consistency theorem (see Section 9.2.1), Ferguson (1973) showed that such a process exists. The DP is denoted as $\mathcal{DP}(cS^*)$, with c the precision and S^* the centering distribution or initial guess. The interpretation of these parameters is much the same as for the first approach. The realized S has again a discrete nature, with $\mathsf{E}\{S(t)\} = S^*(t)$

and $\mathsf{var}\{S(t)\} = S^*(t)\,\{1 - S^*(t)\}/(c+1)$. Hence, again the realizations vary around the initial guess with variance decreasing with c. Note that we have assumed that the survival times are not censored. The case of censoring will be briefly treated below and in the next chapters.

Because of the discrete nature of the realized S, we can write $S(t) = \sum_{h=1}^{\infty} w_h \delta_{m_h}(t)$, with $w_1, w_2, \ldots$ probability weights for the locations m_h and $\delta_{m_h}(t) = 1$ when $t = m_h$ and zero elsewhere. Sethuraman (1994) proposed a sampling procedure to construct the DP. This procedure consists of taking i.i.d. draws m_h from S^*, and with weights in the unit interval, as follows: $w_1 \in [0,1]$, $w_2 \in [0, 1 - w_1]$, $w_3 \in [0, 1 - w_1 - w_2]$, etc. Since this resembles breaking a unit-sized stick into two pieces at a random location of the stick, then breaking the remaining part of the stick at a random location into two sticks, etc. one speaks of the *stick breaking construction.* It can be shown that, when $w_h = v_h \prod_{l<h}(1 - v_l)$ with i.i.d. $v_h \sim \text{Beta}(1, c)$ and i.i.d. $m_h \sim S^*$, w_h and m_h independent, then $S = \sum_{h=1}^{\infty} w_h \delta_{m_h}$ defines a $\mathcal{DP}(cS^*)$. In Figure 9.2 we have shown 10 samples from a $\mathcal{DP}\{c\,\text{Weibull}(1.5, 7)\}$, with $c = 5, 50$ making use of the stick-breaking algorithm. We notice the same dependence of the prior on c as for the first BNP approach, but now there is no fixed grid, rather the jumps are made at random locations.

The DP defines a prior on the survival distributions S for i.i.d. survival times of the sample, i.e., $T_i \mid S \sim S,\ i = 1, \ldots, n$. Ferguson (1973) showed that the DP is conjugate with respect to i.i.d. sampling. This simply implies that if a priori $S \sim \mathcal{DP}(cS^*)$, and a sample $t_1, \ldots, t_n$ is observed from S, then

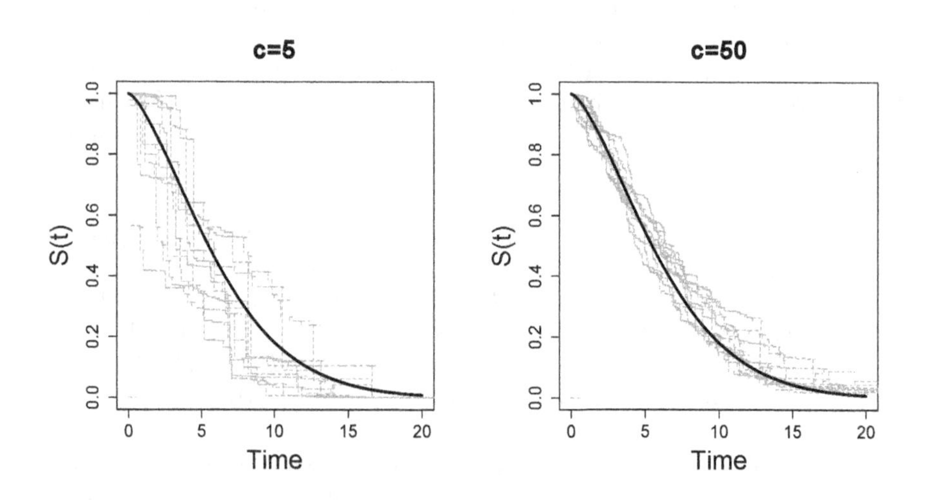

FIGURE 9.2: Illustration of the Dirichlet process: Ten realizations of $\mathcal{DP}\{c\,\text{Weibull}(1.5, 7)\}$ for $c = 5, 50$.

a posteriori we have learned about S and its posterior is again a DP. Formally:

Let $T_1, T_2, \ldots, T_n$ i.i.d. survival times from S and $S \sim \mathcal{DP}(cS^*)$, then $S \mid T_1 = t_1, \ldots, T_n = t_n \sim \mathcal{DP}\{cS^* + \sum_{i=1}^n (1 - \delta_{t_i})\}$.

The previous property implies that

$$\mathsf{E}(S \mid t_1, t_2, \ldots, t_n) = \frac{c}{c+n} S^* + \frac{n}{c+n} S_n,$$

with $S_n = 1 - F_n$ whereby $F_n(t) = \sum_{i=1}^n \delta_{t_i}(t)$ is the empirical cdf. This defines the nonparametric Bayesian estimator of the survival function denoted by $\overline{S}$. Thus the DP prior shrinks the empirical estimator S_n to the parametric model S^*. Shrinkage is severe when c is large and minimal in the other extreme. In addition, the posterior predictive distribution is given by $T_{n+1} \mid t_1, \ldots, t_n \sim \frac{c}{c+n} S^* + \frac{n}{c+n} S_n$, i.e., the same expression as for the posterior distribution. This expression leads to a sampling procedure from the DP, called the *Pólya Urn Scheme.*

It can be shown (Ferguson, 1973; Kalbfleisch and Prentice, 2002) that if the partition $\{B_1, B_2, \ldots, B_k\}$ is defined by the grid $[s_0 = 0, s_1), [s_1, s_2), \ldots, [s_{k-1}, s_k = \infty)$, then for a DP prior it implies that $q_i \sim \text{Beta}(\gamma_{i-1} - \gamma_i, \gamma_i)$ $(i = 1, \ldots, k)$, with $\gamma_i = c\,S^*(s_i)$ $(i = 0, 1, \ldots, k)$. Thus, for the DP independent beta priors are given on the discrete hazard contributions, while in the first approach independent gamma priors are given for the cumulative hazard contributions.

Many variants and extensions of the DP approach have been suggested. Indeed, with the DP prior only discrete solutions for $\overline{S}$ are generated and this is seen as an important restriction of the approach. The discreteness of the solution also creates difficulties when dealing with censored observations. In Section 9.4 we illustrate the use of the DP prior with right-censored survival times. It will be shown that the solution is not a DP process anymore. Further, it does not seem possible to base a semiparametric PH model on the DP approach, rather now the AFT model is the natural choice. However, analytical developments with the DP prior with censored observations in the presence of covariates pose formidable difficulties; see Christensen and Johnson (1988), Johnson and Christensen (1989).

Other extensions of the DP approach have conquered the nonparametric Bayesian world. Here are some examples: *Dirichlet Process Mixture (DPM), Mixture of Dirichlet Processes (MDP)* and *Pólya Trees (PT).* We will illustrate the DPM approach in Chapter 10. The MDP approach, suggested in Hanson and Johnson (2004) will be treated in Chapter 12. Note the difference between the DPM and the MDP approach. The MDP is a direct generalization of the DP; it arises by assigning a prior distribution on the base measure of the DP prior. Therefore, it generates discrete distributions and allows to center the DP prior around a family of distributions instead of a single probability distribution. In the DPM one uses a DP or a MDP as prior for the mixing distribution in a mixture model. The resulting model can be continuous (depending on the kernel), but one loses the nice centering property. For the other BNP and BSP approaches, we refer to Müller et al. (2015).

9.3 Bayesian software

9.3.1 WinBUGS and OpenBUGS

The BUGS (**B**ayesian inference **U**sing **G**ibbs **S**ampling) is a project which started in 1989 in the MRC Biostatistics Unit and initially led to the classical BUGS program. The stand alone WinBUGS program (Lunn et al., 2000) equipped with GUI capabilities has further been developed in cooperation with the Imperial College School of Medicine at St. Mary's, London. The successor of WinBUGS, OpenBUGS (`http://www.openbugs.net`), is being developed at the University of Helsinki and offers more sampling and other modelling tools than WinBUGS. Moreover, next to the MS Windows platforms, it can easily be installed on machines running Linux and with some difficulties also on the MAC OS X systems. More on the history of the BUGS project can be found in Lunn et al. (2009).

Furthermore, it is possible to call a BUGS model, summarize inferences and convergence in tables and graphs, and save the simulation results for easy future access in R using the packages R2OpenBUGS/R2WinBUGS (Sturtz et al., 2005) or BRugs (Thomas et al., 2006) and we will use this capability throughout the book as well. The OpenBUGS examples will also run in WinBUGS (possibly with small changes). Hence for simplicity, we will refer to BUGS.

As we have explained in Section 9.1, interval censoring is naturally tackled in Bayesian models by treating the unobserved event times as additional model parameters. In WinBUGS the I() statement is used to mark an interval-censored observation, while in OpenBUGS also the C() statement can be used. Now follows an example to illustrate the use of the C() command in OpenBUGS.

***Example 9.2 Log-normal model with censored data in* BUGS**

Suppose that the data consist of four observations (exact event time, right-censored, left-censored and interval-censored time):

$$3;\ \lfloor 4,\ \infty);\ (0,\ 2\rfloor;\ \lfloor 2,\ 3\rfloor,$$

and that we assume a log-normal model used in Example 9.1. On the log-scale, the observations are (rounded to 4 decimal places):

$$1.0986;\ \lfloor 1.3863,\ \infty);\ (-\infty,\ 0.6931\rfloor;\ \lfloor 0.6931,\ 1.0986\rfloor.$$

The BUGS datafile is as follows.

```
list(
  logt = c(1.0986, NA, NA, NA),
  logt.low = c(NA, 1.3863, NA, 0.6931),
  logt.upp = c(NA, NA, 0.6931, 1.0986)
)
```

That is, the vector logt contains event times if these are exactly observed and missing values (NA) if these are censored. The vectors logt.low and logt.upp contain missing values for exactly observed event times and limits of corresponding intervals for censored event times with NA replacing $\pm\infty$.

The model of Example 9.1 with the prior hyperparameters $m = 0$, $s = 1\,000$, $h_1 = 1$, $h_2 = 0.005$ is specified in BUGS as

```
model{
  ### Exact observation
  logt[1] ~ dnorm(mu, tau2);

  ### Right-censored observation
  logt[2] ~ dnorm(mu, tau2)C(logt.low[2], );

  ### Left-censored observation
  logt[3] ~ dnorm(mu, tau2)C(, logt.upp[3]);

  ### Interval-censored observation
  logt[4] ~ dnorm(mu, tau2)C(logt.low[4], logt.upp[4]);

  ### Prior
  tau2 ~ dgamma(1, 0.005);
  prec.mu <- 0.001 * tau2;
  mu ~ dnorm(0, prec.mu);
  sigma2 <- 1 / tau2;
}
```

See also Section 10.1.1 for a more complex example.

□

9.3.2 JAGS

JAGS (Just Another Gibbs Sampler) is BUGS-like software for fitting Bayesian models using MCMC methods. JAGS is platform independent and is written in C++ by Martyn Plummer. At the time of writing this book, version 4.2.0 was used (released on February 19, 2016; `http://mcmc-jags.sourceforge.net/`) but the JAGS manual was not yet adapted. JAGS provides some extra functions like mexp() (matrix exponential), sort() (for sorting elements) and %*% (matrix multiplication). Just as with BUGS, truncation and censoring are allowed. Censoring is handled with the function dinterval, which use is illustrated below. For details of how to specify a model in JAGS we refer to its manual, which also details the compilation step, the loading data step and the execution step.

One of the advantages of JAGS over WinBUGS is its platform independence. Therefore it is part of many repositories of Linux distributions making it easily extendable through user-written (C++) modules. Practice also shows that JAGS is in general faster than WinBUGS and OpenBUGS. The R packages rjags (Plummer, 2016) and R2jags (Su and Yajima, 2015) provide an interface between JAGS and R. For most of the BUGS-like examples JAGS will be used via the R package runjags (Denwood, 2016). The package runjags allows also for parallel processing using multiple cores on the computer and will be used throughout this book to conduct the JAGS analyses.

JAGS handles interval-censored data via the function dinterval and specification of a helping ordinal observation $z \in \{0, \ldots, M\}$ together with cut points $c_1 < \ldots < c_M$. In general, if $\boldsymbol{c} = (c_1, \ldots, c_M)'$ is a vector of the cut points, depending on the value z of the helping ordinal observation, a statement z ~ dinterval(t, c) specifies the following for the interval-censored variable t:

$z = 0$:	$t \leq c_1,$
$z = m$:	$c_m < t \leq c_{m+1}, \quad m = 1, \ldots, M-1,$
$z = M$:	$t > c_M.$

It follows from here, that a left-censored observation can be indicated by $z = 0$ while providing a single cut point being equal to the left-censored time (upper limit of the interval). Analogously, a right-censored observation can be indicated by $z = 1$ while providing again a single cut point, now being equal to the right-censored time (lower limit of the interval). Also a genuine interval-censored observation can be indicated by $z = 1$, nevertheless, complemented by a vector of two cut points being equal to both limits of the observed interval. An exact observation is treated separately without usage of the dinterval function, see Example 9.3.

Example 9.3 *Log-normal model with censored data in* JAGS

We take the same four observations and the same prior hyperparameters as in Example 9.2. Here we show one way on how to specify interval-censored observations in JAGS. The data list can be practically the same as in the BUGS analysis, nevertheless, now complemented by a vector of helping ordinal observations (its value for an exactly observed event time can be arbitrary):

```
list(
  logt = c(1.0986, NA, NA, NA),
  z    = c(0, 1, 0, 1),
  logt.low = c(NA, 1.3863, NA, 0.6931),
  logt.upp = c(NA, NA, 0.6931, 1.0986)
)
```

The JAGS model is specified as

```
model{

  ### Exact observation
  logt[1] ~ dnorm(mu, tau2);

  ### Right-censored observation, z[2] = 1
  z[2] ~ dinterval(logt[2], logt.low[2]);
  logt[2] ~ dnorm(mu, tau2);

  ### Left-censored observation, z[3] = 0
  z[3] ~ dinterval(logt[3], logt.upp[3]);
  logt[3] ~ dnorm(mu, tau2);

  ### Interval-censored observation, z[4] = 1
  z[4] ~ dinterval(logt[4], c(logt.low[4], logt.upp[4]));
  logt[4] ~ dnorm(mu, tau2);

  ### Prior
  tau2 ~ dgamma(1, 0.005);
  prec.mu <- 0.001 * tau2;
  mu ~ dnorm(0, prec.mu);
  sigma2 <- 1 / tau2;
}
```

Note that while running JAGS, we must provide initial values for the censored values that satisfy the censoring constraints to avoid the cryptic error message `Observed node inconsistent with unobserved parents at initialization`. For more details, see the accompanying R script which calls JAGS through the runjags package. A more complex example can also be found in Section 10.1.

□

9.3.3 R software

In the remainder of the book we will also use a handful of R packages that focus on a specific statistical Bayesian approach, e.g., ICBayes (Pan et al., 2015) for semiparametric Bayesian modelling of interval-censored data, R package dynsurv (Wang et al., 2017) for dynamic Bayesian PH modelling with a piecewise constant baseline hazard, DPpackage (Jara, 2007; Jara et al., 2011) a general package for nonparametric Bayesian analyses, mixAK (Komárek, 2009; Komárek and Komárková, 2014) for analyses of models involving mixtures, or bayesSurv (Komárek and Lesaffre, 2006, 2007, 2008; Komárek et al., 2007) for Bayesian inference for a variety of regression models with interval-censored data. We will indicate what is required as input for the respective package. Undoubtedly, we will have ignored many of the R packages that have been recently developed or are under development, since the development of R packages has seen an explosion in the last decade.

9.3.4 SAS procedures

Only the SAS procedure LIFEREG has an option for Bayesian survival analysis of interval-censored data, we refer to Section 10.1 for details on how interval-censored observations must be entered. For the SAS procedure MCMC, the likelihood must be specified by the user. We illustrate its use in subsequent chapters.

9.3.5 Stan software

Recently, the Stan software (Carpenter et al., 2017) is becoming an increasingly popular tool of Bayesian analysis. Most analyses shown in this book could be performed using the Stan software as well. Nevertheless, to maintain the amount of material covered by this book within reasonable limits, we will not cover any Stan analysis here.

9.4 Applications for right-censored data

In this section we illustrate the use of Bayesian software for a survival analysis with a right-censored data set. We start with parametric Bayesian analyses and a BUGS analysis from R using the R2WinBUGS/R2OpenBUGS packages. Then we illustrate the use of SAS on the same data set. While the SAS procedures PHREG and LIFEREG both have a Bayesian option, we will ignore the first procedure since it does not allow for interval-censored data. The SAS procedure MCMC will be illustrated in the next chapters. To illustrate the nonparametric Bayesian analyses we make use of self-written

R programs. In the first program, we estimate nonparametrically the survival function with the DP approach. In the second R program Clayton's approach has been implemented making use of a call to BUGS. This program allows for covariates. The use of covariates with the DP approach will be illustrated in later chapters. In all illustrations below we have used the homograft data set introduced in Section 1.6.1.

9.4.1 Parametric models

Example 9.4 Homograft study

In Example 2.4, we found that a Weibull AFT model gave the best fit to the homograft data and that the significant effect of donor type disappears when controlling for age. We now provide a Bayesian analysis of the same model. That is, it is assumed that the survival time T of the homograft satisfies

$$\log(T) = \beta_1 X_1 + \beta_2 X_2 + \varepsilon,$$

where $X_1 \in \{0, 1\}$ indicates whether the homograft is of a pulmonary type ($X_1 = 0$) or aortic ($X_1 = 1$), and X_2 is the patient's age in years. The error term ε is now assumed to follow a type I least extreme value distribution with location β_0 and scale σ. That is, $Y = \log(T)$ has location $\mu = \beta_0 + \beta_1 X_1 + \beta_2 X_2$ and scale σ. Consequently, the event time T is assumed to follow a Weibull distribution $\mathcal{W}(\gamma, \alpha)$, where $\alpha = \exp(\beta_0 + \beta_1 X_1 + \beta_2 X_2)$ is the scale of the Weibull distribution and $\gamma = \sigma^{-1}$ is its shape. See also Section B.3 in Appendix B for our parametrization of the Weibull distribution and its link to the type I least extreme value distribution.

Weakly informative prior distributions were considered for the model parameters. Namely, a priori $\beta_j \sim \mathcal{N}(0, 10^6)$ $(j = 0, 1, 2)$, $\gamma \sim \mathcal{U}(0, 10)$. Three MCMC chains were generated by the BUGS program, each with 10 000 burn-in and subsequent 50 000 iterations. Thinning of 1:10 was used to decrease autocorrelations of the chains. This means that only 10% of the sampled chain elements are retained for processing. Satisfactory convergence of the MCMC procedure was concluded (see Section 9.4.1.1 for more details) and the posterior inference was based on the MCMC sample of length 15 000 (= 3 × 5 000).

The mean and standard deviation of the posterior samples, a 95% HPD interval for the model parameters (Intercept – β_0, Aortic donor graft – β_1 and Age – β_2) and the contour probabilities of the Weibull AFT model are shown in Table 9.1. Since weakly informative prior distributions were used, the posterior mean, the posterior standard deviation and the 95% HPD interval, respectively, are very close to their frequentist counterparts (MLE, its standard error and the 95% confidence interval, respectively) reported in Example 2.4. Among other things, a parametric Bayesian Weibull analysis confirmed that in the presence of age, type of the graft has no real effect as the 95%

TABLE 9.1: Homograft study. Bayesian Weibull AFT model with means and standard deviation (SD) of the posterior samples and the highest posterior density (HPD) interval for the model parameters obtained with OpenBUGS.

	Posterior			Contour
Parameter	**Mean**	**SD**	**95% HPD interval**	**probability**
Intercept	2.3131	0.1276	(2.0640, 2.5930)	< 0.0001
Aortic donor graft	−0.2593	0.1474	(−0.5571, 0.0324)	0.0729
Age	0.0444	0.0101	(0.0227, 0.0640)	< 0.0001

TABLE 9.2: Homograft study. DIC for three Bayesian AFT models.

Model	**DIC**
Weibull	445.699
Log-logistic	455.418
Log-normal	480.740

HPD interval for its effect covers zero, and the contour probability is greater than 0.05.

To compare the Weibull model to the two parametric competitors, we performed similar Bayesian analyses while assuming the log-logistic or the log-normal distribution for the event times. The values of DIC for the fitted models are reported in Table 9.2. They again confirm preference for the Weibull model. □

9.4.1.1 BUGS solution

Example 9.4 in BUGS

The R *packages* R2OpenBUGS *and* R2WinBUGS *allow to call* BUGS *(either* OpenBUGS *or* WinBUGS*) using a script file while returning an* R *object that can be manipulated later. The advantage of this instead of directly using* BUGS *is that the program can be run in batch mode. In addition, the data can be prepared much easier than within* BUGS *and the same applies for the initial values of the stochastic nodes (parameters which are given a prior). This is especially useful in survival applications. Indeed, it appears that* BUGS *has sometimes difficulties with the censored survival times when they are not given initial values. As with other packages, survival data in* BUGS *consist of the survival times and censor indicators. Here we have only right-censoring. Two vectors are given as input to the program: (1) survival times, which are put to* NA *when censored, and (2) censoring times, which are put to* 0 *(or any arbitrary value) when there is no censoring.*

Below we show part of the R *program that makes use of the* R2OpenBUGS *package and its* bugs *command to call* OpenBUGS. *Note that the data has been split up into two parts: (1) uncensored survival times and corresponding covariates (indicated by* _un*) and (2) censored survival times and corresponding covariates (indicated by* _ce*). This appeared necessary to provide the censored survival times with appropriate initial values (*time_init *slightly greater than the censoring time* cens_ce*). The covariates (treatment, age) are contained in the design matrices* X_un *and* X_ce*. Variables* num_un *and* num_ce *contain the number of uncensored and censored observations, respectively.*

First, the BUGS *program must be available in a form of the external* txt *file (here named as* Homograft-Bayes_ParamAFT_Weib.txt*). The* bugs *function in* R *then needs as input: data, initial values and the list of parameters to monitor out of those mentioned in the* BUGS *program. As additional parameters, we specified that 3 chains should be initiated, the total number of iterations should be 60 000 with thinning factor equal to 10, hence only 10% of the sampled chain elements are retained. Further, by default, half of the chain is taken as burn-in. The argument* n.burnin *(here set to 10 000) allows to change the default value. Finally, note that basically the same commands are required when using* WinBUGS *except few changes as indicated in the code below.*

Cautionary note: *When the survival times are censored,* BUGS *code requires the survival times to be coded as* NA*. Hence, when all survival times are censored (as will mostly be the case for interval-censored data) a list object with all values equal to* NA *must then be specified in the data part. When using* BUGS*-like software from* R*, there are two correct ways to specify the censored survival times: (1) not specifying them at all in the data part. After all, all survival times are now parameters or (2) the survival times are specified in* R *as* time_ce = as.numeric(rep(NA, n))*. Without the* as.numeric *function, the elements of the* time_ce *vector are considered as logical, and is not accepted in* BUGS *software.*

```
## List of variables with data for BUGS
##   (all mentioned variables have previously been defined
##    in the R working environment)
> dataBUGS <- list("num_un", "num_ce", "time_un", "cens_ce",
+   "X_un", "X_ce")

## Initial values for censored event times
##   (cens_ce holds censored event times)
> time_init <- cens_ce + 0.5

## Initial parameters: 3 lists for 3 chains
##   (time_ce are the censored event times
##   augmented by Bayesian Data Augmentation)
initsWeib <- list(
```

```
+   list(beta0 = 0,     beta = rep(0, 2),
+        time_ce = time_init, gamma = 2),
+   list(beta0 = -0.5, beta = rep(-0.5, 2),
+        time_ce = time_init, gamma = 1.5),
+   list(beta0 = 0.5,  beta = rep(0.5, 2),
+        time_ce = time_init, gamma = 2.5))

## Parameters to monitor
##   (all mentioned variables are defined by the BUGS model)
> parameters <- c("beta0", "beta", "gamma", "sigma")

## Running OpenBUGS
> library("R2OpenBUGS")
> graft.Weib <- bugs(data = dataBUGS, inits = initsWeib,
+     parameters.to.save = parameters,
+     model.file = "Homograft-Bayes_ParamAFT_Weib.txt",
+     n.chains = 3, n.burnin = 10000, n.iter = 60000, n.thin = 10,
+     clearWD = FALSE, debug = TRUE)

## Running WinBUGS
> library("R2WinBUGS")
> graft.Weib <- bugs(data = dataBUGS, inits = initsWeib,
+     parameters.to.save = parameters,
+     model.file = "Homograft-Bayes_ParamAFT_Weib.txt",
+     n.chains = 3, n.burnin = 10000, n.iter = 60000, n.thin = 10,
+     bugs.directory = "C:/Programs/WinBUGS14",
+     working.directory = NULL, codaPkg = TRUE, debug = TRUE)
```

If WinBUGS *is used for the analysis, the argument* bugs.directory = "C:/Programs/WinBUGS14" *specifies where the* WinBUGS *executable is located. It is necessary to edit the bugs directory when* WinBUGS *does not reside in the classical 'Program Files' directory. The option* working.directory = NULL *signifies that a temporary working directory via tempdir is used. The option* codaPkg = TRUE *is specified for easy access by the* coda *package through function* read.bugs *(see the supplementary code). Finally,* debug = TRUE *implies that the* WinBUGS *requires the user to end the* WinBUGS *run. This allows to inspect first the standard* WinBUGS *output before proceeding with the additional* R *commands. Note that with this option the user must close the* WinBUGS *window in order to proceed with the subsequent* R *commands.*

The BUGS *model which is assumed to be stored in the txt file* `Homograft-Bayes_ParamAFT_Weib.txt` *is as follows.*

```
model{
  ### Uncensored observations
  for(i in 1:num_un){
    mu_un[i]       <- beta0 + inprod(beta[1:2], X_un[i, 1:2])
    invalpha_un[i]  <- exp(-mu_un[i])
    lambda_un[i] <- pow(invalpha_un[i], gamma)
    time_un[i] ~ dweib(gamma, lambda_un[i])
  }

  ### Right-censored observations
  for(i in 1 : num_ce) {
    mu_ce[i] <- beta0 + inprod(beta[1:2], X_ce[i, 1:2])
    invalpha_ce[i]  <- exp(-mu_ce[i])
    lambda_ce[i] <- pow(invalpha_ce[i], gamma)
    time_ce[i] ~ dweib(gamma, lambda_ce[i])I(cens_ce[i],)
  }

  ### Priors
  gamma ~ dunif(0, 10)
  beta0 ~ dnorm(0.0, 0.000001)
  for (i in 1:2){
    beta[i] ~ dnorm(0.0, 0.000001)
  }

  ### Derived parameters (AFT scale parameter)
  sigma <- 1 / gamma
}
```

Note that we now model the survival time distribution on the original time scale since the corresponding distribution on the log-scale, the type I least extreme value distribution, is not available in BUGS. *The scalar product* $\beta_1 X_{i1} + \beta_2 X_{i2}$ *is computed with the* BUGS *command* inprod. *Further, note that the* BUGS *parametrization of the Weibull distribution differs from that used primarily in this book. Namely,* `dweib`(γ, λ) *in* BUGS *corresponds to our* $\mathcal{W}(\gamma, \alpha)$ *with* $\lambda = \left(\frac{1}{\alpha}\right)^{\gamma}$, *see also Appendix B. Variables* `mu_un/mu_ce`, `invalpha_un/invalpha_ce` *and* `lambda_un/lambda_ce`, *respectively, then provide (i) the values of* $\mu_i = \beta_0 + \beta_1 X_{i1} + \beta_2 X_{i2}$, *which is the location parameter of the extreme value distribution of* $\log(T_i)$, *(ii)* $\alpha_i^{-1} = \exp(-\mu_i)$ *which is the inverted scale of the Weibull distribution of* T_i *and* $\lambda_i = (1/\alpha_i)^{\gamma}$ *which is the second parameter of the Weibull parametrization in* BUGS. *Vague normal and independent normal priors with mean zero and precision* 10^{-6} *are* given for the regression coefficients. Note that in BUGS software, the second parameter of the dnorm function is the precision (= 1/variance) rather than the variance. A relatively vague uniform prior is given for the shape parameter γ. Restricting its upper value was necessary to achieve convergence.

Upon running OpenBUGS the sampled values are already available in R inside the object graft.Weib. For further processing, we will use the coda package. To do so, it is necessary to get the sampled values stored in a form of the mcmc coda object (below, it is named as codaWeib). This can be achieved as follows.

```
> library("coda")
> codaChains <- list()
> for (ch in 1:3)
+   codaChains[[ch]] <- mcmc(graft.Weib$sims.array[,ch,],
+             start = graft.Weib$n.burnin / graft.Weib$n.thin + 1,
+             end   = graft.Weib$n.iter / graft.Weib$n.thin, thin = 1)
> codaWeib <- mcmc.list(codaChains[[1]], codaChains[[2]],
+                       codaChains[[3]])
> rm(list = "codaChains")
```

When running WinBUGS by the code mentioned above, the sampled values are stored as external text files on the hard drive. To read them into R for use by the coda package, the following command is used.

```
> codaWeib <- read.bugs(graft.Weib)
```

The next step is to evaluate the quality of the chains. In other words, we need to check whether sampling is done from the posterior distribution. Much can be concluded from the trace plot. We have used the coda function traceplot for this. In Figure 9.3 we show the trace plot of the regression coefficient of type of homograft (β_1). The plot is in fact composed out of three overlaying trace plots, one for each chain. By default each chain is displayed in a different color, thereby clearly showing whether there is good mixing of the chains or not. This cannot be observed from the black and white Figure 9.3, though. The autocorrelation plot shows the correlation (Y-axis) of subsequent values in the chain depending on their lag (X-axis). This plot cannot indicate convergence, but pinpoints how independent the sampled values of the chain are. The more independent, the more efficient the posterior distribution is sampled. Remember here that with thinning = 10 only 10% of the sampled values are examined thereby decreasing the original autocorrelation in the chain. The BGR test and plot verifies the mixing of the three chains. If so, then one might conclude that the three chains are samples from the same distribution, i.e., the posterior distribution. Two curves are plotted. The solid line estimates the mixing of the chains in a cumulative manner and is based on an ANOVA principle. That is, based on cumulative windows the ratio of the *overall variability to the within-chain variability is computed. Good mixing is concluded when this ratio and the pointwise upper one-sided 97.5% confidence bounds (dashed line) are close to 1 (rule of thumb less than 1.1 or 1.2), as can be seen here. Other convergence diagnostics are also supported by the* coda *package. Finally, note that the above diagnostics checked only the β_1 chain. To address all parameters we replace in all commands* codaWeib[,"beta[1]"] *with* codaWeib. *An alternative way to obtain convergence diagnostics is to use the* mcmcplot *function from the package* mcmcplots. *The function produces diagnostics for each parameter in a separate figure object in the working directory.*

When convergence is not contra-indicated, the chain values allow to estimate (marginal) posterior densities of all parameters. A rule of thumb is that the Monte Carlo error for the posterior mean is less than 5% of the posterior standard deviation of the parameter. The command densplot *provides an*

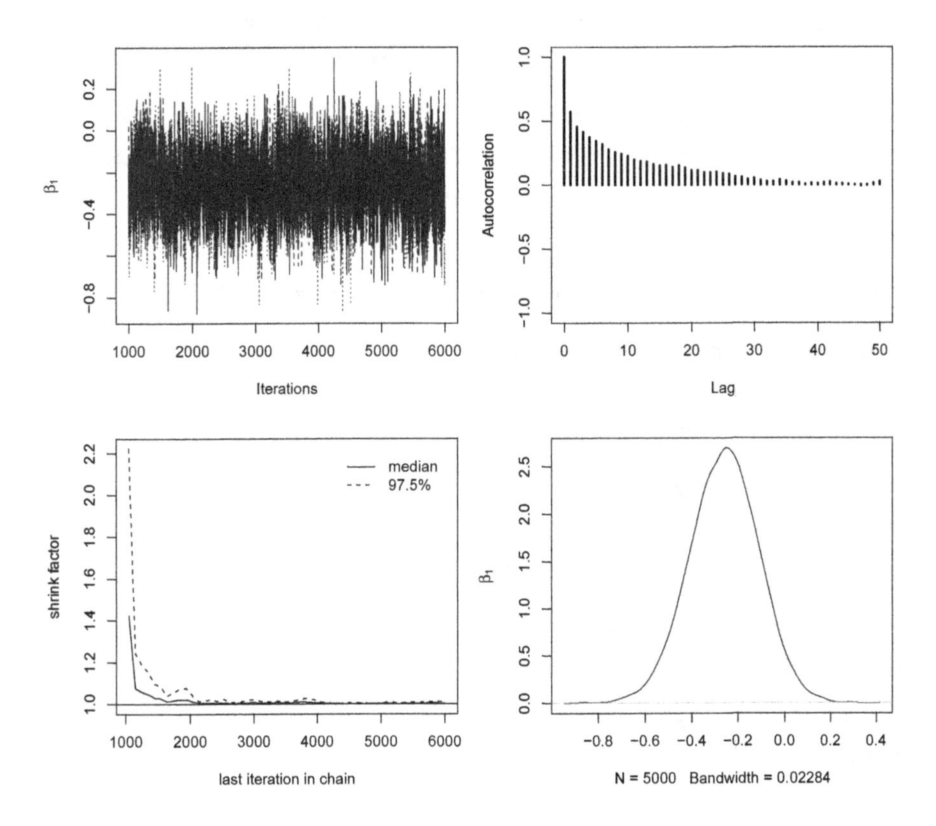

FIGURE 9.3: Homograft study. MCMC diagnostics of regression coefficient of treatment: trace plot (Top, left), autocorrelation plot (Top, right), BGR plot (Bottom, left), and estimate of marginal posterior distribution (Bottom, right), obtained from R package coda.

estimate of the marginal posterior density of the monitored parameters. The command summary *provides the posterior summary measures of these parameters, while the command* HPDinterval *provides 95% HPD intervals (output not shown).*

```
### Gelman convergence diagnostics
> gelman.diag(codaWeib[, "beta[1]"])

### Generic series of plots
> plot(codaWeib)

### Posterior plots for selected parameter
> par(mfrow=c(2, 2))

  ### Traceplots
> traceplot(codaWeib[,"beta[1]"], smooth = FALSE, type = "l",
+     ylab = expression(beta[1]), col = "grey20")

  ### Autocorrelation plots
> autocorr.plot(codaWeib[,"beta[1]"][1], lag.max = 50, ask = FALSE,
+     col = "black", lwd = 2, auto.layout = FALSE)

  ### Gelman plots
> gelman.plot(codaWeib[, "beta[1]"], auto.layout = FALSE)

  ### Marginal posterior distribution
> densplot(codaWeib[, "beta[1]"], ylab = expression(beta[1]),
+     col = "grey20", show.obs = FALSE)

### Plots using the mcmcplots package
> library("mcmcplots")
> mcmcplot(codaWeib, dir = getwd())

### Posterior summary statistics
> summary(codaWeib)

### HPD credible intervals
> HPDinterval(codaWeib)
```

□

9.4.1.2 SAS solution

The Bayesian parametric AFT model can also be fitted in SAS with the LIFEREG procedure. The parameter estimates and corresponding standard errors can be obtained using the following code.

Example 9.4 in SAS

```
PROC LIFEREG DATA=icdata.graft;
MODEL timeFU*homo_failure(0)=Hgraft age/DIST=weibull;
BAYES SEED=1;
RUN;
```

The BAYES *statement specifies that Gibbs sampling should be performed on the Weibull model shape parameter and the prior distribution for the shape parameter (default prior is* $\mathcal{G}(10^{-4},\ 10^{-4})$ *and a flat prior (*COEFFPRIOR=UNIFORM *by default) be used for the regression coefficients. Via the option* WEIBULLSHAPEPRIOR=GAMMA(a,b) *with* a *the shape and* b *the inverse-scale parameter of the gamma distribution, the default values for the prior distribution of the shape parameter may be altered. The keyword* COEFFPRIOR=NORMAL<(options)> *requests a normal prior (by default* $\mathcal{N}(0,\ 10^6\mathbb{I})$ *with* $\mathbb{I}$ *the identity matrix) for the regression coefficients. With the suboption* INPUT=SAS-data-set *the specific mean and covariance information of the normal prior may be provided. For more details, we refer to the* SAS *documentation. The* SEED= *argument guarantees reproducible results.*

By default, three types of diagnostics are produced (not shown). These are autocorrelations at lags 1, 5, 10 and 50, Carlin's estimate of the effective sample size (ESS) and Geweke's diagnostics, which are named here Geweke's spectral density diagnostics. Summary statistics for the posterior sample are displayed in the Fit Statistics *(not shown),* Posterior Summaries, Posterior Intervals *and* Posterior Correlation Matrix *(not shown).*

Posterior Summaries

			Standard		Percentiles	
Parameter	N	Mean	Deviation	25%	50%	75%
Intercept	10000	2.3255	0.1369	2.2313	2.3205	2.4124
Hgraft	10000	-0.2640	0.1508	-0.3619	-0.2626	-0.1644
age	10000	0.0435	0.0104	0.0359	0.0432	0.0502
Scale	10000	0.5118	0.0561	0.4723	0.5081	0.5464

Posterior Intervals

(Cont.)

Parameter	Alpha	Equal-Tail Interval		HPD Interval	
Intercept	0.050	2.0705	2.6109	2.0636	2.6014
Hgraft	0.050	-0.5682	0.0243	-0.5622	0.0286
age	0.050	0.0247	0.0650	0.0237	0.0638
Scale	0.050	0.4146	0.6329	0.4051	0.6214

Finally, trace, density, and autocorrelation plots for the model parameters are created (not shown).

□

9.4.2 Nonparametric Bayesian estimation of a survival curve

Susarla and Van Ryzin (1976) proposed a nonparametric Bayesian approach to compute the survival function for possibly right-censored survival times $T_1, \ldots, T_n$. They used a Dirichlet process prior $\mathcal{DP}(cS^*)$ for the survival function S. From Section 9.2 we know that S^* represents an initial guess of the survival function and $c > 0$ expresses the prior (un)certainty that we attribute to our initial guess. A large value of c indicates that we strongly believe in S^*, while a small value means that our guess is very rough. With uncensored survival times the posterior is again $\mathcal{DP}$. But, when censoring is involved Susarla and Van Ryzin (1976) showed that the posterior is a mixture of beta distributions. The estimated posterior mean of the survival function, i.e., $\overline{S}(t) = \mathsf{E}\{S(t) \,|\, \boldsymbol{\mathcal{D}}\}$, with $\boldsymbol{\mathcal{D}}$ given by the n observed possibly right-censored survival times $t_1, \ldots, t_n$. More specifically, with $\omega(t) = cS^*(t)$:

$$\overline{S}_\omega(t) = \frac{\omega(t) + N^+(t)}{\omega(R^+)} \times \prod_{i=1}^{n} \left\{ \frac{\omega(t_i-) + N^+(t_i) + d_i}{\omega(t_i-) + N^+(t_i)} \right\}^{(\delta_i=0, t_i \leq t)},$$

with t_i- the value just before the observed survival time t_i, $N^+(t)$ the number of (censored or uncensored) survival times in the sample greater than t, d_i equal to the total number of (censored or uncensored) survival times in the sample equal to t_i, δ_i equal to zero for a censored survival time and equal to 1 otherwise. Note that a small c will now result in the classical Kaplan-Meier estimate, except possibly for the upper end of the time scale, see Susarla and Van Ryzin (1976). The authors argue that their estimator is more efficient than the Kaplan-Meier estimator. They also show that their estimator is consistent in a frequentist sense.

Example 9.5 Homograft study

To illustrate the behavior of the nonparametric Bayesian estimator, we have taken the subgroup of aortic homograft patients. We have chosen a Weibull $\mathcal{W}(\gamma, \alpha)$ (in parametrization of Appendix B.3) distribution for S^*. The MLEs of the Weibull parameters for the aortic homograft patients are $\widehat{\gamma} = 2.38$ and

$\widehat{\alpha} = 10.05$. The nonparametric Bayesian solutions with $c = 0.1$ and $c = 100$ are shown in Figure 9.4. As expected, for a low value of c we obtain basically the Kaplan-Meier estimate, except for the right end of the time scale. For a large value of c, the estimate is almost the Weibull fit.

□

9.4.2.1 R solution

A self-written R function NPbayesSurv included in the icensBKL package fits the nonparametric Bayesian solution to the right-censored survival times and plots the solution together with the Kaplan-Meier solution. The function is based on Susarla and Van Ryzin (1976) and Susarla and Van Ryzin (1978).

Example 9.5 in R

```
> library("icensBKL")
> data("graft", package = "icensBKL")
> graft.AH <- subset(graft, Hgraft == "AH")   # aortic homog. patients
```

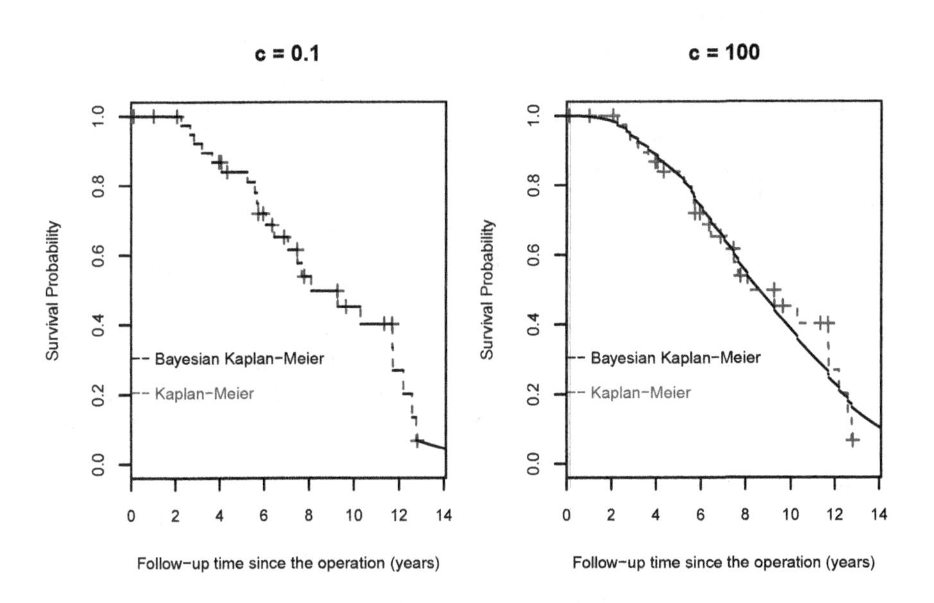

FIGURE 9.4: Homograft study (aortic homograft patients). Nonparametric Bayesian estimate of survival function based on Dirichlet process priors with $c = 100$ (close to fitted Weibull $\mathcal{W}(2.38, 10.05)$ distribution) and $c = 0.1$ (close to Kaplan-Meier estimator) together with Kaplan-Meier estimator, obtained from the R package icensBKL.

```
> time <- graft$timeFU[graft$Hgraft == "AH"]
> censor <- graft$homo.failure[graft$Hgraft == "AH"]
> NPbayesSurv(time, censor, "weibull", c = 0.1,
+     xlab = "Follow-up time since the operation (years)",
+     maintitle = "c = 0.1")
```

The first two arguments in the function NPbayesSurv *are the vector of survival times and the vector of censor indicators (0 = censored, 1 = observed). The third argument chooses S^*. One can choose (a) exponential, (b) Weibull, (c) log-normal distribution. The fourth argument represents c. The final two arguments are used on the plot as the x-axis label and the main title, respectively.*

□

9.4.3 Semiparametric Bayesian survival analysis

While the Dirichlet process (and its extensions) is the most popular nonparametric Bayesian approach, Ibrahim et al. (2001) indicated that the method is difficult to extend to involve covariates. An alternative approach was suggested by Kalbfleisch (1978), who suggested a Bayesian version of the Cox PH model by making use of a gamma process. Clayton (1991) formulates the Bayesian version of the Cox model in a counting process framework introduced by Andersen and Gill (1982). The BUGS program in Examples Volume 1 (Spiegelhalter et al., 1996) for the analysis of leukemia data is based on Clayton's approach. Here follows a brief description of this approach, much inspired by the explanation in the leukemia example.

For subjects $i = 1, \ldots, n$, $N_i(t)$ expresses the number of events occurred in the interval $(0, t]$. $\{N_i(t), t \geq 0\}$ is a counting process, which means that: (1) $N_i(t) \geq 0$, (2) $N_i(t)$ takes integer values and (3) $s < t$, $N_i(t) - N_i(s)$ represents the number of events that occurred in time interval $(s,\ t]$. The rate of a new event for the ith subject is $I_i(t) = Y_i(t)\, \hbar_i(t)$, where $Y_i(t)$ is the at risk indicator, i.e., 1 when ith subject is still at risk at time t, and zero otherwise. Furthermore, $\hbar_i(t) = \hbar(t \mid \boldsymbol{X}_i)$ is the hazard rate for the ith subject characterized by the covariates $\boldsymbol{X}_i$. $I_i(t)$ can be seen as the probability that the event occurs in a small interval $[t, t + \mathrm{d}t)$, given that it has not happened before. Thus, with $\mathrm{d}N_i(t) = N_i(t+\mathrm{d}t-) - N_i(t-)$ the increment of $N_i(t)$ over the interval $[t,\ t + \mathrm{d}t)$, $\mathrm{d}N_i(t) \approx I_i(t)\mathrm{d}t$. More formally,

$$I_i(t)\mathrm{d}t = \mathsf{E}\{\mathrm{d}N_i(t) \,|\, \boldsymbol{\mathcal{D}}_i(t-)\},$$

with $\boldsymbol{\mathcal{D}}_i(t-)$ standing for the history (data collected) of the ith patient just before time t. Further, $\mathsf{E}\{\mathrm{d}N_i(t) \,|\, \boldsymbol{\mathcal{D}}_i(t-)\}$ equals the probability of subject i failing in the interval $[t, t + \mathrm{d}t)$. For the Cox model, proportionality of the hazard functions is assumed. This implies here

$$I_i(t) = Y_i(t)\, \hbar(t) \exp(\boldsymbol{X}_i'\boldsymbol{\beta}).$$

The observed data up to time t is $\mathcal{D}(t) = \{N_i(t), Y_i(t), \boldsymbol{X}_i, (i = 1, \ldots, n)\}$. The unknown parameters are the regression coefficients $\boldsymbol{\beta}$, and the hazard function up to time t, i.e., the cumulative hazard function $H(t) = \int_0^t \hbar(u)\,du$. Under independent noninformative censoring, the total likelihood of the sample (apart from the part that depends on the censoring process) is given by

$$\prod_{i=1}^{n}[I_i(t)]^{\mathrm{d}N_i(t)} \exp\left(\int_{t\geq 0} I_i(u)\,du\right),$$

whereby the subjects contribute to the likelihood only up to the time they are at risk. Note that the above expression uses convention $0^0 = 1$. The above likelihood resembles that of a Poisson with counts now being the increments $\mathrm{d}N_i(t)$. Hence, for the sake of analysis we might assume that the counting process increments $\mathrm{d}N_i(t)$ in the time interval $[t, t + \mathrm{d}t)$ are independent Poisson random variables with means $I_i(t)\mathrm{d}t$:

$$\mathrm{d}N_i(t) \sim \text{Poisson}(I_i(t)\mathrm{d}t).$$

The Poisson mean can be expressed in the cumulative hazard function: $I_i(t)\mathrm{d}t = Y_i(t)\mathrm{d}H(t)\exp(\boldsymbol{X}_i'\boldsymbol{\beta})$, with $\mathrm{d}H(t)$ the increment or jump in the integrated baseline hazard function occurring during the time interval $[t, t+\mathrm{d}t)$. Thus this model could be considered as a nonhomogeneous Poisson process. Bayes' theorem allows to derive the joint posterior distribution for the above model, i.e.,

$$p(\boldsymbol{\beta}, H \mid \mathcal{D}) \propto p(\mathcal{D} \mid \boldsymbol{\beta}, H)\, p(\boldsymbol{\beta})\, p(H).$$

A vague prior for the regression coefficients is often chosen. The prior for the cumulative hazard function H determines the prior for the survival function. If we wish to mimic a semiparametric analysis as in the original Cox model, a nonparametric prior for H seems more appropriate. With the likelihood of the increments of a Poisson type, it would be convenient if $H(\cdot)$ were a process in which the increments $\mathrm{d}H(\cdot)$ are distributed according to gamma distributions. This is exactly the assumption in Kalbfleisch (1978), namely

$$\mathrm{d}H(t) \sim \text{Gamma}(\mathrm{d}H^*(t), c).$$

As with the Dirichlet process, $\mathrm{d}H^*$ represents a prior guess but now of the unknown hazard function. The constant c represents the degree of confidence in this guess, with small values of c corresponding to weak prior beliefs.

Example 9.6 Homograft study

The semiparametric Bayesian approach requires a prior guess for H^* or equivalently for the baseline distribution S^*. As before we have chosen for a Weibull distribution $\mathcal{W}(\gamma, \alpha)$ for which the cumulative hazard function is $H(t) = \left(\frac{t}{\alpha}\right)^{\gamma}$, $t > 0$. Now the best fit based on maximum likelihood estimation for the entire group of patients gives $\widehat{\gamma} = 1.97 \approx 2$ and $\widehat{\alpha} = 15.19$. Hence,

$H^*(t) \approx 0.005\,t^2$ and we have chosen $c = 0.001$. The latter implies that we do not strongly believe in the parametric assumption.

First, we considered a model with the indicator of the homograft type (pulmonary versus aortic graft) as the only covariate. Figure 9.5 shows the Bayesian estimates of the survival functions in the two groups based on the semiparametric PH model. Note that with a large value of c the estimates closely approximate the Weibull template (output not shown).

In our previous parametric analysis (Example 9.4), we found that the two homograft groups did not differ significantly with respect to survival when corrected for age (see Table 9.1). Such a correction can also be applied in a semiparametric Bayesian PH model. In our other analysis, we included age as an additional covariate on top of the indicator of the homograft type. We also varied a value of c and considered 0.001, 100 and 10 000 to examine the dependence of our conclusion on our prior belief of the assumed survival distribution. We evaluated for each setting the predictive ability of the model using DIC and p_D. For the model without age we obtained for the respective values of c: $p_D = 67.6, 34.9, 2.7$ and DIC $= 752.0, 707.3, 680.8$, while the corresponding values for the model with age included are: $p_D = 68.8, 35.8, 3.3$ and DIC $= 723.0, 700.6, 677.9$. We may conclude that the parametric Weibull fit is

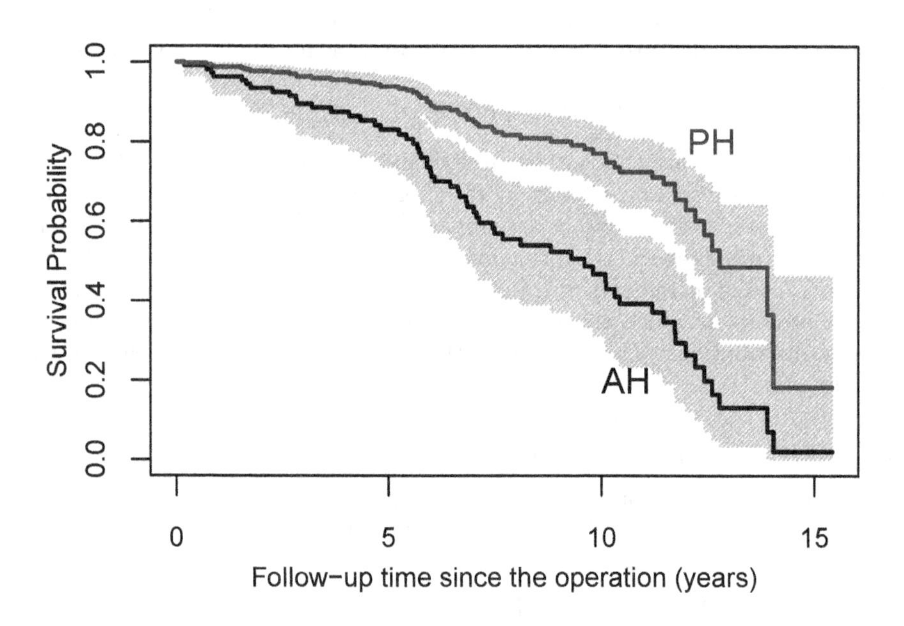

FIGURE 9.5: Homograft study. Semiparametric PH model Bayesian estimate (and pointwise 95% credible intervals) of the pulmonary (PH) and aortic (AH) homograft survival functions based on a gamma process with Weibull as initial guess, obtained from BUGS and R package icensBKL.

appropriate, and that the model with age is somewhat better than the model without age. The improvement due to age is seen especially for small values of c, but appears less important for large values of c. For all values of c, though, the regression coefficient for the type of homograft lost its importance (95% CI: -0.11 - 1.00 for $c = 0.001$), while the regression coefficient for age proved to be always important (95% CI: -0.15 - -0.06) for $c = 0.001$).

□

9.4.3.1 BUGS solution

We have adapted the BUGS program that analyzes the leukemia data from BUGS Examples Volume 1 to the homograft data while including a single binary covariate (type of homograft). R was used for the graphical presentation of the results. The accompanying R program can be easily adapted for other choices of the prior guess of H^* and of c. The same is true for other data sets without changing the BUGS model files that are available in the supplementary materials. However, when a change in H^* is envisaged, then the BUGS program must be adapted. The BUGS program is based on the approach of Clayton (1991) who incorporated the counting process framework in a Bayesian setting.

Example 9.6 in BUGS

Preparation of data and initial values to start the MCMC sampling provided by BUGS proceeds in R in practically the same way as in Example 9.4. The corresponding code can be found in the supplementary materials. Now, we only show the main BUGS code for the model involving a single binary covariate:

```
model
{
  ### Set up data
  for(i in 1:N) {
    for(j in 1:T) {
      # risk set = 1 if obs.t >= t
      Y[i,j] <- step(obs.t[i] - t[j] + eps)

      # counting process jump = 1 if obs.t in [ t[j], t[j+1] )
      #                       i.e. if t[j] <= obs.t < t[j+1]
      dN[i, j] <- Y[i, j] * step(t[j + 1] - obs.t[i] - eps) * fail[i]
    }
  }

  ### Model
  for(j in 1:T) {
    for(i in 1:N) {
      dN[i, j] ~ dpois(Idt[i, j])                    # Likelihood
```

```
      Idt[i, j] <- Y[i, j] * exp(beta * Z[i]) * dH[j]   # Intensity
    }

    dH[j] ~ dgamma(mu[j], c)
    mu[j] <- dH.star[j] * c    # dH.star[j] = prior mean hazard

    ### Survivor function = exp(-Integral{h(u)du})^exp(beta*z)
    S.AH[j] <- pow(exp(-sum(dH[1 : j])), exp(beta *  0.5))
    S.PH[j] <- pow(exp(-sum(dH[1 : j])), exp(beta * -0.5))

  }

  ### Priors
  for (j in 1 : T) {
    dH.star[j] <- pow((t[j + 1] / alpha), gamma)
                  - pow((t[j] / alpha), gamma)
  }
  beta ~ dnorm(0.0, 0.000001)
}
```

The first part of the BUGS *program creates the data in a counting process format. There are two time vectors:* obs.t *consists of all observed (uncensored and censored) event times, while* t *is a fixed grid of time points, here equal to the 59 uncensored event times. The function* step(a) *equals one when* a ≥ 0 *and zero otherwise. Hence,* Y[i,j] *is equal to 1 when at time* t[j] *the ith individual is still at risk (*eps *is a small positive value fixed in the* R *program calling the* BUGS *code). Then,* dN[i,j] *equals one when the ith subject fails before the next time on the grid, and zero otherwise. In the second part of the program, the gamma process model is described according to the theory described above. Note that $Z[i]$ is the indicator for the homograft group, coded -0.5 if the ith subject belongs to the PH group and equal to 0.5 otherwise (AH group). Note that the coding of the binary variable is taken from the original BUGS program. This coding may give more stable parameter estimates. The hazard increment at time* t[j] *is given by* dH[j] *and is assumed to follow a gamma distribution with parameters* mu[j] *and* c*. The mean of the gamma distribution is equal to* mu[j]/c = dH.star[j]*. The parameter* c *is an expression of the variability around the mean curve, since the variance of a gamma distribution is here given by* mu[j]/c^2 = dH.star[j]/c*. For small values of* c *the variance will be high. Finally, an initial guess of the hazard function, corresponding to the increment* dH.star[j] *of the cumulative hazard function $H^\star$, is given. Remember that for the Weibull distribution $\mathcal{W}(\gamma, \alpha)$ the cumulative hazard is $\left(\frac{t}{\alpha}\right)^\gamma$, $t > 0$. Prior parameters* gamma *and* alpha *are specified in the* R *program (coming from the ML Weibull fit to the whole data set). The* R *program can be further adapted to obtain extra output. Code for* *a model involving more covariates (homograft type and age in our case) can be found in the supplementary materials.*

□

Mostafa and Ghorbal (2011) propose to replace the nonparametrically defined baseline survival function by a polygonal approximation. They argue that their approach is flexible enough in practice but simpler to implement than the nonparametric Bayesian approaches. They applied their approach to the leukemia data and compared their result with those from the approach in Clayton (1991). The authors provided also a BUGS program.

9.5 Concluding remarks

In this chapter we have introduced the basic concepts of the Bayesian approach. Bayesian methods were not of interest for the statistical consultant for about 250 years. This has changed dramatically with the development of MCMC techniques and software such as BUGS. In this chapter we have illustrated how to work with Bayesian software, but we have focused on right-censored data. For right-censored data, the Cox PH model takes a central position because of the elegant partial likelihood approach allowing not to estimate the baseline hazard function. The unique central position of the Cox PH procedure is lost with interval-censored data because then for all approaches the baseline hazard must be estimated together with the regression parameters. In the next chapters, we will look what Bayesian approaches can offer to the statistical consultant when faced with interval-censored data.

Chapter 10

Bayesian estimation of the survival distribution for interval-censored observations

We introduce the reader in this chapter to some simple Bayesian analyses for interval-censored survival times. We will limit ourselves to the estimation of the survival density. In Section 10.1 parametric Bayesian analyses are performed on the breast cancer data set using JAGS in combination with R. Also the use of SAS procedures LIFEREG and MCMC will be illustrated. The R package mixAK is used in Section 10.2 to fit a smooth function to the density. Two illustrations of the Bayesian nonparametric (BNP) approach are given in Section 10.3. One illustration is based on a self-written R program using an extension of the Dirichlet Process (DP) to interval-censored survival times. The second illustration makes use of the R package DPpackage and is based on the Dirichlet Process Mixture, which is a further extension of the DP. Bayesian regression models for interval-censored data are treated in the subsequent two chapters.

10.1 Bayesian parametric modelling

In Section 3.2 we introduced parametric modelling of the survival function in a frequentist way. Here, we focus on estimating the survival function in a Bayesian manner.

Example 10.1 Signal Tandmobiel study

We revisit Example 3.2 and estimate again the distribution of the emergence times of tooth 44 (permanent right mandibular first premolar) of boys using specific parametric choices for the survival distribution, now in a Bayesian manner. In particular, we consider the three survival distributions introduced in Sections B.1 – B.3: log-normal, log-logistic and Weibull which are all parametrized using the shape parameter γ and the scale parameter α such that $\mu = \log(\alpha)$ and $\sigma = \gamma^{-1}$ are the location and the scale parameter, respectively, of the distribution of the logarithm of the event time. Vague priors,

TABLE 10.1: Signal Tandmobiel study. Bayesian parametric modelling of emergence of tooth 44 of boys, posterior mean (posterior standard deviation (SD)) of the model parameters μ, σ, γ, α, the estimated mean emergence time $\widehat{\mathsf{E}(T)}$ and median emergence time $\widehat{\text{med}(T)}$ using runjags.

Model	Parameter			
	μ	σ	γ	α
Log-normal	2.352 (0.009)	0.115 (0.007)	8.731 (0.532)	10.512 (0.091)
Log-logistic	2.353 (0.009)	0.066 (0.004)	15.111 (1.017)	10.522 (0.090)
Weibull	2.400 (0.007)	0.092 (0.006)	10.963 (0.710)	11.021 (0.082)

Model	Parameter	
	$\widehat{\mathsf{E}(T)}$	$\widehat{\text{med}(T)}$
Log-normal	10.581 (0.093)	10.512 (0.091)
Log-logistic	10.599 (0.092)	10.522 (0.090)
Weibull	10.523 (0.084)	10.657 (0.083)

$\mu \sim \mathcal{N}(0,\, 10^6)$ and $\sigma \sim \mathcal{U}(0,\, 100)$, have been specified. Two MCMC chains were generated by the JAGS program, each with 3 000 burn-in and subsequent 10 000 iterations that were sufficient to conclude a satisfactory convergence of the MCMC procedure.

The posterior mean and the posterior standard deviation for the model parameters, as well as for the model based emergence time expectation and the median emergence time (see Equations B.4 – B.9) are shown in Table 10.1. Due to vague prior distributions, the posterior means of the model parameters only negligibly differ from the corresponding MLEs reported in Table 3.2 and at the same time, the posterior standard deviations are practically the same as the standard errors of the MLEs.

□

10.1.1 JAGS solution

BUGS can be used to fit parametric models with interval-censored data in a similar way as with right-censored data, see Section 9.4.1.1. Nevertheless, from now on, we use JAGS for most Bayesian analyses when referring to BUGS-like software.

Example 10.1 in JAGS

Below we show the most relevant R *commands to call* JAGS *using the* runjags *package and to fit the log-normal model. All details and also code to fit the Weibull as well as the log-logistic model can be found in the supplementary materials.*

To call the JAGS *program, we have to create several* R *objects. First, we prepare a* list *object* Data.logt *which holds data related information.*

```
> Data.logt <- list(
+   n.left     = n.left,
+   n.right    = n.right,
+   n.interval = n.interval,
+   z.ord      = z.ord,
+   logt.lims  = cbind(log(t.low), log(t.upp)))
```

In particular, n.left, n.right, n.interval *are the numbers of left-, right-, and genuine interval-censored observations in the data set. Further,* z.ord *(a help ordinal variable to indicate the censoring status) is equal to 0 for left-censored observations and 1 for right- and interval-censored observations, see Section 9.3.2. Finally,* logt.lims *is a two-column matrix with n rows providing the logarithms of the limits of the observed intervals (vectors* t.low *and* t.upp*). Analogous to our previous* BUGS *analyses, data are assumed to be sorted such that the left-censored observations come first, followed by the right-censored observations and the genuine interval-censored observations. For left-censored observations,* t.low *is equal to the missing value (*NA*), for right-censored observations,* t.upp *is equal to* NA.

Second, two lists *with initial values for the primary model parameters (μ and σ) as well as the augmented log-event times are specified:*

```
> initLogN1 <- list(mu = 2.0, sigma = 0.1, logt = logt.init1)
> initLogN2 <- list(mu = 3.0, sigma = 0.2, logt = logt.init2)
```

With the above specification, the first chain starts from $\mu = 2.0$ and $\sigma = 0.1$, the second chain starts from $\mu = 3.0$ and $\sigma = 0.2$. The initial values for the log-event times (vectors logt.init1 *and* logt.init2*) were prepared in advance by taking a random value being consistent with the observed intervals.*

Now, we specify the JAGS *program. This can be stored in a separate text file in the same manner as with the* BUGS *analysis (see, e.g., Section 9.4.1.1). An alternative option, used here, is to make the* JAGS *program a* character *variable in* R *(here named* JAGS_LogN*):*

```
> JAGS_LogN <- "
+ model {
+
+   ### Prior distributions
+   mu ~ dnorm(0, 0.000001);
+   sigma ~ dunif(0, 100);
+
+   ### Useful derived parameters
+   sigma2 <- sigma * sigma;
```

```
+    invsigma2 <- 1 / sigma2;
+    gamma <- 1 / sigma;
+    alpha <- exp(mu);
+
+    meanT <- alpha * exp(0.5 * sigma2);
+    sdT <- alpha * sqrt(exp(1.5 * sigma2) - exp(0.5 * sigma2));
+    medT <- alpha;
+
+    ### Left-censored observations
+    for (i in 1:n.left){
+      z.ord[i] ~ dinterval(logt[i], logt.lims[i, 2]);
+      logt[i] ~ dnorm(mu, invsigma2);
+    }
+
+    ### Right-censored observations
+    for (i in (n.left+1):(n.left+n.right)){
+      z.ord[i] ~ dinterval(logt[i], logt.lims[i, 1]);
+      logt[i] ~ dnorm(mu, invsigma2);
+    }
+
+    ### Interval-censored observations
+    for (i in (n.left+n.right+1):(n.left+n.right+n.interval)){
+      z.ord[i] ~ dinterval(logt[i], logt.lims[i, ]);
+      logt[i] ~ dnorm(mu, invsigma2);
+    }
+ }
+ "
```

We first specify the prior distribution for the primary model parameters μ and σ, then specify a set of derived parameters that we may wish to monitor (σ^2, σ^{-2}, $\gamma = \sigma^{-1}$, $\alpha = \exp(\mu)$) and also characteristics of the survival distribution (its mean, standard deviation and median), all calculated using formulas from Section B.1. Finally, the log-normal model (normal distribution for the log-event times) is specified in the same way as in Example 9.3.

The JAGS *program can now be called from* R *using the command* run.jags *from the package* runjags*. If a multicore processor is available, two or more chains can be generated in parallel. To do so, the following option is specified upon loading the package into* R.

```
> library("runjags")
> runjags.options(method = "rjparallel")
```

We can now generate two chains, each with 1 000 (default value) plus 2 000 burn-in iterations and additional 5 000 iterations which are kept for inference:

```
> ModelLogN <- run.jags(model = JAGS_LogN,
+     monitor = c("mu", "sigma", "alpha", "gamma", "meanT",
+                 "sdT", "medT"),
+     data = Data.logt,
+     inits = list(initLogN1, initLogN2),
+     burnin = 2000, sample = 5000, thin = 1,
+     n.chains = 2)
```

By the argument monitor, *we specify which parameters out of those previously defined in the* JAGS *program should be stored. The sampled chains along with some additional information are stored in the object* ModelLogN. *Later we can extend each of the sampled chains by additional 5 000 values by using the command* extend.jags *as follows:*

```
> ModelLogN <- extend.jags(ModelLogN, sample = 5000, thin = 1)
```

Basic posterior summary together with some convergence diagnostics is obtained using

```
> print(ModelLogN)
```

```
JAGS model summary statistics from 20000 samples
(chains = 2; adapt+burnin = 3000):

      Lower95  Median Upper95    Mean        SD Mode       MCerr
mu     2.3348   2.352  2.3683  2.3521 0.0085556   --  0.00008267
sigma 0.10145 0.11466 0.12852 0.11501  0.006971   -- 0.000099062
alpha  10.327  10.506  10.678  10.508  0.089928   --  0.00086904
gamma  7.7133  8.7212  9.7698  8.7267    0.5255   --   0.0074133
meanT  10.396  10.575  10.758  10.578   0.09241   --  0.00095217
sdT    1.0715  1.2125  1.3722  1.2167  0.077284   --   0.0011048
medT   10.327  10.506  10.678  10.508  0.089928   --  0.00086904

      MC%ofSD SSeff       AC.10    psrf
mu          1 10710 -0.0018076 0.99997
sigma     1.4  4952 0.00074696  1.0021
alpha       1 10708 -0.0017416 0.99997
gamma     1.4  5025 0.00017789  1.0021
meanT       1  9419 -0.0013072 0.99995
sdT       1.4  4894  0.0014327  1.0019
medT        1 10708 -0.0017416 0.99997
```

(Cont.)

```
Total time taken: 6.2 seconds
```

For example, columns Mean *and* Median *provide the posterior mean and median which can be considered as point estimates for the quantities of interest. Further, columns* Lower95 *and* Upper95 *give the limits of the 95% equal-tail credible intervals. The 95% HPD credible interval based on 20 000 sampled values from both chains is directly available as the* hpd *component of the object* ModelLogN*:*

```
> print(ModelLogN$hpd)
```

```
         Lower95    Median    Upper95
mu     2.3347553  2.351966  2.3682647
sigma  0.1014458  0.114663  0.1285195
alpha 10.3265540 10.506205 10.6784664
gamma  7.7132504  8.721210  9.7698111
meanT 10.3956027 10.575340 10.7579608
sdT    1.0715035  1.212513  1.3722430
medT  10.3265540 10.506205 10.6784664
```

A set of basic plots (traceplot, posterior CDF, histogram and autocorrelation plot) for each parameter and a crosscorrelation plot are created by calling (output not shown):

```
> plot(ModelLogN)
```

The sampled values are stored as a component mcmc *of the object* ModelLogN *which can directly be used with all capabilities of the* coda *package to perform additional posterior calculations or convergence diagnostics. For example, the 95% HPD credible intervals (now calculated) separately for each chain are obtained in the same way as in our previous* BUGS *analyses (e.g., Section 9.4.1.1) using (output not shown)*

```
> library("coda")
> HPDinterval(ModelLogN$mcmc)
```

□

10.1.2 SAS solution

SAS also allows to fit parametric survival models in a Bayesian way. For interval-censored data, we can either use the Bayesian option in the SAS/STAT procedure LIFEREG or the general Bayesian procedure MCMC. For reasons that will become clear below, we prefer the LIFEREG procedure over the MCMC procedure for this class of models.

Example 10.1 in SAS (PROC LIFEREG)

We only present the code for the log-normal *model, the other models are dealt with by adapting the distribution.*

```
ODS GRAPHICS;
PROC LIFEREG DATA=icdata.tandmob;
WHERE gender=0;                        /* select boys */
MODEL (L44,R44)= /D=LNORMAL;
BAYES SEED=1 OUTPOST=postout;
RUN;
ODS GRAPHICS OFF;
```

The BAYES *statement instructs the procedure to choose for a Bayesian analysis based on a MCMC algorithm, here Gibbs sampling. The* SEED= *argument guarantees reproducible results. The posterior samples for the model parameters are stored in a* SAS *data set for further investigation. You can specify the* SAS *data set by using the option* OUTPOST=. *Alternatively, the output data set can be created by specifying:*

```
ODS OUTPUT PosteriorSample=postout;
```

The following output is then generated.

```
The LIFEREG Procedure

Bayesian Analysis

Uniform Prior for Regression Coefficients

Parameter    Prior

Intercept    Constant


          Independent Prior Distributions for Model Parameters

             Prior
Parameter    Distribution                  Hyperparameters

Scale        Gamma             Shape       0.001    Inverse Scale      0.001
```

(Cont.)

```
        Initial Values of the Chain

   Chain          Seed    Intercept       Scale

       1             1     2.351827     0.11315

                 Fit Statistics

DIC (smaller is better)                  619.994
pD (effective number of parameters)        1.996

                         Posterior Summaries

                                Standard            Percentiles
Parameter         N       Mean  Deviation        25%        50%       75%

Intercept     10000     2.3521    0.00853     2.3464     2.3519    2.3578
Scale         10000     0.1146    0.00695     0.1097     0.1144    0.1191


                       Posterior Intervals

Parameter     Alpha     Equal-Tail Interval        HPD Interval

Intercept     0.050      2.3359       2.3694      2.3359      2.3694
Scale         0.050      0.1018       0.1288      0.1017      0.1286


    Posterior Correlation Matrix

Parameter       Intercept        Scale

Intercept          1.0000       0.1771
Scale              0.1771       1.0000

              Posterior Autocorrelations

Parameter       Lag 1        Lag 5      Lag 10       Lag 50

Intercept      0.0443       0.0117     -0.0018       0.0075
Scale          0.0303      -0.0030      0.0060       0.0033


       Geweke Diagnostics

Parameter               z     Pr > |z|
```

(Cont.)

```
Intercept      -0.8429       0.3993
Scale          -0.2914       0.7707


               Effective Sample Sizes

                        Autocorrelation
Parameter          ESS             Time     Efficiency

Intercept       9186.9           1.0885         0.9187
Scale           9314.6           1.0736         0.9315
```

By default, a flat prior is assumed for the intercept and a noninformative gamma prior for the scale parameter σ. Note that in contrast to BUGS *software, improper priors (area-under-the-curve is infinite) can be chosen in* SAS.

Three types of diagnostics are produced. These are autocorrelations at lags 1, 5, 10 and 50, Carlin's estimate of the effective sample size (ESS) and Geweke's convergence diagnostics. Summary statistics for the posterior sample are displayed in the "Fit Statistics", "Posterior Summaries", "Posterior Intervals" and "Posterior Correlation Matrix". Since noninformative prior distributions were used, these results are similar to the MLEs. The ODS GRAPHICS *statement generates (diagnostic) plots. Trace, autocorrelation and density plots, grouped together for each model parameter, are created by default. Figure 10.1 displays the diagnostic plots for the intercept. A similar graph is created for the scale parameter.*

□

Example 10.1 in SAS (PROC MCMC)

Alternatively, the MCMC *procedure can be used. The same analysis as with the* LIFEREG *procedure can be fitted but now sampling is done with the Metropolis-Hastings algorithm, see Section 9.1.3. However, in the* MCMC *procedure you need to fully specify the model. For this reason, the* LIFEREG *procedure is preferred for standard survival models. The log-normal model can be fitted using the following code.*

```
PROC MCMC DATA=icdata.tandmob OUTPOST=postout
          SEED=1 NMC=20000 MISSING=AC;
WHERE gender=0;                   /* select boys */
PARMS mu 0 sigma 1;
PRIOR mu ~ NORMAL(0, SD=1000);
PRIOR sigma ~ GAMMA(SHAPE=0.001, ISCALE=0.001);

IF (L44^=. AND R44^=. AND L44=R44) THEN
```

```
        llike = LOGPDF('normal',LOG(R44),mu,sigma);
ELSE IF (L44^=. and R44=.) THEN
        llike = LOGSDF('normal',LOG(L44),mu,sigma);
      ELSE IF (L44=. and R44^=.) THEN
              llike = LOGCDF('normal',LOG(R44),mu,sigma);
           ELSE
              llike = LOG(SDF('normal',LOG(L44),mu,sigma) -
                      SDF('normal',LOG(R44),mu,sigma));
MODEL GENERAL(llike);
RUN;
```

The output (not shown) is similar to that of the LIFEREG *procedure but* MCMC *is less accurate (ESS is often much lower). Other survival models can be fitted simply by changing the distribution.*

□

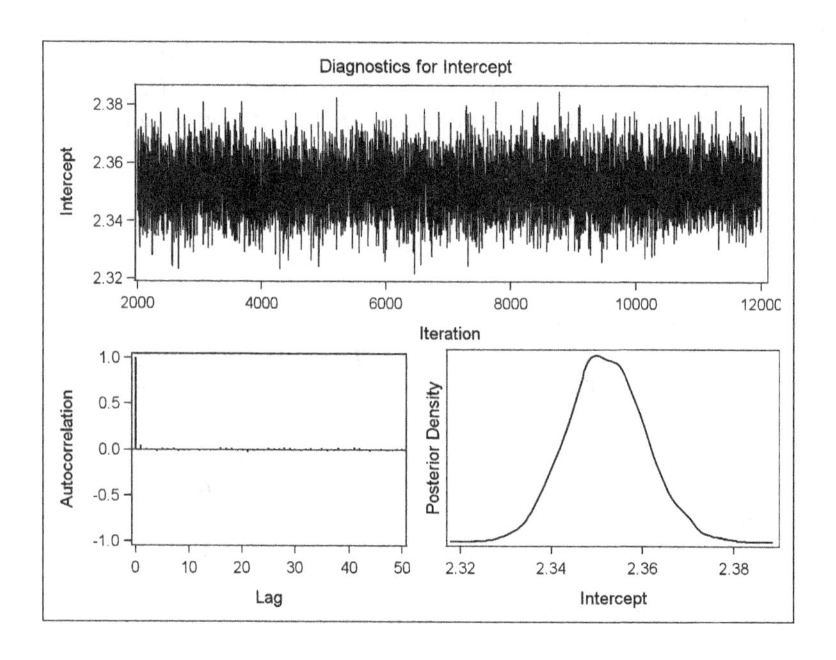

FIGURE 10.1: Signal Tandmobiel study. Summary measures of a Bayesian analysis using the SAS procedure LIFEREG (trace plot, autocorrelation plot and posterior density of the intercept).

10.2 Bayesian smoothing methods

10.2.1 Classical Gaussian mixture

The classical Gaussian mixture model was introduced in Section 3.3.2 without estimating the model parameters. In this section we focus on the Bayesian approach which is quite natural for this type of model, especially in combination with interval-censored data. Now we primarily model the density $g(y)$ of $Y = \log(T)$. As suggested in Section 3.3.2, a flexible model is provided by a (shifted and scaled) Gaussian mixture:

$$g(y) = g(y;\, \boldsymbol{\theta}) = \frac{1}{\tau} \sum_{k=1}^{K} w_k\, \varphi_{\mu_k, \sigma_k^2}\Big(\frac{y-\alpha}{\tau}\Big), \tag{10.1}$$

where $\alpha \in \mathbb{R}$ is the fixed shift parameter and $\tau > 0$ is the fixed scale parameter. The survival density f is then given as

$$f(t) = f(t;\, \boldsymbol{\theta}) = t^{-1}\, g\big\{\log(t);\, \boldsymbol{\theta}\big\}, \qquad t > 0. \tag{10.2}$$

It is indeed theoretically sufficient to consider $\alpha = 0$ and $\tau = 1$. However, for numerical stability of the MCMC procedures described in this section, it is useful to choose α and τ such that the shifted and scaled distribution of Y, i.e., the distribution of $\tau^{-1}(Y - \alpha)$ has approximately zero mean and unit variance. The vector of unknown parameters of Model (10.1) is composed of K, the number of mixture components, $\boldsymbol{w} = (w_1, \ldots, w_K)'$, the mixture weights, $\boldsymbol{\mu} = (\mu_1, \ldots, \mu_K)'$, the mixture means, and $\boldsymbol{d} = \big(\sigma_1^2, \ldots, \sigma_K^2\big)'$ the mixture variances. That is, $\boldsymbol{\theta} = (K, \boldsymbol{w}', \boldsymbol{\mu}', \boldsymbol{d}')'$.

Bayesian estimation

It was explained in Section 3.3.2 that maximum likelihood methods for estimation of parameters of the mixture Model (10.1) pose two main difficulties. First, the dimension of the parameter vector $\boldsymbol{\theta}$ depends on one of its (unknown) elements, namely the number of mixture components K. This violates one of the basic regularity conditions needed for the validity of the classical maximum likelihood theory. Second, the related likelihood is unbounded if $K \geq 2$ and a heteroscedastic mixture is allowed. On the other hand, the *Bayesian* methodology offers a unified framework to estimate both the number of mixture components K and heteroscedastic Gaussian mixtures in the same way as any other unknown parameters, i.e., using posterior summaries.

A breakthrough in Bayesian analysis of models with a parameter space of varying dimension is the introduction of the *reversible jump* Markov chain Monte Carlo (RJ-MCMC) algorithm by Green (1995) which allows to explore a joint posterior distribution of the whole parameter vector $\boldsymbol{\theta}$ from Model (10.1), including the number of mixture components K. Application of the

RJ-MCMC algorithm to estimate a Gaussian mixture from uncensored data was summarized in full generality by Richardson and Green (1997). Extension to interval-censored data is straightforward within the Bayesian framework and it is discussed in more detail by Komárek (2009) and implemented in the R package mixAK. In the remainder of this subsection, we briefly explain the concept of Bayesian estimation of the Gaussian mixture but concentrate mainly on practical issues.

As described in Section 9.1, the vector $\boldsymbol{\theta}$ of model parameters is augmented by suitable auxiliary variables $\boldsymbol{\psi}$ in the spirit of Bayesian data augmentation and a joint prior distribution is specified for $\boldsymbol{\theta}$ and $\boldsymbol{\psi}$ using the decomposition $p(\boldsymbol{\theta}, \boldsymbol{\psi}) = p(\boldsymbol{\psi} \mid \boldsymbol{\theta})\, p(\boldsymbol{\theta})$, where $p(\boldsymbol{\psi} \mid \boldsymbol{\theta})$ is determined by the used model, i.e., a Gaussian mixture, see Komárek (2009) for details. In situations when no prior information is available, it is useful to use weakly informative prior distributions for the original parameters $\boldsymbol{\theta}$, that is, $p(\boldsymbol{\theta})$ is locally flat. For mixtures, a suitable weakly informative prior distribution is described by Richardson and Green (1997) and we refer the reader therein. Further, the prior can also be chosen such that the problem of unbounded likelihood of a heteroscedastic mixture is solved. It is only necessary to guarantee that the related posterior distribution is a proper one. Inference is then based on the converged sample $\boldsymbol{\theta}^{(1)}, \ldots, \boldsymbol{\theta}^{(M)}$ from the corresponding posterior distribution obtained from the RJ-MCMC algorithm (see Richardson and Green, 1997 or Komárek, 2009 for details) as was explained in Sections 9.1.3 and 9.1.4.

Checking convergence of the (RJ-)MCMC algorithm for mixture problems is complicated by the varying dimension of the parameter vector $\boldsymbol{\theta}$ and also because the posterior distribution is invariant against the switching of the labelling of mixture components (Jasra et al., 2005). Nevertheless, the basic convergence diagnostic can be based on quantities which are invariant to label switching. These are, among others, the overall mean and the variance of the mixture (10.1) which are (given $\boldsymbol{\theta}$) equal to

$$\mathsf{E}(Y) = \alpha + \tau \sum_{k=1}^{K} w_k\, \mu_k, \tag{10.3}$$

$$\mathsf{var}(Y) = \tau^2 \sum_{k=1}^{K} w_k \left\{ \sigma_k^2 + \Bigl(\mu_k - \sum_{j=1}^{K} w_j\, \mu_j\Bigr)^2 \right\}. \tag{10.4}$$

Also invariant to label switching is the mean of the survival distribution which is obtained from (10.1) and (10.2) and equals

$$\mathsf{E}(T) = \sum_{k=1}^{K} w_k \exp\Bigl(\alpha + \tau\, \mu_k + \frac{\tau^2 \sigma_k^2}{2}\Bigr). \tag{10.5}$$

Posterior predictive density and survival function

In our context, we are primarily interested in the estimation of the survival distribution, i.e., in the estimation of the survival function S or the survival density f of the original random sample $T_1, \ldots, T_n$. A suitable estimate of the density f is the *posterior predictive density* defined as

$$\widetilde{f}(t) = \mathsf{E}_{\boldsymbol{\theta}|\boldsymbol{\mathcal{D}}}\{f(t;\,\boldsymbol{\theta})\} = \mathsf{E}_{\boldsymbol{\theta}|\boldsymbol{\mathcal{D}}}\Big[t^{-1}g\{\log(t);\,\boldsymbol{\theta}\}\Big]$$
$$= \mathsf{E}_{\boldsymbol{\theta}|\boldsymbol{\mathcal{D}}}\left[t^{-1}\tau^{-1}\sum_{k=1}^{K} w_k\,\varphi_{\mu_k,\,\sigma_k^2}\left\{\frac{\log(t)-\alpha}{\tau}\right\}\right], \tag{10.6}$$

whose MCMC estimate is obtained using Expression (9.7) and is equal to

$$\widehat{f}(t) = \frac{1}{M}\sum_{m=1}^{M}\left[t^{-1}\tau^{-1}\sum_{k=1}^{K^{(m)}} w_k^{(m)}\,\varphi_{\mu_k^{(m)},\,\sigma_k^{(m)2}}\left\{\frac{\log(t)-\alpha}{\tau}\right\}\right], \tag{10.7}$$

where $K^{(m)}$, $w_1^{(m)}, \ldots, w_{K^{(m)}}^{(m)}$, $\mu_1^{(m)}, \ldots, \mu_{K^{(m)}}^{(m)}$, ${\sigma_1^{(m)}}^2, \ldots, {\sigma_{K^{(m)}}^{(m)}}^2$ are elements of the parameter vector $\boldsymbol{\theta}^{(m)}$ sampled at the m-th MCMC iteration. Analogously, the MCMC estimate of the survival function is given by

$$\widehat{S}(t) = \frac{1}{M}\sum_{m=1}^{M}\left[\sum_{k=1}^{K^{(m)}} w_k^{(m)}\left[1-\Phi_{\mu_k^{(m)},\,\sigma_k^{(m)2}}\left\{\frac{\log(t)-\alpha}{\tau}\right\}\right]\right]. \tag{10.8}$$

Example 10.2 ***Signal Tandmobiel study***

We now estimate the distribution of the emergence times of tooth 44 (permanent right mandibular first premolar) of boys using the classical Gaussian mixture. For all parameters, a weakly informative prior distribution has been specified using the guidelines given by Richardson and Green (1997) and Komárek (2009). The MCMC was run with 1:10 thinning, and a burn-in of 10 000 iterations followed by 20 000 iterations kept for inference. The traceplots of sampled K, $\mathsf{E}(T)$, $\mathsf{E}\big(\log(T)\big)$, $\mathsf{SD}\big(\log(T)\big)$, not shown here, did not indicate any serious convergence problems.

Detailed exploration of the posterior distribution of the number of mixture components reveals that a posteriori $K = 1$ with probability 0.916, $K = 2$ with probability 0.071 and $K = 3$ with probability 0.010. The chain visited also the models with $K = 4, \ldots, 8$, however, all of them in less than 0.25% of the cases.

Posterior distributions and summary statistics for characteristics of the fitted distribution are shown in Table 10.2 and in the upper panel of Figure 10.2, respectively. We conclude that the mean emergence time is 10.6 years of age with 95% uncertainty between 10.4 and 10.8 years of age. Note that the same results were obtained with the posterior median and the 95% HPD interval. Further, these results closely correspond to results of Example 10.1. This is not surprising when we realize that the one component Gaussian mixture (which has a posterior probability of more than 0.9) is in fact the log-normal model for the emergence time T.

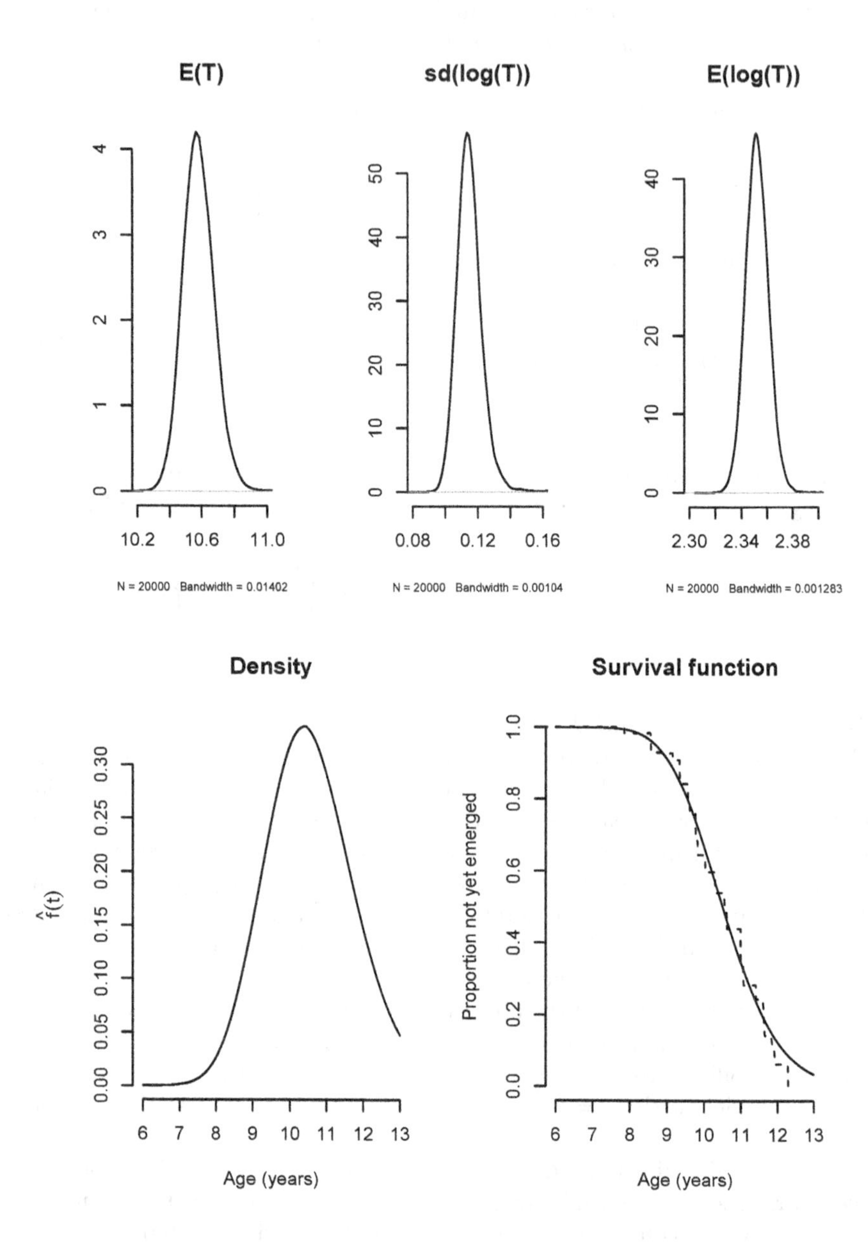

FIGURE 10.2: Signal Tandmobiel study (boys, emergence of tooth 44). Upper panel: posterior densities (their kernel estimates from the MCMC sample) of selected parameters; lower panel: posterior predictive density, and posterior predictive survival function of the emergence time compared to NPMLE (dashed line), all obtained with the R package mixAK.

TABLE 10.2: Signal Tandmobiel study (boys). Posterior summary statistics for the parameters of the emergence distribution of tooth 44 obtained with the R package mixAK.

	Posterior		95% Equal-Tail
Parameter	**Mean**	**SD**	**CI**
$\mathsf{E}(T)$	10.590	0.193	(10.403, 10.791)
$\mathsf{E}(\log(T))$	2.353	0.011	(2.336, 2.371)
$\mathsf{SD}(\log(T))$	0.116	0.017	(0.102, 0.134)

Finally, the lower panel of Figure 10.2 shows the posterior predictive density (Equation (10.7)) and the posterior predictive survival function (Equation (10.8)) compared to the NPMLE.

□

10.2.1.1 R solution

Bayesian estimation of classical Gaussian mixtures using the RJ-MCMC algorithm allowing also for censored data has been implemented in the R package mixAK (Komárek, 2009). The primary function which invokes the RJ-MCMC algorithm is NMixMCMC and results in an object which contains sampled values and several basic posterior summary statistics. Additional methods have been implemented and can be used to process the sampled values. See Appendix D.7 for more details.

Example 10.2 in R

As usual, data must be prepared first. Now three data related vectors need to be created: Y0, Y1 *and* delta*. Vector* Y0 *contains logarithms of the lower limits of observed intervals for genuine interval-censored and also for right-censored observations. On the other hand, the logarithm of the upper limit of the observed interval is included in the vector* Y0 *for left-censored observations. Vector* Y1*, of the same length as vector* Y0*, contains logarithms of the upper limits of the observed intervals for genuine interval-censored observations and an arbitrary value for left- and right-censored observations. Vector* delta *indicates the type of interval according to the convention in Table 1.1, i.e., it is equal to 0 for right-censored, equal to 2 for left-censored and equal to 3 for interval-censored observations.*

In the next step, we create a list which specifies the prior distribution of the Bayesian model. The minimal specification resulting in the default values of most parameters of the prior distribution leading to a weakly informative prior distribution is the following:

```
> Prior <- list(priorK = "uniform", Kmax = 20)
```

With the code above, we specified that a prior distribution for K is uniform on a set $\{1, 2, \ldots, 20\}$. Full user-controlled specification of the prior distribution, with values equal to the default choices, can be achieved by using the following code:

```
> Prior <- list(priorK = "uniform", Kmax = 20,
+               delta = 1,
+               priormuQ = "independentC", xi = -0.319, D = 25.668,
+               zeta = 2, g = 0.2, h = 0.390)
```

This additionally specifies that given K (the number of mixture components), the vector of mixture weights $\boldsymbol{w}$ follows a priori a Dirichlet distribution with parameter $\delta = 1$, the mixture means $\mu_1, \ldots, \mu_K$ are all a priori $\mathcal{N}(\xi, D)$ distributed with $\xi = -0.319$ and $D = 25.668$ and finally the mixture precisions $q_1 = \sigma_1^{-2}, \ldots, q_K = \sigma_K^{-2}$ are all a priori $\mathcal{G}(\zeta, \gamma)$ distributed, where $\zeta = 2$ and γ is a random hyperparameter following $\mathcal{G}(g, h)$ prior with $g = 0.2$, $h = 0.390$. Particular values of the prior hyperparameters that lead to a weakly informative prior distribution are chosen according to the procedure described in Komárek (2009).

Upon loading the package mixAK, it is now possible to run the MCMC simulation and store the results in object mod using:

```
> library("mixAK")
> mod <- NMixMCMC(y0 = Y0, y1 = Y1, censor = delta,
+        prior = Prior,
+        nMCMC = c(burn = 10000, keep = 20000, thin = 10, info = 1000))
```

In the above call, suitable values of the shift α and scale τ are derived automatically from the data. They can also be explicitly specified by adding an argument scale = list(shift = 2.331, scale = 0.148) into the above call to the NMixMCMC function.

Used values of the parameters of the prior distribution and values of the shift and scale parameters can be seen (output not shown) using:

```
> print(mod$prior)
> print(mod$scale)
```

The chain for the mean $\mathsf{E}(T)$ of the survival distribution is added to the resulting object using:

```
> mod <- NMixChainsDerived(mod)
```

Basic posterior summary statistics of the fitted model are obtained with

```
> print(mod)
```

```
    Normal mixture with at most 20 components estimated using RJ-MCMC
    ==================================================================
Posterior distribution of K:
----------------------------
      1       2       3       4       5       6       7
0.91575 0.07145 0.00985 0.00235 0.00025 0.00025 0.00005
      8
0.00005

Posterior summary statistics for moments of mixture for original data:
----------------------------------------------------------------------
Mean:
      Mean    Std.Dev.        Min.        2.5%     1st Qu.
2.35277436 0.01106800 2.30791108 2.33574809 2.34643575
    Median     3rd Qu.       97.5%        Max.
2.35222786 2.35819318 2.37100998 2.65325232

Standard deviation:
       Mean     Std.Dev.         Min.         2.5%     1st Qu.
0.116056768 0.017496965 0.004089886 0.101613880 0.109756964
     Median      3rd Qu.        97.5%         Max.
0.114328629 0.119286071 0.134292862 0.658129100

Posterior summary statistics for mean of exp(data):
---------------------------------------------------
      Mean    Std.Dev.        Min.        2.5%     1st Qu.
10.5902180  0.1929784 10.1986843 10.4034876 10.5157477
    Median     3rd Qu.       97.5%        Max.
10.5784169 10.6441858 10.7911856 19.6680009
```

Section Posterior summary statistics for moments of mixture for original data *shows posterior summary statistics of* $\mathsf{E}\big(\log(T)\big)$ *(Equation (10.3)) and the standard deviation* $\mathsf{SD}\big(\log(T)\big)$ *(square root of Expression (10.4)). Section* Posterior summary statistics for mean of exp(data) *gives posterior summary statistics of* $\mathsf{E}(T)$ *(Equation (10.5)), which are the basis for the reported values in Table 10.2. The items* 2.5%, 97.5% *determine the 95% equal-tail credibility intervals.*

The sampled values of model parameters are stored as components of the object mod *and we can further process them using the* coda *package. To do so, we first have to store the sampled values as* mcmc coda *objects. This is done (chains stand for* K, $\mathsf{E}(T)$, $\mathsf{E}(Y) = \mathsf{E}\big(\log(T)\big)$, $\mathsf{SD}(Y) = \mathsf{SD}\big(\log(T)\big)$*) using (arguments* start *and* end *are shown only in the first call to the* mcmc *function):*

```
> library("coda")
> K <- mcmc(mod$K,
+            start = mod$iter - length(mod$K) + 1, end = mod$iter)
> ET <- mcmc(mod$chains.derived$expy.Mean.1)
> EY <- mcmc(mod$mixture$y.Mean.1)
> sdY <- mcmc(mod$mixture$y.SD.1)
```

Standard capabilities of the coda package can now be used. For example, more extensive posterior summary and the HPD credible interval for the survival expectation $\mathsf{E}(T)$ are obtained using (output not shown):

```
> summary(ET)
> HPDinterval(ET, prob=0.95)
```

The traceplot and estimated posterior density (not shown) for $\mathsf{E}(T)$ are drawn using:

```
> traceplot(ET, ylab = "E(T)")
> densplot(ET, show.obs = FALSE, main = "E(T)")
```

In the same spirit, the coda package capabilities can be used for convergence diagnostics in a usual fashion.

As was mentioned above, some methods implemented in the mixAK package compute and visualize posterior predictive densities and cumulative distribution functions. First, we specify a grid of values in which we evaluate the posterior predictive density and cdf:

```
> tgrid <- seq(6, 13, length=100)
> ygrid <- log(tgrid)
```

The posterior predictive density of $Y = \log(T)$ is computed and its basic plot (not shown) drawn using:

```
> pdensY <- NMixPredDensMarg(mod, grid=ygrid)
> plot(pdensY)
```

The posterior predictive density of the emergence time T is obtained together with its basic plot using:

```
> pdensT <- Y2T(pdensY)
> plot(pdensT)
```

Similarly, the posterior cumulative distribution functions of Y and T and their basic plots are obtained with:

```
> pcdfY <- NMixPredCDFMarg(mod, grid=ygrid)
> plot(pcdfY)
> pcdfT <- Y2T(pcdfY)
> plot(pcdfT)
```

The lower panel of Figure 10.2 was drawn using (code drawing the NPMLE not shown):

```
> plot(pdensT$x[[1]], pdensT$dens[[1]], type="l",
+        main="Density",
+        xlab="Age (years)", ylab=expression(hat(f)(t)))
> plot(pcdfT$x[[1]], 1 - pcdfT$cdf[[1]], type="l",
+        ylim=c(0, 1), main="Survival function",
+        xlab="Age (years)", ylab="Proportion not yet emerged")
```

□

10.2.2 Penalized Gaussian mixture

Also the penalized Gaussian mixture (PGM) introduced in Section 3.3.3 can be estimated in a Bayesian way. When doing so, the penalty term (3.21) is translated into a prior distribution for the transformed coefficients $\boldsymbol{a}$. Furthermore, the Bayesian approach allows also for direct estimation of the smoothing parameter λ as this is treated in the same way as all other unknown parameters of the model. For more details, we refer to Section 12.3 where we estimate a regression AFT model with the PGM in the error distribution. If only an intercept is included in the AFT model, the survival distribution can directly be fitted.

10.3 Nonparametric Bayesian estimation

10.3.1 The Dirichlet Process prior approach

In Section 9.2.2 we have introduced the Dirichlet process prior, and in Section 9.4.2 we have applied the approach to estimate nonparametrically the survival function $S(t)$ in the presence of right-censored event times. In this section, this approach is extended to estimate the survival function in the presence of interval-censored event times, making use of the approach given in Calle and Gómez (2001).

The DP approach is based on a prior on the function space of S. The DP prior samples around an initial guess of the shape of the survival function, denoted as S^*. But, since there is uncertainty with respect to the actual survival function, the DP prior samples (discrete) distributions around S^* with precision expressed by a positive value c. Namely, with c small there is a wild variation around the initial guess, while a large value of c puts a lot of prior belief in S^*. This prior is classically denoted as $\mathcal{DP}(cS^*)$.

Following Ferguson's work, Susarla and Van Ryzin (1976, 1978) derived analytically the nonparametric Bayesian estimator for the survival function $S(t)$ based on right-censored data. They derive the posterior estimate of $S(t)$ after having observed right-censored survival times $t_1, \ldots, t_n$ and show that it approximates the Kaplan-Meier estimate for c small (see also Section 9.4.2). But, as Calle and Gómez (2001) point out, their approach cannot be applied for interval-censored survival times. The approach of Calle and Gómez (2001) exploits MCMC techniques to derive the Bayesian nonparametric posterior estimate of $S(t)$ after having observed interval-censored survival times. They argue that the Bayesian estimator of $S(t)$ can be interpreted as a way of 'shrinking' Turnbull's nonparametric estimator towards a smooth parametric family.

Suppose the grid of observed time points $0 = s_0 < s_1 < \cdots < s_J = +\infty$ defines a partition on the positive real line. This grid is assumed here to be taken from all the left- and right endpoints of the interval-censored observations. The survival function is known if its increments $w_j = \mathsf{P}(T \in (s_{j-1}, s_j]) = S(s_{j-1}) - S(s_j)$ $(j = 1, \ldots, J)$ are known. The Dirichlet process prior enters the estimation by assuming that the distribution of $\boldsymbol{w}$ is Dirichlet:

$$p(\boldsymbol{w} \mid \omega_1, \ldots, \omega_J) = \frac{\Gamma(\omega_1 + \ldots + \omega_J)}{\Gamma(\omega_1) \cdots \Gamma(\omega_J)} \left(\prod_{j=1}^{J-1} w_j^{\omega_j - 1} \right) \left(1 - \sum_{j=1}^{J-1} w_j \right)^{\omega_J - 1},$$

with $\omega_j = c(S^*(s_{j-1}) - S^*(s_j))$. If it was known to which the true survival time T_i of the ith individual belongs, one could construct an indicator vector $\boldsymbol{\delta}_i$ that pinpoints the interval $(s_{j-1}, s_j]$ to which T_i belongs, i.e., $\delta_{ij} = 1$ if $T_i \in (s_{j-1}, s_j]$ and zero otherwise. The distribution of δ_i given $\boldsymbol{w}$ is then a multinomial of size 1, namely

$$p(\boldsymbol{\delta}_i \mid \boldsymbol{w}) = \prod_{j=1}^{J} w_j^{\delta_{ij}}, \quad \text{where} \quad \sum_{j=1}^{J} \delta_{ij} = 1.$$

Calle and Gómez (2001) used the Gibbs sampler to sample from the posterior distribution. Sampling from two conditionals gives ultimately a sample from the posterior distribution of $\boldsymbol{w}$. To this end, they used data augmentation by assuming the latent vector $\boldsymbol{n} = (n_1, \ldots, n_J)'$, containing the frequencies that the true survival times are in the intervals defined by the grid. The two conditionals are $p(\boldsymbol{n} \mid \boldsymbol{w}, \boldsymbol{\mathcal{D}})$ and $p(\boldsymbol{w} \mid \boldsymbol{n}, \boldsymbol{\mathcal{D}})$. The first conditional is obtained from the conditional $p(\boldsymbol{\delta}_i \mid \boldsymbol{\mathcal{D}}_i, \boldsymbol{w})$, whereby $\boldsymbol{\mathcal{D}}_i = \lfloor l_i, u_i \rfloor$, equal to $w_1^{\delta_{i1}} \cdots w_J^{\delta_{iJ}} / \sum_{j | T_i \in (s_{j-1}, s_j]} w_j$ and zero otherwise. The conditional distribution $p(\boldsymbol{n} \mid \boldsymbol{w}, \boldsymbol{\mathcal{D}})$ is then obtained by taking the sum $\boldsymbol{n} = \sum_{i=1}^{n} \boldsymbol{\delta}_i$. On the other hand, $p(\boldsymbol{w} \mid \boldsymbol{n}, \boldsymbol{\mathcal{D}})$ is a Dirichlet distribution with parameters $\omega_j + n_j \, (j = 1, \ldots, J)$.

Example 10.3 Breast cancer study

We illustrate the nonparametric Bayesian solution on the early breast cancer patients treated with radiotherapy alone. Together with the origin and the "infinity" value, the grid is based on 33 points, and hence involves $r = 32$ intervals. Based on an initial guess of a Weibull distribution whereby its parameters were estimated using MLE (shape $\widehat{\gamma} = 1.1$, scale $\widehat{\alpha} = 57.5$), the nonparametric solution is shown in Figure 10.3. We have taken c equal to $\sqrt{n} = 6.8$, which is proved to represent a consistent solution of the survival function, see Calle and Gómez (2001). In the MCMC procedure, we have taken one chain with 10 000 iterations and 5 000 burn-in iterations. Furthermore, we have obtained the 95% CIs at the grid points.

We note a small difference between the Turnbull and the Bayesian estimator. Simulations in Calle and Gómez (2001) indicate a better frequentist performance of the nonparametric Bayesian estimate under the quadratic loss function than the Turnbull estimator.

□

10.3.1.1 R solution

The R function NPICbayesSurv included in the icensBKL package implements the approach of Calle and Gómez (2001) while providing the posterior sample of the weight vector $\boldsymbol{w}$, latent vector $\boldsymbol{n}$, values of the survival function evaluated at the grid of the observed time points $0 = s_0 < s_1 < \ldots < s_{J-1}$ and related pointwise posterior means as well as the credible intervals.

Example 10.3 in R

The main calculations implemented in the function NPICbayesSurv *are invoked by the following command:*

```
> library("icensBKL")
> Samp <- NPICbayesSurv(low, upp, choice = "weibull",
+                       n.sample = 5000, n.burn = 5000)
```

In the above code, low *and* upp *are vectors with the lower and upper limits of the observed intervals with our usual convention of having* low *equal to the missing value indicator* NA *for left-censored observations and analogously having* upp *equal to* NA *for right-censored observations. The argument* choice *determines the distribution of the initial guess S^* (alternatives are* "exp" *and* "lnorm" *for exponential and log-normal distribution, respectively). Finally, arguments* n.sample *and* n.burn *determine the length of the burn-in period* *of the MCMC procedure and the number of subsequent iterations used for posterior calculations, respectively.*

The sampled weights $w_1, \ldots, w_J$ as well as the sampled values of the survival function at the finite grid points $s_0 < \ldots < s_{J-1}$ are available as elements w *and* Ssample *of the* list *object* Samp *and can be used for convergence diagnostics (not shown here) or other posterior calculations. Posterior means of the values of $S(s_0), \ldots, S(s_{J_r-1})$ as well as related credible intervals are available as the* S *element of the object* Samp:

```
> print(Samp$S)
```

```
     t      Mean     Lower     Upper
1    0 1.0000000 1.0000000 1.0000000
2    4 0.9651370 0.8817327 0.9999868
3    5 0.9473143 0.8653850 0.9952490
```

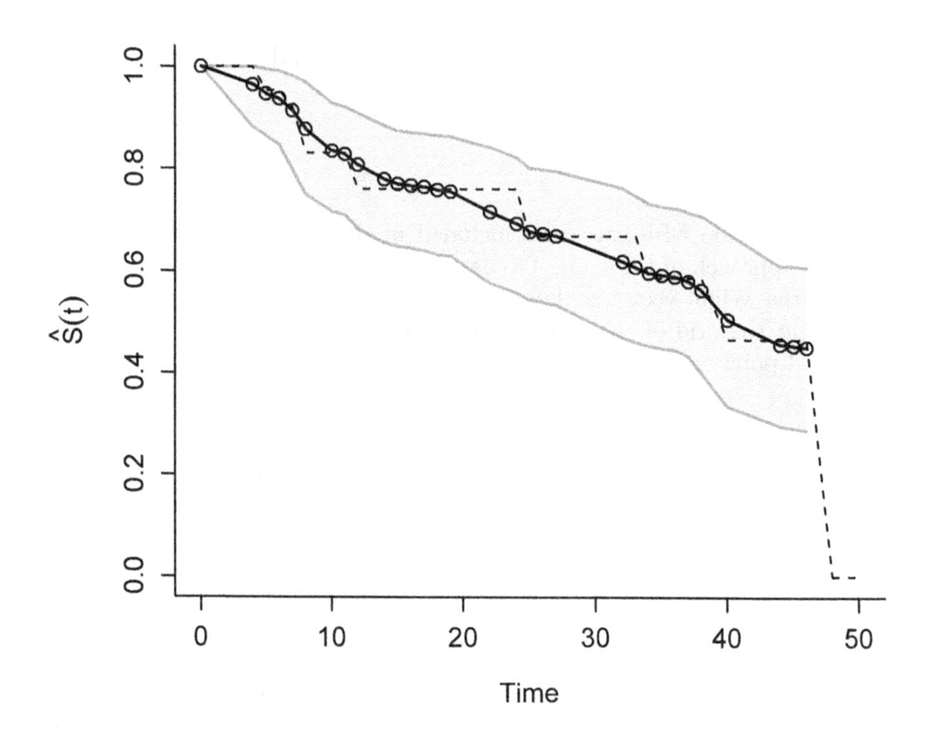

FIGURE 10.3: Breast Cancer study of the radiotherapy-only group. Nonparametric Bayesian estimate of survival function (solid line), together with the NPMLE of the survival function (dashed line), both obtained with the R package icensBKL. The gray area represents the 95% pointwise credible intervals. The grid of time points used for the Bayesian solution is indicated by circles.

(Cont.)

```
4    6 0.9378508 0.8475914 0.9919168
...
30 45 0.4534894 0.2905785 0.6091534
31 46 0.4503830 0.2860806 0.6062566
32 48 0.4240502 0.2164857 0.5930606
```

Column t *provides the finite grid points $s_0 < \ldots < s_{J-1}$. The above output can further be used, e.g., to draw a plot of the estimated survival function as depicted on Figure 10.3 (code not shown).*

□

10.3.2 The Dirichlet Process Mixture approach

While the MCMC posterior estimate of a DP often looks pretty smooth, the result of a DP is in fact a discrete distribution. This is in contrast to the classical assumption that the true survival distribution is continuous. The *Dirichlet Process Mixture (DPM)* assumes a continuous kernel for the survival times, but with a DP for the parameters of the kernel. Adapted to the survival context, we assume for a DPM that $Y_i = T_i$ or $Y_i = \log(T_i)$ have a density with parameters that have a DP prior. For n independent Y_i this implies:

$$\begin{aligned} Y_i \mid \boldsymbol{\theta}_i &\sim f_{\boldsymbol{\theta}_i} \quad (i = 1, \ldots, n), \\ \boldsymbol{\theta}_i \mid G &\sim G \quad (i = 1, \ldots, n), \end{aligned} \tag{10.9}$$

and $G \sim DP(cG^*)$, with G^* an initial guess of the distribution function of $\boldsymbol{\theta}_i$ and c is, as before, a measure of initial belief in G^*. This extension of the DP was first suggested by Antoniak (1974). Combined with observed data y_i $(i = 1, \ldots, n)$ the posterior of a DPM is a mixture of DP's. Formally, Antoniak (1974) derived that for n independent Y_i with $Y_i \mid \boldsymbol{\theta}_i \sim f_{\boldsymbol{\theta}_i}$ and independent $\boldsymbol{\theta}_i \mid G$ and $G \sim DP(cG^*)$, for a sample of observed (log) survival times $\boldsymbol{y} = \{y_1, \ldots, y_n\}$ the posterior of G is given by:

$$G \mid \boldsymbol{y} \sim \int DP(cG^* + \sum_{i=1}^{n} \delta_{\boldsymbol{\theta}_i})\, \mathrm{d}p(\boldsymbol{\theta}_1, \ldots, \boldsymbol{\theta}_n \mid \boldsymbol{y}).$$

From this result we can conclude that the posterior of G has a discrete nature determined by the $\boldsymbol{\theta}_1, \ldots, \boldsymbol{\theta}_n$ and with the most plausible $\boldsymbol{\theta}_i$ (given the data) having most impact. From a general result on DPs, it is known that the posterior of a DP has a nonzero probability on ties. In this sense, one says that the DPM model induces a probability model on clusters. There are therefore only $k \leq n$ unique values of $\boldsymbol{\theta}_i$ in the posterior sample, see further details in Müller et al. (2015, Section 2.3).

Example 10.4 ***Signal Tandmobiel study***
We now estimate the density, survival distribution and hazard function of the

emergence times of tooth 44 (permanent right mandibular first premolar) of boys using a continuous Gaussian mixture on the density. This subset of the data set contains 256 interval-censored emergence times with values between 6 and 14 years of age. In contrast to the analysis in Section 10.2.1, we work here with the original emergence times. The analysis on the log-scale gives here basically the same result as on the original scale since there is almost a linear relation between T and $\log(T)$ when T runs from 6 to 14. The analysis on the original scale avoids extra calculations to obtain the results for T given those for $\log(T)$. The density, survival distribution and hazard function are depicted in Figure 10.4.

□

10.3.2.1 R solution

As an illustration, we will estimate the density of the emergence times of tooth 44 of boys using a DPM with normal kernel. For this, we will use the function DPMdencens from the R package DPpackage.

Example 10.4 in R
The DPM implemented in DPMdencens *based on a choice of c and F^* is specified as :*

$$\begin{aligned} Y_i \mid \mu_i, \Sigma_i &\sim \mathcal{N}(\mu_i, \Sigma_i) \quad (i = 1, \ldots, n), \\ (\mu_i, \Sigma_i) \mid G &\sim G \quad (i = 1, \ldots, n), \\ G \mid c, G^* &\sim DP(cG^*). \end{aligned} \tag{10.10}$$

The user needs to specify the template G^* (also called baseline distribution) and express how much belief (s)he has in it expressed by c. Here, G^* corre-

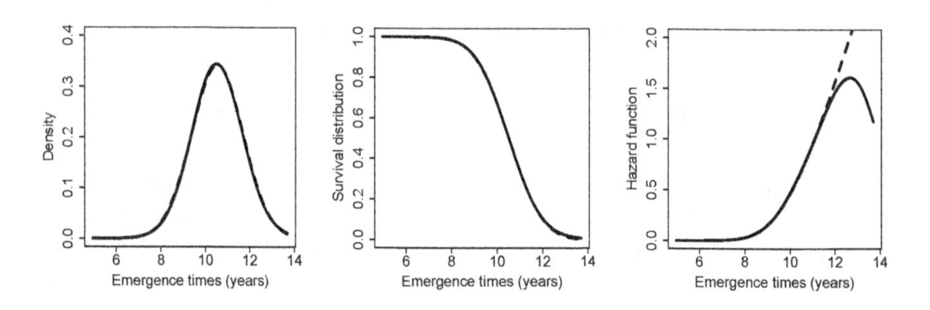

FIGURE 10.4: Signal Tandmobiel study. Estimated density, survival distribution and hazard function from DPMdencens of the emergence time for tooth 44 of boys for two precision parameters: $c = 1$ (solid), $c \sim \mathcal{G}(0.01, 0.01)$ (dashed).

sponds to an initial guess of a distribution for the means μ_i and variances Σ_i. In the function DPMdencens *a conjugate normal-inverse-Wishart distribution is chosen. This is given (in the notation of the* R *package* DPpackage*) by:*

$$G^*(\mu, \Sigma) = \mathcal{N}(\mu \mid m_1, (1/k_0)\Sigma) IW(\Sigma \mid \nu_1, \Psi_1),$$

where m_1, k_0, ν_1, Ψ_1 is chosen by the user as also c. Alternatively, these parameters could each be given a hyperprior. In the package, the following hyperpriors can be chosen:

$$\begin{aligned} c \mid a_0, b_0 &\sim \mathcal{G}(a_0, b_0), \\ m_1 \mid m_2, s_2 &\sim \mathcal{N}(m_2, s_2), \\ k_0 \mid \tau_1, \tau_2 &\sim \mathcal{G}(\tau_1/2, \tau_2/2), \\ \Psi_1 \mid \nu_2, \psi_2 &\sim IW(\nu_2, \Psi_2). \end{aligned}$$

The choice of the prior parameters should preferably be inspired by the problem at hand (not the data!). In the package, c is referred to alpha*. Relatively small values of c imply large uncertainty around the template, see Section 9.2.2. Here, m_2 corresponds to the mean of the emergence times, roughly around 10. When s_2 is large, the prior is vague. The parameters ν_1 and ν_2 correspond to the degrees of the inverse Wishart distributions. When ν_1 is close to the dimension of the scale matrix Ψ_1, here equal to 1, $IW(\Sigma \mid \nu_1, \Psi_1)$ is a vague prior. Note that instead of choosing for a particular Ψ_1, in* DPMdencens *Ψ_1 is given an inverse Wishart prior.*

The question of interest here is what the most likely density of the emergence times is, taking into account that the observed emergence times are interval-censored. For this we need to load DPpackage *to access function* DPMdencens*. We also load* icensBKL *to access the emergence times and* coda *to produce convergence diagnostics.*

The R *object* tandmobDMFboys *contains the emergence times of tooth 44 for the boys. From this data set we extract the left and right endpoints of the interval-censored emergence times, see the commands below. In the first analysis, we fix c to 1. This implies a large prior uncertainty in the assumed template. For instance, with this choice of c the 95% prior uncertainty for $S(t) = 0.2$ runs from 0.05 to 0.50. The objects* m2_1 *and* s2_1 *are the mean and variance of the template. They have been chosen here to reflect roughly the average age and spread of emergence times of tooth 44. The object* psiinv2_1 *corresponds to the scale of the inverse Wishart (here inverse gamma because the dimension of the observations is 1) and its small value implies a vague prior for the variance of the emergence times. In a next step all prior choices are collected in a list object called* prior1*, which contains also the other prior parameters. Note that* nu1 $= \nu_1$, nu2 $= \nu_2$, tau1 $= \tau_1$, tau2 $= \tau_2$, *with $\nu_1, \nu_2, \tau_1, \tau_2$ defined above. Then, the parameters of the MCMC sampling exercise are specified: (1)* nburn*: number of burn-in iterations, (2)* nsave*: number of iterations to be saved, (3)* nskip*: thinning factor and (4)* ndisplay*: frequency* of alerting the user of the progress in MCMC sampling. Finally, the DPMdencens is accessed and an object called fit1 of class DPMdencens is created. Note status = TRUE means that a fresh MCMC chain has been started up, when equal to FALSE a previous MCMC chain is continued. Only one chain can be started; if one wishes to initiate more than one chain, then one has to do this by hand and in a later stage combine the sampled values of the different chains.

```
> N <- length(tandmobDMFboys$IDNR)
> left <- tandmobDMFboys$L44
> right <- tandmobDMFboys$R44
> c_1 <- 1              # first analysis fixing alpha
> m2_1 <- 10; s2_1 <- 25; psiinv2_1 <- 1/10000
>  prior1 <- list(alpha = c_1, nu1 = 2, nu2 = 2, s2 = s2_1,
+        m2 = m2_1, psiinv2 = psiinv2_1, tau1 = 0.01, tau2 = 0.01)

# MCMC parameters
> nburn <- 5000; nsave <- 5000; nskip <- 10; ndisplay <- 100
> mcmc <- list(nburn = nburn, nsave = nsave,
+        nskip = nskip, ndisplay = ndisplay)

# Fitting the model
> fit1 <- DPMdencens(left = left, right = right, ngrid = 100,
+           prior = prior1, mcmc = mcmc, state = state, status = TRUE)
```

Before inspecting the posterior summary measures, one needs to check for convergence. This can be done by using the R package coda as illustrated below. We have omitted the diagnostic information which shows no specific deviation from convergence.

```
> coda.obj <- mcmc(fit1$save.state$thetasave)
> traceplot(coda.obj, ask = FALSE, auto.layout = FALSE)
> autocorr.plot(coda.obj, ask = FALSE, auto.layout = FALSE)
> geweke.diag(coda.obj, frac1 = 0.1, frac2 = 0.5)
```

Upon convergence, we look at the summary measures, see the next three commands. The first is the default summary, which provides standard MCMC output. Note that with alpha $= 1$, ncluster ≈ 5, which implies that the MCMC analysis recognizes about 5 different μ_i, Σ_i values. The DPpackage function DPrandom provides estimates for the mean and variance of each true interval-censored emergence time (output not shown). Because individual times are allocated to one of the 5 clusters, these means and variances differ across boys. When adding predictive = TRUE to the DPrandom command, the predictive means and variances of μ_i and Σ_i are given. See below for output of the first and last command.

```
> summary(fit1)
> DPrandom(fit1)
> DPrandom(fit1, predictive = TRUE)
```

```
DPM model for interval-censored data

Call:
DPMdencens.default(left = left, right = right, ngrid = 100,
    prior = prior1, mcmc = mcmc, state = state, status = TRUE)

Baseline distribution:
                Mean       Median     Std. Dev.  Naive Std.Error
m1-var1         1.067e+01  1.057e+01  1.316e+00  1.861e-02
k0              3.486e+01  5.950e+00  7.666e+01  1.084e+00
psi1-var1:var1  5.153e-01  3.553e-01  7.023e-01  9.932e-03
                95%HPD-Low  95%HPD-Upp
m1-var1         8.646e+00   1.340e+01
k0              1.576e-05   1.710e+02
psi1-var1:var1  4.435e-02   1.392e+00

Precision parameter:
          Mean    Median  Std. Dev.  Naive Std.Error  95%HPD-Low
ncluster  4.7394  5.0000  1.9944     0.0282           1.0000
          95%HPD-Upp
ncluster  8.0000

Number of Observations: 256
Number of Variables: 1
```

```
> DPrandom(fit1,predictive = TRUE)
```

```
Random effect information for the DP object:

Call:
DPMdencens.default(left = left, right = right, ngrid = 100,
    prior = prior1, mcmc = mcmc, state = state, status = TRUE)


Predictive distribution:
          Mean      Median    Std. Dev.  Naive Std.Error
mu-var1   10.68744  10.54669  2.27374    0.03216
```

(Cont.)

```
var-var1    1.56711    1.25400    7.29335    0.10314
           95%HPD-Low  95%HPD-Upp
mu-var1     9.71266    11.18424
var-var1    0.20930     2.09153
```

In the second analysis, the precision parameter c (alpha) is given a prior itself. This is done by specifying values for a_0 and b_0. First, we have taken $a_0 = b_0 = 0.01$, see below. Although not specified in the package, here the gamma representation with shape and scale was taken. This means that we assume a prior mean and variance for c equal to 0.001 and 0.0001. Hence, a very weak belief in the baseline distribution. We have varied the values of a_0 and b_0 up to each 100. The posterior value for c depends in a nontrivial manner on the choice of a_0 and b_0 but always pointed to a small value below 0.5.

```
> prior2 <- list(a0 = 0.01, b0 = 0.01, nu1 = 2, nu2 = 2, s2 = s2_1,
+       m2 = m2_2, psiinv2 = psiinv2_2, tau1 = 0.01, tau2 = 0.01)
```

With $a_0 = b_0 = 0.1$, the posterior of alpha points to only one cluster (output not shown). With increasing a_0 and b_0 the number of clusters remains very low. In Figure 10.4 we have compared two solutions ($c = 1$, $c \sim \mathcal{G}(0.01, 0.01)$) for the density, the survival distribution and the hazard function of the emergence times. The density is obtained directly from the R package, the other estimates are easily derived from the density (see the accompanied R program). We observe that the density and survival distributions are quite alike,

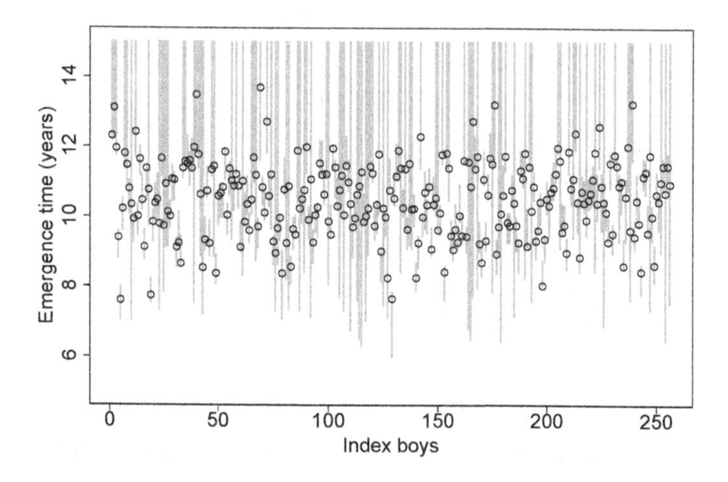

FIGURE 10.5: Signal Tandmobiel study. Imputed emergence times for tooth 44 of boys from 2nd solution obtained from the R package DPpackage together with observed intervals truncated at [6, 15] years.

but that the hazard functions differ around ages 12 to 14. It appears that for all considered values of a_0 *and* b_0 *the survival distribution and density are much alike, but that the hazard varies between ages 12 and 14. This has to do probably with the relatively large number of right-censored emergence times (39%).*

Finally, we show in Figure 10.5 the index plot of the imputed emergence times from the second analysis together with the observed intervals truncated at 6 and 15.

□

Note that the classical Gaussian mixture approach in Section 10.2.1 is similar in spirit as the current BNP approach, i.e., both approaches assume a mixture of Gaussian densities for the emergence density. While they are technically different, both approaches appear to agree on the estimated survival distribution, density and hazard function. They also appear to agree on a relatively simple (approximate) Gaussian structure for the (log) emergence times (see comment above).

10.4 Concluding remarks

This chapter reviewed a number of Bayesian approaches to estimate the true survival distribution despite the fact that the survival times are observed only as intervals. We are now ready for Bayesian regression models, either in a parametric or nonparametric way, which is of main interest in basically every statistical analysis. The aim of this chapter was partly also to prepare the reader for the way interval-censored data are handled in current Bayesian software.

Chapter 11

The Bayesian proportional hazards model

In the parametric case, the Bayesian treatment of a proportional hazards model for interval-censored data is similar to that for right-censored data. This is illustrated in Section 11.1 with a JAGS analysis and using the SAS procedure MCMC. We also illustrate model selection and model checking in a Bayesian way. The restrictive assumptions can be relaxed by replacing the parametrically defined baseline hazard function by a smooth version, which is illustrated in Section 11.2 with the R package ICBayes. The aforementioned approaches are illustrated on the breast cancer data. In Section 11.2 we also introduce and illustrate an approach that is based on slowly varying piecewise constant baseline hazard functions allowing also for a dynamic analysis, i.e., regression coefficients that vary in time. This is provided with the R package dynsurv and is applied to emergence times of tooth 44 recorded in the Signal Tandmobiel study. There is no fully developed Bayesian semiparametric model and definitely no software available, so Section 11.3 is quite short.

11.1 Parametric PH model

The proportional hazards (PH) assumption was introduced in Section 2.3.1. Recall that this implies for the hazard function at time t for a subject with covariate vector $\boldsymbol{X}$:

$$\hbar(t \mid \boldsymbol{X}) = \hbar_0(t) \exp(\boldsymbol{X}'\boldsymbol{\beta}), \tag{11.1}$$

where $\hbar_0(t)$ represents a baseline hazard function and $\boldsymbol{\beta}$ a vector of regression parameters. We look here at the case when the baseline hazard, or equivalently the baseline survival function $S_0(t) = \exp\Big\{-\int_0^t \hbar_0(s)\,\mathrm{d}s\Big\}$, is defined parametrically. Estimation of the model parameters using the maximum likelihood method was illustrated in Section 5.1. By revisiting Example 5.1, we show here how the model may be fitted in a Bayesian manner.

Example 11.1 Breast cancer study

As in Example 5.1, we use the PH model to compare the two treatments (*radiotherapy only* and *radiotherapy and chemotherapy*) for the early breast cancer study, i.e., $\boldsymbol{X} \equiv X_1 \in \{0, 1\}$ is the treatment group indicator and $\boldsymbol{\beta} \equiv \beta_1$ is the logarithm of the hazard rate. The analysis will now proceed using a Bayesian approach. As before, we consider for the baseline hazard function $\hbar_0(t)$ a Weibull distribution. In Appendix B.9, we consider two parametrizations (for $t > 0$)

$$\hbar_0(t) = \frac{\gamma}{\alpha_0}\left(\frac{t}{\alpha_0}\right)^{\gamma-1} = \lambda_0\,\gamma\,t^{\gamma-1}, \tag{11.2}$$

$$S_0(t) = \exp\left\{-\left(\frac{t}{\alpha_0}\right)^{\gamma}\right\} = \exp\bigl(-\lambda_0\,t^{\gamma}\bigr),$$

where γ and α_0 are unknown shape and scale parameters, respectively, which give rise to parameter $\lambda_0 = \alpha_0^{-\gamma}$ which is the second parameter of the Weibull parametrization in BUGS and JAGS. When the PH assumption (11.1) is combined with the Weibull baseline hazard function (11.2), it is seen that a subject with covariate vector $\boldsymbol{X}$ follows a Weibull distribution with the shape parameter γ (the same as for the baseline distribution) and the scale parameter $\alpha(\boldsymbol{X})$ or parameter $\lambda(\boldsymbol{X})$ (in the BUGS/JAGS parametrization) being given as

$$\alpha(\boldsymbol{X}) = \alpha_0\,\exp\left(-\frac{1}{\gamma}\boldsymbol{X}'\boldsymbol{\beta}\right), \qquad \lambda(\boldsymbol{X}) = \lambda_0\,\exp(\boldsymbol{X}'\boldsymbol{\beta}). \tag{11.3}$$

In the following, we additionally transform parameter λ_0 into the model intercept $\beta_0 = \log(\lambda_0)$. That is, the primary model parameters will be the intercept term β_0, the regression parameters $\boldsymbol{\beta}$ leading to $\lambda(\boldsymbol{X}) = \exp\bigl(\beta_0 + \boldsymbol{X}'\boldsymbol{\beta}\bigr)$ and the Weibull shape parameter γ. For all model parameters, vague priors are specified, namely, $\beta_0 \sim \mathcal{N}(0,\,10^6)$, $\boldsymbol{\beta} \equiv \beta_1 \sim \mathcal{N}(0,\,10^6)$. The shape parameter γ is assigned an exponential prior distribution with mean (and scale) equal to 1 000. Three MCMC chains were generated by the JAGS program, each with 6 000 (×10) burn-in and subsequent 10 000 (×10) iterations. Thinning of 1:10 was used leading to 30 000 sampled values used for inference. Table 11.1 shows the posterior means, posterior standard deviations and 95% HPD intervals for the model parameters as well as the treatment hazard ratio ($= e^{\beta_1}$) and the median event times in the two treatment groups which are calculated from the model parameters using Expression (B.9). As vague priors were used, the posterior means only negligibly differ from the MLEs of the corresponding parameters reported in Example 5.1.

□

11.1.1 JAGS solution

Both BUGS and JAGS can be used to fit a parametric PH model in a Bayesian manner. As we already announced in Section 10.1, we concen-

TABLE 11.1: Breast cancer study. Bayesian Weibull PH model, posterior means, posterior standard deviations, and 95% HPD intervals of the model parameters, treatment hazard ratio and the median event times in the two treatment groups obtained from a runjags analysis.

Parameter	Posterior mean	Posterior SD	95% HPD interval
Intercept (β_0)	−6.424	0.728	(−7.845, −4.990)
Treatment (β_1)	0.938	0.284	(0.377, 1.487)
Shape (γ)	1.643	0.194	(1.269, 2.023)
Baseline scale (α_0)	50.70	7.55	(37.58, 65.75)
Hazard ratio (radio + chemo vs. radio only)	2.66	0.78	(1.35, 4.22)
Median time (radio only)	40.4	5.7	(29.8, 51.6)
Median time (radio + chemo)	22.6	2.5	(18.0, 27.8)

trate on JAGS and show how to fit the considered models by combining JAGS with the R package runjags.

Example 11.1 in JAGS

The most relevant R *commands to call* JAGS *using the* runjags *package and to fit the Weibull PH model are shown below. Full code is available in the supplementary materials.*

Analogously to our previous JAGS *analysis in Section 10.1.1, several* R *objects with data and initial values must be prepared first. The corresponding steps are in fact more or less the same as in Example 10.1. We now create objects* N1, N2, N3 *(numbers of left-, right- and genuine interval-censored observations),* lim1, lim2, lim3 *(two-column matrices with limits of observed intervals for left-, right- and interval-censored observations, respectively),* trtRCH1, trtRCH2, trtRCH3 *(vectors with zero-one treatment group indicators for the left-, right- and interval-censored observations) and also censoring vectors* z.ord1, z.ord2, z.ord3 *corresponding to the left-, right- and interval-censored observations, respectively, with all values equal to zero (*z.ord1*) or one (*z.ord2, z.ord3*). All objects are stored in a* list *object* data.breast*:*

```
> data.breast <- list(
+    N1 = N1, lim1 = lim1, z.ord1 = z.ord1, trtRCH1 = trtRCH1,
+    N2 = N2, lim2 = lim2, z.ord2 = z.ord2, trtRCH2 = trtRCH2,
+    N3 = N3, lim3 = lim3, z.ord3 = z.ord3, trtRCH3 = trtRCH3)
```

The next three commands specify the initial values of the stochastic nodes. It is advisable to provide initial values not only for the model parameters but also for all censored observations. After all, they are considered as parameters in the JAGS program. Note that three chains will be initiated. It is recommended to choose three widely spread starting values for all parameters. We have done this only for the model parameters but not for the censored observations:

```
> ### Initials for augmented times (will be the same for all 3 chains)
> time1_init <- lim1[, 2] / 2                      # left-censored
> time2_init <- lim2[, 1] + 0.1                    # right-censored
> time3_init <- (lim3[, 1] + lim3[, 2]) / 2        # interval-censored
```

```
> ### All initials put in the list to be passed to JAGS
> inits.breast <- list(
+    list(time1 = time1_init, time2 = time2_init, time3 = time3_init,
+         beta0 = rnorm(1, 0, 1), beta1 = rnorm(1, 0, 1),
+         gamma = runif(1, 0, 1)),
+    list(time1 = time1_init, time2 = time2_init, time3 = time3_init,
+         beta0 = rnorm(1, 2, 1), beta1 = rnorm(1, 2, 1),
+         gamma = runif(1, 0, 1)),
+    list(time1 = time1_init, time2 = time2_init, time3 = time3_init,
+         beta0 = rnorm(1, -2, 1), beta1 = rnorm(1, -2, 1),
+         gamma = runif(1, 0, 1)))
```

The parameters to be monitored by JAGS are stored in the R object parameters.breast1:

```
> parameters.breast1 <- c("beta0", "beta1", "gamma", "lambda0",
+                         "alpha0", "HR", "median0", "median1")
```

Now we discuss the JAGS program that we specify (analogously to Example 10.1) as a character variable (named JAGS_Cox_Weib) in R:

```
> JAGS_Cox_Weib <- "
+ model {
+
+   ### Model for left-censored observations
+   for(i in 1 : N1) {
+     lambda1[i] <- exp(beta0 + beta1 * trtRCH1[i])
+     z.ord1[i] ~ dinterval(time1[i], lim1[i, 2])
+     time1[i] ~ dweib(gamma, lambda1[i])
+     time1.rep[i] ~ dweib(gamma, lambda1[i])
+   }
+
+   ### Model for right-censored observations
```

```
+   for(i in 1 : N2) {
+     lambda2[i] <- exp(beta0 + beta1 * trtRCH2[i])
+     z.ord2[i] ~ dinterval(time2[i], lim2[i, 1])
+     time2[i] ~ dweib(gamma, lambda2[i])
+     time2.rep[i] ~ dweib(gamma, lambda2[i])
+   }
+
+   ### Model for interval-censored observations
+   for(i in 1 : N3) {
+     lambda3[i] <- exp(beta0 + beta1 * trtRCH3[i])
+     z.ord3[i] ~ dinterval(time3[i], lim3[i, ])
+     time3[i] ~ dweib(gamma, lambda3[i])
+     time3.rep[i] ~ dweib(gamma, lambda3[i])
+   }
+
+   ### Useful derived parameters
+   ### (median survival times in the two groups)
+   ### median0 = radio only, median1 = radio+chemo
+   median0 <- pow(log(2) * exp(-beta0), 1 / gamma)
+   median1 <- pow(log(2) * exp(-beta0 - beta1), 1 / gamma)
+
+   lambda0 <- exp(beta0)
+   alpha0 <- exp(-beta0 / gamma)
+
+   ### Priors
+   beta0 ~ dnorm(0, 1.0E-6)
+   beta1 ~ dnorm(0, 1.0E-6)
+   gamma ~ dexp(0.001)
+ }
+ "
```

Two sorts of variables define the censored outcomes: the z.ord *(helping ordinal observations) and the* time *(augmented event times) variables, one for each type of censoring. Moreover, we added a* time.rep *variable (replicated event time) for each type, to be used later for a model check. As indicated above, the Weibull distribution is assumed to have the same shape parameter (*gamma*) for the two treatments, but the second parameter of the* JAGS *Weibull parametrizations (*lambda*) depends on covariates as dictated by (11.3). Finally, we are also interested in the median survival times of the* two treatments (median0 and median1) which are derived from the model parameters by Expression (B.9).

At this stage, we are ready to call JAGS from R. This is accomplished with the run.jags command, which produces an object of class runjags here called jagsfit. Note that we could have used also the autorun.jags program, which samples until convergence is met with a maximal running time, see the documentation of the package for further details and for other options not used in this analysis. To allow for parallel sampling of the three chains (if at least three CPUs are available), we also modify the runjags options.

```
> runjags.options(method = "rjparallel")
> jagsfit <- run.jags(model = JAGS_Cox_Weib,
+     monitor = parameters.breast1,
+     data = data.breast, inits = inits.breast,
+     n.chains = 3, adapt = 1000, thin = 10,
+     burnin = 5000, sample = 10000)
```

The following arguments are specified in the run.jags call:

- model = JAGS_Cox_Weib: locates the character variable with the JAGS model;
- monitor = parameters.breast1, data = data.breast, inits = inits.breast: parameters to be monitored, data to be used and specification of initial values, respectively;
- n.chains = 3, adapt = 1000, burnin = 5000, thin = 10, sample = 10000: three chains are initiated, the size of the adaptation phase of the sampler (part of chain where sampler is being optimized) is equal to 1 000 iterations, burn-in part consists of 5 000 iterations, and ten times 10 000 values are sampled since we asked for a thinning factor of 10, respectively.

The next step is to explore the quality of the fitted model. We should start with checking the convergence of the chain(s). This is done in the two R commands below. The first command asks for some default plots provided by the runjags package, see Figure 11.1 for the regression parameter β_1. The second command checks whether there is good mixing of the three chains. This is done with the BGR diagnostic or the Geweke diagnostic, both supported by the coda package.

```
> plot(jagsfit, layout = c(2, 2))
> gelman.diag(jagsfit, multivariate = TRUE)
```

Upon convergence of the chains, we explore the posterior summary measures. This is done with the default command summary. The output is shown below, now produced with the print statement, which is a bit more sophisticated than the summary command allowing to control the number of reported *digits (=3). The columns* Lower95 *and* Upper95 *provide an equal-tail 95% credible interval and other standard posterior summary measures (*AC.100 *is autocorrelation at lag 100). Of interest is the effective sample size, which shows that for some parameters the equivalent chain size is around 2 000 despite the fact that actually 30 000 values were generated. Note, that the* R *object* jagsfit *contains additional information. It is advised to look via* attributes *what other information can be retrieved. For instance, HPD intervals can be obtained from the second command below.*

```
> summary(jagsfit)
```

```
  JAGS model summary statistics from 30000 samples
  (thin = 10; chains = 3; adapt+burnin = 6000):

           Lower95  Median Upper95    Mean      SD Mode    MCerr
beta0        -7.85   -6.39   -4.99   -6.42   0.728   --   0.0168
beta1        0.377   0.934    1.49   0.938   0.284   --  0.00252
gamma         1.27    1.63    2.02    1.64   0.194   --   0.0045
lambda0 0.000144 0.00168 0.00507 0.00208 0.00157   -- 3.24e-05
alpha0        37.6    49.7    65.7    50.7    7.55   --   0.0541
```

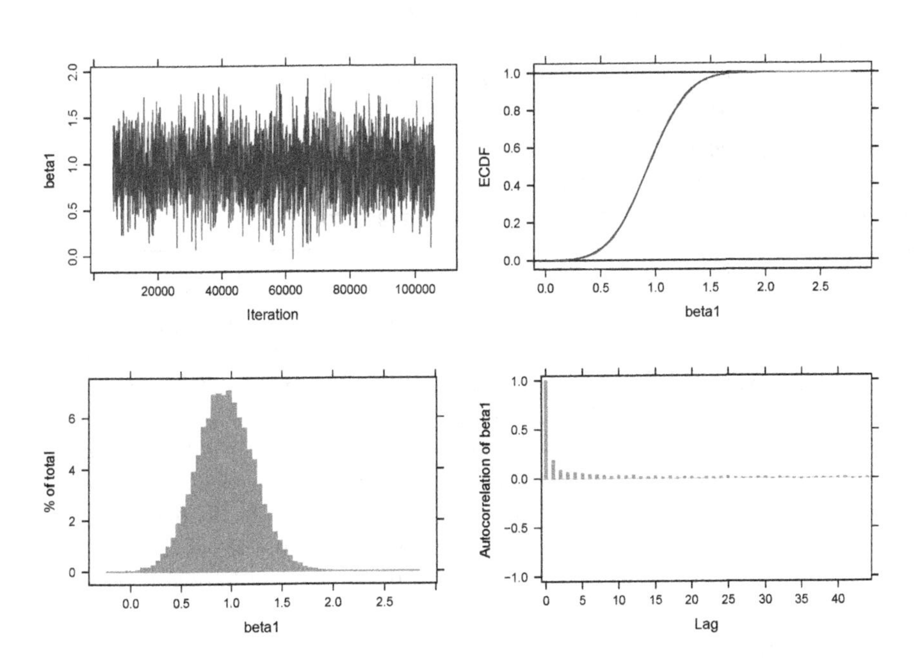

FIGURE 11.1: Breast cancer study. Diagnostic plots for parameter β_1 (Top, left: trace plot; Top, right: empirical cdf; Bottom, left: density; Bottom, right: autocorrelation plot) produced by the R package runjags.

(Cont.)

```
HR            1.35     2.55     4.22     2.66    0.783     --   0.00713
median0       29.8     39.7     51.6     40.4     5.74     --    0.0374
median1         18     22.5     27.8     22.6     2.51     --    0.0172

         MC%ofSD SSeff  AC.100 psrf
beta0        2.3  1890   0.282    1
beta1        0.9 12724  0.0317    1
gamma        2.3  1852   0.284    1
lambda0      2.1  2351   0.202    1
alpha0       0.7 19430  0.0142    1
HR           0.9 12084  0.0327    1
median0      0.7 23583 0.00824    1
median1      0.7 21344  0.0147    1

Total time taken: 59.5 seconds
```

```
> jagsfit$hpd
```

```
               Lower95       Median      Upper95
beta0   -7.8453649534 -6.390191839 -4.989778313
beta1    0.3770626218  0.934273834  1.486893714
gamma    1.2689486208  1.633798052  2.022930106
lambda0  0.0001436445  0.001677934  0.005071897
alpha0  37.5791198883 49.657268441 65.746418184
HR       1.3490855321  2.545364430  4.219748852
median0 29.8186003024 39.739399833 51.573689438
median1 17.9561570258 22.493457430 27.760837077
```

Extra information can be retrieved using the R command extract.runjags, e.g., to highlight which samplers have been used. This command provides also DIC (or versions thereof), but this is not available with JAGS for censored data. Some extra R commands provide the estimated $p_D = 2.96$ and $DIC = 292.5$, close to what is obtained in WinBUGS and SAS.

Since we have achieved convergence, we can now check whether the fitted model is appropriate for the data at hand. For this, we need the true but latent survival times, but also replicated survival times from the fitted Weibull distribution. To obtain this, we continue our sampling but now monitoring additional parameters for a limited number of chain values. This is achieved below.

```
> parameters.breast2 <- c("time1", "time2", "time3",
+                          "time1.rep", "time2.rep", "time3.rep")
> jagsfit2 <- extend.jags(jagsfit, add.monitor = parameters.breast2,
+                          sample = 1000)
```

Extra programming then produces the two Q-Q plots in Figure 11.2. The values on the X-axis represent the sorted values of the posterior medians of the latent true survival times taking into account the observed censoring. On the Y-axis, we have plotted the estimated quantiles using the expression $\widehat{F}((i-0.5)/(n+0.5))$, $(i=1,\ldots,n)$, with $\widehat{F}$ the estimated Weibull distribution based on the posterior mean of the parameters. From the figure, we may conclude that the Weibull assumption is not perfect, but not too wild of an assumption either. Further exploration of this assumption appears to be necessary.

Next, we verified the Weibull assumption using posterior predictive checks (PPCs). In Section 9.1.5 we have introduced the PPC. It involves the comparison of a characteristic, also called summary measure, of the observed data with the same characteristic of replicated data under the parametric assumption, taking into account the (posterior) uncertainty of the model parameters. The characteristic of the data can either involve only the observed data (test statistic) or also the unknown parameters (discrepancy measure). Here, the observed data are intervals. However, it is complicated to work with intervals. That is why we compared the characteristics of the sampled true latent survival times conditioning on the observed intervals with the unconditional replicated survival times from the same Weibull distribution, taking into account the posterior uncertainty of the model parameters. For a test statistic this results in a histogram of generated summary measures, which is then contrasted

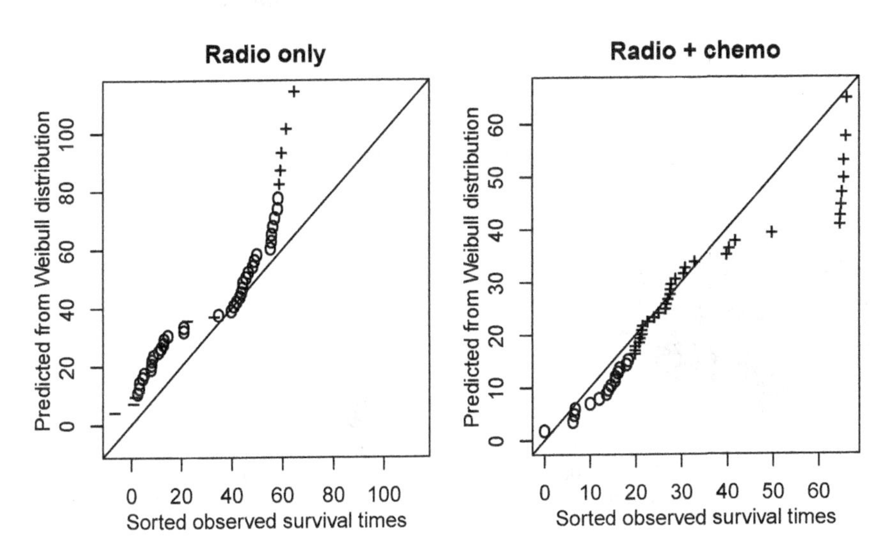

FIGURE 11.2: Breast cancer study. Q-Q plots to contrast the "true latent" survival times with the "model-based replicated" survival times per treatment group. Notation: left-censored (symbol −), interval-censored (symbol o), right-censored (symbol +).

with the observed test statistic. For a discrepancy measure, a $X-Y$ plot is generated, with one axis showing the generated discrepancy measures for the observed data and the other axis the generated discrepancy measures based on the replicated data. In both cases, an exceedance probability is computed, called the posterior predictive P-value (PPP)-value. The PPP-value quantifies how extreme the observed data are in view of the parametric assumption based on the chosen characteristic of the data. A large or a small PPP-value, say below 0.05 or above 0.95, is a sign that the parametric assumption is violated. Note, however, that the PPC is a conservative procedure because the data are used both for estimating as well as testing, see, e.g., Lesaffre and Lawson (2012).

There is no unique recipe for choosing the characteristic of the data. Here we have chosen two discrepancy measures: (1) range: difference between maximum and minimum of the sample; (2) maximal gap: maximum of the gaps in the sample, where the gaps are defined as the differences $t_{(i+1)} - t_{(i)}$ $(i = 1, \ldots, n-1)$ of the sorted survival times $t_{(i)}$ $(i = 1, \ldots, n)$. Note that we could not use the classical definition of a PPC, since here both the true latent as well as the replicated survival times are generated under the Weibull assumption, differing in the sense that the former takes into account the censoring interval while the latter does not. This type of PPC is reminiscent of the proposal in Gelman et al. (2005). In Figure 11.3 we show the two corresponding plots. The PPP-value for the range test expressing that the observed range is greater than under the Weibull assumption is 0.35, while the

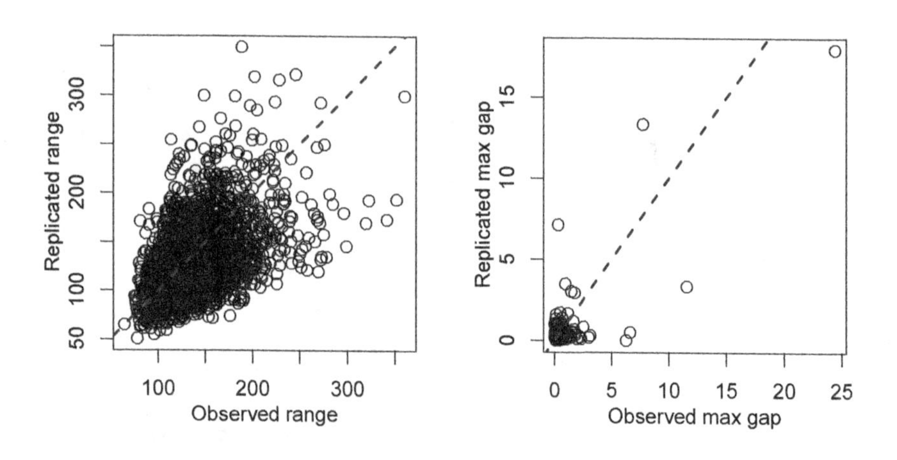

FIGURE 11.3: Breast cancer study. PPCs corresponding to the range (left) and max gap (right) test. The PPP-value for the range test is 0.35, while for the max gap test, the PPP-value is 0.48.

PPP-value for the max gap test is 0.48. We stress, however, that these tests are conservative for two reasons: (1) the PPC is in general conservative and (2) the "observed" and "replicated" survival times are both sampled under the Weibull assumption.

We have not tested the PH assumption here. This could be done by fitting a Weibull model for each of the treatment groups, add the two DICs and compare this to the DIC obtained above. When the difference in DICs is large enough (say larger than 5 to 10), one then would conclude that the PH assumption is violated. However, this is a naïve approach, which works only for binary covariates. A better approach will be shown in Section 11.2.2 using the R *package* dynsurv.

Finally, we have seen in Section 9.1 that it is standard to perform a sensitivity analysis in a Bayesian analysis. This involves varying the distributional assumptions of the data and the choice of the (noninformative) prior. In the SAS *analysis below two other survival distributions are fitted to the data, with again similar results for the treatment effect.*

□

11.1.2 SAS solution

In SAS, the PH model with a parametric baseline can be fitted in a Bayesian way using the procedure MCMC.

Example 11.1 in SAS (PROC MCMC)

First the data are prepared using the preliminary data step shown in Section 5.1.3. As in the frequentist way using the NLMIXED *procedure, the log-likelihood function must be explicitly specified. In addition, priors must be specified for the parameters. The following code fits a PH model with a Weibull baseline hazard.*

```
PROC MCMC DATA=breast OUTPOST=postout SEED =1234 NBI=10000 NMC=50000
          MISSING=AC DIC STATISTICS = (SUMMARY INTERVAL)
          MONITOR=(alpha beta0 beta1 gamma median0 median1);
PARMS gamma=1 beta0=0.69 beta1=0;
PRIOR beta0 ~ NORMAL (0, SD =1000);
PRIOR beta1 ~ NORMAL (0, SD =1000);
PRIOR gamma ~ GAMMA(SHAPE=1,ISCALE=0.001);
BEGINNODATA;
alpha=exp(-beta0/gamma);
median0 = (LOG(2)*EXP(-beta0))**(1/gamma);          /*radio only */
median1 = (LOG(2)*EXP(-beta0 - beta1))**(1/gamma); /* radio+chemo*/
ENDNODATA;
S_1=SDF('WEIBULL',lower,gamma,alpha)**EXP(beta1*treat);
```

```
S_r=SDF('WEIBULL',upper,gamma,alpha)**EXP(beta1*treat);
     IF MISSING(lower) AND NOT MISSING (upper) THEN lik = 1 - S_r;
ELSE IF NOT MISSING(lower) AND MISSING (upper) THEN lik = S_l;
ELSE lik = S_l - S_r;
llik=log(lik);
MODEL GENERAL(llik);
RUN;
```

The number of burn-in iterations and the number of MCMC iterations excluding the burn-in iterations are specified with the options NBI and NMC, respectively. The option MISSING=AC (all cases) is essential because the default value is CC (complete cases) which discards all records that have missing or partial missing values (i.e., left- and right-censored observations) before carrying out the simulation. The option DIC requests the deviance information criterion which may be used to compare models with each other (see Section 9.1.5 and below). Using the STATISTICS option, the posterior means, standard deviations, the 25th, 50th, and 75th (by default) percentile points (value SUMMARY), and 95% (by default) equal-tail and HPD credible intervals (value INTERVAL) for each variable are requested. In the MONITOR option, we list all the parameters and additional quantities of interest like the median survival time in both treatment groups (median0 and median1) that must be monitored. The posterior samples of all these items will be stored in the output data set defined using the OUTPOST option. If the MONITOR option is omitted, only the model parameters specified in the PARMS statement are monitored and outputted by default. Rather than specifying all model parameters explicitly, one can also use the abbreviation _PARMS_. By specifying the calculation of alpha, median0 and median1 between BEGINNODATA and ENDNODATA, PROC MCMC calculates these values without stepping through the entire data set. The programming statements are executed only twice, namely at the first observation of the data set to ensure proper values are used in the subsequent computations and at the last observation of the data set when PROC MCMC also adds the log of the prior density to the log of the posterior density, hereby speeding up the calculations by avoiding unnecessary observation-level computations.

The MCMC Procedure

Posterior Summaries

Parameter	N	Mean	Standard Deviation	Percentiles 25	50	75
alpha	50000	51.2748	7.9683	45.7183	50.2777	55.7457
gamma	50000	1.5870	0.1939	1.4541	1.5798	1.7146
beta1	50000	0.9186	0.2879	0.7242	0.9199	1.1082

(Cont.)

```
                         Posterior Intervals

Parameter    Alpha    Equal-Tail Interval         HPD Interval

alpha        0.050    38.6108     69.7944     37.1929     66.9635
gamma        0.050     1.2208      1.9815      1.2163      1.9734
beta1        0.050     0.3526      1.4934      0.3399      1.4735
```

The results are similar to the results of the frequentist approach with a hazard ratio of 2.53 (= exp(0.9277)) (95% HPD interval: 1.45 – 4.45*). This is expected because we used noninformative priors on the parameters. These must be specified using* PRIOR *statements. For* beta0 *and* beta1 *we used a normal prior with mean zero and a standard deviation of 1 000. For the shape parameter of the Weibull distribution (*gamma*) the noninformative prior is a gamma distribution $\mathcal{G}(a, b)$ with $a = 1$ and $b = 10^{-3}$. Using the options* SHAPE=a *and* ISCALE=b *within parentheses the shape and inverse-scale parameter can be changed. Note that because the shape parameter is equal to 1, an exponential distribution is in fact used.*

In addition to the posterior statistics shown above, the output also contains information on the number of observations used, the parameters and their given priors, convergence diagnostics, DIC and diagnostic plots for all parameters (all not shown).

Other distributions can be easily fitted by changing the distribution in the SDF *function and adapting the parameters accordingly. For a log-normal baseline hazard, the survival function at the lower and upper limit may be calculated as follows.*

```
S_l = SDF('LOGNORMAL', lower, mu, sigma)**EXP(beta1*treat);
S_r = SDF('LOGNORMAL', upper, mu, sigma)**EXP(beta1*treat);
```

For a log-logistic baseline hazard, the code is

```
S_l = SDF('LOGISTIC', LOG(lower), mu, sigma)**EXP(beta1*treat);
S_r = SDF('LOGISTIC', LOG(upper), mu, sigma)**EXP(beta1*treat);
```

In addition, for the models with a log-normal or log-logistic baseline hazard a noninformative normal prior will be used for mu *and a noninformative gamma prior for* sigma.

The different models may be compared using DIC. The best fitting model with the lowest DIC was the Weibull model (DIC = 292.636), which did slightly better than the log-logistic (DIC = 292.951) and log-normal (DIC = 293.668) model. The warning relating to the use of the GENERAL *function and DIC may be ignored as the comparison is meaningful in this example.*

□

11.2 PH model with flexible baseline hazard

11.2.1 Bayesian PH model with a smooth baseline hazard

Lin et al. (2015) proposed a general Bayesian approach for fitting a PH model with a smooth baseline hazard. To this end, they made use of integrated splines (I-splines). The estimation of the model parameters is made tractable and efficient by using the relationship of the PH model with a latent nonhomogeneous Poisson process (see Section 9.4.3) and data augmentation. The approach is implemented in the R package ICBayes (Pan et al., 2015). The baseline cumulative hazard function $H_0(t) = \int_0^t \hbar_0(s)\,ds$ is estimated smoothly using a monotone spline basis function of order d (degree + 1) and γ_ℓ nonnegative spline coefficients to ensure that $H_0(t)$ is nondecreasing. We refer to Appendix F.3 for more technical details on the smoothing splines. The number and position of the knots can be chosen by comparing different choices via a model selection criterion such as DIC. In the ICBayes package, the log pseudo marginal likelihood (LPML) was chosen as criterion (larger is better), see Section 9.1.5.

It can be shown that if $N(t)$ is a nonhomogeneous Poisson process (see also Section 9.4.3) with a cumulative intensity function $H_0(t)\exp(\boldsymbol{X}'\boldsymbol{\beta})$, then for T, the first occurrence in the Poisson process: $\mathsf{P}(T > t) = \mathsf{P}\{N(t) = 0\} = \exp\Big\{-H_0(t)\exp(\boldsymbol{X}'\boldsymbol{\beta})\Big\}$ what proves that T follows the PH model. Lin et al. (2015) then suggest to use data augmentation which essentially generates independent Poisson random variables that are compatible with the above defined Poisson process. Data augmentation considerably simplifies the MCMC process for estimating the parameters. Vague Gaussian priors are taken for the regression parameters, while an hierarchical prior is taken for the spline coefficients. That is, independent exponential priors are given for the γ parameters and a hyper gamma prior connects them and controls the over-fitting.

We have applied the approach to the breast cancer data. Incidentally, this data set is also used in the documentation of the ICBayes package.

Example 11.2 Breast cancer study

The early breast cancer study is one of the two data sets that are delivered with the package ICBayes (version 1.0). While this data set can be imported from ICBayes for the analysis, we transformed the breast cancer data set from the package icensBKL to be suitable for ICBayes. The same model as in Example 11.1 is considered here with the only difference that a smooth baseline hazard of Lin et al. (2015) is exploited.

Several runs with each 20 000 iterations (and 10 000 burn-in iterations) were performed varying the number of inner knots by varying the length of the equal-sized intervals (see below). The maximum LPML was obtained for length = 4. This is the analysis that we show below. The estimated treatment effect is now 0.847 (95% equal tail CI: 0.308 – 1.392) with hazard ratio of 2.33 (95% equal tail CI: 1.36 – 4.02), which is somewhat lower than the estimates obtained before by a parametric analysis (see Table 11.1).

□

11.2.1.1 R solution

The R package ICBayes contains functions that fit the proportional hazards model, the proportional odds model and the probit model for interval-censored observations using a Bayesian approach. Here we show only its use for the PH model and for the general case of interval-censored observations. Note that the package can also be used for current status data.

Example 11.2 in R

First the packages ICBayes *and* icensBKL *are loaded. Then we create the variable* time, *which consists of* NA *for all patients. For the left-censored observations, we changed the left endpoints from* NA *into* 0, *which is required by the package (although not specified in its documentation). The right endpoints can be simply taken over from the original data set, as done here. A* status *variable (actually censoring variable) needs to be created with values* 0, 1, 2 *for a left-, right- and interval-censored observation, respectively. The variable* trtRCH *is a matrix with columns representing the covariates in the model. All created variables are then combined into a* data.frame, *which is then ready to be used for the analysis.*

```
> library("ICBayes")
> library("icensBKL")
> data("breastCancer", package = "icensBKL")
>
> N <- length(breastCancer$treat)
> time <- rep(NA, N)
> L  <- breastCancer$low
> R  <- breastCancer$upp
> status <- rep(0, N)
> status <- ifelse(is.na(R), 2, 1)
> status[is.na(L)] <- 0
> L[is.na(L)] <- 0
> trtRCH <- 1 * (breastCancer$treat == "radio+chemo")
> breast <- data.frame(cbind(L, R, status, trtRCH))
```

The command ICBayes *below requires the following input. First, the argument* model = "case2ph" *specifies that we are dealing with general interval-*

censored observations. As indicated above, the package can also handle other types of the survival times and other models. Then follows L = breast[, "L"], R = breast[, "R"], status = breast[, "status"], xcov = breast[, "trtRCH"], *which specify the data. The option* x_user = c(0, 1) *tells* ICBayes *to estimate survival curves for* $X = 0$ *and* $X = 1$ *(here the two treatment groups). With the argument* knots = seq(0.1, 60.1, length = 4) *the left and right endpoints of the interval of survival times are specified along with the length of the intervals that regulates the number of knots. By specifying the argument* grids = seq(0.1, 60.1, by = 1) *we ask the program to evaluate the estimated survival function at a fine grid of values, i.e.,* $0.1, 1.1, 2.1, \ldots$. *The last two arguments regulate the number of iterations and the burn-in size. We have varied the length parameter in the* knots *argument, i.e., 12, 10, 8, 6, 4, and 3. The computed LPML values were* -149.78, -149.25, -148.96, -148.46, -147.03, -147.23, *respectively, showing that the model specified below appears best. However, the difference in LPML is not great.*

```
> breastICB <- ICBayes(model = "case2ph",
+   L = breast[, "L"], R = breast[, "R"], status = breast[, "status"],
+   xcov = breast[, "trtRCH"], x_user = c(0, 1),
+   knots = seq(0.1, 60.1, length = 4),
+   grids = seq(0.1, 60.1, by = 1),
+   niter = 20000, burnin = 10000)
> print(breastICB)
```

For the chosen model, we can obtain the parameter estimates with a print *command, which provides the posterior mean of regression coefficient of* trtRCH *(= 0.847), its posterior SD (= 0.278) and the 95% equal-tail CI (= (0.308, 1.392)) and LPML equal to* -147.03.

The default plot.ICBayes *command produces a plot of the estimated baseline survival function. In the code below, we have taken the first* ngrid *elements from* breastICB$S_m *to plot the survival function for the "radio only" group, and the last* ngrid *elements to plot the survival function for the "radio+chemo" group, see Figure 11.4. Recall that to find the parameter estimates in the* R *object, the command* str(breastICB) *can be used. In this figure we have additionally overlaid the parametric estimates from a Weibull fit obtained in Section 11.1.1 (code not shown). We can observe that the two solutions are quite close, providing evidence of the goodness of fit of the Weibull distribution. Unfortunately, the* ICBayes *package does not provide confidence bounds.*

```
> ngrid <- length(breastICB$S0_m)
> plot(breastICB$grids, breastICB$S_m[1:ngrid], type = "l",
+      lty = "solid", xlab = "Survival times (months)",
+      ylab = "Estimated survival distributions", ylim = c(0, 1))
```

```
> lines(breastICB$grids, breastICB$S_m[(ngrid+1):(2*ngrid)],
+       lty = "dashed")
```

The estimated hazard ratio (2.33) and related 95% equal tail credible interval (1.36, 4.02) are not automatically available, but are easily obtained from the ICBayes *object as shown by the last four lines of coding below.*

```
> HR <- exp(breastICB$coef)
> HR.CI <- exp(breastICB$coef_ci)
> print(HR)
> print(HR.CI)
```

□

We have illustrated above that ICBayes can be used to check the goodness of fit of an assumed parametric model. This smooth fit might also be used to compute the PPC as done in Section 11.1.1. There we have used the Weibull assumption twice, once for the observed intervals to create the

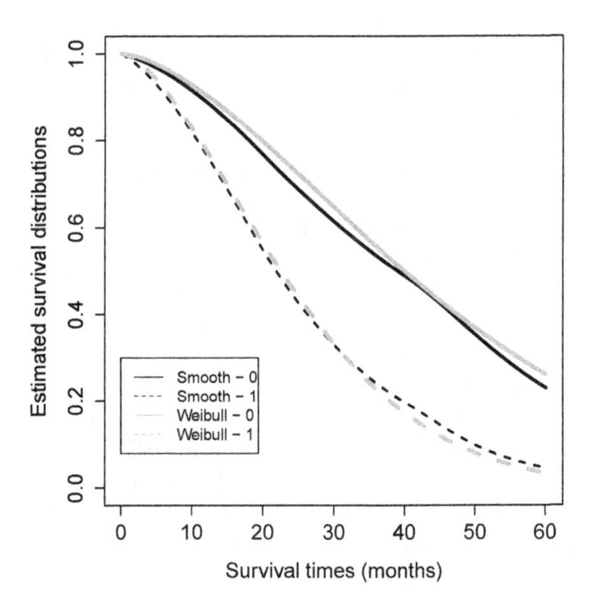

FIGURE 11.4: Breast cancer: Smooth (black) and Weibull (gray) survival functions for treatment group "radio only" (0) and "radio+chemo" (1) obtained from the output of the R package ICBayes.

"true latent" survival times and for the replicated survival times under the Weibull assumption. Clearly, this must give a conservative test. But, if we had computed the "true latent" survival times using the smooth fit the comparison would have been definitely less conservative. However, once having fitted the baseline survival function in a smooth manner, one will probably not be interested anymore in the parametric fit.

Finally, we note that the package ICBayes expects that all observations are censored. Hence an exactly observed survival time should be replaced by a small interval around the observed survival time and treated as an interval-censored observation.

11.2.2 PH model with piecewise constant baseline hazard

Wang et al. (2013) suggested a Bayesian PH model with a piecewise constant baseline hazard. They also allow the effect of the regression coefficients to vary over time, which in fact produces a dynamic survival model. In this way their method provides a check for the PH assumption. The parameters are estimated via a reversible jump MCMC procedure, which was introduced in Section 10.2.1. The sampling distribution is made tractable by making use of data augmentation. Their approach is implemented in the R package dynsurv (Wang et al., 2017). We now sketch this approach.

A fine time grid is chosen, say $0 = s_0 < s_1 < s_2 < \ldots < s_K < \infty$. This prespecified grid may be defined in many ways and it contains all potential jump points of the curves. The Cox model with time-varying regression coefficients replaces Expression (11.1) with:

$$\hbar(t \mid \boldsymbol{X}) = \hbar_0(t) \exp\{\boldsymbol{X}'\boldsymbol{\beta}(t)\}$$

with again $\hbar_0(t)$ an arbitrary unspecified baseline hazard function, but now the regression coefficients may vary with time. It is assumed that $\hbar_0(t)$ and each regression coefficient are left continuous step functions, with a random number of jumps with random locations to be estimated from the data at hand. However, here the potential jump points are limited to the grid points. Because it is assumed that the (hazard and regression coefficients) functions can vary only at the jump points, the notation $\hbar_k = \hbar_0(s_k)$ and $\boldsymbol{\beta}_k = \boldsymbol{\beta}(s_k)$ for $k = 1, 2, \ldots, K$ is justified. To simplify the MCMC process, data augmentation is used. Two variables $\mathrm{d}N_{ik}$ and Y_{ik} are created for each interval $(k = 1, 2, \ldots, K)$ and for each subject i $(i = 1, \ldots, n)$. Namely for the ith subject with an interval-censored survival $(l_i, u_i]$: (1) $\mathrm{d}N_{ik}$ is an indicator variable, which is equal to 1 only when the latent, but true, survival time T_i belongs to interval $(s_{k-1}, s_k]$, otherwise $\mathrm{d}N_{ik}$ is equal to 0 and (2) and then the at-risk indicator $Y_{il} = 1$ for all $l < k$ (at-risk period), while $Y_{il} = 0$ for all $l > k$ (not any more at risk) and $Y_{ik} = \dfrac{T_i - s_{k-1}}{\Delta_k}$ with $\Delta_k = s_k - s_{k-1}$ (partially at risk). For subject i with a right-censored interval (l_i, ∞): $\mathrm{d}N_{ik} = 0$ for all k and $Y_{ik} = I(s_k \leq l_i)$.

The likelihood contribution for the ith subject, having either an interval-censored or right-censored survival time, is equal to:

$$L_i = \prod_{k=1}^{K} \{\hbar_k \exp(\boldsymbol{X}_i'\boldsymbol{\beta}_k)\}^{\mathrm{d}N_{ik}} \exp\{-\Delta_k \hbar_k \exp(\boldsymbol{X}_i'\boldsymbol{\beta}_k) Y_{ik}\}$$

The total likelihood from the augmented data is then $\prod_{i=1}^{n} L_i$, and this likelihood will be used in the MCMC procedure. The marginal likelihood (integrated over the latent true survival times), based on the observed data (intervals) and given by (1.13), is used to evaluate model performance.

The authors considered three models, which can be fitted with the R package dynsurv. Firstly, the software allows to fit a classical PH model with a piecewise constant hazard with jump points at all prespecified grid points (Model M_1). For the second model also the regression coefficients are allowed to make jumps at (all) the grid points (Model M_2). The third model (Model M_3) involves the selection of the jump points with the RJ-MCMC procedure. We illustrate here the use of Model M_1 and Model M_3.

For Model M_1 the priors are

$$\begin{aligned} \hbar_j &\sim \mathcal{G}(\delta_j,\, \gamma_j) \quad (j = 1, \ldots, K) \text{ independent;} \\ \beta_s &\sim \mathcal{N}(0, \omega_s) \quad (s = 1, \ldots, p). \end{aligned}$$

When the grid is dense, a large number of parameters need to be estimated, especially for Model M_3. In addition, with independent priors the hazard function may vary wildly. In the R package dynsurv two priors for Model M_3 have been implemented in combination with the RJ-MCMC procedure. The number of jumps J is assumed to follow a discrete uniform distribution ranging from 1 to K. For a fixed J, the jump times $0 = s_0 < s_1 < s_2 < \ldots < s_{J-1}$ are randomly selected but smaller than $s_J \equiv s_K$ (the original maximal jump point). Give J and the jump points, we describe here two options for the priors of Model M_3 available in dynsurv. We refer to Wang et al. (2013) and the documentation of the package for other choices. The first prior consists of independent gamma priors for the baseline hazard components $\hbar_k$, while for the regression coefficients of the p covariates $\beta_{k1}, \ldots, \beta_{kp}$ $(k = 1, \ldots, K)$ an hierarchical first-order autoregressive (AR(1)) prior, also referred to as a Markov type process prior, is chosen. This prior is given by:

$$\begin{aligned} \hbar_j &\sim \mathcal{G}(\delta_j,\, \gamma_j) && (j = 1, \ldots, J) \text{ independent;} \\ \beta_{1,s} &\sim \mathcal{N}(0,\, a_0\omega_s) && (s = 1, \ldots, p); \\ \beta_{j,s} \mid \beta_{j-1,s}, \omega_s &\sim \mathcal{N}(\beta_{j-1,s},\, \omega_s) && (j = 2, \ldots, J,\ s = 1, \ldots, p); \\ \omega_s &\sim \mathcal{IG}(\alpha_0,\, \eta_0) && (s = 1, \ldots, p). \end{aligned}$$

Further, a_0 is a hyperparameter and $\mathcal{IG}(\alpha_0, \eta_0)$ represents an inverse gamma distribution with shape parameter α_0 and scale parameter η_0 with mean $\eta_0/(\alpha_0 - 1)$. The independent gamma prior is related to the gamma process prior introduced by Kalbfleisch (1978), see Chapter 7 in Wang et al. (2012).

For the second prior, the logarithm of the baseline hazard is incorporated in the regression coefficients as an intercept and its prior is then given by:

$$\begin{aligned}\log(\hbar_1) &\equiv \beta_{1,0} & &\sim \mathcal{N}(0,\ a_0\omega_0); \\ \log(\hbar_j) &\equiv \beta_{j,0} \mid \beta_{j-1,0},\ \omega_0 & &\sim \mathcal{N}(\beta_{j-1,0},\ \omega_0), \qquad j = 2, \ldots, J.\end{aligned}$$

That now also a Markov process prior is chosen for the baseline hazard jump, implies a better control of the jump fluctuations.

The sampling algorithm combines the RJ-MCMC procedure with data augmentation generating the true but latent event times T_i given all other model parameters and thereby also the event indicators $\mathrm{d}N_{ik}$ and at-risk indicators Y_{ik}. More technical details on the sampling procedure can be found in Wang et al. (2013).

Checking convergence of a RJ-MCMC run is not easy. Wang et al. (2013) suggest to check the stability of the jumps at particular chosen grid points. Model comparison is done using LPML and DIC using the marginal likelihood.

Example 11.3 Signal Tandmobiel study

We use the R package dynsurv to estimate the distribution of the emergence time of tooth 44 (permanent right mandibular first premolar) as a function of gender and whether or not there is a history of caries on the deciduous

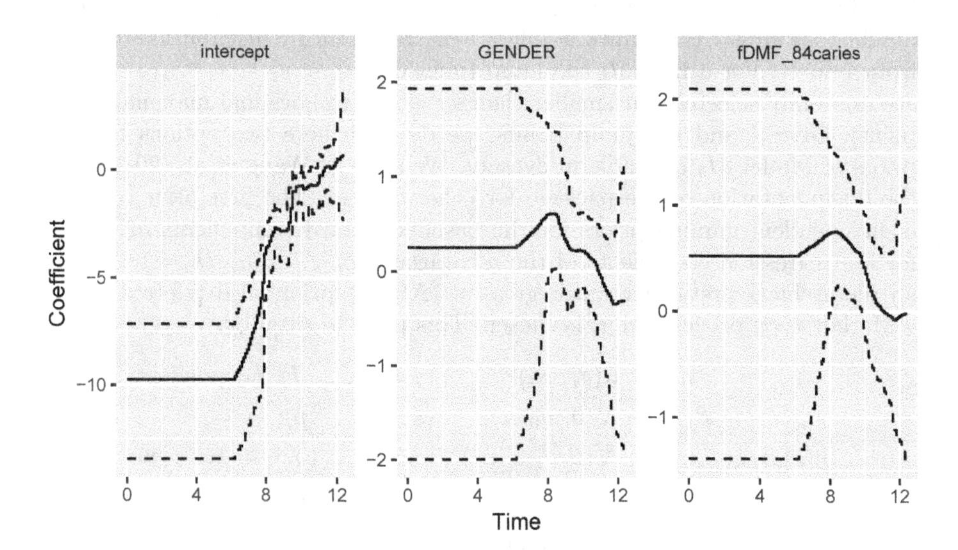

FIGURE 11.5: Signal Tandmobiel study. Estimated piecewise constant dynamic regression coefficients (model M_3 with the second prior) obtained from the R package dynsurv.

TABLE 11.2: Signal Tandmobiel study. Comparison of the estimated Models M_1 and M_3 (combined with two priors, see text for details) with R package dynsurv according to LPML and DIC with p_D.

Model	LPML	DIC (p_D)
M_1	−549.1	1 096.7 (16.5)
M_3 prior 1	−545.3	1 088.7 (19.9)
M_3 prior 2	−559.1	1 111.3 (9.1)

predecessor, i.e., tooth 84. We first fit a classical Bayesian PH model (Model M_1), with a piecewise constant baseline hazard. In a second step, the PH assumption is relaxed via Model M_3. That is, we then allow the effect of gender and caries on tooth 84 to vary over time. Two priors are considered. For the first prior, the hazard jumps are given independent priors, while for the second prior the hazard jumps are more restricted. The comparison of the results on Model M_1 and Model M_3 allows to check whether the PH assumption holds.

According to Model M_1, gender has no significant impact on the emergence of tooth 44, while a history of caries on tooth 84 accelerates the emergence of tooth 44. The latter result was also seen in previous analyses and is plausible since caries at a deciduous tooth often leads to its extraction making the way clear here for tooth 44 to emerge. For Model M_3 the effect of covariates cannot be summarized anymore in a single value, but we need to look at the evolution of the regression coefficients over time. This is seen in Figure 11.5 for Model M_3 with the second prior (same graph for the regression coefficients is seen for the first prior). The figure highlights that the effect of both covariates is constant and low initially (when there is no emergence at all), then increases and later on decreases. The two figures confirm that on average the effect of gender is relatively low compared to that of a history of caries on tooth 84. That their effect first increases and then decreases has possibly to do with (1) that permanent teeth on girls emerge indeed earlier (seen in the full data set) and (2) that when the permanent tooth is ready to emerge (early period) its emergence will be accelerated by caries on the predecessor which will be extracted for that reason, while the opposite occurs when the permanent tooth is far from ready to emerge. Then early extraction of the deciduous tooth will let the gingiva close the gap making it harder for the permanent tooth to break through. From the model selection criteria in Table 11.2, we conclude that the PH assumption should be rejected in favor of a dynamic model. That the second prior produces a better fitting model is remarkable, but illustrated in Figure 11.6. Indeed, the estimated hazard seems to have preferable ages for tooth 44 to emerge, with gaps in-between. This result is not immediately obvious and needs to be confirmed in the larger data set.

□

11.2.2.1 R solution

The R package dynsurv consists of a number of R functions that fit time-varying coefficient models for interval- and right-censored survival data. We consider here only the function bayesCox. Note that for exactly observed survival times, they need to be replaced (as with the package ICBayes) by small intervals around the observed times.

Example 11.3 in R

The tandmob *data set from the package* icensBKL *needs to be adapted slightly for use with the* bayesCox *function from the* dynsurv *package. Further, the grid* $0 = s_0 < s_1 < s_2 < \ldots < s_K < \infty$ *can be specified by the user or can be left unspecified. In the former case, the documentation of the package provides*

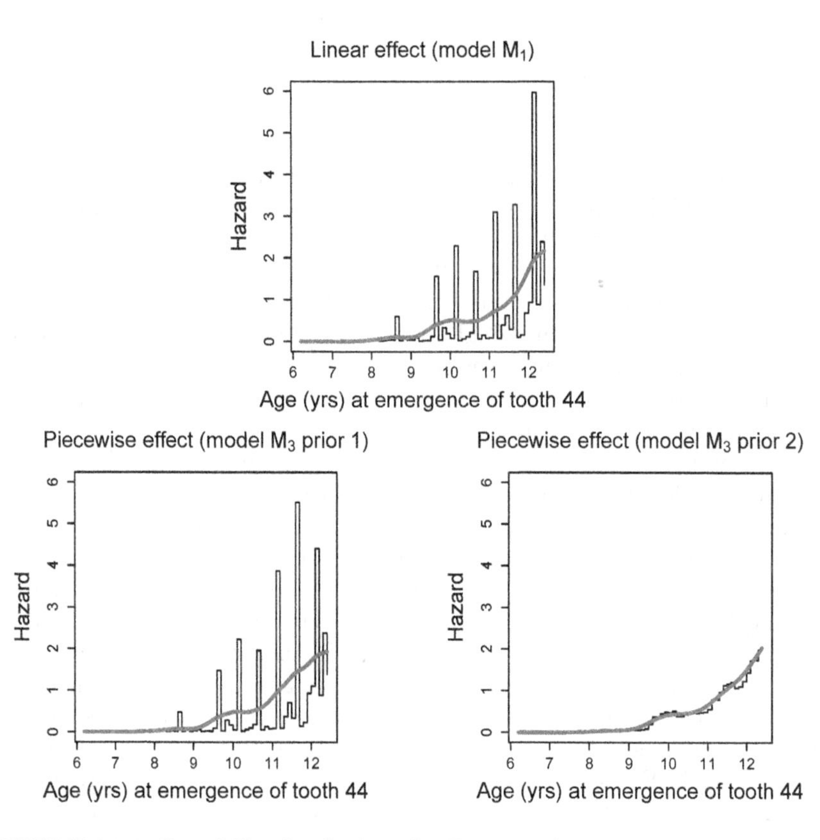

FIGURE 11.6: Signal Tandmobiel study. Estimated piecewise constant baseline hazard functions obtained from the R package dynsurv corresponding to Model M_1 (top), Model M_3 prior 1 (lower left) and Model M_3 prior 2 (lower right). Smooth estimates of the hazard functions using smooth.spline with df = 10 are overlaid.

some guidelines to construct the grid. Details of data preparation are provided in the R script accompanying the book. Below, we show the call to bayesCox for a classical PH model (i.e., with regression coefficients not depending on time), but with a piecewise baseline hazard at all prespecified grid points. This model is referred to as Model M_1.

```
> fit1 <- bayesCox(myformula, mydata, mygrid, out = "tiCox.txt",
+           model = "TimeIndep",
+           base.prior = list(type = "Gamma", shape = 0.1, rate = 0.1),
+           coef.prior = list(type = "Normal", mean = 0, sd = 10),
+           gibbs = list(iter = 50000, burn = 15000, thin = 10,
+                        verbose = TRUE, nReport = 100))
```

What follows is a brief explanation of the arguments of the function:

- myformula, mydata, mygrid: an R formula describing the regression model, which is here: Surv(L44, R44_Inf, type = "interval2") ~ GENDER + fDMF_84. Then the data (containing variables appearing in the model formula) and the grid are specified, respectively. All must have been defined before in the program;

- out = "tiCox.txt": a txt file containing MCMC samples will be deposited at the working directory, to be used for constructing, e.g., trace plots;

- model = "TimeIndep": all priors (for the piecewise baseline hazard and regression coefficients) are independent;

- base.prior = list(type = "Gamma", shape = 0.1, rate = 0.1): a vague gamma prior $\mathcal{G}(0.1, 0.1)$ is chosen for each $\hbar_k$;

- coef.prior = list(type = "Normal", mean = 0, sd = 10): a vague normal prior $\mathcal{N}(0, 10^2)$ is chosen for β_1 (gender) and β_2 (dmf status tooth 84 = 0, when there is no caries experience and = 1, otherwise);

- gibbs = list(iter = 50000, burn = 15000, thin = 10, verbose = TRUE, nReport = 100): details on the Gibbs procedure, e.g., nReport = 100 means that after each 100 iterations, an alert is produced.

The estimated posterior medians of the hazard ratios (95% equal tail CI) for Model M_1 are for gender: 1.17 (0.92, 1.47) and for caries experience on tooth 84: 1.44 (1.12, 1.85). These are obtained by taking the exponent of the corresponding regression coefficients. This result shows that boys and girls do not seem to differ in age at emergence of tooth 44, but that this tooth emerges earlier when there is a history of caries on tooth 84. The extensions of the PH model provided in dynsurv allow to check whether the impact of gender and caries on tooth 84 is constant over time. In other words, can we trust the PH assumption in Model M_1? The first extension is illustrated in the accompanying R program and corresponds to Model M_2 in Wang et al. (2013), but is skipped here. We look immediately at Model M_3 with the two priors described above. Both priors allow for time-varying behavior regression coefficients of gender and dmf status tooth 84. But the second prior is likely to give less fluctuating jumps. The call to bayesCox for the first prior is:

```
> fit3.1 <- bayesCox(myformula, mydata, mygrid, out = "dynCox1.txt",
+            model = "Dynamic",
+            base.prior = list(type = "Gamma", shape = 0.1, rate = 0.1),
+            coef.prior = list(type = "HAR1", shape = 2, scale = 1),
+            gibbs = list(iter = 50000, burn = 15000, thin = 10,
+                         verbose = TRUE, nReport = 100))
```

The model argument specifies "Dynamic", meaning that we ask for piecewise constant coefficient vectors β_{k1} and β_{k2} with the hierarchical priors. The "HAR1" prior corresponds to the above-described hierarchical AR prior for this part of the model. The independent gamma priors of the ω_s are defined by their shape and scale parameter in the above call. Note that the value of a_0 is by default 100 aiming for a vague prior. For the second prior, we specify:

```
> fit3.2 <- bayesCox(myformula, mydata, mygrid, out = "dynCox2.txt",
+            model = "Dynamic",
+            base.prior = list(type = "Const"),
+            coef.prior = list(type = "HAR1", shape = 2, scale = 1),
+            gibbs = list(iter = 50000, burn = 15000, thin = 10,
+                         verbose = TRUE, nReport = 100),
+            control = list(intercept = TRUE))
```

Note the change in program. Now we specify base.prior = list(type = "Const"), which means that there is no real baseline hazard anymore. Argument control = list(intercept = TRUE) stipulates that the logarithm of the baseline hazard is considered part of the vector of regression coefficients and given the hierarchical AR prior. For the jump points, a uniform prior on the chosen grid is taken.

To check convergence, it is standard to look at trace plots. But, bayesCox does not provide trace plots for the coefficients. Except for Model M_1, they are difficult to interpret as the number of jumps and the location of jumps vary in the MCMC process. In that case, the following R commands could be used:

```
> plotJumpTrace(jump(fit3.2))
> plotJumpHist(jump(fit3.2))
```

The first command asks for a trace plot of the number of jumps for each *covariate for Model M_3 with the second prior, see Figure 11.7. Apparently, the MCMC procedure samples mostly between 2 and 15 possible jump points from the original 63 grid points for both regression coefficients. There is wider variability for the intercept with the MCMC procedure that varies roughly between 5 and 40 jump points. The second command asks to display the probability of jumping at each grid point (output not shown).*

The estimated piecewise constant dynamic regression coefficients for Model M_3 with the second prior are displayed in Figure 11.5. The plot was obtained using:

```
> plotCoef(coef(fit3.2, level = 0.95))
```

We notice that the hazard for emergence increases with age. We also notice that around 6–10 years of age, girls are more prone for their tooth 44 to emerge, then (evidently) afterward probability of emergence of tooth 44 is higher for boys. Finally, throughout the whole age range the probability of emergence of tooth 44 is increased with a history of caries on tooth 84, but more at an earlier age. Note, however, that there is quite some uncertainty about these estimated effects.

When comparing the estimated baseline hazard functions for Model M_1 and under the two priors for Model M_3, we notice the most stability for the

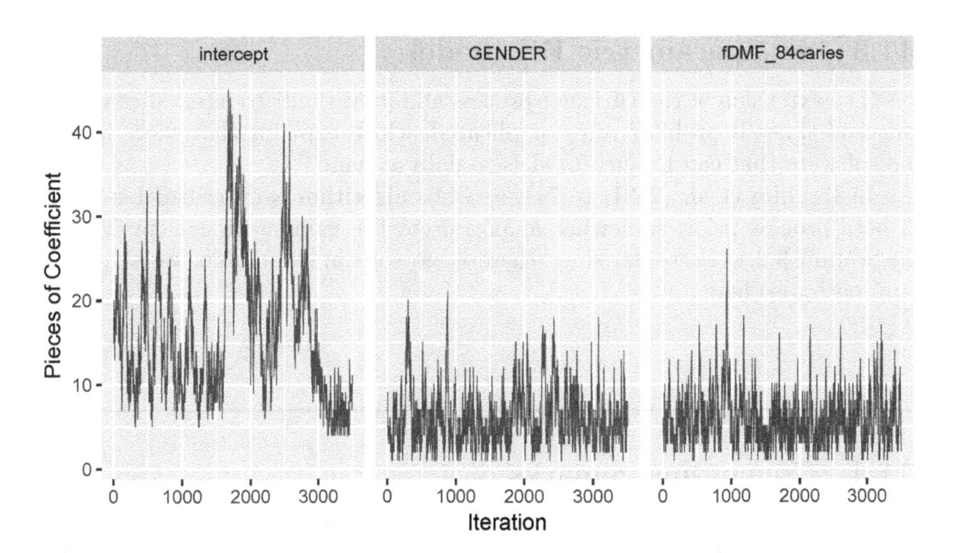

FIGURE 11.7: Signal Tandmobiel study: Frequency of jump points obtained from R package dynsurv with the second specification of prior under Model M_3 for the intercept, regression coefficient of gender and regression coefficient of caries experience at predecessor.

second prior of Model M_3, see Figure 11.6. Its last panel related to Model M_3 with the second prior was obtained using:

```
> plot(fit3.2$grid, exp(fit3.2$est$beta[,1]), type = "s",
+      xlab = "Age (yrs) at emergence of tooth 44", ylab = "Hazard")
> haz3.2.spl <- smooth.spline(fit3.2$grid, exp(fit3.2$est$beta[,1]),
+                             df = 10)
> lines(haz3.2.spl, lwd = 3, col = "grey60")
```

Finally, Table 11.2 displays the model selection criteria LPML and DIC. For both criteria, Model M_3 with the first prior is the winning model. This seems to indicate that there are preferred times for tooth 44 to emerge, a result that is not immediately understood and needs to be confirmed in the complete data set.

The R *package* dynsurv *allows for a comparison of two fitted survival functions either by overlaying the two curves or by computing the difference. Such a comparison shows that girls with a history of caries on its predecessor have tooth 44 earlier emerged than boys with a sound predecessor (plot not shown).*

□

11.3 Semiparametric PH model

The extension of the Dirichlet process and its variants to interval-censored survival times is prohibitively complicated, which explains that there is also no software that can fit such models to data at hand.

In Ibrahim et al. (2001, p. 71), a Gibbs algorithm is described based on a beta process prior, somewhat analogous to the gamma process prior seen in Section 9.2.1. However, since there is no software available, we skip this approach also here.

11.4 Concluding remarks

In this chapter, we have given an overview of the available Bayesian software that allows for PH regression models for interval-censored outcomes. We can observe that recently most of the software developments consist in focused packages dealing with a particular type of model and using a very specific estimation approach. The number of packages has grown exponentially. It is therefore not unlikely that, at the time of publication of this book, additional software for interval-censored survival times will have appeared.

Chapter 12

The Bayesian accelerated failure time model

In Chapter 6 we discussed the accelerated failure time (AFT) model with interval-censored data and showed how to perform frequentist inference on its parameters. In this chapter, we show how the AFT model can be fitted with Bayesian software. The parametric Bayesian model can be fitted with BUGS-like and SAS software in very much the same manner as seen in the previous three chapters. Hence we will quickly go over parametric Bayesian AFT modelling. The AFT counter part of the PH models with a flexible baseline hazard is found in the R package bayesSurv. This package fits a smooth AFT model in a Bayesian way using either the RJ-MCMC or the Bayesian penalization techniques. We will deal with this package in detail here. We have seen in the previous chapter that Bayesian nonparametric software for PH modelling with interval-censored data is basically non-existing. In contrast, with the R package DPpackage practitioners have several programs at their disposal for fitting Bayesian AFT models in a nonparametric manner.

12.1 Bayesian parametric AFT model

The AFT model was first introduced in Section 2.3.2. For the AFT model, the event time T is related to the covariate vector $\boldsymbol{X}$ as

$$\log(T) = \boldsymbol{X}'\boldsymbol{\beta} + \varepsilon, \tag{12.1}$$

where $\boldsymbol{\beta}$ is a vector of regression coefficients and ε an error random variable. In Section 6.1 we discussed maximum likelihood inference for a parametrically specified distribution of ε, or equivalently, the baseline survival distribution S_0 with the hazard function $\hbar_0$ of the event time $T_0 = \exp(\varepsilon)$. Remember from Section 2.3.2 that the AFT Model (12.1) implies the following hazard function and survival function for a subject with covariate vector $\boldsymbol{X}$:

$$\begin{aligned} \hbar(t \mid \boldsymbol{X}) &= \hbar_0\{\exp(-\boldsymbol{X}'\boldsymbol{\beta})\,t\}\,\exp(-\boldsymbol{X}'\boldsymbol{\beta}), \\ S(t \mid \boldsymbol{X}) &= S_0\{\exp(-\boldsymbol{X}'\boldsymbol{\beta})\,t\}. \end{aligned} \tag{12.2}$$

The maximum likelihood based inference for parameters of Model (12.1) under parametric assumptions was discussed in Section 6.1. We now revisit Example 6.1 and show key aspects of the Bayesian analysis of the parametric AFT model using JAGS.

Example 12.1 Signal Tandmobiel study

We refit the parametric AFT model which relates the emergence time of tooth 44 (permanent right mandibular first premolar) to gender ($X_1 \in \{0, 1\}$ with 0 for boys and 1 for girls), the DMF status of tooth 84 (primary mandibular right first molar, $X_2 \in \{0, 1\}$ with 0 for a sound primary tooth and 1 for a primary tooth with caries history). As in Example 6.1, an interaction term will be considered to allow for a possibly different effect of the DMF status of tooth 84 on the emergence distribution of tooth 44 for boys and girls. Hence $\boldsymbol{X} = \big(X_1, X_2, X_1 \cdot X_2\big)'$ and $\boldsymbol{\beta} = \big(\beta_1, \beta_2, \beta_3\big)'$ in Model (12.1).

In Example 6.1, a log-normal distribution of the emergence times (normal distribution for the error term ε) was assumed. For illustrative purposes, another distribution will be assumed now. Namely, we will consider the Weibull distribution for T (type I least extreme value distribution for ε). Note that the Weibull distribution was assumed also in Example 11.1, combined with the PH model. By contrasting the current example to Example 11.1, differences between the PH and AFT model can again be seen.

In the following expressions, two parametrizations of the Weibull distribution (see Appendix B.9) are given. Namely, the baseline hazard and survival functions are assumed to take, for $t > 0$, the following form:

$$\begin{aligned} \hbar_0(t) &= \frac{\gamma}{\alpha_0}\left(\frac{t}{\alpha_0}\right)^{\gamma-1} &&= \lambda_0\,\gamma\,t^{\gamma-1}, \\ S_0(t) &= \exp\left\{-\left(\frac{t}{\alpha_0}\right)^{\gamma}\right\} &&= \exp\big(-\lambda_0\,t^{\gamma}\big), \end{aligned} \tag{12.3}$$

where γ and α_0 are unknown shape and scale parameters, respectively, which leads to parameter $\lambda_0 = \alpha_0^{-\gamma}$, the second parameter of the Weibull parametrization in BUGS and JAGS. Combination of the baseline hazard function (12.3) with the AFT assumption (12.2) shows that a subject with covariate vector $\boldsymbol{X}$ follows a Weibull distribution with the shape parameter γ (the same as for the baseline distribution) and the scale parameter $\alpha(\boldsymbol{X})$ or parameter $\lambda(\boldsymbol{X})$ (in the BUGS/JAGS parametrization) given by

$$\alpha(\boldsymbol{X}) = \alpha_0\,\exp\big(\boldsymbol{X}'\boldsymbol{\beta}\big), \qquad \lambda(\boldsymbol{X}) = \lambda_0\,\exp\big(-\gamma\,\boldsymbol{X}'\boldsymbol{\beta}\big).$$

If we define $\mu = \log(\alpha_0)$ and $\sigma = 1/\gamma$, we can rewrite Model (12.1) as

$$\log(T) = \mu + \boldsymbol{X}'\boldsymbol{\beta} + \sigma\,\varepsilon^*, \tag{12.4}$$

where $\varepsilon^* = \sigma^{-1}(\varepsilon - \mu)$ is the standardized error term following a standard

TABLE 12.1: Signal Tandmobiel study. Bayesian Weibull AFT model for emergence of tooth 44, posterior means, posterior standard deviations, and 95% HPD intervals of the model parameters, acceleration factors (AF) due to DMF = 1 for boys and girls, median emergence times for the four groups obtained with R package runjags.

Parameter	Posterior mean	Posterior SD	95% HPD interval
Intercept (μ)	2.42	0.01	(2.40, 2.44)
Gender[girl] (β_1)	−0.0258	0.0138	(−0.0527, 0.0011)
DMF[1] (β_2)	−0.0524	0.0170	(−0.0854, −0.0187)
Gender:DMF (β_3)	0.0311	0.0243	(−0.0166, 0.0784)
Scale (σ)	0.099	0.004	(0.090, 0.108)
AF_{boys}(DMF)	0.95	0.02	(0.92, 0.98)
AF_{girls}(DMF)	0.98	0.02	(0.95, 1.01)
Median time: boys, DMF = 0	10.80	0.11	(10.59, 11.01)
boys, DMF = 1	10.25	0.14	(9.98, 10.53)
girls, DMF = 0	10.53	0.10	(10.33, 10.73)
girls, DMF = 1	10.31	0.15	(10.02, 10.62)

type I least extreme value distribution with Density (6.7). That is, μ (the model intercept) is the location and σ the scale of the original error term ε (see also Section 6.1 for relationships between parameters of the event time distribution and parameters of the distribution of the error terms of the AFT model).

In the Bayesian analysis, vague priors are considered for all model parameters. In particular, $\mu \sim \mathcal{N}(0, 10^6)$, $\beta_j \sim \mathcal{N}(0, 10^6)$ ($j = 1, 2, 3$) and $\sigma \sim \mathcal{U}(0, 100)$. For inference, three MCMC chains were generated by the JAGS program, each with 3 000 burn-in and 10 000 (×5) iterations. Thinning of 1:5 was used leading to 30 000 sampled values used for inference. Principal posterior summary statistics for primary model parameters, as well as several derived quantities are shown in Table 12.1.

The effect of decayed primary tooth 84 (DMF = 1) on emergence of the permanent tooth 44 is best seen on the posterior summary statistics for the corresponding acceleration factors AF(DMF; boys) = e^{β_2} and AF(DMF; girls) = $\text{e}^{\beta_2+\beta_3}$, respectively. For both genders, emergence of the permanent tooth is accelerated if the primary tooth was decayed, nevertheless, only for boys, this effect is of statistical importance (95% HPD credible interval being fully below a value of one). Finally, Table 12.1 shows posterior summary statistics for the median emergence times for the four combinations of the covariate values. Those were calculated from samples derived from the primary model parameters using Expression (B.9).

Finally, we note that with the maximum likelihood analysis in Example 6.1, the log-normal distribution seemed to fit the emergence times best among the three parametric models (see Table 6.1). The same conclusion is drawn if DIC is used as a model fit statistic with the Bayesian analysis. The Weibull model leads to a DIC of 1 299.25, the log-logistic model to a DIC of 1 261.70 and the log-normal model to the best value of 1 257.24.

□

12.1.1 JAGS solution

Fitting the parametric AFT model in BUGS or JAGS is almost the same as in the case of a parametric PH model (see Section 11.1.1). In particular, preparation of data as well as post-processing of the sampled chains is exactly the same. The only difference is seen in the JAGS model file, which is discussed below in the example.

Example 12.1 in JAGS

The JAGS program, specified as a character variable (named JAGS_Weib_AFT) in R used to sample from the posterior distribution of the parameters of the Weibull AFT model is as follows:

```
> JAGS_Weib_AFT <- "
+ model {
+   ### Prior distributions
+   mu ~ dnorm(0, 0.000001);
+   beta.gender ~ dnorm(0, 0.000001);
+   beta.dmf ~ dnorm(0, 0.000001);
+   beta.gen.dmf ~ dnorm(0, 0.000001);
+   sigma ~ dunif(0, 100);
+
+   ### Shape parameter of the Weibull distribution
+   ### calculated from sigma
+   gamma <- 1 / sigma;
+
+   ### Left-censored observations
+   for (i in 1:n.left){
+     eta[i] <- mu + beta.gender * gender[i] + beta.dmf * dmf[i] +
+                beta.gen.dmf * gender[i] * dmf[i];
+     alpha[i] <- exp(eta[i]);
+     lambda[i] <- pow(1 / alpha[i], gamma);
+     z.ord[i] ~ dinterval(t[i], t.lims[i, 2]);
+     t[i] ~ dweib(gamma, lambda[i]);
```

```
+   }
+
+   ### Right-censored observations
+   for (i in (n.left+1):(n.left+n.right)){
+     eta[i] <- mu + beta.gender * gender[i] + beta.dmf * dmf[i] +
+               beta.gen.dmf * gender[i] * dmf[i];
+     alpha[i] <- exp(eta[i]);
+     lambda[i] <- pow(1 / alpha[i], gamma);
+     z.ord[i] ~ dinterval(t[i], t.lims[i, 1]);
+     t[i] ~ dweib(gamma, lambda[i]);
+   }
+
+   ### Interval-censored observations
+   for (i in (n.left+n.right+1):(n.left+n.right+n.interval)){
+     eta[i] <- mu + beta.gender * gender[i] + beta.dmf * dmf[i] +
+               beta.gen.dmf * gender[i] * dmf[i];
+     alpha[i] <- exp(eta[i]);
+     lambda[i] <- pow(1 / alpha[i], gamma);
+     z.ord[i] ~ dinterval(t[i], t.lims[i, ]);
+     t[i] ~ dweib(gamma, lambda[i]);
+   }
+
+   ### Acceleration factors due to dmf = 1 for boys and girls
+   AF.dmf.boy <- exp(beta.dmf);
+   AF.dmf.girl <- exp(beta.dmf + beta.gen.dmf);
+
+   ### Characteristics of the survival distribution of T
+   ### for the four combinations of covariates
+   alpha.boy.0 <- exp(mu);
+   alpha.boy.1 <- exp(mu + beta.dmf);
+   alpha.girl.0 <- exp(mu + beta.gender);
+   alpha.girl.1 <- exp(mu + beta.gender + beta.dmf + beta.gen.dmf);
+
+   elg.1.sigma <- exp(loggam(1 + sigma))
+   meanT.boy.0  <- alpha.boy.0 * elg.1.sigma;
+   meanT.boy.1  <- alpha.boy.1 * elg.1.sigma;
+   meanT.girl.0 <- alpha.girl.0 * elg.1.sigma;
+   meanT.girl.1 <- alpha.girl.1 * elg.1.sigma;
```

```
+
+    sqrt.etc.sigma <- sqrt(exp(loggam(1 + 2 * sigma)) -
+                           exp(2 * loggam(1 + sigma)));
+    sdT.boy.0  <- alpha.boy.0 * sqrt.etc.sigma;
+    sdT.boy.1  <- alpha.boy.1 * sqrt.etc.sigma;
+    sdT.girl.0 <- alpha.girl.0 * sqrt.etc.sigma;
+    sdT.girl.1 <- alpha.girl.1 * sqrt.etc.sigma;
+
+    pow.log2.gamma <- pow(log(2), 1/gamma)
+    medT.boy.0  <- alpha.boy.0 * pow.log2.gamma;
+    medT.boy.1  <- alpha.boy.1 * pow.log2.gamma;
+    medT.girl.0 <- alpha.girl.0 * pow.log2.gamma;
+    medT.girl.1 <- alpha.girl.1 * pow.log2.gamma;
+ }
+ "
```

The JAGS *program first specifies prior distributions for the primary model parameters and then calculates the Weibull shape parameter γ. Analogously to Example 11.1, the model is specified first for left-censored then for right-censored and finally for interval-censored observations. Only one censoring vector* z.ord *of length n is now created. Its elements are sorted such that left-censored observations come first, right-censored observations second and interval-censored observations last. The same ordering is assumed for rows of a two-column matrix* t.lims *providing limits of the observed intervals (with a missing value indicator for lower limits of left- and upper limits of right-censored observations) and also for vectors* gender *and* dmf *with the covariate values. The number of left-, right- and interval-censored observations are specified by the variables* n.left, n.right *and* n.interval, *respectively. Variables* eta[i], alpha[i] *and* lambda[i] *specify, for each observation with the covariate vector $\boldsymbol{X}_i$ $(i = 1, \ldots, n)$, a value of the linear predictor $\mu + \boldsymbol{X}_i'\boldsymbol{\beta}$, the Weibull scale parameter $\alpha(\boldsymbol{X}_i)$ and the value of the second parameter of the Weibull parametrization in* JAGS, *respectively. Finally, derived parameters (acceleration factors due to DMF, means, standard deviations and medians of the emergence time distribution for the four covariate combinations) are specified to let* JAGS *generate the corresponding samples.*

Additional details, including calculation of DIC or JAGS *scripts for fitting the log-normal and the log-logistic AFT models can be found in the code available in the supplementary materials.*

□

12.1.2 SAS solution

In the Bayesian framework, AFT models can be fitted using the SAS/STAT procedures LIFEREG and MCMC. We prefer the LIFEREG procedure over the MCMC procedure for reasons explained in Section 10.1.2.

Example 12.1 in SAS (PROC LIFEREG)

Adding the BAYES *statement below to the* LIFEREG *program for Example 6.1 (see Section 6.1.3) instructs the procedure to choose for a Bayesian analysis based on a MCMC algorithm, here Gibbs sampling. Note that in contrast to the previous* JAGS *analysis, a different prior, namely* $\mathcal{G}(0.001, 0.001)$ *is assumed for the scale parameter* σ.

```
BAYES NBI=5000 NMC=20000 SEED=1 OUTPOST=postout;
```

With the option INITIALMLE, *the MLE estimates are taken as initial values. Here, the initial values are taken from the mode of the posterior distribution. Note that to obtain the posterior mode no sampling is involved, only the maximisation of the posterior (in fact the product of the likelihood with the prior). The option* NBI=5 000 *increases the number of burn-in iterations from 2 000 (default value) to 5 000 iterations. The option* NMC=20 000 *(10 000 by default) determines the number of iterations after the burn-in. Thinning with parameter* k *(only every* k*th chain is retained) is obtained with the option* THINNING=k, *where* k *is the number for which every* k*-th iteration should be kept (default* $k = 1$*).*

The default prior for the regression parameter is a uniform, but with the option COEFFPRIOR=NORMAL *a normal prior is taken on the regression parameters which is by default* $\mathcal{N}(0, c\mathbb{I})$ *with* $c = 10^6$ *and* $\mathbb{I}$ *the identity matrix. Using options within parentheses after* NORMAL *allows to change the variance constant, use a diagonal matrix with diagonal elements equal to the variances of the corresponding ML estimator multiplied by a constant or provide the mean and covariance information of the normal prior in a data set. For further details, we refer to the* SAS *documentation manual.*

The default prior for the scale parameter is a gamma distribution $\mathcal{G}(a, b)$ *with* $a = 10^{-3}$ *and* $b = 10^{-3}$*. Using the options* SHAPE=a *and* ISCALE=b *within parentheses, the shape and inverse-scale parameter can be changed. The option* RELSHAPE=c *specifies the use of a* $\mathcal{G}(c\widehat{\sigma}, c)$ *where* $\widehat{\sigma}$ *is the MLE of the scale parameter and* $c = 10^{-3}$ *by default.*

Three diagnostics are automatically produced: (1) autocorrelations (at lags 1, 5, 10 and 50), (2) (Carlin's estimate of the) effective sample size (ESS) and (3) Geweke's spectral density diagnostics. One can request specific diagnostics by specifying the DIAGNOSTICS= *or* DIAG= *option. For the first diagnostic, the requested autocorrelations are given by* LAGS= *list for each parameter, with default value of the keyword being* AUTOCORR *(*LAGS=1, 5, 10, 50*). The keyword* ESS *computes Carlin's ESS, the correlation time and the efficiency* *of the chain for each parameter. The keyword* GELMAN *computes the Gelman and Rubin convergence diagnostics. The keyword* MCSE *or* MC *computes the Monte Carlo standard error for each parameter. The keyword* GEWEKE *computes the Geweke spectral density diagnostics. Other convergence diagnostics are obtained with the keywords* HEIDELBERGER *and* RAFTERY*. For more details about these diagnostics, we refer to the* SAS *documentation or to Lesaffre and Lawson (2012). Finally, the keywords* ALL *and* NONE *produce all available or no diagnostics, respectively. The results (not shown) are similar to the frequentist results (see Section 6.1.3).*

□

Example 12.1 in SAS *(PROC MCMC)*

In procedure MCMC *the log-normal model can be fitted using the following code (output not shown). For other models, the code is easily adapted.*

```
PROC MCMC DATA=icdata.tandmob OUTPOST=postout
          SEED=1 NMC=20000 MISSING=AC;
WHERE gender=0;                      /* select boys */
PARMS mu 0 sigma 1;
PRIOR mu ~ NORMAL(0, SD=1000);
PRIOR sigma ~ GAMMA(SHAPE=0.001,ISCALE=0.001);

IF (L44^=. AND R44^=. AND L44=R44) THEN
     llike = LOGPDF('normal',LOG(R44),mu,sigma);
ELSE IF (L44^=. and R44=.) THEN
     llike = LOGSDF('normal',LOG(L44),mu,sigma);
ELSE IF (L44=. and R44^=.) THEN
     llike = LOGCDF('normal',LOG(R44),mu,sigma);
ELSE
     llike = LOG(SDF('normal',LOG(L44),mu,sigma) -
                 SDF('normal',LOG(R44),mu,sigma));
MODEL GENERAL(llike);
RUN;
```

□

12.2 AFT model with a classical Gaussian mixture as an error distribution

In Section 10.2.1, a classical Gaussian mixture was introduced as a flexible model for an unknown distribution. In this section, we incorporate it in the error term of the AFT model. The methods shown here generalize the methodology of Section 10.2.1 by incorporating covariates in the model. This section stems largely from Komárek and Lesaffre (2007) where the model is even further generalized by inclusion of random effects in the AFT Expression (12.1).

The model

To robustify the AFT Model (12.1) towards misspecified distributional assumptions, a density g of the error term ε is now specified as a (shifted and scaled) Gaussian mixture, i.e.,

$$g(e) = \frac{1}{\tau} \sum_{k=1}^{K} w_k \, \varphi_{\mu_k, \sigma_k^2}\Big(\frac{e - \alpha}{\tau}\Big). \tag{12.5}$$

Hence, we assume that the AFT model with a classical Gaussian mixture as an error distribution (CGM AFT model) expresses the relationship of the response Y to covariates $\boldsymbol{X}$.

Analogously to Section 10.2.1, $\alpha \in \mathbb{R}$ and $\tau > 0$ in (12.5) are the fixed shift parameter and the fixed scale parameter, respectively, included in the model to improve numerical stability of the MCMC procedures. Without any loss of generality, we can set $\alpha = 0$ and $\tau = 1$. The unknown parameters in (12.5) are: the number of mixture components K, the mixture weights $\boldsymbol{w} = (w_1, \ldots, w_K)'$ ($w_k > 0$, $k = 1, \ldots, K$, $\sum_{k=1}^{K} w_k = 1$), the mixture means $\boldsymbol{\mu} = (\mu_1, \ldots, \mu_K)'$ and the mixture variances $\boldsymbol{d} = (\sigma_1^2, \ldots, \sigma_K^2)'$ ($\sigma_k^2 > 0$, $k = 1, \ldots, K$). The vector $\boldsymbol{\theta}$ of the unknown parameters of the CGM AFT model contains the unknown parameters from (12.5) and additionally regression coefficients vector $\boldsymbol{\beta}$, i.e., $\boldsymbol{\theta} = (K, \boldsymbol{w}', \boldsymbol{\mu}', \boldsymbol{d}', \boldsymbol{\beta}')'$.

Standardized error term

In Section 12.1 (Expression 12.4) and in Chapter 6 as well, we wrote the AFT model also using the standardized error term $\varepsilon^* = \sigma^{-1}(\varepsilon - \mu)$ as

$$Y = \log(T) = \mu + \boldsymbol{X}'\boldsymbol{\beta} + \sigma\varepsilon^*,$$

where $\mu \in \mathbb{R}$ was the location (model intercept) and $\sigma > 0$ the scale parameter of the distribution of the error term ε. To allow for easy comparison of the fitted CGM to (standardized) parametric distributions, we take

$$\mu = \mathsf{E}(\varepsilon) = \alpha + \tau \sum_{k=1}^{K} w_k \mu_k,$$

$$\sigma = \mathsf{SD}(\varepsilon) = \tau \sqrt{\sum_{k=1}^{K} w_k \Big\{ \sigma_k^2 + \Big(\mu_k - \sum_{j=1}^{K} w_j \mu_j \Big)^2 \Big\}},$$

which leads to a zero-mean and unit-variance standardized error term ε^* having a density

$$g^*(e^*) = \sigma\, g(\mu + \sigma\, e^*) = \frac{\sigma}{\tau} \sum_{k=1}^{K} w_k\, \varphi_{\mu_k, \sigma_k^2}\Big(\frac{\mu + \sigma\, e^* - \alpha}{\tau} \Big). \tag{12.6}$$

Bayesian estimation

The likelihood of the CGM AFT model is given by general Equations (6.8) and (6.9), now with densities g and g^* given by Expressions (12.5) and (12.6), respectively. The prior distribution of the regression coefficients vector $\boldsymbol{\beta}$ is assumed to be independent of the mixture parameters K, $\boldsymbol{w}$, $\boldsymbol{\mu}$, $\boldsymbol{d}$. Consequently, the prior distribution of all parameters $\boldsymbol{\theta}$ of the CGM AFT model is specified using a decomposition

$$p(\boldsymbol{\theta}) = p(K,\, \boldsymbol{w},\, \boldsymbol{\mu},\, \boldsymbol{d},\, \boldsymbol{\beta}) = p(K,\, \boldsymbol{w},\, \boldsymbol{\mu},\, \boldsymbol{d})\, p(\boldsymbol{\beta}). \tag{12.7}$$

For the first factor in the decomposition (12.7) we can use the same weakly informative prior as in Section 10.2.1. This is based on the additional decomposition

$$p(K,\, \boldsymbol{w},\, \boldsymbol{\mu},\, \boldsymbol{d}) = p(K)\, p(\boldsymbol{w} \,|\, K)\, p(\boldsymbol{\mu} \,|\, K)\, p(\boldsymbol{d} \,|\, K), \tag{12.8}$$

see Example 12.2 below and Komárek and Lesaffre (2007) for details on specification of factors in decomposition (12.8). For $p(\boldsymbol{\beta})$ in (12.7), a multivariate normal prior $\mathcal{N}_p(\boldsymbol{\beta}_0,\, \boldsymbol{\Sigma}_0)$ is chosen. Typically, a weak informative prior for these regression parameters is taken by setting $\boldsymbol{\beta}_0 = \mathbf{0}$ and $\boldsymbol{\Sigma}_0$ to a diagonal matrix with sufficiently high values.

In Example 12.2 inference is based on the converged sample $\boldsymbol{\theta}^{(1)}, \ldots, \boldsymbol{\theta}^{(M)}$ from the posterior distribution obtained with the *reversible jump* Markov chain Monte Carlo (RJ-MCMC) algorithm of Green (1995). Details on this are given in Komárek and Lesaffre (2007).

Posterior predictive survival distribution for a new subject

As in Chapter 6, we can estimate the survival distribution for a subject with covariate vector, say, $\boldsymbol{X}_{pred}$. Let $S_{pred}(t;\, \boldsymbol{\theta})$, $f_{pred}(t;\, \boldsymbol{\theta})$ and $\hbar_{pred}(t;\, \boldsymbol{\theta})$

denote the survival, the density and the hazard function, respectively corresponding to $\boldsymbol{X}_{pred}$. Given the CGM AFT model and the parameter vector $\boldsymbol{\theta}$ (and hence K), the above functions are given by

$$\left.\begin{aligned} S_{pred}(t;\,\boldsymbol{\theta}) &= \sum_{k=1}^{K} w_k \left\{1 - \Phi_{\mu_k,\sigma_k^2}\left(\frac{\log(t) - \alpha - \boldsymbol{X}'_{pred}\boldsymbol{\beta}}{\tau}\right)\right\}, \\ f_{pred}(t;\,\boldsymbol{\theta}) &= \frac{1}{\tau\, t}\sum_{k=1}^{K} w_k\,\varphi_{\mu_k,\sigma_k^2}\left(\frac{\log(t) - \alpha - \boldsymbol{X}'_{pred}\boldsymbol{\beta}}{\tau}\right), \\ \hbar_{pred}(t;\,\boldsymbol{\theta}) &= \frac{f_{pred}(t;\,\boldsymbol{\theta})}{S_{pred}(t;\,\boldsymbol{\theta})}. \end{aligned}\right\} \tag{12.9}$$

Estimates of the above quantities are obtained from a converged Markov chain $\boldsymbol{\theta}^{(m)}$ $(m = 1, \ldots, M)$, i.e., by

$$\widehat{S}_{pred}(t) = \widehat{\mathsf{E}}_{\boldsymbol{\theta}|\boldsymbol{\mathcal{D}}}\big\{S_{pred}(t;\,\boldsymbol{\theta})\big\} = \frac{1}{M}\sum_{m=1}^{M} S_{pred}\big(t;\,\boldsymbol{\theta}^{(m)}\big), \tag{12.10}$$

and $\widehat{f}_{pred}(t) = \widehat{\mathsf{E}}_{\boldsymbol{\theta}|\boldsymbol{\mathcal{D}}}\big\{f_{pred}(t;\,\boldsymbol{\theta})\big\}$ and $\widehat{\hbar}_{pred}(t) = \widehat{\mathsf{E}}_{\boldsymbol{\theta}|\boldsymbol{\mathcal{D}}}\big\{\hbar_{pred}(t;\,\boldsymbol{\theta})\big\}$ calculated analogously.

Additionally, the pointwise equal-tail credible intervals for the values of $S_{pred}(t;\,\boldsymbol{\theta})$, $f_{pred}(t;\,\boldsymbol{\theta})$, or $\hbar_{pred}(t;\,\boldsymbol{\theta})$ can easily be calculated by taking appropriate empirical quantiles.

Posterior predictive error density

The *posterior predictive density* provides an estimate of the density of the (standardized) error term $\varepsilon^\star$. An MCMC estimate of this posterior predictive density stems from Equation (12.6) and it is given by $\widehat{g^\star}(e^\star) = \widehat{\mathsf{E}}_{\boldsymbol{\theta}|\boldsymbol{\mathcal{D}}}\{g^\star(e^\star)\}$, again calculated analogously to Expression (12.10).

Example 12.2 Signal Tandmobiel study

We revisit Examples 6.1 and 12.1 by fitting the same AFT model to the emergence times of tooth 44, except for the error distribution which is now a CGM. When fitting the model, the shift and scale parameters were set to $\alpha = 0$ and $\tau = 1$.

TABLE 12.2: Signal Tandmobiel study. AFT model for emergence of tooth 44 with a CGM error distribution. Posterior summary statistics for the most important model parameters (AF=acceleration factor) obtained with the R package bayesSurv.

	Posterior		**95% Equal Tail**	**Contour**
Parameter	**mean**	**SD**	**CI**	**probability**
Intercept (μ)	2.38	0.02	(2.36, 2.41)	**<0.0001**
Gender[girl] (β_1)	−0.0460	0.0146	(−0.0750, −0.0178)	**0.0012**
DMF[1] (β_2)	−0.0751	0.0182	(−0.1107, −0.0399)	**<0.0001**
Gender:DMF (β_3)	0.0479	0.0265	(−0.0042, 0.1005)	0.0710
Scale σ	0.125	0.050	(0.108, 0.150)	
AF(DMF; boys)	0.93	0.02	(0.90, 0.96)	**<0.0001**
AF(DMF; girls)	0.97	0.02	(0.94, 1.01)	0.1598

Prior distributions

We have taken the following priors dictated by decompositions (12.7) and (12.8):

$$\left.\begin{array}{c} K \sim \mathcal{U}\{1,\ldots,20\}, \\ \boldsymbol{w} \mid K \sim \mathcal{D}ir_K(1,\ldots,1), \qquad \boldsymbol{\mu} \mid K \sim \prod_{k=1}^{K} \mathcal{N}(2.4,\,10^2), \\ (\sigma_1^{-2},\ldots,\sigma_K^{-2}) \mid K \sim \prod_{k=1}^{K} \mathcal{G}(1.5,\,h), \qquad h \sim \mathcal{G}(0.8,\,100), \\ \boldsymbol{\beta} \sim \mathcal{N}_3\big(\boldsymbol{0},\,\mathsf{diag}(10^2,\,10^2,\,10^2)\big). \end{array}\right\} \tag{12.11}$$

The choice of the hyperparameters of the above prior distributions is inspired by the guidelines described in Komárek and Lesaffre (2007). The intention is to choose a reasonable and at the same time weakly informative prior distribution. For example, the prior mean of 2.4 in the distribution $\boldsymbol{\mu} \mid K$ is approximately equal to the MLE of the intercept of the AFT model with a normal error distribution, see Table 6.2. The prior is indeed weakly informative since the prior standard deviation of 10 is several orders of magnitude higher than the posterior standard deviation, see Table 12.2. The same holds also for the prior standard deviations in the prior distribution of the regression coefficients $\boldsymbol{\beta}$.

Posterior predictive error distribution

A single Markov chain was run with 1:10 thinning, and a burn-in of 10 000 iterations followed by 10 000 iterations kept for inference. Standard convergence diagnostics were performed. The posterior distribution of K reveals that a posteriori $K = 1$ is visited with probability 0.8767, $K = 2$ with probability 0.1185, $K = 3$ with probability 0.0046, $K = 4$ and $K = 5$ both with probability of 0.0001. Values $K \geq 6$ were not visited by our chain and hence we conclude that $\mathsf{P}\big(K \leq 5 \,\big|\, \boldsymbol{\mathcal{D}}\big) = 1$. Figure 12.1 then shows posterior predictive density of the standardized error term compared to a standard normal density. It shows, among other things, that values of K above one that are represented in the posterior distribution with probability $1 - 0.8767 = 0.1233$ mainly serve to capture a mild skewness of the error distribution.

Posterior summary statistics

Posterior summary statistics (averaged over all sampled values, irrespective of K) for the regression coefficients, intercept and scale parameters and also for the acceleration factors of emergence due to decayed primary predecessors, i.e., due to DMF = 1 are shown, separately for boys and girls, in Table 12.2.

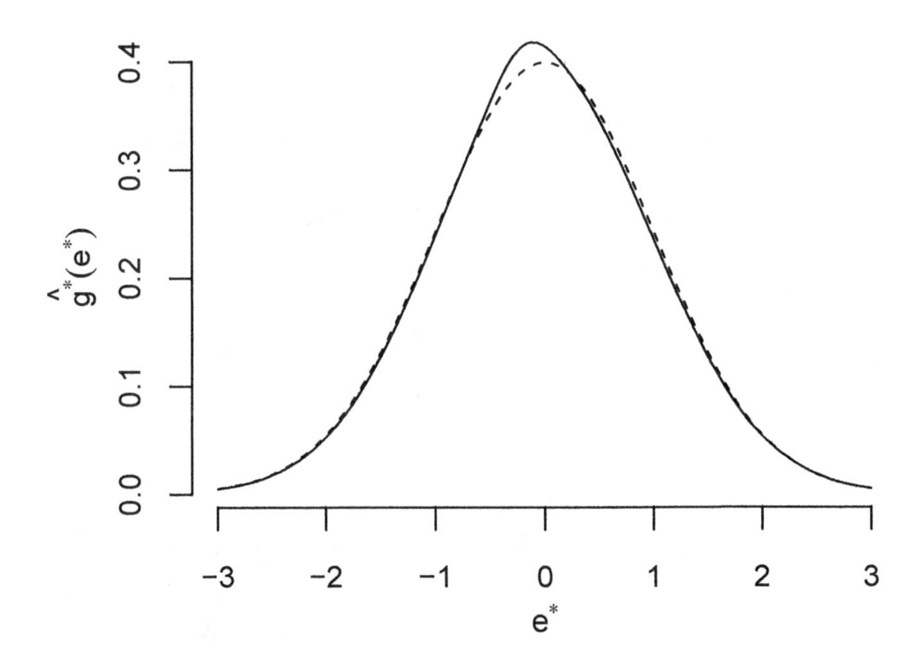

FIGURE 12.1: Signal Tandmobiel study. Posterior predictive density of the standardized error term in the AFT model for emergence of permanent tooth 44 (solid line) compared to a standard normal density (dashed line), obtained with the R package bayesSurv.

The acceleration factors are defined in the same way as in Example 12.1, that is, AF(DMF; boys) = e^{β_2}, AF(DMF; girls) = $e^{\beta_2+\beta_3}$.

Since the estimated error distribution shows only mild deviations from normality (see Figure 12.1) and a weakly informative prior distribution was used, it is not surprising to see that the results shown in Table 12.2 are close to the results shown in Table 6.2 based on maximum likelihood estimation with a normal distribution for the error term. Consequently also all conclusions stated in Examples 6.1 and 12.1 (where, however, type I least extreme value distribution was assumed for the error term) can (basically) be drawn also now.

Posterior predictive survival distributions

Finally, we calculated the posterior predictive survival and hazard functions for the combinations of the covariate values corresponding to (i) boys with DMF = 0, (ii) boys with DMF = 1, (iii) girls with DMF = 0, (iv) girls with DMF = 1 using the procedure described above. The resulting survival curves, their pointwise 95% credible intervals and their NPMLE for comparison are shown for boys in Figure 12.2. The hazard functions together with

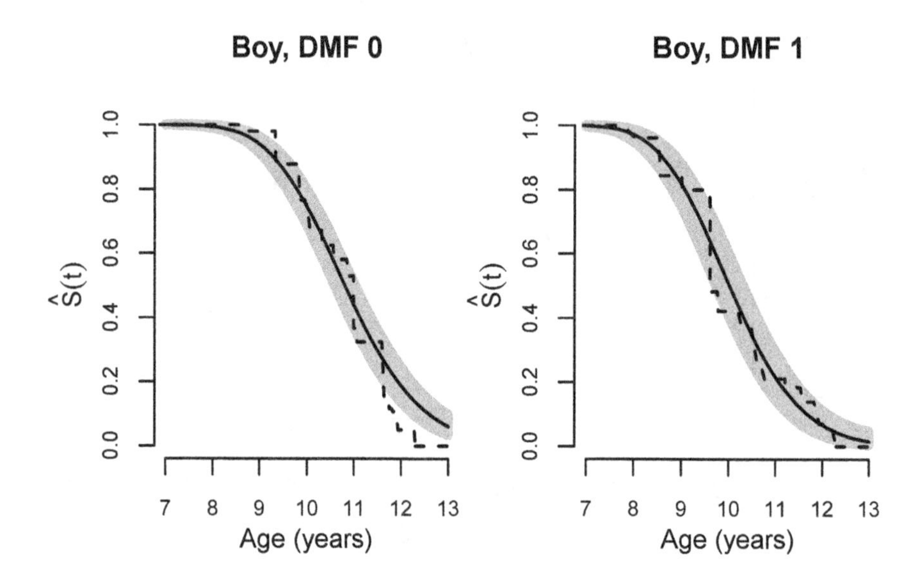

FIGURE 12.2: Signal Tandmobiel study. Posterior predictive survival functions for emergence of permanent tooth 44 by DMF groups for boys based on the AFT model with the CGM error distribution (solid lines) compared to NPMLE (dashed lines), obtained with the R package bayesSurv. The filled area shows pointwise 95% credible intervals for the values of the survival functions.

the pointwise 95% credible intervals are shown for boys in Figure 12.3. The results for girls (not shown) are similar.

A comparison of the posterior predictive survival curves to their NPMLE in Figure 12.3 reveals that even though the error distribution was modelled in a flexible way, there is not yet a perfect agreement between the two curves, especially for boys with DMF = 0. For girls (not shown) the agreement was better. Possible reasons for this disagreement are (i) the survival distributions in the four groups are not all of the same shape, (ii) the survival distributions in the four groups are of the same shape but of different variability. Note that neither (i) nor (ii) can be captured by a model of this section.

□

12.2.1 R solution

Bayesian inference for the AFT model with the classical normal mixture in the error distribution has been implemented in the R package bayesSurv. MCMC sampling is provided by function bayessurvreg1, which generates the posterior samples and stores them in a form of the txt files in a directory specified by the user. Additional functions have been implemented and can be

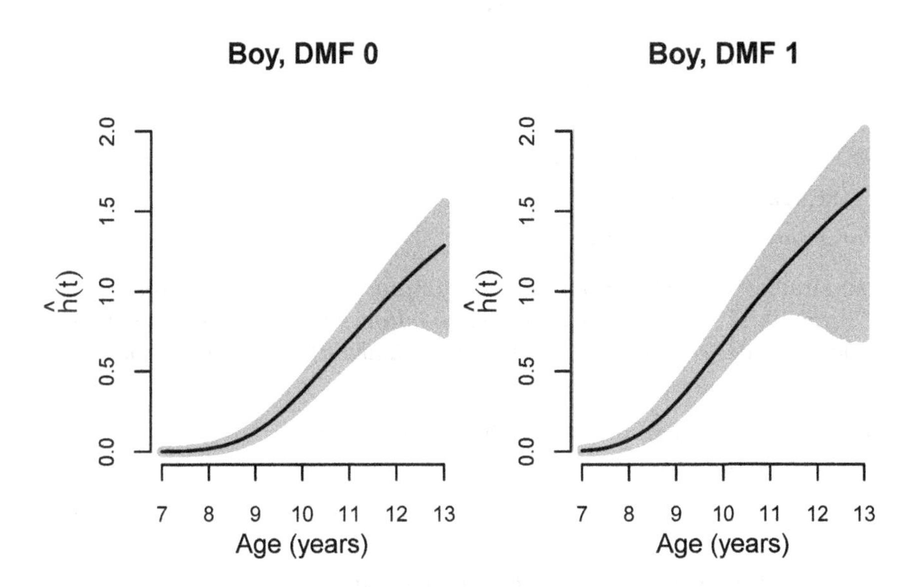

FIGURE 12.3: Signal Tandmobiel study. Posterior predictive hazard functions for emergence of permanent tooth 44 in DMF groups for boys based on a AFT model with the CGM error distribution (solid lines), obtained with the R package bayesSurv. The filled area shows pointwise 95% credible intervals for the values of the hazard functions.

used to process the posterior samples to get the final results. See Appendix D.8 for more details.

Example 12.2 in R

We start by loading the data and creating a data frame tandDMF84 *containing only those children with nonmissing values for all covariates included in the model.*

Prior distribution

In the next steps, we create lists that specify the prior distribution of the Bayesian model with the values of the hyperparameters as indicated by Expressions (12.11). First, we create a list prior.eps *specifying prior hyperparameters for the mixture parameters* K, $\boldsymbol{w}$, $\boldsymbol{\mu}$, $\sigma_1^{-2}, \ldots, \sigma_K^{-2}$. *The priors for the number of mixture components* K *and the mixture means* $\boldsymbol{\mu}$ *are specified as follows.*

```
> prior.eps <- list()
>
> ## Prior and its hyperparameters for K
> prior.eps$kmax <- 20
> prior.eps$k.prior <- "uniform"
>
> ## Prior hyperparameters for mu
> prior.eps$mean.mu <- 2.4
> prior.eps$var.mu <- 10^2
```

Hyperparameters of the priors for the mixture weights $\boldsymbol{w}$ *and inverted mixture variances* $\sigma_1^{-2}, \ldots, \sigma_K^{-2}$ *can be chosen automatically by the* bayessurvreg1 *function below. Nevertheless, the user may change them by an obvious modification of the following code:*

```
> ## Prior hyperparameter of the Dirichlet prior
> ## for mixture weights
> prior.eps$dirichlet.w <- 1
>
> ## Prior hyperparameters of the gamma prior
> ## for mixture precisions sigma_1^{-1}, ..., sigma_K^{-2}
> prior.eps$shape.invsig2 <- 1.5
> prior.eps$shape.hyper.invsig2 <- 0.8
> prior.eps$rate.hyper.invsig2 <- 100
```

Second, a list prior.beta *is specified giving the hyperparameters of the prior for the regression coefficients* $\boldsymbol{\beta}$.

```
> nregres <- 3          ## number of regression parameters
> prior.beta <- list()
> prior.beta$mean.prior <- rep(0, nregres)
> prior.beta$var.prior  <- rep(10^2, nregres)
```

Initial values

To start the MCMC simulation, reasonable initial values must be specified for the model parameters. These can be taken, e.g., from a parametric maximum-likelihood fit (see Table 6.2). In the following, we create a list init *containing the initial values in the required format. The following inits will be considered:* $K = 1$, $w_1 = 1$, $\mu_1 = 2.4$ *(approximately estimated intercept from Table 6.2),* $\sigma_1^2 = 0.12^2$ *(0.12 being estimated scale from Table 6.2),* $\boldsymbol{\beta} = (-0.046, -0.075, 0.048)'$ *(estimated values of regression coefficients from Table 6.2).*

```
> init <- list()
> init$mixture <- c(1,      ## initial number of mixture components
+      1, rep(0, prior.eps$kmax - 1),           ## initial weights
+      2.4, rep(0, prior.eps$kmax - 1),         ## initial means
+      0.12^2, rep(0, prior.eps$kmax - 1))      ## initial variances
> init$beta <- c(-0.046, -0.075, 0.048)
```

Posterior simulation

After having loaded the package bayesSurv, *the MCMC simulation is done with the* bayessurvreg1 *function. We point out that generated samples are stored in* txt *files in a directory specified by the argument* dir *(directory* /home/BKL/AFT_CGM/ *on the currently active drive in the example below). Note that the corresponding directory must physically exist before starting the simulation.*

```
> library("bayesSurv")
> simCGM <- bayessurvreg1(
+       Surv(L44, R44, type = "interval2")
+         ~ fGENDER + fDMF_84 + fGENDER:fDMF_84,
+       data = tandDMF84,
+       dir = "/home/BKL/AFT_CGM/",
+       nsimul = list(niter = 20000, nthin = 10, nburn = 10000,
+                     nwrite = 1000),
+       prior = prior.eps, prior.beta = prior.beta, init = init,
+       store = list(y = FALSE, r = FALSE, u = FALSE))
```

Chosen values of the prior hyperparameters and initial values of the model parameters to start the MCMC simulation can be displayed (output not shown) using:

```
> attr(simCGM, "prior")
> attr(simCGM, "prior.beta")
> attr(simCGM, "init")
```

Results can be further processed by the R *package* coda. *To do that, we first use function* files2coda *from the package* bayesSurv *and read the sampled values into* R.

```
> library("coda")
> parCGM <- files2coda(dir = "/home/BKL/AFT_CGM/")
```

Object parCGM *is the* coda *object of class* mcmc. *Its most important columns include:* k *(sampled values of parameter K),* Intercept *(sampled values of the model intercept μ),* fGENDERgirl, fDMF_84caries, fGENDERgirl.fDMF_84caries *(sampled values of the regression coefficients $\boldsymbol{\beta}$) and* Scale *(sampled values of the scale parameter σ). For example, the traceplots of the sampled values of parameter K, the number of mixture components, can easily be drawn using:*

```
> traceplot(parCGM[, "k"], ylab = "K")
```

Posterior predictive error distribution

The posterior distribution of the number of mixture components is shown using:

```
> tabK <- table(parCGM[, "k"])
> print(ptabK <- prop.table(tabK))
```

```
     1      2      3      4      5
0.8767 0.1185 0.0046 0.0001 0.0001
```

The posterior predictive density of the (standardized) error term is produced by the function bayesDensity *from the package* bayesSurv. *Its argument* stgrid *determines a grid of $e^\star$ values in which the density is to be evaluated.*

```
> epsstargrid <- seq(-3, 3, length = 500)
> gepsstar <- bayesDensity(dir = "/home/BKL/AFT_CGM/",
                           stgrid = epsstargrid)$standard
```

The function bayesDensity *calculates by default posterior predictive densities of both the standardized error term $\varepsilon^\star$ and the unstandardized error term ε. In the above code, we stored only values of the posterior predictive density of the standardized error term in object* gepsstar. *This is a data frame with the following first 6 rows and first 11 columns:*

```
        grid unconditional         k = 1        k = 2
1 -3.000000    0.004211694 0.004431848 0.002686984
2 -2.987976    0.004366105 0.004594301 0.002785409
3 -2.975952    0.004525543 0.004762020 0.002887179
4 -2.963928    0.004690145 0.004935148 0.002992395
5 -2.951904    0.004860053 0.005113830 0.003101163
6 -2.939880    0.005035409 0.005298216 0.003213590
        k = 3        k = 4         k = 5 k = 6 k = 7 k = 8 k = 9
1 0.001630512 0.001034489 0.002811724     0     0     0     0
2 0.001696045 0.001101214 0.002986556     0     0     0     0
3 0.001764211 0.001171795 0.003170185     0     0     0     0
4 0.001835113 0.001246424 0.003362910     0     0     0     0
5 0.001908857 0.001325298 0.003565033     0     0     0     0
6 0.001985555 0.001408625 0.003776854     0     0     0     0
```

The first column grid *provides a grid of values where the posterior predictive density was evaluated, the second column* unconditional *provides values of the posterior predictive density* $\widehat{g^{\star}}(e^{\star}) = \widehat{\mathsf{E}}_{\boldsymbol{\theta}|\boldsymbol{\mathcal{D}}}\{g^{\star}(e^{\star})\}$. *Those columns were also used to produce Figure 12.1. Additional columns labeled* k = 1, k = 2, ... *provide estimated values of* $\widehat{\mathsf{E}}_{\boldsymbol{\theta}|\boldsymbol{\mathcal{D}}}\{g^{\star}(e^{\star}) \,|\, K = 1\}$, $\widehat{\mathsf{E}}_{\boldsymbol{\theta}|\boldsymbol{\mathcal{D}}}\{g^{\star}(e^{\star}) \,|\, K = 2\}$, ... *that we do not discuss here.*

Posterior summary statistics

The posterior summary statistics for the model parameters shown in Table 12.2 are obtained using:

```
> summary(parCGM)
```

The contour probabilities in Table 12.2 are calculated using:

```
> PARAMS <- c("Intercept", "fGENDERgirl", "fDMF_84caries",
+             "fGENDERgirl.fDMF_84caries")
> PVal <- numeric(length(PARAMS))
> names(PVal) <- PARAMS
> for (i in 1:length(PARAMS)){
+    PVal[i] <- 2 * min(sum(parCGM[, PARAMS[i]] < 0),
+                       sum(parCGM[, PARAMS[i]] > 0)) / nrow(parCGM)
+ }
```

To calculate the posterior summary statistics for the acceleration factors of emergence due to decayed primary predecessors, we first create an object with the sampled chains of the acceleration factors. When doing so, remember $AF(DMF;\ boys) = \mathrm{e}^{\beta_2}$, $AF(DMF;\ girls) = \mathrm{e}^{\beta_2+\beta_3}$.

```
> AFsample <- mcmc(data.frame(
+              Boy  = exp(parCGM[, "fDMF_84caries"]),
+              Girl = exp(parCGM[, "fDMF_84caries"]
+                         + parCGM[, "fGENDERgirl.fDMF_84caries"])))
> colnames(AFsample) <- c("Boy", "Girl")
```

Posterior summary statistics and contour probabilities shown in Table 12.2 are then calculated by applying the same set of commands as was done above with the primary model parameters.

Posterior predictive survival distributions

Finally, we calculate the posterior predictive survival and hazard curves including their pointwise 95% credible intervals using the function predictive of the package bayesSurv. We start by creating a data frame pdata with the covariate combinations for which we want to calculate the quantities of the posterior predictive distribution:

```
> x.label <- paste(rep(c("Boy", "Girl"), each = 2), ", ",
+                  rep(c("DMF 0", "DMF 1"), 2), sep = "")
> pdata <- data.frame(fGENDER = factor(rep(c(0, 1), each = 2),
+                                      labels = c("boy", "girl")),
+                     fDMF_84 = factor(c(0, 1, 0, 1),
+                                      labels = c("sound", "caries")),
+                     L44     = 0.001,
+                     R44     = as.numeric(NA))
> rownames(pdata) <- x.label
> print(pdata)
```

```
             fGENDER fDMF_84   L44 R44
Boy, DMF 0       boy   sound 0.001  NA
Boy, DMF 1       boy  caries 0.001  NA
Girl, DMF 0     girl   sound 0.001  NA
Girl, DMF 1     girl  caries 0.001  NA
```

Columns fGENDER and fDMF_84 correspond to the original covariates used in the call to the bayessurvreg1 function which generated the posterior sample. They must have the same name and format as the original covariates. Columns L44 and R44 correspond to the original variables with the observed intervals. They must be included in the data frame pdata; nevertheless the L44 column may contain any positive value and the R44 column even the missing data value as shown in the example above. Further, we create a numeric vector tgrid *containing a time (age) grid in which we want to evaluate the posterior predictive survival and hazard functions:*

```
> tgrid <- seq(7, 13, length = 200)
```

Posterior predictive survival and hazard functions are calculated by running the following code:

```
> simCGMp1 <- predictive(Surv(L44, R44, type = "interval2")
+        ~ fGENDER + fDMF_84 + fGENDER:fDMF_84,
+        data = pdata, grid = tgrid,
+        quantile = c(0.025, 0.5, 0.975),
+        predict = list(Et = FALSE, t = FALSE, Surv = TRUE,
+                       hazard = TRUE, cum.hazard = FALSE),
+        store = list(Et = FALSE, t = FALSE, Surv = FALSE,
+                     hazard = FALSE, cum.hazard = FALSE),
+        dir = "/home/BKL/AFT_CGM/")
```

By setting the quantile *argument to a numeric vector of* (0.025, 0.5, 0.975), *we also calculate the pointwise 2.5%, 50% and 97.5% posterior quantiles of the values of the survival and hazard functions that serve as basis for the calculation of the pointwise credible intervals.*

The above call to the predictive *function assumes that the sampled Markov chains of the model parameters generated by the* bayessurvreg1 *function are stored in a directory* /home/BKL/AFT_CGM/. *New TXT files* quantS1.sim, ..., quantS4.sim *and* quanthazard1.sim, ..., quanthazard4.sim *are created which contain calculated values of the posterior predictive survival and hazard function for each covariate combination (on the last row of each file) and requested pointwise posterior quantiles (on the second to the prelast row of each file). The first row of each file contains the chosen time grid. To read the calculated values to* R, *the following code can be used:*

```
> S <- haz <- list()
> for (i in 1:length(x.label)){
+   S[[i]] <- as.data.frame(cbind(tgrid,
+     t(read.table(paste("/home/BKL/AFT_CGM/quantS", i, ".sim",
+                        sep=""),
+                  header = FALSE, skip = 1))))
+   haz[[i]] <- as.data.frame(cbind(tgrid,
+     t(read.table(paste("/home/BKL/AFT_CGM/quanthazard", i, ".sim",
+                        sep=""),
+                  header = FALSE, skip = 1))))
+   colnames(S[[i]]) <- colnames(haz[[i]]) <- c("Time",
```

```
+           paste("Q", c(0.025, 0.5, 0.975)*100, "%", sep=""), "Mean")
+    rownames(S[[i]]) <- rownames(haz[[i]]) <- 1:length(tgrid)
+ }
> names(S) <- names(haz) <- x.label
```

Each component of the list S *contains the posterior predictive survival function and related posterior quantiles for one covariate combination. For example, the quantities related to the first covariate combination (boys with DMF = 0) look as follows (column* Mean *gives the posterior predictive survival function).*

```
> print(S[[1]])
```

```
      Time       Q2.5%       Q50%     Q97.5%       Mean
1 7.000000 0.9995802 0.9998929 0.9999799 0.9998592
2 7.030151 0.9995307 0.9998763 0.9999761 0.9998399
3 7.060302 0.9994703 0.9998575 0.9999718 0.9998180
4 7.090452 0.9994059 0.9998361 0.9999666 0.9997934
5 7.120603 0.9993363 0.9998119 0.9999607 0.9997657
...
```

The list haz *has the same structure, nevertheless it contains the posterior quantities related to the hazard functions.*

Lists S *and* haz *can then be used to draw plots shown in Figures 12.2 and 12.3. To draw the pointwise credible intervals in a form of a filled area, we can exploit the function* cbplot *from the* R *package* mixAK. *For example, Figure 12.3 was produced using the following code:*

```
> library("mixAK")
> par(mfrow=c(2, 2))
> for (i in 1:4){
+    cbplot(haz[[i]]$Time, haz[[i]]$Mean,
+           haz[[i]][,"Q2.5%"], haz[[i]][,"Q97.5%"],
+           col = "black", band.type = "s", scol = "grey80",
+           xlab="Age (years)", ylab=expression(hat(h)(t)),
+           xlim=c(7, 13), ylim=c(0, 2), lwd=1.5)
+    title(main=x.label[i])
+ }
```

□

12.3 AFT model with a penalized Gaussian mixture as an error distribution

In Section 6.2, an alternative to the classical Gaussian mixture (CGM), namely the penalized Gaussian mixture (PGM), was introduced as a suitable semiparametric model for unknown error distribution of the AFT model. In Section 6.2 a frequentist inference using a method of penalized maximum likelihood was discussed. In this section, we show that the penalty term can be translated into a prior distribution for the (transformed) mixture weights while allowing for a full Bayesian inference on the model parameters. In a more general setting of the AFT model for clustered (doubly) interval-censored data, the methodology was introduced by Komárek et al. (2007) and Komárek and Lesaffre (2008) and is implemented by the R package bayesSurv. Here we outline the most important aspects.

The model

We continue assuming the AFT Model (12.1) where a density g of the error term ε is modelled by a shifted and scaled Gaussian mixture that will now be written (in accordance with notation of Section 6.2) as

$$g(e) = \frac{1}{\tau} \sum_{k=-K}^{K} w_k \varphi_{\mu_k,\, \sigma_0^2}\Big(\frac{e-\alpha}{\tau}\Big). \tag{12.12}$$

In contrast to Section 12.2, the prespecified fixed parameters are now: the number of components $2K+1$, the means $\boldsymbol{\mu} = \big(\mu_{-K},\, \dots,\, \mu_K\big)'$ playing the role of knots and the basis standard deviation σ_0. The means (knots) are assumed to be equidistant and centered around zero, i.e., $\mu_k = k\,\delta$, $k = -K, \dots, K$, for some $\delta > 0$. Typical choices for the fixed parameters are $K = 15$, $\delta = 0.3$, $\sigma_0 = (2/3)\delta$ (see Section 6.2 for motivation of those choices).

Next to the regression coefficients $\boldsymbol{\beta}$ from the AFT Expression (12.1), the unknown model parameters, stemming from the Gaussian mixture (12.12) are the shift parameter α, the scale parameter $\tau > 0$ and the weights $\boldsymbol{w} = \big(w_{-K},\, \dots,\, w_K\big)'$ satisfying $0 < w_k < 1$, $k = -K, \dots, K$, $\sum_{k=-K}^{K} w_k = 1$. Analogously to Section 6.2, constraints on the weights are avoided by estimating the transformed weights $\boldsymbol{a} = \big(a_{-K},\, \dots,\, a_K\big)'$, where $a_0 = 0$ and

$$w_k = \frac{\exp(a_k)}{\sum_{j=-K}^{K} \exp(a_j)} \qquad (k = -K, \dots, K).$$

The prior distribution

In the prior distribution, we assume usual independence of the blocks of parameters $\boldsymbol{\beta}$, α, τ and $\boldsymbol{a}$. That is, the prior distribution is factorized as

$$p(\boldsymbol{\beta},\, \alpha,\, \tau,\, \boldsymbol{a}) \;=\; p(\boldsymbol{\beta})\, p(\alpha)\, p(\tau)\, p(\boldsymbol{a}).$$

For both the regression coefficients $\boldsymbol{\beta}$ and the shift parameter α, standard normal priors are considered, namely $\boldsymbol{\beta} \sim \mathcal{N}_p(\boldsymbol{\beta}_0,\, \boldsymbol{\Sigma}_0)$ and $\alpha \sim \mathcal{N}(\alpha_0,\, \sigma_\alpha^2)$. For the scale parameter τ, we consider either a uniform prior on interval $(0,\, h_{\tau,2})$ or a gamma prior for τ^{-2} with shape and rate parameters being $h_{\tau,1}$ and $h_{\tau,2}$. The prior hyperparameters $h_{\tau,1}$ and $h_{\tau,2}$ are assumed to be fixed.

The penalty prior for the transformed mixture weights

To avoid overfitting and identifiability problems, the (minus) penalty term (Expression (3.21)) on the (transformed) mixture weights was introduced in Section 6.2. This penalty term can also be interpreted as an informative log-prior distribution (see, e.g., Silverman, 1985, Section 6) and leads to the following penalty prior for the transformed mixture weights:

$$p(\boldsymbol{a}) \;\propto\; \exp\biggl\{-\frac{\lambda}{2}\sum_{k=-K+s}^{K}\bigl(\Delta^s\, a_k\bigr)^2\biggr\} \;=\; \exp\Bigl(-\frac{\lambda}{2}\,\boldsymbol{a}'\boldsymbol{P}_s'\boldsymbol{P}_s\boldsymbol{a}\Bigr), \qquad (12.13)$$

where Δ^s denotes a difference operator of order s (see Section 3.3.3 for details) and $\boldsymbol{P}_s$ the corresponding difference operator matrix. As in Section 6.2, the hyperparameter λ controls the smoothness of the resulting density of the AFT error term.

In principle, the prior (12.13) can also be interpreted as a (singular) multivariate normal distribution with zero mean and covariance matrix $\lambda^{-1}\bigl(\boldsymbol{P}_s'\boldsymbol{P}_s\bigr)^{-}$, where $\bigl(\boldsymbol{P}_s'\boldsymbol{P}_s\bigr)^{-}$ denotes a generalized inverse of the matrix $\boldsymbol{P}_s'\boldsymbol{P}_s$. This distribution is also known as a Gaussian Markov random field (GMRF) and is extensively used in spatial statistics. Although the prior (12.13) is improper (the matrix $\boldsymbol{P}_s'\boldsymbol{P}_s$ has a deficiency of s in its rank), the resulting posterior will be proper as soon as there is enough information in the data, see Besag et al. (1995).

In Section 6.2, the smoothing hyperparameter λ was chosen by a grid search to find the minimal AIC. However, the subsequent inference did not take into account the uncertainty in estimating λ. In the Bayesian approach, λ can be introduced as an additional model parameter which is assigned a prior distribution. Its selection as well as the evaluation of the uncertainty of this selection is then implicitly included in the joint posterior distribution of all model parameters.

It is seen from (12.13) that the prior distribution for λ determines the variability of the transformed weights $\boldsymbol{a}$. Hence, classical prior choices for this type of parameters can be exploited. As for the scale parameter τ, we consider either a uniform prior on interval $(0,\, h_{\lambda,2})$ for $1/\sqrt{\lambda}$ or a gamma prior with shape and rate parameters $h_{\lambda,1}$ and $h_{\lambda,2}$ for λ.

Model intercept and scale parameters

In Section 6.2 we imposed two more constraints on the mixture weights (Expressions (6.14) and (6.15)) which ensured that the mean and the variance of the standardized error term $\varepsilon^* = \tau^{-1}\,(\varepsilon - \alpha)$ were equal to zero and one, respectively. Analogous constraints could have been incorporated in the Bayesian model by properly constraining the prior distribution (12.13). Nevertheless, this would significantly complicate subsequent MCMC sampling and hence we do not follow this strategy. Consequently, the first two moments of the error distribution are determined jointly by the shift parameter α, the scale parameter τ and the mixture weights $\boldsymbol{w}$. In the following, in agreement with previous sections, we understand by the model intercept μ and the model scale σ the mean and the standard deviation, respectively, of the error distribution, given by

$$\mu = \alpha + \tau \sum_{k=-K}^{K} w_k\,\mu_k, \qquad \sigma = \tau\sqrt{\sum_{k=-K}^{K} w_k\,\bigl(\mu_k^2 + \sigma_0^2\bigr)}. \tag{12.14}$$

Bayesian estimation

Unlike Section 12.2, the parameter space is now of fixed dimension. Standard MCMC procedures can be used to sample from the posterior distribution. For technical details concerning implementations of the MCMC procedures, we refer to Komárek et al. (2007) and Komárek and Lesaffre (2008). On the other hand, as soon as a sample from the posterior distribution is available, characteristics of the posterior predictive survival distribution as well as the posterior predictive error density are calculated using analogous formulas as those shown in Section 12.2.

Example 12.3 Signal Tandmobiel study

The AFT model used in Examples 6.1, 12.1 and 12.2 which relates distribution of the emergence times of tooth 44 to gender and the DMF status of the primary predecessor will now be refitted while assuming a penalized normal mixture in the error distribution.

Prior distributions

The prior distributions were chosen as proposed above with the following hyperparameter choices leading to a weakly informative prior distribution: $\boldsymbol{\beta}_0 = (0,\,0,\,0)'$, $\boldsymbol{\Sigma}_0 = \mathsf{diag}(100,\,100,\,100)$, $\alpha_0 = 0$, $\sigma_\alpha^2 = 100$. That is, apriori

TABLE 12.3: Signal Tandmobiel study. AFT model for emergence of tooth 44 with a PGM error distribution. Posterior summary statistics for the most important model parameters (AF=acceleration factor) obtained with R package bayesSurv.

Parameter	Posterior mean	SD	95% Equal Tail CI	Contour probability
Intercept (μ)	2.39	0.01	(2.36, 2.41)	**<0.0001**
Gender[girl] (β_1)	−0.0577	0.0151	(−0.0867, −0.0277)	**0.0002**
DMF[1] (β_2)	−0.0834	0.0176	(−0.1169, −0.0483)	**0.0001**
Gender:DMF (β_3)	0.0618	0.0251	(0.0121, 0.1103)	**0.0162**
Scale σ	0.117	0.006	(0.107, 0.131)	
AF(DMF; boys)	0.92	0.02	(0.89, 0.95)	**0.0001**
AF(DMF; girls)	0.98	0.02	(0.95, 1.01)	0.2073

$\beta_j \sim \mathcal{N}(0, 100)$, $j = 1, 2, 3$, $\alpha \sim \mathcal{N}(0, 100)$. Further, a uniform prior on interval (0, 100) ($h_{\tau,2} = 100$) was considered for the scale parameter τ.

Furthermore, third order ($s = 3$) differences were considered in the penalty prior (12.13) for the transformed mixture weights. Finally, the smoothing hyperparameter λ was assigned a gamma prior $\mathcal{G}(1, 0.005)$ ($h_{\lambda,1} = 1$, $h_{\lambda,2} = 0.005$).

Posterior summary statistics

Table 12.3 shows posterior summary statistics for the same set of parameters as for a fully parametric AFT analysis (Weibull model, Table 12.1) and an analysis based on the AFT model with a classical Gaussian mixture in the error distribution (Table 12.2). The results now are based on a single Markov chain of length 20 000 obtained with a thinning factor of 50 after 10 000 ($\times$ 50) burn-in iterations. As compared to the AFT model with a CGM in the error distribution, all regression coefficients are now slightly further away from zero. Nevertheless, all summary statistics including the credible intervals for the acceleration factors are practically the same while using both flexible models (CGM in Table 12.2 and PGM in Table 12.3).

Posterior predictive error distribution

The posterior mean and median of the smoothing hyperparameter λ are equal to 469.9 and 414.6, respectively, with a 95% HPD credible interval of (44.6, 1 023.1). Remember from Table 6.3 that an optimal value of λ found in Section 6.2 is equal to $2.7\,n = 2.7 \cdot 496 = 1\,339.2$, which extends even the upper bound of the credible interval. Hence, it is not surprising that the predictive error distribution whose density is shown by a solid line in Figure 12.4 is somehow less smooth than found in the frequentist analysis (see Figure 6.3). Its bimodal shape seems to suggest an omitted covariate. Figure 12.4 further shows the posterior predictive density based on the CGM model (dotted line) from Section 12.2 and also the CGM predictive density conditioned by $K = 4$ mixture components (dashed line) which closely resembles the fitted PGM density.

Posterior predictive survival distributions

Posterior predictive survival functions in DMF groups for boys are plotted on Figure 12.5 (solid lines) along with the pointwise 95% credible intervals depicted by a gray band. For comparison, dashed lines show the predictive survival functions based on the CGM AFT model from Section 12.2. It is seen that despite some differences between the posterior predictive densities of the error terms there is only a negligible difference between the corresponding predictive survival functions. The results for girls (not shown) are similar.

□

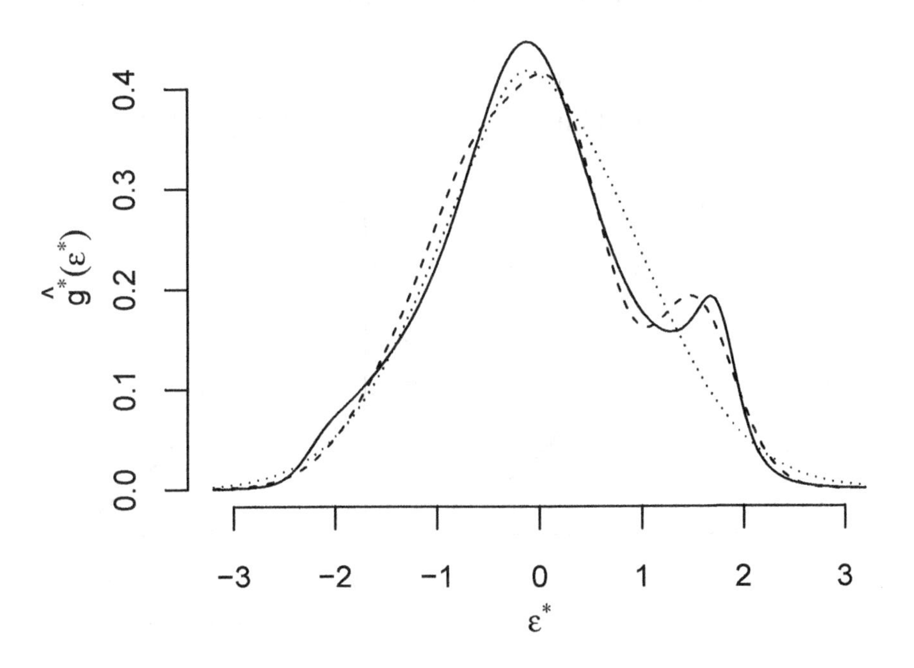

FIGURE 12.4: Signal Tandmobiel study. Posterior predictive density of the standardized error term in the AFT model for emergence of permanent tooth 44 (solid line) compared to a CGM posterior predictive density (dotted line) and a CGM posterior predictive density conditioned by $K = 4$ (dashed line), obtained with the R package bayesSurv.

12.3.1 R solution

Other functions from the R package bayesSurv, namely the functions bayessurvreg2 and bayessurvreg3 implement the MCMC sampling for the AFT model with the penalized Gaussian mixture in the error distribution. Both functions are capable of fitting two different generalizations of the model considered in this section, whereas both lead to identical results for the model used here. In the example below, we use the function bayessurvreg3 while showing the most important aspects. More details can be found in the R script available in the supplementary materials.

Example 12.3 in R

Prior distribution

Initial preparation of data, which results in the data.frame tandDMF84, *proceeds in the same way as in Section 12.2.1. Hence we directly proceed here*

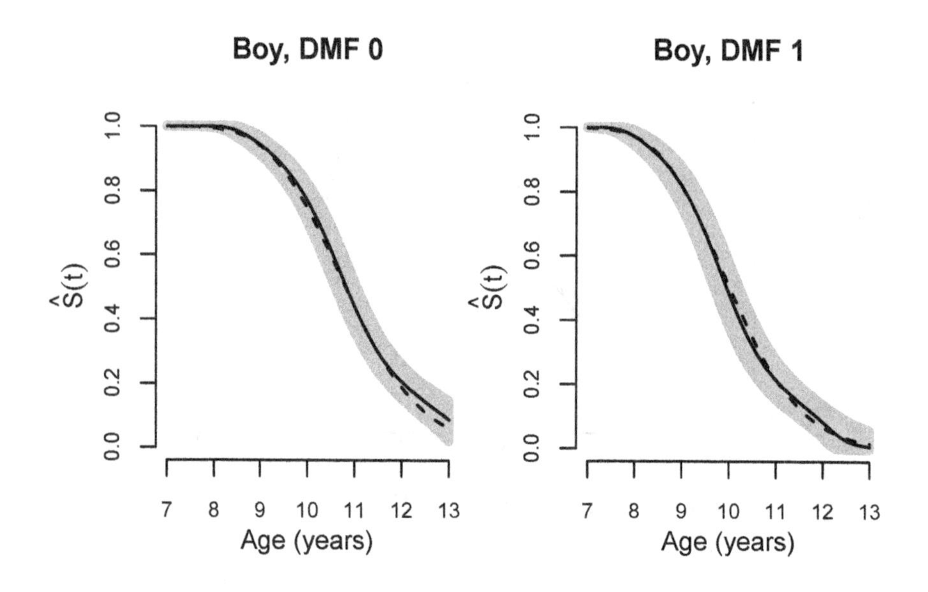

FIGURE 12.5: Signal Tandmobiel study. Posterior predictive survival functions for emergence of permanent tooth 44 in DMF groups for boys based on the AFT model with the PGM error distribution (solid lines) compared the estimates from the AFT model with the CGM error distribution (dashed lines), obtained with the R package bayesSurv. The filled area shows pointwise 95% credible intervals for the values of the survival functions based on the PGM AFT model.

to the specification of the prior hyperparameters stored in the lists prior.PGM *and* prior.beta.

```
>   ### Prior for PGM
> prior.PGM <- list(K = 15,         ## 2*15+1 knots
+   c4delta = 1.5,                  ## delta=1.5*sigma, delta=mu[k+1]-mu[k]
+   order = 3,                      ## GMRF (penalty) of the 3rd order
+   prior.lambda = "gamma",         ## Gamma(1, 0.005) prior on lambda
+     shape.lambda = 1, rate.lambda = 0.005,
+   prior.intercept = "normal", ## N(0, 100) prior on alpha
+     mean.intercept = 0, var.intercept = 100,
+   prior.scale = "sduniform",  ## Unif(0, 100) prior on tau
+     rate.scale = 100)
>
>   ### Prior for beta:
> prior.beta <- list(mean.prior = rep(0, 3),
+                    var.prior = rep(10^2, 3))
```

Posterior simulation

The initial values for the model parameters can be specified analogously to Section 12.2.1. In the following, we assume that they are stored in a list inits. *The MCMC sampling is then provided by the following command:*

```
> simPGM <- bayessurvreg3(
+      Surv(L44, R44, type = "interval2")
+        ~ fGENDER + fDMF_84 + fGENDER:fDMF_84,
+      data = tandDMF84,
+      dir = "/home/BKL/AFT_PGM/",
+      nsimul = list(niter = 30000, nthin = 50, nburn = 10000,
+                    nwrite = 1000),
+      prior = prior.PGM, prior.beta = prior.beta,
+      init = inits,
+      store = list(a = TRUE, y = FALSE, r = FALSE))
```

Note that the meaning of all above arguments of the bayessyrvreg3 *function is analogous to the meaning of the respective arguments of the* bayessurvreg1 *function used in Section 12.2.1. The above command runs in total 30 000 (×50) MCMC iterations. From the last 20 000 (×50) scans, each 50th value is retained for the posterior inference. The sampled values are stored in a form of* txt *files in a directory specified by the argument* dir *(directory* /home/BKL/AFT_PGM/ *here).*

Posterior summary statistics

To read the sampled values back to R*, the following commands (different from those shown in Section 12.2.1) can be used. The code shows reading of the sampled values of the regression coefficients* $\boldsymbol{\beta}$ *and of the moments of the error terms (see Expression 12.14) which are stored in the* data.frame Error.

```
> DIRSIM <- "/home/BKL/AFT_PGM/"
> Beta <- scanFN(paste(DIRSIM, "beta.sim", sep = ""))
> MixMoment <- scanFN(paste(DIRSIM, "mixmoment.sim", sep = ""))
> Error <- data.frame(Intercept  = MixMoment[, "Mean.1"],
+                     Scale      = sqrt(MixMoment[, "D.1.1"]))
```

To get the posterior summary statistics (see Table 12.3), the function give.summary *is available in the* bayesSurv *package (output shown for the regression coefficients* $\boldsymbol{\beta}$*).*

```
> give.summary(Beta)
```

```
        fGENDERgirl fDMF_84caries fGENDERgirl:fDMF_84caries
means       -0.0577       -0.0834                    0.0618
50%         -0.0581       -0.0835                    0.0619
2.5%        -0.0867       -0.1169                    0.0121
97.5%       -0.0277       -0.0483                    0.1103
p.value      0.0002        0.0001                    0.0162
```

Note that the line labeled as p.value *provides the contour probability. The standard* summary *function from the* coda *package can be used as well (output shown for the moments of the error terms).*

```
> summary(mcmc(Error))
```

```
Iterations = 1:20000
Thinning interval = 1
Number of chains = 1
Sample size per chain = 20000

1. Empirical mean and standard deviation for each variable,
   plus standard error of the mean:

            Mean       SD  Naive SE Time-series SE
Intercept 2.3866 0.011456 8.101e-05      0.0004217
Scale     0.1168 0.006201 4.385e-05      0.0003506

2. Quantiles for each variable:
```

(Cont.)

```
              2.5%     25%     50%     75%   97.5%
Intercept 2.3638 2.3790 2.3869 2.3945 2.4088
Scale     0.1068 0.1126 0.1161 0.1199 0.1311
```

Other capabilities of the coda *package can be used for both convergence diagnostics and posterior inference in a usual way.*

Posterior predictive error distribution

To calculate the posterior predictive error density based on the PGM Model (12.12), the function bayesGspline *can be used. For example, the standardized (shifted and scaled to have a zero mean and unity variance) density is calculated as follows.*

```
> stgrideps <- seq(-3.2, 3.2, length = 500)
> sterrDens <- bayesGspline(dir = "/home/BKL/AFT_PGM/",
+     grid1 = stgrideps, version = 32, extens = "",
+     standard = TRUE, nwrite = 1000)
```

The dir *argument specifies a path to the directory with the sampled values of the model parameters, the* grid1 *argument specifies a grid to calculate the density values. Arguments* version *and* extens *must always be set to the values indicated above (i.e.,* version = 32, extens = "" *) if the error density corresponds to the AFT model as described in this section (with other values of those two arguments, posterior predictive densities for other models are calculated). Finally,* standard = TRUE *specifies that a standardized version of the density is to be calculated and the argument* nwrite *determines how often we get informed on the progress of computation. The solid line from Figure 12.4 is then drawn using:*

```
> plot(sterrDens$grid, sterrDens$average, type = "l")
```

Posterior predictive survival distributions

No specific function is available in the bayesSurv *package to calculate the posterior predictive survival or hazard functions. Nevertheless, by using formulas analogous to Expression (12.9), one can calculate all needed quantities using the sampled values of the model parameters. The relevant code used to calculate the posterior predictive survival function including the pointwise credible intervals as shown on Figure 12.5 is available in the supplementary materials.*

□

12.4 Bayesian semiparametric AFT model

Take again the AFT model given by (12.1). In Sections 12.2 and 12.3, the error distribution was specified as a Gaussian mixture. Moreover, the prior distribution of the mixture weights in Section 12.3 could also be considered as a penalty term which ensures a smooth error distribution. In this way, almost a semiparametric approach is realized because, apart from the assumed linear effect of the covariates on the log-scale, the error distribution is only weakly restricted to a parametric family. Such models were analyzed both with a frequentist (Section 6.2) and a Bayesian approach (Section 12.3). In the Bayesian nonparametric (BNP) approach, introduced in Section 9.1.1, the aim is to move further away from a parametric model. As seen in Sections 9.2.2 and 10.3, the Dirichlet Process (DP) prior is a popular choice. In Section 10.3.2, we have also seen the Dirichlet Process Mixture (DPM) generalization to the DP. There are many other generalizations of the basic DP, but here we look at one particular extension and allow for a systematic part. This leads to a *Bayesian semiparametric process (BSP)*.

Recall that the DP prior is based on an initial estimate of a parametric baseline (also referred to as a "template") survival function S^*. This template could be determined in a classical (frequentist) way. The deviation of the actual survival distribution from the template is governed by a parameter c. The smaller c, the more the survival curve is allowed to vary around the parametric template. Instead of fixing the template survival function, one could consider a parametric family of templates and give their parameters also a prior distribution. This leads to the *Mixture of Dirichlet processes* (MDP) prior. Formally, if T_0 is an event time random variable, we say that it follows the MDP if

$$\begin{aligned} T_0 \mid S_0 &\sim S_0, \\ S_0 \mid c, \boldsymbol{\theta} &\sim \mathcal{DP}(c\, S_{\boldsymbol{\theta}}), \\ \boldsymbol{\theta} &\sim p(\boldsymbol{\theta}), \end{aligned} \tag{12.15}$$

where $S_{\boldsymbol{\theta}}$ is a survival distribution coming from a specific parametric family of absolutely continuous survival distributions being indexed by a parameter (vector) $\boldsymbol{\theta}$ and $p(\boldsymbol{\theta})$ representing its prior distribution. Observe the difference with the DPM prior described in Section 10.3.2.

Here we consider the MDP prior applied to an AFT model. A popular MCMC implementation of the MDP prior for an AFT model was suggested by Hanson and Johnson (2004) and is the basis for the R function DPsurvint of the R package DPpackage developed by Jara (2007), see also Jara et al. (2011). Below we largely adopt the description of the MDP prior for an AFT as given in the manual of the DPpackage.

As usual, let the interval-censored survival time related to the covariate vector $\boldsymbol{X}$ be denoted as T. We now write the AFT Model (12.1) as

$$T = \exp(\boldsymbol{X}'\boldsymbol{\beta})\, T_0, \tag{12.16}$$

where $T_0 = \exp(\varepsilon)$ is a random variable representing the baseline survival distribution. Be aware of the fact that in the R package DPpackage, the regression coefficients are assumed to have opposite signs. We will note the difference in the practical analysis, nevertheless, for consistency with the previous sections, we explain the concepts using the AFT parametrization (12.16) used throughout the book.

We now assume that T_0 follows the MDP prior (12.15) where $S_{\boldsymbol{\theta}}(\cdot)$ is a reasonable (template) parametric baseline survival distribution. In Section 10.3, $\boldsymbol{\theta}$ would have been obtained from, say a parametric (classical) AFT analysis yielding a preliminary estimate for $\boldsymbol{\theta}$ and the template survival distribution for the Dirichlet prior. In contrast, here the model parameters $\boldsymbol{\theta}$ are given a prior together with the regression coefficients which leads to the MDP prior. In addition, the dispersion parameter c can be fixed or can also be given a prior distribution.

As before, assume n independent event times T_i $(i = 1, \ldots, n)$ whose relationship to the covariate vectors $\boldsymbol{X}_i$ is given by the AFT Model (12.16). At the same time, the observed data are intervals $\lfloor l_i,\, u_i \rfloor$ that contain the true survival times T_i $(i = 1, \ldots, n)$. Sampling from the posterior distribution based on the above BSP is not straightforward. In Hanson and Johnson (2004) two sampling algorithms are described. The first algorithm is based on only Gibbs sampling, while for the second algorithm Gibbs sampling is combined with a Metropolis-Hastings step. The R function DPsurvint is based on the second algorithm which makes use of the *stick-breaking process*, introduced in Section 9.2.2. The result of the sampling algorithm is an estimate for the survival function in the spirit of Section 9.1.1.

One is also interested in the posterior predictive distribution of the survival times. Remember from (12.2) that given the AFT Model (12.16) where the baseline survival function is S_0, the survival function at time t of a subject with a given covariate vector $\boldsymbol{X}_{pred}$ is $S(t \,|\, \boldsymbol{X}_{pred}) = S_0\Big(\exp(-\boldsymbol{X}'_{pred}\boldsymbol{\beta})\, t\Big)$. Given converged Markov chains $\boldsymbol{\theta}^{(m)}$, $\boldsymbol{\beta}^{(m)}$, $T_i^{(m)}$ $(i = 1, \ldots, n,\ m = 1, \ldots, M)$, where $T_i^{(m)}$ denotes Bayesian augmented event times, an estimate of the predictive survival function for the covariate vector $\boldsymbol{X}_{pred}$ at time t is

$$\begin{aligned}\widehat{S}_{pred}(t) = \frac{1}{M}\sum_{m=1}^{M}\Bigg\{ & c\, S_{\boldsymbol{\theta}^{(m)}}\Big(\exp(-\boldsymbol{X}'_{pred}\boldsymbol{\beta}^{(m)})\, t\Big) \\ & + \frac{1}{c+n}\sum_{i=1}^{n}\Delta_{T_i^{(m)}}(t)\Bigg\},\end{aligned}$$

where $\Delta_{T_i^{(m)}}(t)$ equals one when $T_i^{(m)} > t$ and zero otherwise. The derivation of the above results is, however, quite technical and it is beyond the scope of this book to further elaborate on this. We refer to Hanson and Johnson (2004) for more technical details.

Example 12.4 ***Signal Tandmobiel study***
The data set analyzed in Examples 6.1 and 12.1 – 12.3 is fit here with a semiparametric Bayesian AFT model. Recall that we wish to examine the dependence of the distribution of the emergence time T of the permanent tooth 44 on the gender of a child and the history of caries experience of the deciduous tooth 84 expressed by its dichotomized DMF score. The covariate vector used in the above AFT models is $\boldsymbol{X} = (X_1, X_2, X_3)'$, where now X_1 equals 1 for girls and 0 for boys, X_2 is a dichotomized DMF score where a value of 1 means that the primary predecessor was decayed, missing due to caries or filled and $X_3 = X_1 \cdot X_2$ is the gender by DMF interaction covariate.

Prior distributions

The semiparametric approach assumes a reasonable parametric baseline survival distribution $S_{\boldsymbol{\theta}}$ as template. In the DPsurvint implementation, the log-normal baseline $S_{\boldsymbol{\theta}} \equiv \mathcal{LN}$ is assumed. The R parametrization is considered with $\boldsymbol{\theta} = (\mu, \sigma)'$, where μ and σ, respectively, are parameters called meanlog and sdlog, respectively, in Table B.1. With the approach of Section 10.3, estimates for μ and σ would be obtained from, say, a parametric (classical) AFT analysis and the resulting log-normal distribution would be used as a template in the DP prior. In contrast here, the model parameters μ and σ themselves are given a prior. In DPsurvint the following choices are made:

$$\begin{aligned} \mu &\sim \mathcal{N}(m_0, s_0), \\ \sigma^{-2} &\sim \mathcal{G}(\tau_1/2, \tau_2/2), \end{aligned} \tag{12.17}$$

with m_0, s_0, τ_1 and τ_2 being the fixed hyperparameters. In our analysis, we use $m_0 = 2.3$ (which is motivated by results of previous analyses), $s_0 = 10^2$, $\tau_1 = \tau_2 = 0.1$ (leading to vague prior distributions).

For the dispersion parameter c, DPsurvint allows for a fixed prespecified value or a gamma prior, i.e., $c \sim \mathcal{G}(a_0, b_0)$ with again a_0, b_0 being the fixed hyperparameters. In this analysis, several fixed values for c were applied: 1, 5, 10 and 20.

Finally, a normal prior is assumed for the regression coefficients, i.e.,

$$\boldsymbol{\beta} \sim \mathcal{N}(\boldsymbol{\beta}_0, \boldsymbol{\Sigma}_0), \tag{12.18}$$

where $\boldsymbol{\beta}_0$ and $\boldsymbol{\Sigma}_0$ are fixed hyperparameters. For our application, we have chosen $\boldsymbol{\beta}_0 = (0, 0, 0)'$ and $\boldsymbol{\Sigma}_0 = \mathsf{diag}(10^2, 10^2, 10^2)$, again leading to vague prior distribution.

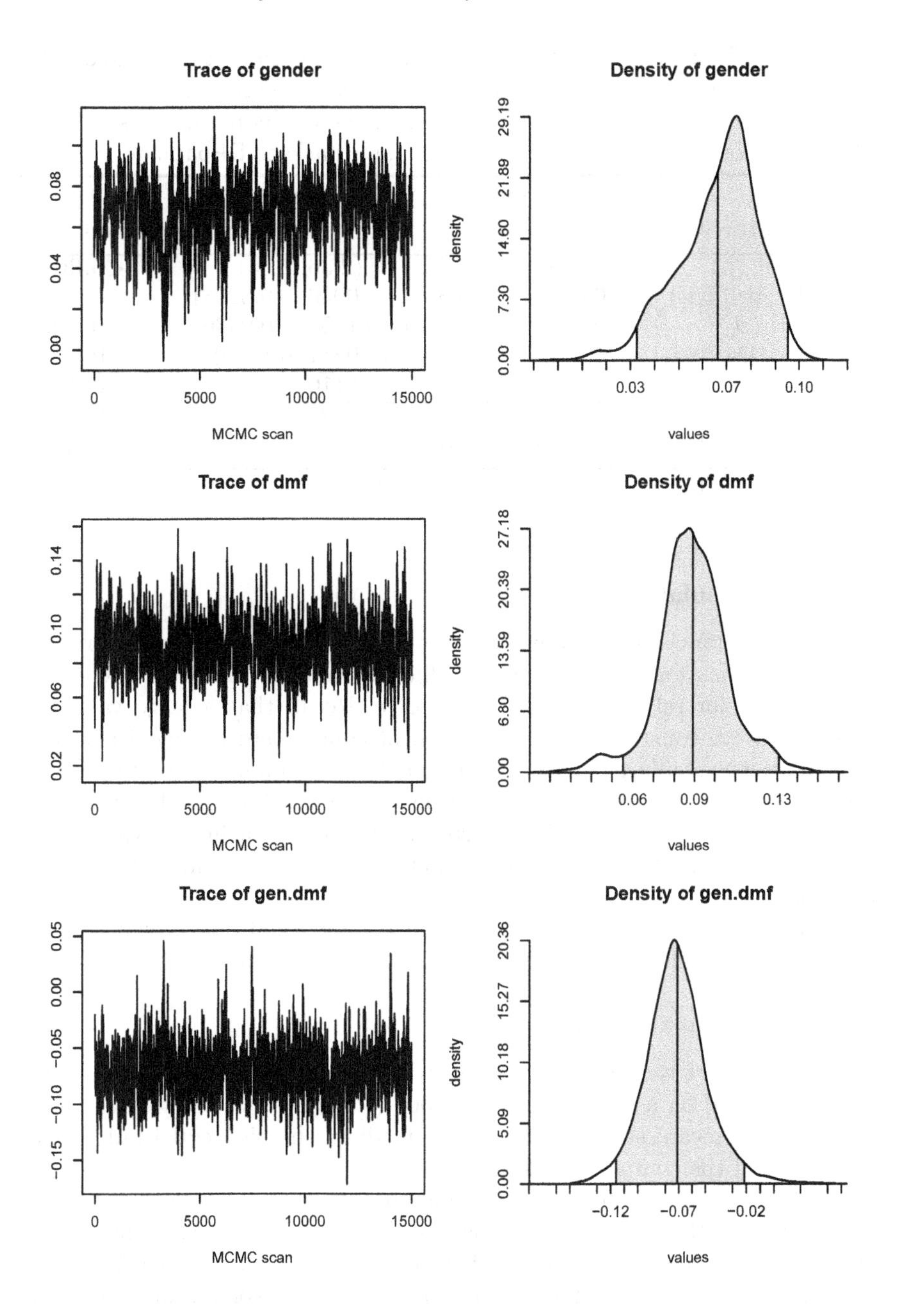

FIGURE 12.6: Signal Tandmobiel study. Semiparametric Bayesian AFT model trace plots and marginal posterior densities for the regression coefficients β_1 (main gender effect), β_2 (main DMF effect) and β_3 (gender by DMF interaction) to predict emergence of permanent tooth 44, obtained with the R package DPpackage.

TABLE 12.4: Signal Tandmobiel study. Semiparametric Bayesian AFT model for emergence of tooth 44 with log-normal baseline survival distribution. Posterior summary measures of the model parameters (AF=acceleration factor), obtained with the R package DPpackage.

	Posterior		**95% Equal Tail**	**Contour**
Parameter	**mean**	**SD**	**CI**	**probability**
Intercept (μ)	2.40	0.02	(2.36, 2.45)	**<0.0001**
Gender[girl] (β_1)	−0.0664	0.0168	(−0.0936, −0.0298)	**0.0012**
DMF[1] (β_2)	−0.0892	0.0172	(−0.1258, −0.0476)	**<0.0001**
Gender:DMF (β_3)	0.0711	0.0232	(0.0211, 0.1158)	**0.0145**
Scale σ	0.143	0.019	(0.110, 0.184)	**<0.0001**
AF(DMF; boys)	0.91	0.02	(0.88, 0.95)	**<0.0001**
AF(DMF; girls)	0.98	0.01	(0.95, 1.01)	0.1776

Posterior simulation

For small values of c, convergence was notoriously slow with very high autocorrelations for the regression coefficients and μ. The convergence was rather quick for relatively large values of c. Recall that a large value of c means that we constraint S close to $S_{\boldsymbol{\theta}}$, of course here letting $\boldsymbol{\theta}$ also vary. The preference of a large c value, likely points to an unidentifiability problem. This problem in the data in combination with a BNP approach was already alluded to in Section 10.3.2. For $c = 20$ convergence was achieved with burn-in = 15 000 values and 15 000 saved chain elements with a thinning factor of 10 (10% of the chain is retained).

In Figure 12.6 we show the trace plot and marginal posterior density for the regression coefficients $\boldsymbol{\beta}$ obtained from a plot statement.

Posterior summary statistics

In Table 12.4 the posterior summary measures of the model parameters are given based on a single chain of 15 000 ($\times$ 10). We calculated, as before, the posterior mean, standard deviation, median, the 95% equal-tail credible intervals and the contour probabilities.

The regression coefficients are similar but greater in magnitude as those obtained with the approaches of Sections 12.2 and 12.3. Now the contour probability for the interaction term is below the "magical" 0.05 level. Also the posterior mean of the AFT scale parameter σ is now somewhat higher than obtained before. As before we computed the acceleration factors for boys and girls. Apparently, the acceleration factor for the boys is a bit smaller and that for the girls a bit higher than obtained before.

Posterior predictive error distribution

The distribution of the unknown baseline event times $T_{0i} = \exp(-\boldsymbol{X}_i'\beta)\, T_i$, $(i = 1, \ldots, n)$, can be obtained from their posterior means computed from a converged Markov chain. The histogram of these "error" latent emergence times is shown in Figure 12.7. Overlaid are two lines: the solid line is a kernel density estimate and the dashed line corresponds to a log-normal distribution whose parameters are obtained from the posterior mean of μ and posterior median of σ. The closeness of the smooth line with the posterior estimate of the survival function indicates that the choice of a log-normal template was reasonable. The closeness of the lines of course also depends on the choice of the bandwidth in the kernel density estimator. In Figure 12.7 we have chosen for a bandwidth equal to five times the standard bandwidth (adjust = 5 in the R function density). When sampling μ and σ from their joint posterior, considerable variability of the estimated log-normal density is seen (plot not shown). Note that the estimated distribution extends here to 16 – 17 years of age. For an individual child, this estimated age must be multiplied with the acceleration or deceleration effect of the covariates gender and DMF.

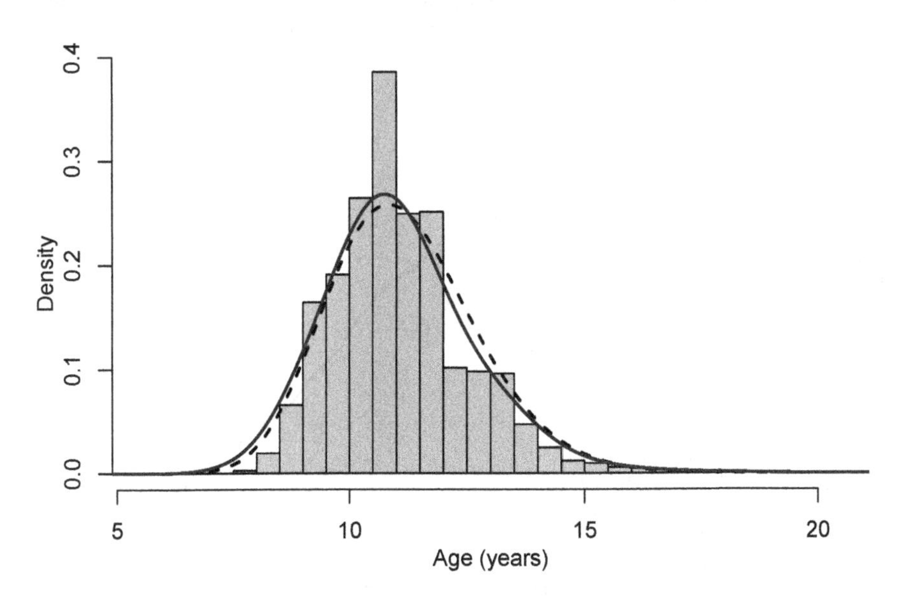

FIGURE 12.7: Signal Tandmobiel study. Semiparametric Bayesian AFT model for emergence of tooth 44 with log-normal baseline survival distribution. Histogram of latent ε_i, together with a kernel density estimate (solid line) and the posterior estimate of the "template" log-normal density (dashed line), obtained with the R package DPpackage.

Posterior predictive survival distributions

The R function predict.DPsurvint produces the posterior predictive survival function and stores it in an R object for a prespecified grid of emergence times. The plot function applied to that R object allows for separate survival plots for each of the covariate combinations, or a figure with the different survival functions overlaid as shown in Figure 12.8. The plot lacks, however, detail. Unfortunately changing the graphical control parameters does not have much effect on the quality of the plot. Alternatively, the user can produce customized plots using basic plot statements directly on the delivered R object.

□

12.4.1 R solution

As already mentioned earlier, the semiparametric Bayesian estimate for an AFT model is implemented in the function DPsurvint from the R package DPpackage. This package provides many R functions to estimate parameters in a semiparametric Bayesian way for a variety of regression models. We refer to the DPpackage manual for a description of the routines of this package. We now show the most relevant commands to produce the results discussed above.

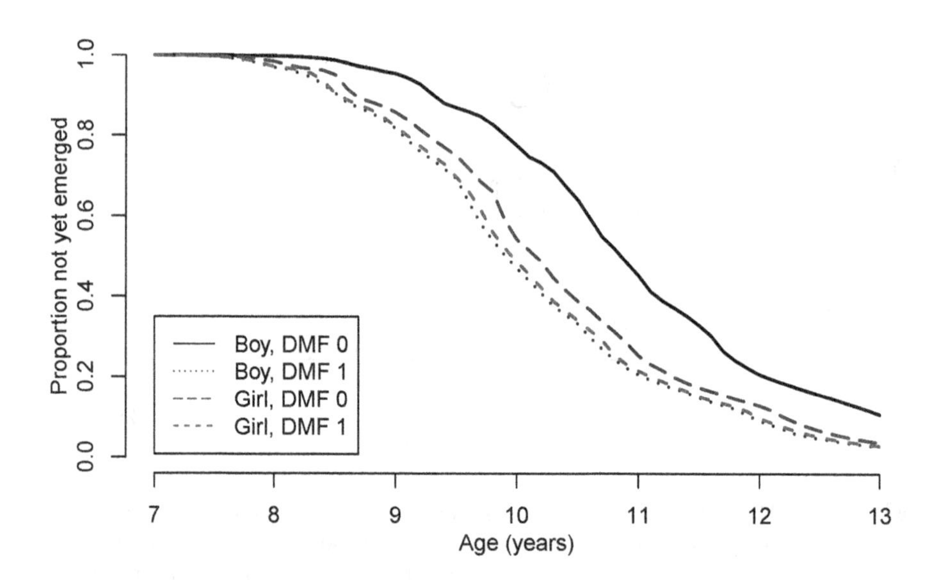

FIGURE 12.8: Signal Tandmobiel study. Survival functions for emergence of permanent tooth 44 in gender by DMF groups based on a semiparametric Bayesian model with MDP priors, obtained with the R package DPpackage.

Example 12.4 in R

We start by loading the data and creating a data frame containing only those children with nonmissing values for the covariates included in the model. As usual, we load needed packages DPpackage, icensBKL *and* coda. *Next, the emergence data set needs to be adapted for* DPsurvint. *To this end, we create a two-column matrix* t.lims *containing the limits of observed intervals. The lower limits must be set to zero for left-censored observations. For the upper endpoints of the intervals corresponding to right-censored observations, a value of* -999 *must be used.*

To fit the AFT model to the emergence data, we further need to specify the MCMC parameters provided as elements of an R list mcmc *which will be used later when calling the function* DPsurvint.

```
> mcmc <- list(nburn = 15000, nsave = 15000, nskip = 10,
+              ndisplay = 1000, tune = 0.25)
```

Above, we specified the thinning factor nskip, *here 10, which implies that only 10% of the generated chain values will be retained. The number of burn-in iterations is provided by* nburn *and the number of extra chain values to be kept for further processing by* nsave. *Both* nburn *and* nsave *need to be multiplied with* nskip *to know the exact number of generated chain values. With* ndisplay *we indicate after how many iterations we wish to obtain a screen message of the sampling progress. Then, we choose the prior distributions by specifying the hyperparameters in a* list prior.

```
> prior <- list(alpha = 20,
+               m0 = 2.3, s0 = 10^2, tau1 = 0.1, tau2 = 0.1,
+               beta0 = rep(0, 3), Sbeta0 = diag(10^2, 3))
```

The dispersion parameter c *is specified as the* alpha *element of this list, here* $c = 20$ *(*alpha $= 20$*) is chosen. Elements* m0, s0, tau1 *and* tau2 *specify the corresponding hyperparameters of the prior distribution (12.17). The prior mean* $\boldsymbol{\beta}_0$ *and the covariance matrix* $\boldsymbol{\Sigma}_0$ *from the prior (12.18) are specified by elements* beta0 *and* Sbeta0. *Nevertheless, be aware of the fact that* beta0 *and* Sbeta0 *specify, in fact, a prior for* $(-\boldsymbol{\beta})$ *since the signs of the regression coefficients are switched in the definition of the AFT model in* DPpackage. *On the other hand, with our normal prior which has a mean* $\boldsymbol{\beta}_0 = (0, 0, 0)'$ *the prior is indeed the same for both* $\boldsymbol{\beta}$ *and* $(-\boldsymbol{\beta})$.

To start sampling, a call to DPsurvint *is initiated with the first parameter a model formula (*t.lims *is a two-column matrix with the limits of the observed intervals created as described above, objects* gender, dmf *and* gen.dmf *are vectors with the values of the three covariates) followed by the* prior *and* mcmc *objects. The parameter* status *must be equal to* TRUE *for a new run, and* FALSE *for a continuation of a previous run. In the latter case,* state *must contain the final sampled values of the parameters.*

```
> fit1 <- DPsurvint(t.lims ~ gender + dmf + gen.dmf, prior = prior,
+                   mcmc = mcmc, state = NULL, status = TRUE)
> summary(fit1)
> plot(fit1, ask = FALSE)
> anova(fit1)
```

Posterior summary measures can be obtained by the summary *function applied to the* fit1 *object. The results were reported in Table 12.4 (remember that the signs of the regression coefficients had to be switched as compared to the* R *output). Note that the acceptance rate for the Metropolis step equals 0.36. The output also shows the average number of clusters (*ncluster*, equal to 71 here), which represents the number of discrete points in the stick-breaking representation of the DP prior, which varies with iteration in the algorithm of Hanson and Johnson (2004). In the plot statement, we ask for traceplots and marginal posterior densities of the model parameters* $\beta_1, \beta_2, \beta_3, \mu, \sigma^2$ *and* ncluster*. In Figure 12.6 the first of three plots are shown. The function* anova *computes contour probabilities to evaluate the "significance" of each of the regression parameters.*

To obtain the sampled values of the model parameters $\boldsymbol{\beta}$*,* μ*,* σ^2*, you need the first command line below, while the sampled values of* T_{0i}*,* $(i = 1, \ldots, n)$ *are obtained by the second line.*

```
> fit1$save.state$thetasave
> fit1$save.state$randsave
```

In particular, the sampled values of the regression coefficients β_1*,* β_2*,* β_3 *are available using the following commands (again, note a change in sign):*

```
> beta1 <- mcmc(-fit1$save.state$thetasave[, "gender"])
> beta2 <- mcmc(-fit1$save.state$thetasave[, "dmf"])
> beta3 <- mcmc(-fit1$save.state$thetasave[, "gen.dmf"])
```

We can also calculate the sampled values of the acceleration factors due to DMF = 1 for boys and girls and store them as mcmc coda *objects using:*

```
> AF1 <- mcmc(data.frame(Boys = exp(as.numeric(beta2)),
+                        Girls = exp(as.numeric(beta2 + beta3))))
```

Posterior summary statistics are then easily obtained using:

```
> summary(AF1)
```

Finally, The R *function* predict *provides the estimated posterior predictive survival functions at a user-specified grid of time points and for given covariate combinations. The* plot *function can then be used to plot these estimates. Below we show two versions of the code towards a plot. In the first command* *(output not shown), we ask for a separate plot for each of the four covariate combinations (*all = FALSE*) and the 95% pointwise credible intervals. In the second command, we ask to superimpose the four survival functions. The plot is similar to the one shown in Figure 12.8.*

```
> xnew  <- matrix(rbind(c(0, 0, 0), c(0, 1, 0),
+                       c(1, 0, 0), c(1, 1, 1)), nrow = 4, ncol = 3)
> rownames(xnew) <- c("Boy (DMF = 0)", "Boy (DMF = 1)",
+                     "Girl (DMF = 0)", "Girl (DMF = 1)")
> colnames(xnew) <- c("gender", "dmf", "gen.dmf")
> grid  <- seq(7, 13, 0.1)
> pred1 <- predict(fit1, xnew = xnew, grid = grid)
> plot(pred1, all = FALSE, band = TRUE, xlim = c(7, 13))
> plot(pred1, xlim = c(7, 13), lwd = 3, col = 1:4)
```

In the accompanied R *script we have also provided additional commands to process the chains produced by* DPsurvint *or to run two parallel chains.*

□

12.5 Concluding remarks

The parametric AFT is fairly trivial to fit once we know how to fit interval-censored data with the software. The addition of the systematic part to the model comes in naturally and poses no difficulties. Things are different when moving away from pure parametric assumptions. This was clearly demonstrated here in the smooth and nonparametric Bayesian approaches. Especially, the BSP approach can be notoriously difficult to fit to the interval-censored data, especially in the presence of a large proportion of right-censored survival times. Note that, at the time of publication of the book, basically only one package can fit such models. The package DPpackage provides a great variety of BNP and BSP approaches to fit all kinds of data. Some functions, as DPsurvint, allow for interval-censored data. The function LDTFPsurvival also allows for interval-censored data using linear dependent tailfree processes, but its treatment is omitted here in this book.

Chapter 13

Additional topics

In this chapter, we treat extensions of the classical PH and AFT regression models but in a Bayesian context. We start in Section 13.1 with the analysis of clustered interval-censored survival times using shared frailty models. Frailty models were introduced in a frequentist context in Chapter 8. We treat parametric and flexible shared frailty models, but as far as we know the semiparametric version has not been pursued yet in a Bayesian context.

In Section 13.2 we review the possible statistical approaches to analyze multivariate interval-censored survival times. Shared frailty models may be an option for this, and this is illustrated here on emergence times recorded in the Signal Tandmobiel study. However, the frailty approach assumes conditional independence of the survival times given the subject and provides regression estimates that have a subject-specific interpretation. For a more complex association structure and a marginal interpretation of the regression coefficients, a general multivariate approach is required. Software to fit multivariate interval-censored survival times is, however, basically nonexisting for dimension $d > 2$. Therefore, the bivariate case is primarily discussed in Section 13.2. We consider parametric and flexible Bayesian approaches including copula models. Software for the semiparametric approach with the R package DPpackage can also be used using a trick, and this will be illustrated here. We end this section with a parametric analysis of a multivariate problem ($d = 7$) on the ranking of emergence times of permanent teeth.

The particular case of doubly interval-censored data is treated in Section 13.3. Both the univariate as well as the multivariate case will be considered. In the univariate case there is essentially one time line, e.g., HIV followed by AIDS, emergence of a tooth followed by caries on the tooth, etc. For multivariate interval-censored survival times, there are several parallel time lines. We illustrate how univariate DI-censored data can be analyzed using JAGS. In the same section, we illustrate how a flexible analysis can be done with the R package bayesSurv and a semiparametric analysis with the R package DPpackage. For the multivariate case, two R packages can handle this: the package bayesSurv for a flexible analysis and the package DPpackage for a semiparametric approach. Since these analyses are relatively easy extensions of the univariate case, we refer to the publications and the manuals of the packages for illustrations.

13.1 Hierarchical models

In Section 8.2.1 we have introduced the (shared) frailty model and the use of frequentist software. Some of the frequentist analyses of Section 8.2.1 will now be replayed with JAGS, R and SAS software. Parametric shared frailty models will be considered first, followed by more flexible frailty models.

13.1.1 Parametric shared frailty models

In the parametric case a particular baseline survival distribution and frailty distribution are chosen. The expressions for the AFT and PH shared frailty models were given in Chapter 8, but are repeated here for completeness.

For the PH frailty model the frailty part of the model acts multiplicatively on the hazard function. Namely for the jth subject in the ith cluster it is assumed that the hazard function at time $t > 0$ is given by:

$$\hbar_{ij}(t \mid \boldsymbol{X}_{ij}, Z_i) = Z_i \, \hbar_0(t) \exp(\boldsymbol{X}_{ij}'\boldsymbol{\beta}^P) \qquad (i = 1, \ldots, n;\ j = 1, \ldots, m_i), \quad (13.1)$$

with $\hbar_0(t)$ the baseline hazard, $\boldsymbol{X}_{ij} = (X_{ij,1}, \ldots, X_{ij,d})'$ the covariates for the ith subject in jth cluster, $\boldsymbol{\beta}^P = (\beta_1^P, \ldots, \beta_d^P)'$ the vector of fixed effects regression coefficients and Z_i the frailty term which follows a distribution g_Z with a fixed location parameter.

The AFT frailty model for the true, yet possibly interval-censored event time T_{ij} of the jth subject in the ith cluster is given by:

$$\log(T_{ij}) = \mu + b_i + \boldsymbol{X}_{ij}'\boldsymbol{\beta}^A + \sigma \, \varepsilon_{ij}^* \quad (i = 1, \ldots, n;\ j = 1, \ldots, m_i), \qquad (13.2)$$

with the same meaning of $\boldsymbol{X}_{ij}$ and $\boldsymbol{\beta}^A = (\beta_1^A, \ldots, \beta_d^A)'$ again a vector of fixed effects regression coefficients. Further, b_i is the random intercept for cluster i with density g_b with a fixed mean, e.g., a normal density with zero mean. The standardized error term ε^* has density g_ε^* with fixed both location and scale parameters, often a standard normal or logistic density.

Recall that the interpretation of a regression coefficient for an AFT model is opposite to that of a PH model. Namely, a positive β_r^P ($r = 1, \ldots, d$) implies that the median survival is shortened with increasing values of the regressor $X_{ij,r}$, while the opposite is true for β_r^A. Also, for both frailty models the meaning of a regression coefficient is conditional on the other regressors but also on the random part but in a different manner, i.e., β_r^P expresses the impact of $X_{ij,r}$ but conditional on the multiplicatively acting frailty Z_i while the interpretation of β_r^P is conditional on the additively acting random intercept b_i. This is referred to the *subject-specific* interpretation of the regression coefficient. In the remainder of this chapter, we will denote the vector of regression coefficients as $\boldsymbol{\beta}$ for both the PH and the AFT model.

TABLE 13.1: Mastitis study. Bayesian shared frailty models: PH Weibull frailty models with log-normal and gamma frailties and AFT model with normal errors as well as random intercept, posterior means and SD obtained from runjags. N.A.= Not Applicable

Model	PH				AFT	
	Weibull baseline hazard				Normal errors	
Frailty	Log-normal		Gamma		Normal random intercept	
	Mean	SD	Mean	SD	Mean	SD
μ	N.A.	N.A.	N.A.	N.A.	0.433	0.144
β_{rear}	0.176	0.123	0.181	0.123	−0.151	0.071
β_{par24}	0.037	0.388	−0.196	0.333	−0.005	0.190
β_{par56}	1.924	0.576	1.465	0.503	−0.777	0.272
γ	2.046	0.117	1.945	0.110	N.A.	N.A.
λ_0	0.246	0.074	0.753	0.200	N.A.	N.A.
α_0	2.050	0.290	1.189	0.157	N.A.	N.A.
θ	2.831	0.616	1.638	0.288	0.617	0.123
σ^2	N.A.	N.A.	N.A.	N.A.	0.416	0.042

We now illustrate the use of JAGS and SAS software to fit parametric shared frailty PH and AFT models to the mastitis data.

Example 13.1 Mastitis study

As in Example 8.2, we check the impact of parity (1, 2 to 4 and > 4 calvings) and the position of the udder quarter (front or rear) on the time to inflammation of the udder by parametric shared frailty models. In the PH case, we assumed a Weibull baseline hazard $\hbar_0$ given by Expression (11.2) with the shape parameter γ and parameter λ_0 (BUGS parametrization is considered here, see Appendix B.9). Remember from Section 11.1 that $\beta_0 = \log(\lambda_0)$ can also be interpreted as the model intercept and $\alpha_0 = \exp\big(-\beta_0/\gamma\big)$ is the scale parameter (see Appendix B.3) of the baseline Weibull distribution.

The Weibull baseline hazard was combined with a log-normal ($\log(Z_i) \sim \mathcal{N}(0, \theta)$) and a gamma ($Z_i \sim \mathcal{G}(\theta^{-1}, \theta^{-1})$) frailty distribution, respectively. Table 13.1 shows the posterior mean and standard deviation for the model parameters while assuming vague prior distributions. Note that for both frailty choices, parameter θ refers to the variance of the corresponding (log-)frailty term. By comparison with Table 8.2, we see that basically the same results were obtained as with frequentist methods.

For AFT Model (13.2), we combined a log-normal distribution for the survival times ($\varepsilon^*_{ij} \sim \mathcal{N}(0, 1)$) with a normal random intercept term ($b_i \sim \mathcal{N}(0, \theta)$).

We did some limited model checking on the Weibull PH model with log-normal frailties, i.e., checking the assumed frailty log-normal distribution. This is shown in the LHS of Figure 13.1. In the RHS of that figure we provide an estimate of the survival distribution of the true but latent survival times. More extensive checking of the assumed model is possible via posterior predictive checking, but was not done here.

□

13.1.1.1 JAGS solution

To fit parametric frailty models, JAGS in combination with the R package runjags can be used in basically the same way as was shown earlier in Sections 11.1.1 and 12.1.1 in case of parametric PH and AFT model, respectively, without frailties. It is only necessary to extend the model description accordingly as we indicate below.

Example 13.1 in JAGS

Scripts to fit the three considered models using JAGS *by the mean of the* R *package* runjags *are available in the supplementary materials. They are in fact only mild modifications of the scripts discussed previously with Examples 11.1 (parametric PH model without frailties) and 12.1 (parametric AFT model without random intercept) and hence we omit the majority of the code here.*

The R *character variable* PHWeibullLN *below defines the PH Weibull model with log-normal frailties.*

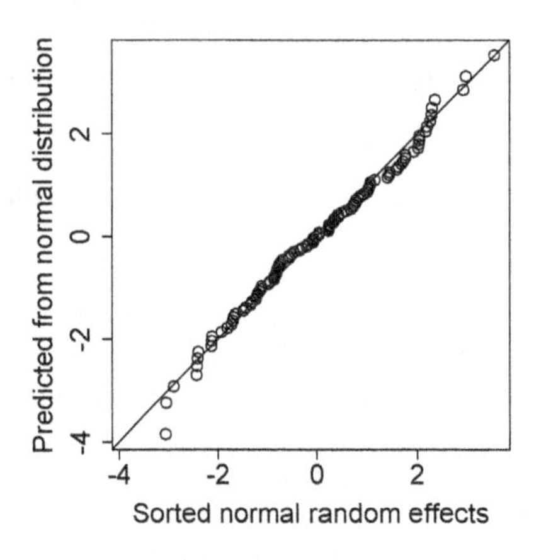

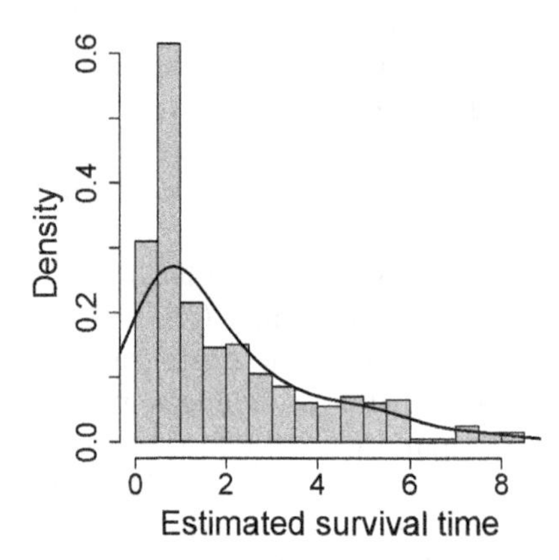

FIGURE 13.1: Mastitis study. Log-normal frailty distribution with Weibull baseline hazard in the PH model: Normal probability plot of posterior mean of $b_i = \log(Z_i)$ (left) and histogram (+ smoothed version) of true but unknown survival times (right), both obtained from runjags.

```
> PHWeibullLN <- "
+   model{
+
+     ### Model for right censored
+     for (i in 1:nr){
+       y[i] ~ dinterval(t[i], lim[i, 1])
+       t[i] ~ dweib(gamma, lambdar[i])
+       lambdar[i] <- exp(b[cow[i]] + beta0 + beta[1]*rear[i]
+                         + beta[2]*par24[i] + beta[3]*par56[i])
+     }
+
+     ### Model for left censored
+     for (i in (nr + 1):(nr + nl)){
+       y[i] ~ dinterval(t[i], lim[i, 2])
+       t[i] ~ dweib(gamma, lambdal[i])
+       lambdal[i] <- exp(b[cow[i]] + beta0 + beta[1]*rear[i]
+                         + beta[2]*par24[i] + beta[3]*par56[i])
+     }
+
+     ### Model for interval censored
+     for (i in (nr + nl + 1):(nr + nl + ni)){
+       y[i] ~ dinterval(t[i], lim[i, ])
+       t[i] ~ dweib(gamma, lambdai[i])
+       lambdai[i] <- exp(b[cow[i]] + beta0 + beta[1]*rear[i]
+                         + beta[2]*par24[i] + beta[3]*par56[i])
+     }
+
+     ### Priors
+     gamma ~ dgamma(1.0, 0.001)
+     beta0 ~ dnorm(0, 0.0001)
+
+     for (i in 1:3){
+       beta[i] ~ dnorm(0, 0.0001)
+     }
+
+     for (i in 1:ncow){
+       b[i] ~ dnorm(0, taub)
+     }
+     sigmab ~ dunif(0, 1000)
```

```
+     taub <- pow(sigmab, -2)
+
+     ### Some derived parameters
+     lambda0 <- exp(beta0)
+     alpha0 <- exp(-beta0 / gamma)
+     theta <- pow(sigmab, 2)
+ }
+ "
```

As was done before, data were divided into three parts that correspond to right-, left- and interval-censored observations. The corresponding numbers of observations are given by the values of the variables nr, nl *and* ni, *respectively, while the limits of observed intervals are supplied in a two-column matrix* lim. *Variables* y *and* t *represent censoring and latent event times, respectively. The covariate values are supplied in vectors* rear, par24 *and* par56, *respectively. The code above also shows that the model parameters were given the following priors:* $\gamma \sim \mathcal{G}(1,\ 0.001)$, $\beta_0 \sim \mathcal{N}(0,\ 10^4)$, $\beta_r \sim \mathcal{N}(0,\ 10^4)$ $(r = 1,\ 2,\ 3)$, $\sigma_b := \sqrt{\theta} \sim \mathcal{U}(0,\ 1\,000)$.

We then call JAGS *via the* run.jags *command. Sampling is again done in three steps: an adaptive phase, the burn-in part and the actual sampling part. The object* jagsPHWeibLN *is based on three chains. The adaptive phase consists of 1 000 iterations, the next 15 000 iterations belong to the burn-in part and finally 30 000 iterations provide the estimated model parameters. Note that data provided in the object* data.mastitis *as well as the initial values for the model parameters (*inits.mastitis*) were prepared in advance in* R *(commands not shown here).*

```
> parameters.mastitis <- c("beta0", "beta", "gamma", "sigmab",
+                          "lambda0", "alpha0", "theta")
> jagsPHWeibLN <- run.jags(model = PHWeibullLN,
+     monitor = parameters.mastitis,
+     data = data.mastitis, inits = inits.mastitis, adapt = 1000,
+     n.chains = 3, n.sims = 3, thin = 1, method = "parallel",
+     burnin = 15000, sample = 30000)
```

Standard diagnostics (not shown) confirmed convergence of all parameters. To estimate (predict) the random effect values b_i *as well as the values of latent event times* T_{ij} $(i = 1, \ldots, n;\ j = 1,\ \ldots,\ m_i)$, *both needed to prepare Figure 13.1, we invoked the function* extend.jags *for an extra 2 000 iterations (1 000 for the adaptive phase and 1 000 iterations for estimation).*

```
> parameters.mastitis2 <- c("b", "t")
> jagsPHWeibLN2 <- extend.jags(jagsPHWeibLN,
```

```
+    add.monitor = parameters.mastitis2,
+    drop.monitor = parameters.mastitis,
+    sample = 1000)
```

The argument add.monitor = parameters.mastitis2 *asks for monitoring a new set of parameters listed in the vector* parameters.mastitis2*, while* drop.monitor = parameters.mastitis *instructs to stop monitoring the original parameters.*

Extra R *commands (not shown here) exploit the structure of the* R *object* jagsPHWeibLN2 *to provide a quick check of the distributional assumption of the random effects on the log-scale for the Weibull log-normal frailty model. A normal probability plot based on the posterior means of the log-frailty terms* $b_i = \log(Z_i)$ *is given in the LHS of Figure 13.1. A histogram (with a kernel density estimate) of the posterior medians of the true survival times is provided in the RHS of Figure 13.1.*

□

JAGS does not provide DIC for (interval-) censored observations. If a choice between frailty models needs to be made, the computation of DIC needs to be done with extra R code outside the JAGS program using the sampled values of the chain. This extra programming part is, however, left to the reader.

13.1.1.2 SAS solution

We now illustrate the analysis of shared frailty models using the SAS procedure MCMC.

Example 13.1 in SAS

We consider again the log-normal and the gamma frailty distribution in combination with a Weibull baseline survival distribution. We refer to Section 8.2.1.2 for the SAS *commands to turn the mastitis data into the correct format.*

Most of the arguments to the MCMC *procedure have been discussed before, except for* MISSING=AC*. With this option* PROC MCMC *neither discards any missing values nor augments them (which means no values will be imputed). The* parms *command line specifies all unknown parameters, and the next three lines provide (vague) prior distributions to these parameters. Subsequent commands specify the (parametric) likelihood contribution for each cow given the random effect. Clearly, here analytical expressions replace data augmentation to compute the contribution of each censored observation to the likelihood. The normal distribution of* $b_i = \log(Z_i)$ *is specified by the* RANDOM *statement.*

```
PROC MCMC DATA=Mastitis SEED=1 NMC=500000 thin=100 MISSING=AC;
parms theta gamma lambda beta1 beta2 beta3;
prior gamma ~ gamma(0.001,is=0.001);
prior theta ~ igamma(0.001,is=0.001);
```

```
prior lambda ~ gamma(0.001,is=0.001);
prior beta1 ~normal(0,var=1e6);
prior beta2 ~normal(0,var=1e6);
prior beta3 ~normal(0,var=1e6);
IF ll>0 THEN
S1 = exp(-exp(b)*lambda*(ll/91.31)**gamma*
          exp(beta1*rear+beta2*par24+beta3*par56));
S2 = exp(-exp(b)*lambda*(ul/91.31)**gamma*
          exp(beta1*rear+beta2*par24+beta3*par56));
     IF event=3 THEN lik=S1-S2;
else IF event=2 THEN lik=1-S2;
else IF event=1 THEN lik=S1;
llik=LOG(lik);
MODEL dummy~GENERAL(llik);  ## Specification of model
RANDOM b~NORMAL(0,var=theta) SUBJECT=cow; ## Lognormal frailty
RUN;
```

Convergence is slower than with runjags*. With 500 000 iterations and keeping 1% (*NMC=500000 thin=100*) convergence was obtained. The thinned chain showed an autocorrelation close to zero at lag 15. The estimated parameters are close to those shown in Table 13.1 and are therefore not shown.*

For the gamma frailty model, 'exp(b)' is replaced by 'b' and the random statement is now:

```
RANDOM  b~GAMMA(shape=1/theta,iscale=1/theta) SUBJECT=cow;
```

With the above statement the variance of b is theta*. We have given* theta *first a gamma distribution and then a uniform distribution as a sensitivity analysis, providing estimates again close to those in Table 13.1. The full programs can again be found in the supplementary materials. Also the* SAS *code for the AFT model with a normal random intercept distribution combined with a normal error distribution is included.*

□

13.1.2 Flexible shared frailty models

The shared frailty models in the previous section assume a particular baseline survival and frailty distribution. Relaxing these assumptions can be achieved by choosing a more flexible baseline survival distribution and/or frailty distribution. Bayesian software for flexible frailty models with interval-censored survival times is scarce, despite the rich statistical literature on smoothing and the many R programs written for flexible frailty models with right-censored survival times. We discuss here two approaches for flexible PH shared frailty models, and one approach for flexible AFT shared frailty models. Currently, it seems that only for the third approach R software is available on CRAN.

In Henschel et al. (2009) a semiparametric PH shared frailty model is proposed with a smooth baseline survival distribution. Smoothness is implemented on the logarithm of the baseline hazard function. Similar as, e.g., in Section 3.3.2, B-splines for $\log(\hbar_0(t))$ are specified on a time grid. A prior distribution on the B-spline coefficients ensures a smooth behavior of the baseline hazard function. This is the Bayesian P-spline approach of Lang and Brezger (2004) relying on the earlier work of Eilers and Marx (1996). Data augmentation allows then for interval-censored survival times. The approach was implemented in the R package survBayes, but currently only an archive and no more maintained version of the package is available on CRAN.

Recently, Çetinyürek Yavuz and Lambert (2016) proposed a semiparametric PH frailty model for clustered interval-censored survival times. The baseline survival distribution is modelled flexibly in combination with a parametric or flexible frailty distribution based on the P-splines approach of Lang and Brezger (2004). The method was applied to the joint distribution of the (left-, right- or interval-censored) emergence times of teeth 14, 15, 24, 25, 34, 35, 44, 45 recorded on a subsample of 300 children of the Signal Tandmobiel study. From a simulation study, the authors concluded that with sufficiently large sample sizes and number of clusters, their approach produces smooth and accurate posterior estimates for the baseline survival function and for the frailty density, and can correctly detect and identify unusual frailty density forms. The R program can be downloaded from the Archives of *Statistical Modelling*, see `http://www.statmod.org/smij/Vol16/Iss5/Yavuz/Abstract.html`. In addition, the simulation study pointed out that the estimation of regression parameters is robust to misspecification of the frailty distribution.

The third approach, proposed in two slightly different contexts by Komárek et al. (2007) and Komárek and Lesaffre (2008), builds on the penalized Gaussian mixture idea, which was introduced in Section 3.3, and further developed in Sections 6.2 and 12.3. To take into account clustering in the data, again the random effects AFT Model (13.2) is considered. For clarity, we will now write it again, nevertheless, in a slightly different way.

Let $\boldsymbol{T}_i = (T_{i1}, \ldots, T_{im_i})'$ be independent random vectors representing times-to-event in the ith cluster which are observed as intervals $\lfloor l_{ij},\, u_{ij} \rfloor$ $(j = 1, \ldots, m_i)$ and $\boldsymbol{X}_{ij}$ be the covariate vector for the jth observation in the ith cluster $(i = 1, \ldots, n;, j = 1, \ldots, m_i)$. The random effects AFT Model (13.2), now written with nonstandardized error terms, assumes that the (i, j)th event time is expressed as

$$\log(T_{ij}) = b_i + \boldsymbol{X}'_{ij}\boldsymbol{\beta} + \varepsilon_{ij} \qquad (i = 1, \ldots, n;\ j = 1, \ldots, m_i), \tag{13.3}$$

where ε_{ij} are (univariately) i.i.d. random errors with a density g_ε and $b_1, \ldots, b_n$ are cluster-specific i.i.d. random effects with a density g_b. As before, $\boldsymbol{\beta}$ is a vector of fixed-effects regression coefficients.

Komárek et al. (2007) consider a more general model where the random intercept terms b_i ($i = 1, \ldots, n$) are replaced by possibly multivariate random effects combined with covariate values in the mood of a linear mixed model of Laird and Ware (1982). On the other hand, only the random intercept AFT Model (13.3) is considered in Komárek and Lesaffre (2008). Nevertheless, the event times are allowed to be doubly interval-censored. In the following, we briefly describe the methodology assuming Model (13.3) for interval-censored data.

To avoid parametric assumptions concerning both the error density g_ε related to the baseline survival distribution and the random intercept density g_b, they can both be expressed as a univariate penalized Gaussian mixture (PGM) defined in Expressions (12.12) and (12.3) with the PGM shift parameter α for g_b fixed to zero for identifiability reasons. Let $\mu_{\varepsilon,k}$ ($k = -K, \ldots, K$) be the PGM means (knots) and $\sigma_{\varepsilon,0}$ the PGM standard deviation related to the model for the density g_ε. Let $\mu_{b,k}$ ($k = -K, \ldots, K$) and $\sigma_{b,0}$ be the analogous parameters of the PGM model for the random intercept density g_b. Further, let $\boldsymbol{\theta} = (\boldsymbol{\beta}', \alpha_\varepsilon, \tau_\varepsilon, \boldsymbol{a}'_\varepsilon, \tau_b, \boldsymbol{a}'_b)'$ represent the unknown model parameters where the subscript ε and b distinguishes the PGM parameters defining the densities g_ε and g_b. Finally, let $\boldsymbol{w}_\varepsilon = \boldsymbol{w}_\varepsilon(\boldsymbol{a}_\varepsilon) = \big(w_{\varepsilon,-K}, \ldots, w_{\varepsilon,K}\big)'$ be the vector of the PGM weights calculated from the vector $\boldsymbol{a}_\varepsilon$ using Expression (12.3). Analogously, let $\boldsymbol{w}_b = \big(\boldsymbol{a}_b\big) = \big(w_{b,-K}, \ldots, w_{b,K}\big)'$ be the PGM weights related to the density g_b being calculated from the transformed weights $\boldsymbol{a}_b$.

Given the random effects b_i, the log-likelihood contributions of the m_i elements in the clusters simply add up, because of the assumed conditional independence. The marginal likelihood of the model involves two integrations: one for the interval-censored survival times and one integrating out the random effect. Therefore, with observed intervals $\lfloor l_{ij}, u_{ij} \rfloor$ ($i = 1, \ldots, n;\ j = 1, \ldots, m_i$), the total marginal log-likelihood is given by:

$$\ell(\boldsymbol{\theta}) = \sum_{i=1}^{n} \log\Bigg[\int_{-\infty}^{\infty} \Bigg\{\prod_{j=1}^{m_i} \int_{l_{ij}}^{u_{ij}} (t_{ij}\,\tau_\varepsilon)^{-1} \sum_{k=-K}^{K} w_{\varepsilon,k}(\boldsymbol{a}_\varepsilon)\, \varphi_{\mu_{\varepsilon,k},\sigma^2_{\varepsilon,0}}\left(\frac{\log(t_{ij}) - \boldsymbol{X}'_{ij}\boldsymbol{\beta} - b_i - \alpha_\varepsilon}{\tau_\varepsilon}\right) \mathrm{d}t_{ij}\Bigg\} \tau_b^{-1} \sum_{k^*=-K}^{K} w_{b,k^*}(\boldsymbol{a}_b)\, \varphi_{\mu_{b,k^*},\sigma^2_{b,0}}\left(\frac{b_i}{\tau_b}\right) \mathrm{d}b_i\Bigg].$$

Due to complexity of the likelihood, the model parameters need to be estimated in a Bayesian way using Markov chain Monte Carlo techniques. To this end, the GMRF penalty prior (12.13) is considered for both sets of the PGM transformed weights $\boldsymbol{a}_\varepsilon$ and $\boldsymbol{a}_b$. Further details can be found in Komárek et al. (2007) and Komárek and Lesaffre (2008).

Now we exemplify this approach on the sample of 500 children from the Signal Tandmobiel study.

Example 13.2 Signal Tandmobiel study

We now extend previous Examples 6.1 and 12.1 – 12.3, and analyze jointly the emergence times of pre-molars 14, 24, 34, 44. More specifically, we check whether the emergence distributions of these teeth differ between boys and girls and whether caries experience on the deciduous predecessor (*DMF* = 1) impacts the emergence of the permanent tooth. The covariate vector $\boldsymbol{X}$ (we left out the subindexes for clarity) in Equation (13.3) is given by $\boldsymbol{X} = (X_1, \ldots, X_{11})'$, where X_1, X_2, X_3 are three dummy variables corresponding to the categorical covariate representing the four *teeth*, X_4 is *gender* equal to one for girls and zero for boys, X_5 is a dichotomized *DMF* score where a value of one means that there has been caries on the predecessor of the permanent tooth, X_6, X_7, X_8 represent the *tooth:gender* interactions and finally, X_9, X_{10}, X_{11} represent the *tooth:DMF* interaction.

We discuss first the PGM-PGM and the Norm-Norm model. The first model assumes a PGM model for both g_ε and g_b, while the second model assumes two normal densities. Table 13.2 contains the posterior summary statistics for both models and for each covariate effect. Importance of covariate effects is judged with contour probabilities (see Section 9.1.4), whereby a simultaneous contour probability is computed for the interaction terms.

For the PGM-PGM model, the individual contour probabilities show that the main effect of *tooth 14* is the same as for *tooth 24*. The same is basically true for the interaction terms of *gender* and *DMF* with *tooth 24*. Horizontal symmetry is therefore not contradicted from these results and the same is true with *teeth 34* and *44*. There is, however, no vertical symmetry since the effect of *tooth 44* is apparently different from that of *tooth 14* with earlier emergence for *tooth 14*. There is a strong *gender* effect, with girls having their permanent teeth emerge earlier. The effect of *gender* does not change with *tooth*, but changes with *DMF* present or absent. There is also a strong *DMF* main effect, i.e., caries experience on the predecessor (*DMF = 1*) significantly accelerates the emergence of the permanent tooth with an effect depending on the tooth. The simultaneous contour probabilities are for the *tooth* effect smaller than 0.001, for the *gender:tooth* interaction term greater than 0.5 and for the *gender:tooth* interaction term smaller than 0.001.

These results lead to consider a simpler model while removing the *gender:tooth* interaction term and replacing the covariate *tooth* by covariate *jaw* which only distinguishes *mandibular* (lower) teeth (34 and 44) from *maxillary* (upper) teeth (12 and 24). Posterior summary statistics (not shown here) then indicate that, as above, the effect of *DMF* is different for both boys and girls. Further, caries experience on the predecessor accelerates the emergence of *maxillary* teeth significantly more than the emergence of *mandibular* teeth.

The results from those of the Norm-Norm model are quantitatively different but qualitatively they lead to basically the same conclusions, as can be seen in Table 13.2. Note that our analysis only involves a sample of 500 subjects from the original data set. Hence, the results reported here differ

TABLE 13.2: Signal Tandmobiel study. Two Bayesian flexible AFT random-effects models applied to emergence times of permanent first premolars (teeth 14, 24, 34, 44), obtained from the R package bayesSurv. The PGM-PGM model assumes a PGM model for both g_ε and g_b, while the Norm-Norm model assumes normal densities. Posterior mean (95% equal-tail CI) (P = contour probability for the effect of covariates).

Effect	Model			
	PGM-PGM		**Norm-Norm**	
tooth = *24*	−0.005	(−0.017, 0.006) P = 0.330	−0.009	(−0.024, 0.006) P = 0.220
tooth = *34*	−0.017	(−0.029, −0.048) P = 0.007	−0.013	(−0.029, 0.002) P = 0.100
tooth = *44*	−0.025	(−0.037, −0.013) P < 0.001	−0.023	(−0.039, −0.008) P = 0.003
gender = *girl*	−0.048	(−0.071, −0.027) P < 0.001	−0.047	(−0.071, −0.023) P < 0.001
DMF = 1	−0.132	(−0.156, −0.103) P < 0.001	−0.133	(−0.156, −0.110) P < 0.001
gender:tooth = *24*	0.0004	(−0.015, 0.016) P = 0.960	0.0024	(−0.017, 0.022) P = 0.810
gender:tooth = *34*	0.0027	(−0.013, 0.018) P = 0.730	−0.0002	(−0.020, 0.020) P = 0.990
gender:tooth = *44*	0.0060	(−0.010, 0.023) P = 0.450	−0.0040	(−0.015, 0.024) P = 0.660
DMF:tooth = *24*	0.021	(0.001, 0.041) P = 0.042	0.018	(−0.006, 0.042) P = 0.138
DMF:tooth = *34*	0.055	(0.027, 0.080) P < 0.001	0.060	(0.036, 0.083) P < 0.001
DMF:tooth = *44*	0.007	(0.046, 0.095) P < 0.001	0.074	(0.051, 0.098) P < 0.001
gender:DMF	0.040	(0.015, 0.064) P = 0.002	0.042	(0.019, 0.065) P = 0.006

somewhat from those reported in the literature, which are based on the full data set of more than 4 000 children.

□

The PGM-PGM model and the Norm-Norm model are the two extreme settings. In-between are the PGM-Norm model and the Norm-PGM model where only one of the densities g_ε and g_b is modelled by the penalized Gaussian mixture. A check on the posterior summary statistics (not shown) reveals that there are only minor differences between the four models. Yet, the Norm-PGM model results are closest to those of the Norm-Norm model, and the PGM-

Norm model results are closest to those of the PGM-PGM model. Hence, with the same error distribution, the results are similar so that we conclude here that the assumed frailty distribution has more effect on the parameter estimates than the error distribution.

Analogously to Section 12.3, both the error density g_ε and the random effects density g_b can be estimated by means of a posterior predictive density. These are given in Figure 13.2 for the Norm-PGM model and the PGM-Norm model. The densities for the Norm-PGM model look about the same as for the Norm-Norm model, while the PGM-Norm densities are close to those of the PGM-PGM model. In addition one can also obtain the predicted marginal survival or hazard function for specific combinations of covariate values. These are given in Figure 13.3 for the PGM-PGM model for tooth 24 and boys. We can observe that the permanent first pre-molars emerge earlier in subjects with a predecessor that showed cavities. While the PGM-PGM model is somewhat overspecified, as discussed in García-Zattera et al. (2016), the marginal survival and hazard curves are identifiable and in fact were basically the same for the four models.

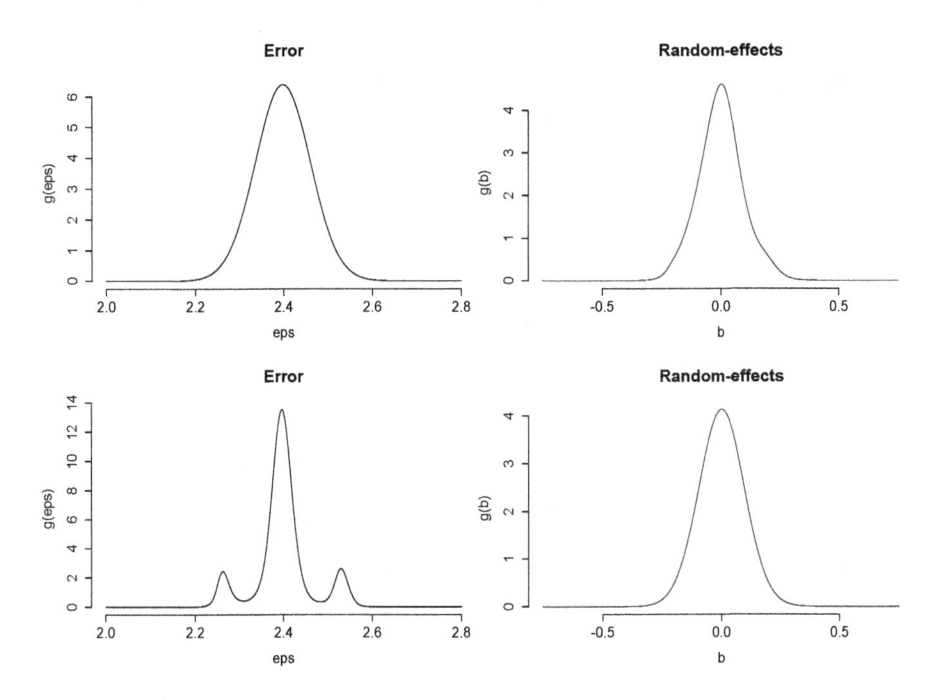

FIGURE 13.2: Signal Tandmobiel study. Estimated error (left) and random effects density (right) for Norm-PGM (top) and PGM-Norm (bottom) models to evaluate the emergence times of teeth 14, 24, 34 and 44, obtained with R package bayesSurv.

13.1.2.1 R solution

Bayesian inference for the random-effects AFT model with the penalized Gaussian mixture in the random effects and error distribution has been implemented in the R package bayesSurv. MCMC sampling is provided by function bayessurvreg3 which generates the posterior samples and stores them in txt files in the same way as was done in a situation of simpler models discussed in Section 12.3.1. The function bayesGspline computes the estimate of the density of the error terms and the random effects, the function predictive2 provides an estimate of the posterior predictive survival function.

Example 13.2 in R

Only selected steps of calculation will be shown here. All details can be found in the R *scripts available in the supplementary materials.*

As it is usual, the R *package* bayesSurv *needs to be loaded. The next step is then to load the data set* tandmob. *The following variables are included: the identification number of the child (*IDNR*), the* TOOTH *factor with levels* 14, 24, 34, 44, *the censoring endpoints* LEFT *and* RIGHT *of the emergence*

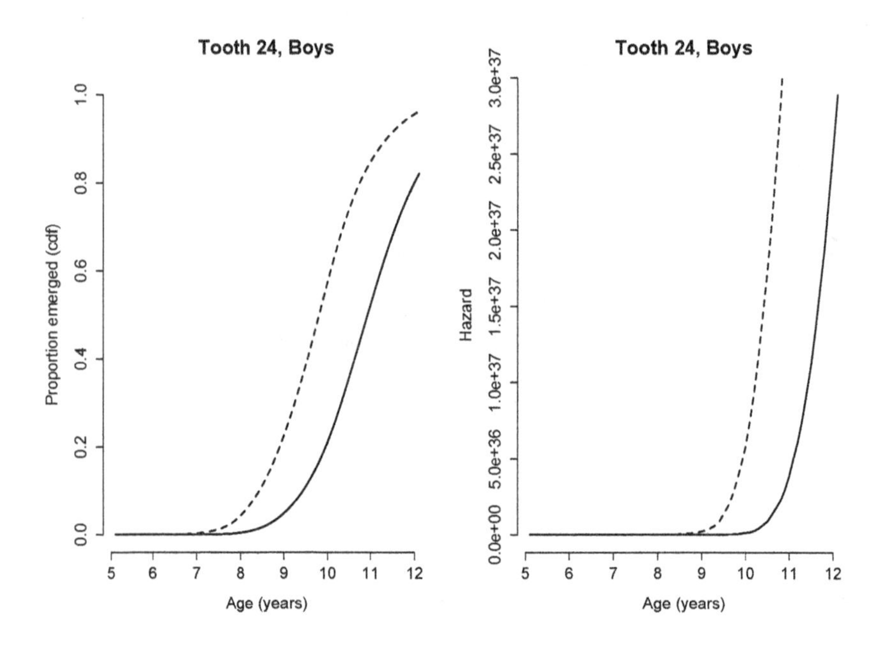

FIGURE 13.3: Signal Tandmobiel study. PGM estimated marginal predictive incidence curve (left) and marginal hazard function (right) from PGM-PGM model to compare emergence times of tooth 24 for boys with caries experience (solid line) on primary predecessor or not (dashed line), obtained with R package bayesSurv.

time, GENDER *(one for girls, zero for boys) and caries experience on the primary predecessor (*DMF = *1 (caries),* DMF = *0 (no caries)). In the original* data.frame tandmob *each line represents a child, but now we need a data set where each row represents a tooth. The following* R *commands create the data set* Tandmob2Analyse *with one row per tooth and teeth removed if no information on caries experience on the predecessor was available.*

```
> library("bayesSurv")
> data("tandmob", package = "icensKBL")
>
> ## Transform tandmob (one row per child)
> ## to Tandmob2Analyse (one row per tooth)
> Tandmob2Analyse <- data.frame(
+     IDNR     = rep(tandmob[, "IDNR"], each = 4),
+     TOOTH    = factor(rep(c(14, 24, 34, 44), nrow(tandmob))),
+     GENDER   = rep(tandmob[, "GENDER"], each = 4),
+     DMF      = as.numeric(rbind(tandmob[, "DMF_54"],
+                    tandmob[, "DMF_64"], tandmob[, "DMF_74"],
+                    tandmob[, "DMF_84"])),
+     LEFT     = as.numeric(rbind(tandmob[, "L14"], tandmob[, "L24"],
+                    tandmob[, "L34"], tandmob[, "L44"])),
+     RIGHT    = as.numeric(rbind(tandmob[, "R14"], tandmob[, "R24"],
+                    tandmob[, "R34"], tandmob[, "R44"]))
> Tandmob2Analyse <- subset(Tandmob2Analyse, !is.na(DMF))
```

The next steps are now documented, i.e., choosing the prior distribution, providing the initial values, calling the MCMC sampler, and post-processing the sampled values. For more explanation, we also refer to Section 12.3.1.

Prior distribution

First, we specify the prior distribution for the penalized Gaussian mixture used in the model for either the error terms or the random intercepts. With K = 15, $2{\cdot}15{+}1$ *knots are chosen spaced* $\delta = 1.5\,\sigma_0$ (c4delta = *1.5) apart. A third order (*order = *3) GMRF penalty is chosen. Further, for* λ*, the smoothing hyperparameter a* $\mathcal{G}(1, 0.005)$ *prior was chosen (*prior.lambda, shape.lambda, rate.lambda*), while a* $\mathcal{N}(0, 10^2)$ *prior on* α *(*prior.intercept, mean.intercept *and* var.intercept*) and a uniform prior on* $(0,\,100)$ *for* τ *(*prior.scale, rate.scale*). Then for the regression parameters (*nBeta = *number of regression parameters) vague normal priors with a variance of 100 are chosen.*

```
> prior.PGM <- list(K = 15, c4delta = 1.5, order = 3,
+     prior.lambda    = "gamma", shape.lambda = 1, rate.lambda = 0.005,
```

```
+     prior.intercept = "normal",
+                           mean.intercept = 0, var.intercept = 100,
+     prior.scale = "sduniform", rate.scale = 100)
> prior.beta <- list(mean.prior = rep(0, nBeta),
+                        var.prior = rep(100, nBeta))
```

For the Norm-PGM and PGM-Norm models also a prior for the normal distribution (= PGM based on $2 \cdot 0 + 1 = 1$ knot) parameters is specified:

```
> prior.Norm <- list(K = 0,
+     order = 0,
+     prior.lambda = "gamma", shape.lambda = 1, rate.lambda = 0.005,
+     prior.intercept = "normal",
+                           mean.intercept = 0, var.intercept = 100,
+     prior.scale = "sduniform", rate.scale = 100)
```

Note that elements prior.lambda, shape.lambda *and* rate.lambda *can be specified arbitrarily as they are effectively ignored when a normal distribution is assumed for the particular random term of the model.*

Initial values

Initial values are required to start the MCMC simulation, here for the PGM parameters of the error distribution (g_ε) and the distribution of the random effect b (g_b) and for the regression parameters. Below we show the initial values for the Norm-PGM model. Some of the PGM initial values in fact determine the PGM itself and remain fixed when sampling. Those are: gamma *and* gamma.b *(position of the middle knots $\mu_{\varepsilon,0}$ and $\mu_{b,0}$, respectively),* sigma *and* sigma.b *(PGM basis standard deviations $\sigma_{\varepsilon,0}$ and $\sigma_{b,0}$, respectively),* alpha.b *(the PGM intercept α_b which is fixed to zero for identifiability reasons). Moreover, the value of* lambda *(λ_ε) is effectively ignored when a normal distribution (PGM with one knot) is assumed for the error terms.*

```
> inits.Norm.PGM <- list(
+    gamma = 0, sigma = 0.2, lambda = 5000,
+    intercept = 2.40, scale = 0.20,
+    gamma.b = 0, sigma.b = 0.2, lambda.b = 1000,
+    intercept.b = 0, scale.b = 0.20)
```

We finalize the specification of the initial values by providing the initial values for the regression coefficients $\boldsymbol{\beta}$ which were motivated by a maximum-likelihood fit of the model without a random intercept, i.e., while ignoring dependencies between the emergence times of the teeth of one child.

```
> inits <- inits.NormPGM
> inits$beta <- c(rep(0, 3), -0.10, -0.10, rep(0, 3), rep(0, 3), 0.04)
```

Posterior simulation

Sampling is done with the bayessurvreg3 *function and the generated samples are stored in txt files in a directory specified by the argument* dir. *Also the number of iterations, the burn-in size, the thinning factor and the number of iterations to refresh the screen (*nwrite*) are specified by the user. The call to* bayessurvreg3 *here is similar to that of* bayessurvreg1 *in Section 12.2 and to* bayessurvreg3 *in Section 12.3. Posterior inference invoked by the code shown below is based on a single chain of 50 000 sampled values obtained from a 1:10 thinned MCMC with 15 000 (×10) burn-in iterations. The* random = ~1 *means that only a random intercept is included in the model corresponding here to the child. The option* a = TRUE *of the argument* store *specifies that the transformed weights* $\boldsymbol{a}$ *of the G-spline defining the error distribution are stored, while with* a.b = TRUE *this applies to the random effects distribution.*

```
> nsimul <- list(niter = 65000, nburn = 15000, nthin = 10,
+     nwrite = 100)
> sampleNormPGM <- bayessurvreg3(
+     Surv(LOWER, UPPER, type = "interval2") ~ TOOTH + GENDER + DMF +
+         GENDER:TOOTH + DMF:TOOTH + GENDER:DMF + cluster(IDNR),
+     random      = ~1,
+     data        = Tandmob2Analyse,
+     dir         = "/home/BKL/AFT_NormPGM",
+     nsimul      = nsimul, prior = prior.Norm,
+     prior.beta = prior.beta, prior.b = prior.PGM,
+     init        = inits, store = list(a = TRUE, a.b = TRUE))
```

Posterior summary information

The R *package* coda *offers the standard Bayesian summary measures. But, one could also use* give.summary *of the package* bayesSurv, *which additionally provides unidimensional contour probabilities. Simultaneous contour probabilities are obtained with the function* simult.pvalue.

An estimate of the error-density g_ε *and of the random effects density* g_b *is obtained using the* R *function* bayesGspline *based on a prespecified grid. This produces Figure 13.2 for the Norm-PGM and PGM-Norm models.*

Posterior predictive information

The R *function* predictive2 *computes the predictive survival curve or hazard function for prespecified values of the covariates. Here we have looked at tooth 24 split up according to gender and caries experience on its primary predecessor. See Figure 13.3 for the PGM-PGM model.*

□

13.1.3 Semiparametric shared frailty models

A Bayesian semiparametric model would involve a nonparametric Bayesian model for the baseline error distribution and the frailty distribution possibly combined with a PH or an AFT assumption. Nevertheless, as far as we know, this approach does not seem to have been pursued in the literature.

13.2 Multivariate models

In Section 13.1.2, a shared frailty AFT model with interval-censored responses was fitted. The model assumed conditional independence of the outcomes given a random intercept representing the subjects. With Gaussian responses this corresponds to a multivariate Gaussian distribution for the log-event times with a compound symmetry covariance matrix. However, often a more general covariance matrix is required, in some cases even depending on covariates. Or even a more general class of multivariate distributions is needed, e.g., with flexible marginal distributions and/or with a non-Gaussian association structure. One possibility is then to consider copula models which were introduced in Section 7.3.

Statistical software to fit multivariate distributions for dimension $d > 2$ is scarce, let alone combined with interval censoring. Below we discuss some Bayesian approaches for bivariate models. We first show that a simple trick allows BUGS or JAGS and SAS software to handle interval censoring for some standard bivariate distributions. Then we illustrate the use of the R package bayesSurv to fit smooth bivariate densities with interval-censored responses. However, for the general multivariate case and Bayesian copula models we limit ourselves to give a brief review of the literature. The R function LDPDdoublyint from the package DPpackage was developed to handle multivariate doubly interval-censored observations. By taking the censoring interval for the first event close to zero, we can use this function to model multivariate interval-censored observations in a semiparametric way. The methods are illustrated using the Signal Tandmobiel study.

13.2.1 Parametric bivariate models

For some models, standard Bayesian software can be used to fit bivariate interval-censored responses. This is possible when the bivariate density is factorized into a marginal and a conditional density and both can be sampled by the software. This is the case for the bivariate Gaussian distribution, but also for the bivariate t-distribution, as will be illustrated below. We estimated the joint distribution of interval-censored emergence times of teeth 14 and 24 recorded in the Signal Tandmobiel study with JAGS and SAS software. Only a basic analysis is shown here.

Example 13.3 Signal Tandmobiel study
The emergence distribution of permanent teeth in Flanders was extensively explored in Bogaerts (2007). Marginally the log-logistic distribution proved to fit the emergence data best for most teeth except for some teeth where the log-normal distribution was best. But, always the log-normal fit was close to that of the log-logistic. Since it is not obvious to fit a bivariate log-logistic distribution to data using JAGS software, we opted to fit a bivariate log-normal distribution to interval-censored emergence times of premolars 14 and 24. In Figure 13.4 we show a scatterplot of the estimated latent predicted emergence times of teeth 14 and 24 obtained from the JAGS analysis.

□

13.2.1.1 JAGS solution

Example 13.3 in JAGS
As before, a complete R *script used to prepare the data and call* JAGS *by means of the* runjags *package is available in the supplementary materials. The* JAGS *command changes with the type of censoring (interval-, left- or right-censoring). For a bivariate distribution, this implies nine possible combinations and hence nine (slightly) different sections within the* JAGS *model*

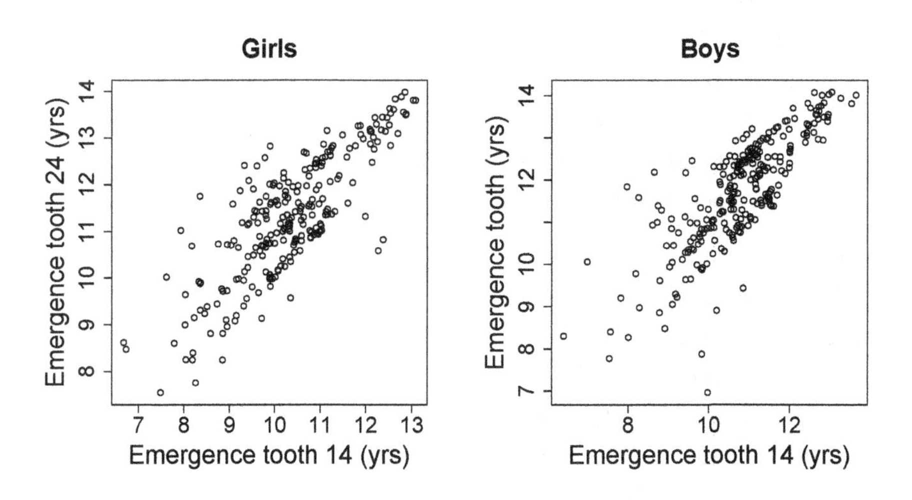

FIGURE 13.4: Signal Tandmobiel study. Imputed emergence times of teeth 14 and 24 assuming a bivariate log-normal distribution obtained using runjags.

description. The first step is to split up the total data set into the nine possible combinations. The combinations left-left censoring and right-left censoring did not occur in the sample of 500 children leaving only seven combinations to consider. After sorting the subjects according to their censoring pattern, the logarithm is taken of all left- and right endpoints to be combined with the JAGS *function* dnorm. *An alternative is to work on the original scale and use the* JAGS *function* dlnorm. *The next step is to specify the initial values of the parameters and then to call* JAGS *via* run.jags. *Below we show part of the* JAGS *model specification to fit a bivariate distribution to the log emergence times of teeth 14 and 24 for boys and girls.*

```
model{
  ### interval-interval case
  for (i in (cindex[1]+1):cindex[2]){
    mu1[i]    <- a1 + alpha1*gender[i]
    mu2[i]    <- a2 + alpha2*gender[i] +
          beta1*(emer1[i] - a1 - alpha1*gender[i])
    y1[i]      ~ dinterval(emer1[i], lim1[i,])
    emer1[i] ~ dnorm(mu1[i], tau[1])
    y2[i]      ~ dinterval(emer2[i], lim2[i,])
    emer2[i] ~ dnorm(mu2[i], tau[2])
  }
......

  ### MVN parameters
  csig.21 <- 1/tau[2]
  sig.11  <- 1/tau[1]
  sig.22  <- csig.21 + sig.11*pow(beta1,2)
  rho     <- beta1*sqrt(sig.11/sig.22)
  sig.12  <- beta1*sig.11

  ### priors
  alpha1 ~ dnorm(0.0, 1.0E-6)
  alpha2 ~ dnorm(0.0, 1.0E-6)
  beta1 ~ dnorm(0.0, 1.0E-6)
  tau[1] ~ dgamma(0.001, 0.001)
  tau[2] ~ dgamma(0.001, 0.001)
  a1 ~ dnorm(0.0, 1.0E-6)
  a2 ~ dnorm(0.0, 1.0E-6)
}
```

There are seven loops in the BUGS *code, one for each observed combination of censoring but we have only shown the first combination. The vector* cindex *contains the start and end index of the respective censoring combinations. The vectors* y1 *and* y2 *are the censoring observations being equal to 0 for left-censoring, and 1 for right- and interval censoring. The bivariate normal fit is based on the well-known factorization of the bivariate Gaussian distribution into a marginal and a conditional Gaussian distribution.*

After the first MCMC run (results of which are assumed to be stored in the R *object* jagsfitlog1*), where we only monitored the primary model parameters (specified by the character vector* parameters.tand1*), a second run can be initiated using the* extend.jags *function to sample 1 000 predicted survival times specified in* parameters.tand2*, using the following code:*

```
> parameters.tand2 <- c("emer1", "emer2")
> jagsfitlog2 <- extend.jags(jagsfitlog1,
+       drop.monitor = parameters.tand1,
+       add.monitor = parameters.tand2, sample = 1000)
```

For diagnostic testing and summarizing the posterior information, standard R *functions can be invoked. We were now only interested to obtain the predicted emergence times. This was done by computing the posterior means of the emergence times with the* R *function* apply *and taking the exponent of sampled log-emergence times. The scatterplots shown in Figure 13.4 are based on these predicted means, but jittering was applied proportional to the variability of the sampled emergence times. See the* R *file for the code.*

□

13.2.1.2 SAS solution

The above JAGS analysis was repeated with the SAS procedure MCMC.

Example 13.3 in SAS

As before, the SAS *program computes directly the likelihood contribution of each child, instead of using data augmentation. Below we show the* SAS *code to compute the likelihood contribution of teeth pairs with an interval-censored emergence time for tooth 14 combined with an interval-censored, left-censored or right-censored emergence time for tooth 24. To compute the likelihood contributions, the* SAS *functions* PROBNORM *and* PROBNRM *are needed to compute the cdf of the univariate standard normal density and the standard bivariate cumulative distribution function, respectively. The complete* SAS *program can be found in the supplementary materials. The predicted emergence times are established with extra* SAS *code (not shown).*

```
PROC MCMC DATA=icdata.tandmob outpost=out1
          SEED=1 NMC=20000 THIN=10 MISSING=AC;
```

```
 .....

mu1=a1 + alpha1*gender;
mu2=a2 + alpha2*gender;

.....

/* interval 1 interval censored */
ELSE IF (L14^=. and R14^=. and L15^=. and R15^=.) THEN
   like =  PROBBNRM((log(R14)-mu1)/sigma1,(log(R15)-mu2)/sigma2,rho)
          -PROBBNRM((log(R14)-mu1)/sigma1,(log(L15)-mu2)/sigma2,rho)
          -PROBBNRM((log(L14)-mu1)/sigma1,(log(R15)-mu2)/sigma2,rho)
          +PROBBNRM((log(L14)-mu1)/sigma1,(log(L15)-mu2)/sigma2,rho);
ELSE IF (L14^=. and R14^=. and L15=. and R15^=.) THEN
   like =  PROBBNRM((log(R14)-mu1)/sigma1,(log(R15)-mu2)/sigma2,rho)
          -PROBBNRM((log(L14)-mu1)/sigma1,(log(R15)-mu2)/sigma2,rho);
ELSE IF (L14^=. and R14^=. and L15^=. and R15=.) THEN
   like =  PROBNORM((log(R14)-mu1)/sigma1)
          -PROBNORM((log(L14)-mu1)/sigma1)
          -PROBBNRM((log(R14)-mu1)/sigma1,(log(L15)-mu2)/sigma2,rho)
          +PROBBNRM((log(L14)-mu1)/sigma1,(log(L15)-mu2)/sigma2,rho);
llike=log(like);
MODEL GENERAL(llike);
RUN;
```

□

Note that this example also illustrates that Bayesian software, such as JAGS and SAS procedure MCMC, can handle interval-censored covariates, as long as the covariates are assumed stochastic and their distribution can be sampled. Also (log) t-distributed emergence times can be analyzed with JAGS and SAS via the well-known relationship between the multivariate normal and multivariate t-distribution. We refer to the supplementary materials for an example.

13.2.2 Bivariate copula models

Bivariate models based on copulas were introduced in Section 7.3. Most of the developments in copula modelling have taken place in the frequentist paradigm. Contributions to the Bayesian approach can be found in Romeo et al. (2006) and Silva and Lopes (2008), and in references therein.

Copula models admit the combination of flexible marginal distributions with a variety of association structures, which makes these models attractive from a modelling perspective. Parameter estimation could be done with a 2-step or 1-step procedure. For a 2-step procedure, the marginal model parameters are first estimated and then fixed into the joint model to determine the association parameters. With a 1-step procedure all model parameters are estimated jointly. Simulation studies done by Romeo et al. (2006) and Silva and Lopes (2008) revealed that the 2-step procedure is sufficient for point estimation, but the 1-step procedure is preferred for appropriately taking into account all uncertainty. Bayesian bivariate copula modelling with interval-censored responses was examined in Cecere and Lesaffre (2008). Some computational procedures were suggested to perform the 1-step procedure and were compared to the 2-step procedure, ending up in the same conclusions as found in Romeo et al. (2006) and Silva and Lopes (2008).

There seems to be no publicly available software to fit Bayesian copula models to data. Further, it is not immediately clear how to program in general Bayesian copula models with JAGS or SAS.

13.2.3 Flexible bivariate models

Komárek and Lesaffre (2006) proposed to fit a flexible bivariate AFT model for interval-censored survival times. The methodology can also be used with doubly interval-censored data. Let $\boldsymbol{T}_i = (T_{i1},\, T_{i2})'$ $(i = 1, \ldots, n)$ be independent bivariate random vectors representing times-to-event of the ith pair observed as intervals $\lfloor l_{il},\, u_{il} \rfloor$ with covariate vectors $\boldsymbol{X}_{ij}$ $(i = 1, \ldots, n;\ j = 1, 2)$. We now assume an AFT model for the (i, l)-th event time given by

$$\log(T_{ij}) = \boldsymbol{X}_{ij}'\boldsymbol{\beta} + \varepsilon_{ij} \qquad (i = 1, \ldots, n;\ j = 1, 2), \tag{13.4}$$

where $\boldsymbol{\varepsilon}_i = (\varepsilon_{i1},\, \varepsilon_{i2})'$ are bivariate i.i.d. random errors with a density $g(\boldsymbol{\varepsilon})$. Komárek and Lesaffre (2006) suggested to use a bivariate PGM as a model for this density. The suggested model is given by

$$g(\boldsymbol{\varepsilon}) = (\tau_1\, \tau_2)^{-1} \sum_{k_1=-K_1}^{K_1} \sum_{k_2=-K_2}^{K_2} w_{k_1,k_2} \\ \varphi_{\mu_{1,k_1},\sigma_0^2}\Big(\frac{\varepsilon_1 - \alpha_1}{\tau_1}\Big)\, \varphi_{\mu_{2,k_2},\sigma_0^2}\Big(\frac{\varepsilon_2 - \alpha_2}{\tau_2}\Big), \tag{13.5}$$

where $\boldsymbol{\mu}_1 = \{\mu_{1,k_1}\}_{k_1=-K_1}^{K_1}$ and $\boldsymbol{\mu}_2 = \{\mu_{2,k_2}\}_{k_2=-K_2}^{K_2}$ are fixed grids of knots for the first and second margin, $\boldsymbol{w} = \{w_{k_1,k_2}\}_{k_1=-K_1,\ldots,K_1,\, k_2=-K_2,\ldots,K_2}$ are weights to be estimated and $\boldsymbol{\alpha} = (\alpha_1,\, \alpha_2)'$ and $\boldsymbol{\tau} = (\tau_1,\, \tau_2)'$ are intercept and scale parameters to be estimated as well. As for the univariate case (see Sections 6.2 and 12.3), one could work with transformed weights $\boldsymbol{a} = \{a_{k_1,k_2}\}_{k_1=-K_1,\ldots,K_1,\, k_2=-K_2,\ldots,K_2}$ being linked to the weights by expressions analogous to (12.3). Note that, although, the mixture components in (13.5) are all uncorrelated, the correlation of the resulting mixture depends on weights $\boldsymbol{w}$ and it is not necessarily equal to zero. Based on this model the log-likelihood of the model with parameters $\boldsymbol{\theta} = (\boldsymbol{\beta}',\, \boldsymbol{\alpha}',\, \boldsymbol{\tau}',\, \boldsymbol{a}')'$ is equal to

$$\ell(\boldsymbol{\theta}) = \sum_{i=1}^{n} \log\Biggl\{ \int_{l_{i1}}^{u_{i1}} \int_{l_{i2}}^{u_{i2}} (t_1\, t_2\, \tau_1\, \tau_2)^{-1} \sum_{k_1=-K_1}^{K_1} \sum_{k_2=-K_2}^{K_2} w_{k_1,k_2}(\boldsymbol{a}) \\ \varphi_{\mu_{1,k_1},\sigma_0^2}\biggl(\frac{\log(t_1) - \boldsymbol{X}_{i1}'\boldsymbol{\beta} - \alpha_1}{\tau_1}\biggr) \\ \varphi_{\mu_{2,k_2},\sigma_0^2}\biggl(\frac{\log(t_2) - \boldsymbol{X}_{i2}'\boldsymbol{\beta} - \alpha_2}{\tau_2}\biggr)\, \mathrm{d}t_2\, \mathrm{d}t_1 \Biggr\}.$$

The model parameters can be estimated using penalized maximum-likelihood by maximizing $\ell_P(\boldsymbol{\theta};\ \lambda_1,\, \lambda_2) = \ell(\boldsymbol{\theta}) - q(\boldsymbol{a};\ \lambda_1,\, \lambda_2)$, where λ_1 and λ_2 are the values of the smoothing hyperparameters. Now, the penalty on the transformed mixture weights $\boldsymbol{a}$ is given according to the margins as follows

$$\begin{aligned} q(\boldsymbol{a};\ \lambda_1,\, \lambda_2) \;&=\; \frac{\lambda_1}{2} \sum_{k_1=-K_1}^{K_1} \sum_{k_2=-K_2+s}^{K_2} (\Delta_1^s a_{k_1,k_2})^2 \\ &\quad + \frac{\lambda_2}{2} \sum_{k_2=-K_2}^{K_2} \sum_{k_1=-K_1+s}^{K_1} (\Delta_2^s a_{k_1,k_2})^2, \end{aligned} \tag{13.6}$$

where Δ_j^s ($j = 1, 2$) denotes the backward difference operator of order s for the j-th margin, e.g., $\Delta_1^3 a_{k_1,k_2} = a_{k_1,k_2} - 3a_{k_1,k_2-1} + 3a_{k_1,k_2-2} - a_{k_1,k_2-3}$. However, maximization of the penalized log-likelihood is computationally difficult.

On the other hand, analogously to the univariate case of Section 12.3, the factor $\exp\bigl\{-q(\boldsymbol{a};\ \lambda_1,\, \lambda_2)\bigr\}$ can be interpreted, up to a normalizing constant, as the prior density of the transformed weights $\boldsymbol{a}$. Komárek and Lesaffre (2006) suggested a Bayesian approach in combination with MCMC procedures to estimate all parameters. Unless prior information is available, vague but proper priors are used for the model parameters. That is, normal priors with a large variance for the regression coefficients $\boldsymbol{\beta}$ and for the intercept parameters $\boldsymbol{\alpha}$; gamma priors with small values of the shape and rate parameters for the PGM inverse variances τ_1^{-2} and τ_2^{-2} and for smoothing hyperparameters λ_1 and λ_2, or alternatively, a uniform prior for the standard deviations τ_1 and τ_2.

This approach was implemented in the R function bayesBisurvreg of the package bayesSurv. In the next example, we exemplify the use of the R function to model the bivariate emergence distribution of teeth 14 and 24 (two horizontally symmetric teeth), but also of teeth 24 and 34 (vertically symmetric teeth). This modelling exercise allows to test horizontal and vertical symmetry, respectively of the emergence distributions, based on data collected in Flanders.

Example 13.4 Signal Tandmobiel study

We test horizontal and vertical symmetry in the emergence distribution of permanent teeth with the above-described bivariate AFT models (13.4). As before, we additionally evaluated whether boys have a different emergence distribution than girls, and whether a predecessor with caries experience (DMF = 1) has an impact on the emergence distribution. The test for horizontal symmetry was done by comparing teeth 14 and 24 and is investigated in model H, while testing vertical symmetry involved teeth 24 and 34 and is investigated in model V. In both analyses, we also evaluated the residual association between the emergence times, i.e., the association between ε_1 and ε_2.

In the following, let the covariate tooth represent the indicator for tooth 24 in model H and tooth 34 in model V. Covariates included in the model are gender (X_1), DMF (X_2) and the interactions gender:tooth (X_3), DMF:tooth (X_4), gender:DMF (X_5). For an explanation of the covariates, we also refer to Section 13.2. Note that the main effect of tooth is not included in the covariate vector $\boldsymbol{X}$ as it is included implicitly in the model as a difference of means $\mathsf{E}(\varepsilon_2)-\mathsf{E}(\varepsilon_1)$ of the error terms, where $\mathsf{E}(\varepsilon_1) = \alpha_1 + \tau_1 \sum_{k_1=-K_1}^{K_1} \sum_{k_2=-K_2}^{K_2} w_{k_1,k_2}\mu_{1,k_1}$ and similarly for $\mathsf{E}(\varepsilon_2)$.

Posterior summary statistics for the covariate effects are shown in Table 13.3 (the row labeled as tooth represents posterior summary statistics for $\mathsf{E}(\varepsilon_2) - \mathsf{E}(\varepsilon_1)$). In model H, the contour probabilities of the main effect of tooth and the interactions gender:tooth and DMF:tooth do not contradict horizontal symmetry. A strong impact is seen of gender and DMF with earlier emergence seen with girls, and when there has been caries experience on the predecessor, even when taking into account the significant interaction term. Both association measures indicate a strong (residual) relationship between the two emergence processes. The results for model V are quite similar supporting also vertical symmetry, and with similar effects for the other terms except for the interaction DMF:tooth which appears more important now. While the associations between the two emergence processes are smaller now, they are still quite high.

We note that both considered models were fitted and results shown also by Komárek and Lesaffre (2009). There, however, the emergence times of all (more than 4 000 children) were analyzed (whereas only a subsample of 500 children is considered here). Not all of the above results are supported by the analysis shown in Komárek and Lesaffre (2009), however. There the dmf:tooth interaction was not significant for teeth 14–24, but was significant for teeth 24–34 (P = 0.034). For the gender:dmf interaction, significant interactions were found in Komárek and Lesaffre (2009). Finally, in Komárek and Lesaffre (2009) no (residual) correlation was found for teeth 24 and 34, while here we found an important correlation. In retrospect, we discovered that the original analysis in Komárek and Lesaffre (2009) contained an error.

Posterior predictive calculations provide the cumulative distribution functions of the emergence for teeth 14, 24 and 34 in both models (graphs not shown). In Figure 13.5 the estimated bivariate density of the error terms for the emergence times is shown for teeth 14, 24 and for teeth 24, 34. In both

TABLE 13.3: Signal Tandmobiel study. Bivariate AFT models applied to emergence times of horizontally symmetric teeth 14, 24 (model H) and vertically symmetric teeth 24, 34 (model V), respectively using R package bayesSurv. Posterior mean (95% equal-tail credible interval) for the effect of covariates and for the residual association, P = pseudo-contour probability for the effect of covariates.

Effect	Model H (teeth 14–24)		Model V (teeth 24–34)	
tooth	−0.007	(−0.022, 0.008) P = 0.40	−0.005	(−0.023, 0.0124) P = 0.55
gender = *girl*	−0.053	(−0.077, −0.030) P < 0.001	−0.047	(−0.070, −0.023) P < 0.001
DMF = 1	−0.123	(−0.152, −0.094) P < 0.001	−0.108	(−0.136, −0.081) P < 0.001
gender:tooth	0.001	(−0.018, 0.020) P = 0.92	−0.000	(−0.021, 0.021) P = 0.98
DMF:tooth	0.022	(−0.005, 0.046) P = 0.06	0.041	(0.014, 0.068) P = 0.002
gender:DMF	0.041	(0.005, 0.077) P = 0.03	0.030	(−0.003, 0.064) P = 0.07
	Residual association			
Pearson correlation	0.776	(0.697, 0.847)	0.639	(0.550, 0.730)
Kendall's tau	0.519	(0.463, 0.571)	0.421	(0.362, 0.477)

cases, the figures suggest that normality is only approximately true for the emergence distributions. The strong (residual) association between the two emergence processes found in Table 13.3 is confirmed graphically.

□

13.2.3.1 R solution

The bivariate AFT Model (13.4) with the error distribution specified as the penalized Gaussian mixture (13.5) is implemented in the R package bayesSurv. The MCMC sampling is provided by the function bayesBisurvreg. Additional functions related specifically to the posterior calculations include bayesGspline, marginal.bayesGspline, predictive2 and sampled.kendall.tau.

Example 13.4 in R

Complete R scripts to perform the analysis of both model H and V are available in the supplementary materials. Many programming steps are similar to those seen in Section 13.1.2.1, so we will be brief on these.

The lists that specify the priors for the model parameters are as follows.

```
> prior.PGM <- list(K = c(5, 5), c4delta = c(1.5, 1.5),
+     neighbor.system = "uniCAR", order = 3,
```

```
+     equal.lambda = FALSE, prior.lambda = c("gamma", "gamma"),
+     shape.lambda = c(1, 1), rate.lambda = c(0.005, 0.005),
+     prior.intercept = c("normal","normal"),
+     mean.intercept = c(0, 0), var.intercept = c(100, 100),
+     prior.scale = c("sduniform","sduniform"),
+     rate.scale = c(100, 100))
> prior.beta <- list(mean.prior = rep(0, 5), var.prior = rep(100, 5))
```

The R *code shows that the same prior distribution is given for the emergence processes of both teeth involved in the analysis. For example,* K = c(5, 5) *means that $K_1 = K_2 = 5$ and there are $2 \cdot 5 + 1$ knots in each margin. On the other hand, the argument* equal.lambda = FALSE *signifies that there are possibly different smoothing hyperparameters in the margins. The argument* neighbor.system = "uniCAR" *stands for "univariate conditional autoregression", which is a prior based on squared differences of a given order (see argument* order*) in each margin as indicated by Expression (13.6). For both τ_1 and τ_2 twice a uniform prior on $(1, 100)$ is chosen on the standard deviation scale, see argument* prior.scale = c("sduniform", "sduniform"), rate.scale = c(100, 100). *Also the effect of an alternative prior is examined as a sensitivity analysis (results not shown).*

Note that we have chosen here for $2 \times 5 + 1$ knots for each margin to reduce the number of parameters. For the specification of the initial values of the model parameters, we refer to the R *program.*

Sampling is done with the bayesBisurvreg *function and the generated samples are again stored in txt files in a pre-specified directory. The call*

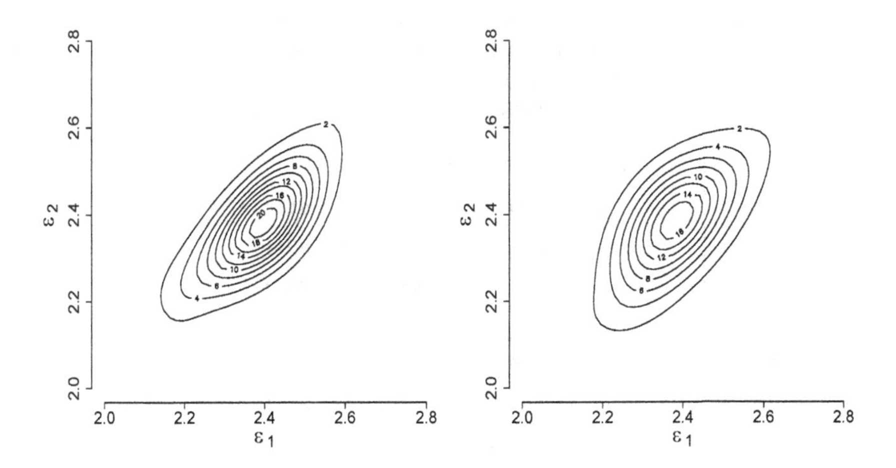

FIGURE 13.5: Signal Tandmobiel study. Contour plots of the estimated error distributions for emergence times of teeth 14 and 24 (left) and teeth 24 and 34 (right), obtained with R package bayesSurv.

to bayesBisurvreg *is similar to that of* bayessurvreg3 *in Sections 12.3.1 and 13.1.2.1:*

```
> nsimul <- list(niter = 10000, nburn = 500, nthin = 1000,
+     nwrite=100)
> sample14.24 <- bayesBisurvreg(
+     Surv(LOWER, UPPER, type="interval2") ~ GENDER + DMF +
+         GENDER:TOOTH + DMF:TOOTH + GENDER:DMF + cluster(IDNR),
+   data = Tandmob2Analyse, dir = "/home/BKL/AFT_Bivar",
+ nsimul = nsimul,  prior = prior.PGM, prior.beta = prior.beta,
+ init = inits, store = list(a = FALSE))
```

The parameter estimates are now based on a single chain of 9 500 values with 500 (×1 000) burn-in iterations and thinning factor 1 000. A large number of iterations is necessary for the parameters of the error densities, convergence is reached much earlier with the regression parameters. The large thinning factor reduces considerably the processing time later on. The R *code above provides sampling from the posterior distribution for the analysis of the emergence of teeth 14 and 24.*

In the next R *commands, characteristics of the error density are computed. Use is made of the function* scanFN*, which reads the needed sampled values. Next, the* data.frame Error *is created which contains the sampled values of (i) the means of the marginal distributions of the error terms* ε_1 *and* ε_2 *(*Intercept1 *and* Intercept2*), (ii) the standard deviations of* ε_1 *and* ε_2 *(*Scale1 *and* Scale2*), (iii) the Pearson correlation (column* Correlation*) and (iv) Kendall's tau for which the sampled values were first obtained by the mean of the function* sampled.kendall.tau*. Note that calculation of sampled values of Kendall's tau may be quite time consuming with large chains. The arguments* skip *and* by *can be specified in the function* sampled.kendall.tau *by the user to omit initial chain values and allow for additional thinning in the computation.*

```
> ChainDir <- "/home/BKL/AFT_Bivar"
> MixMoment <- scanFN(paste(ChainDir, "mixmoment.sim", sep = ""))
> KendallTau <- sampled.kendall.tau(dir = ChainDir, K = c(5, 5),
+                                   nwrite = 1000)
> Error <- data.frame(
+     Intercept1 = MixMoment$Mean.1, Intercept2 = MixMoment$Mean.2,
+     Scale1 = sqrt(MixMoment$D.1.1), Scale2 = sqrt(MixMoment$D.2.2),
+     Correlation = MixMoment$D.2.1 / sqrt(MixMoment$D.1.1
+                                          * MixMoment$D.2.2),
+     KendallTau = KendallTau)
```

The accompanied R script further reports on the posterior summary information, the joint and marginal error densities and the survival and hazard curves for one tooth (not shown here).

□

13.2.4 Semiparametric bivariate models

Jara et al. (2010) suggested a Bayesian semiparametric approach for multivariate doubly interval-censored responses. The method was implemented in the function LDPDdoublyint of the R package DPpackage. When the onset times are replaced by small intervals around the onset times, the function LDPDdoublyint could also be used for the analysis of bivariate interval-censored observations. This approach is another generalization of the DP prior, called the *Poisson-Dirichlet (PD) process prior*. The method is not based on any AFT or any PH assumption, and therefore the impact of covariates cannot be summarized in a regression coefficient. In addition, each onset time and gap time (remember that we are dealing with multivariate survival times) are modelled jointly without the independence assumption. Basically, it is assumed that the logarithm of the onset and gap time have a continuous mixture of Gaussian distributions as distribution. Joint modelling the onset and gap time is not necessary for multivariate interval-censored observations, but is done to relax the independence assumption of onset and gap time for DI-censored survival times. More details to follow in Section 13.3.

Example 13.5 Signal Tandmobiel study
We look again at the emergence distribution of permanent teeth 14 and 24. Here the use of the R function LDPDdoublyint is sketched to analyze jointly the emergence times of teeth 14 and 24 from the Signal Tandmobiel study. Because the impact of a covariate cannot be expressed as a simple numerical value, we compared the effect of covariates graphically by contrasting the different estimated emergence distributions. Here, the two marginal emergence distributions are displayed. These are given in Figure 13.6. We see again that both teeth emerge earlier for girls and when the deciduous predecessor had a caries experience. In the current version of the program, no genuine bivariate comparisons are possible. It would be, say, interesting to see the distribution of the (pairwise) difference in emergence times. The interested reader is referred to the DPpackage manual for more details on the function LDPDdoublyint.

□

13.2.4.1 R solution

The R script to perform the Bayesian semiparametric analysis of the emergence distributions of teeth 14 and 24 can be found in the supplementary materials. We base our analysis on the same data set as in Section 13.4. Below we have described parts of the R program focusing on those parts that serve as input to the function LDPDdoublyint of the package DPpackage.

Example 13.5 in R

We start (as usual) with preparing the data, i.e., the two interval-censored responses and the covariates. Below, the R *code is given for the specification of the two onset (*first*) and event (*second*) responses and the covariates* xfirst *and* xsecond *(all of them exploit* R *objects defined earlier in the script). The dimensions are p and q for the covariates of the onset and event time, respectively.*

```
> first <- cbind(l1_14, r1_14, l1_24, r1_24)
> second <- cbind(l2_14, r2_14, l2_24, r2_24)
> intercept <- rep(1, n)
> p <- 1
> xfirst <- cbind(Intf1 = intercept, Intf2 = intercept)
> q <- 3
> xsecond <- cbind(Ints1 = intercept, x1_14, x2_14,
+                  Ints2 = intercept, x1_24, x2_24)
```

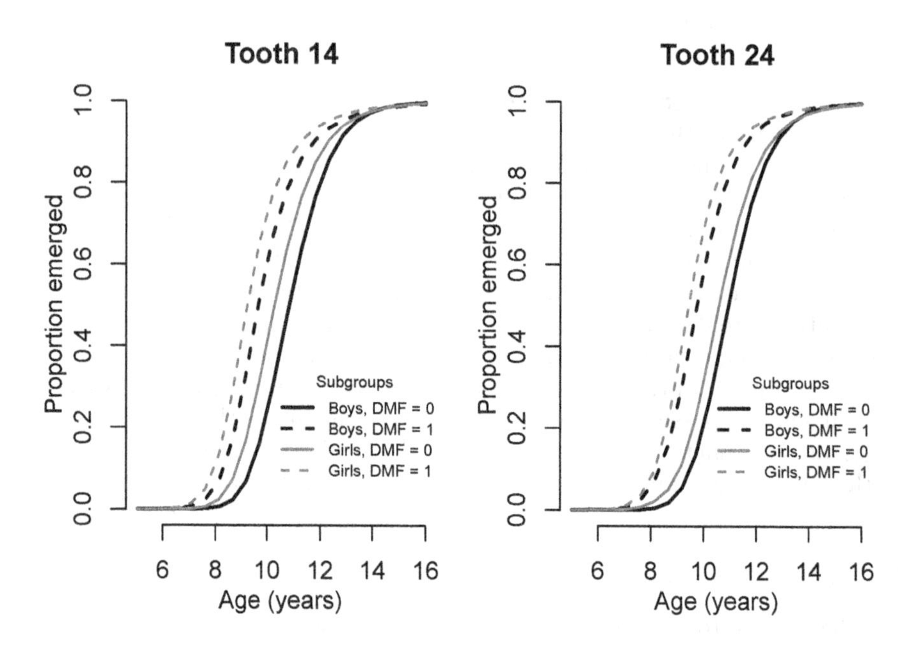

FIGURE 13.6: Signal Tandmobiel study. Emergence distributions for tooth 14 and tooth 24 for four covariate combinations, obtained with the R package DPpackage.

The vectors l1_14 *and* l1_24 *(the lower limits for the onset times) are vectors of zeros. The vectors* r1_14 *and* r1_24 *(the upper limits for the onset times) were randomly sampled from a uniform distribution on interval* $(0, 0.1)$. *That is, the observed intervals for the onset times are all very short intervals above zero. Vectors* l2_14, l2_24 *and* r2_14, r2_24 *provide the lower and the upper limits, respectively, for the event times. Since both teeth 14 and 24 never emerge before 5 years of age, those values are equal to the corresponding observed limits minus five. By doing that, we avoid fitting a zero probability mass of the emergence distribution on interval* $(0, 5)$ *hereby avoiding numerical problems. When interpreting final results, we only have to take into account the fact that a bivariate distribution of* $(T_{i,1} - 5, T_{i,2} - 5)$ $(i = 1, \ldots, n)$ *is fitted.*

Further, we need to choose also a grid of time values for the predicted survival/hazard functions. Note that this input is now essential for this approach, in contrast with previous approaches where prediction is optional. The reason is that for the semiparametric approach the effect of covariates cannot be summarized in a single regression coefficient, but must be evaluated graphically. The matrix object grid *contains for each of the two onset and event times the time points at which prediction should be evaluated. We have chosen for a vector starting at 0 and ending at 11. As all event times were shifted by 5 years, this means that prediction of the emergence distributions for a time span from 5 till 16 years of age will be performed. However, note that the results from about 12 year onwards are extrapolated results since the children were only examined up to that age.*

```
> grid <- matrix(c(rep(seq(0, 11, length = 22), 1),
+                  rep(seq(0, 11, length = 22), 1),
+                  rep(seq(0, 11, length = 22), 1),
+                  rep(seq(0, 11, length = 22), 1)),
+                nrow = 4, byrow = TRUE)
```

The matrix object xpred *includes the covariate values of the subgroups for which prediction is asked for. First the covariate values of the two onset times are given. For the onset times the model contains only intercepts (columns 1 and 2). Then follows the description of the subgroups for the two event times (columns 3 to 8).*

```
> xpred <- matrix(
+     c(1, 1, 1, 0, 0, 1, 0, 0,   # boys,  DMF = 0
+       1, 1, 1, 0, 1, 1, 0, 1,   # boys,  DMF = 1
+       1, 1, 1, 1, 0, 1, 1, 0,   # girls, DMF = 0
+       1, 1, 1, 1, 1, 1, 1, 1),  # girls, DMF = 1
+     ncol = 8, byrow = TRUE)
> colnames(xpred) <- colnames(cbind(xfirst, xsecond))
```

Then we start with the preparation of the nonparametric analysis. The first R command below specifies that the MCMC run starts from scratch. Another option is that one updates a previous run, then the object state should contain the output of the last value of the parameters necessary to restart the analysis.

```
> state <- NULL
>
> prior<-list(a0 = 0.5, b0 = 1, q = 0.5, mub = 10, sigmab = 200,
+       nu = 10, tinv = diag(1, 4), m0 = rep(0, 8), S0 = diag(100, 8),
+       nub = 10, tbinv = diag(1, 8), maxm = 40)
>
> mcmc <- list(nburn = 1000, nskip = 9, ndisplay = 100, nsave = 15000,
+          tune1 = 0.25, tune2 = 1)
>
> fit.14.24 <- LDPDdoublyint(onset = first, failure = second,
+              p = p, xonset = xfirst, q = q, xfailure = xsecond,
+              xpred = xpred, grid = grid, prior = prior, mcmc = mcmc,
+              state = state, status = TRUE, work.dir = work.dir)
```

It would go too far to detail all components of the prior object. Here follows some documentation on some of the parameters in the R codes above. Some other details will be provided in Section 13.3.3.1. The distribution of the logarithmically transformed onset and gap time is a continuous mixture of Gaussian distributions $\mathcal{N}(\boldsymbol{X}_i'\boldsymbol{\beta}_i, \boldsymbol{\Sigma})$. The parameters (nu = 10, tinv = diag(1, 4)) represent the degrees of freedom and the scale matrix of the inverse Wishart distribution of $\boldsymbol{\Sigma}$. The mixture is taken over $\boldsymbol{\beta}_i$ $(i = 1, \ldots, n)$ with a Poisson-Dirichlet (PD) process prior with parameters a, b and $F_{\boldsymbol{\beta},0} = \mathcal{N}(\boldsymbol{m}_{\boldsymbol{\beta}}, \boldsymbol{S}_{\boldsymbol{\beta}})$. The objects (m0 = rep(0, 8), S0 = diag(100, 8)) are the Gaussian mean and covariance prior of the regression mean $\boldsymbol{m}_{\boldsymbol{\beta}}$ with dimension equal to the sum of dimensions of the number of regression parameters of the onset and gap time. The objects (nub = 10, tbinv = diag(1, 8)) are the parameters of the inverse Wishart prior of $\boldsymbol{S}_{\boldsymbol{\beta}}$. The parameters a0, b0, q, mub, sigmab all relate to a and b of the PD process, but it would again go too far to explain them.

The mcmc list object contains classical typical parameters that direct the MCMC sampling (except for tune1 and tune2 which relate to a and b). We ask for 15 000 iterations with 1 000 burn-in iterations, with a 10% thinning factor. After each 1 000 iterations, a message is sent to the screen. Note, however, that there is no formal way to check convergence here. No trace plots are provided, nor is it easy to extract them from the output. One can only run the program with an increasing number of iterations to check stability of the results. To this end, one restarts the calculations with the object state containing the relevant information to restart the sampling. See the manual of the package for details.

The parameters in the call to LDPDdoublyint are self-explanatory. The re-

sult of the MCMC run is then stored in the object fit.14.24. *Then, based on object* fit.14.24 *we construct the predictions from which the marginal survival and hazard functions can be constructed either without or with 95% CI bands, see below the* R *codes. The* plot *statement creates several plots containing the survival functions for the onset and the gap time for each covariate combination specified in* xpred. *However, we have written some extra* R *commands to produce Figure 13.6 and other more illustrative plots, see the accompanied* R *program for details.*

```
> ### Without CI bands and intervals
> fit.14.24.pred <- predict(fit.14.24)
> plot(fit.14.24.pred)
>
> ### With CI bands and intervals
> fit.14.24.pred <- predict(fit.14.24, compute.band = TRUE)
> plot(fit.14.24.pred)
```

□

13.2.5 Multivariate case

The extension of the parametric bivariate approach to the multivariate case is conceptually relatively easy, but not in practice. We are aware only of Bayesian software for modelling the multivariate normal model, and because of the well-known relationship of the t-distribution with the normal model, also for the (multivariate) t distribution. Moreover, only the Archimedian copula model can be extended to the multivariate case, see, e.g., Nelsen (1998). But for interval-censored observations, there is basically no software available in the Bayesian context.

While the multivariate Gaussian distribution is often too restrictive for a problem, it was used successfully in Cecere et al. (2006) to model 7-dimensional emergence distributions of teeth and their ranking based on data collected in the Signal Tandmobiel study. Since the problem examined in that paper is uncommon but interesting, we briefly describe below the research questions and the approach followed.

Adequate knowledge of timing and pattern of permanent tooth emergence is useful for diagnosis and treatment planning in pediatric dentistry and orthodontics and is essential in forensic dentistry. The dependence of the emergence distribution of permanent teeth on caries experience in deciduous teeth has been illustrated in previous examples. A more challenging research question is to estimate the probability of the emergence rankings of permanent teeth. This was most often done in the literature on small data sets. The Signal Tandmobiel study allows for a more accurate estimation of these probabilities at least for children in Flanders and to determine what factors impact these rankings. Note that there are $7! = 5\,040$ possible rankings to consider within one quadrant of the mouth.

Since the marginal emergence distributions were approximately normal (both on an original scale as well as on a logarithmic scale) a multivariate Gaussian distribution was chosen for the seven emergence times of a quadrant. Covariates may affect not only the marginal emergence distributions, but also their association structure. Thus, besides admitting interval censoring, also the covariance matrix should be made dependent on covariates. A condition for this dependence is that the covariance matrix remains positive definite (pd) for all possible values of the covariates. Pourahmadi (1999) proposed to factorize the covariance matrix into a unit lower triangular matrix $\boldsymbol{T}$ and a diagonal matrix $\boldsymbol{D}$, making use of the modified Cholesky decomposition of the inverse of the covariance matrix. By letting the elements of $\boldsymbol{T}$ and $\boldsymbol{D}$ depend on covariates, this parametrization ensures the pd condition. The elements of $\boldsymbol{T}$ can be interpreted as regression coefficients of conditional regression models regressing response Y_s on previous responses, whereas the diagonal elements of $\boldsymbol{D}$ represent the (conditional) variances in these regression models.

The Bayesian approach based on the modified Cholesky decomposition was implemented into an R program calling a C++ program for the calculation of multivariate normal integrals. The program can be downloaded from the Archives of the Statistical Modelling journal (`http://statmod.org/smij/`).

Based on the fitted multivariate Gaussian distribution, it was possible to estimate the probability of the emergence rankings, i.e., the following probabilities (and for all permutations of the teeth indexes) were estimated

$$P(Y_{i_1} < Y_{i_2} < \ldots < Y_{i_7}) = \int_S \varphi_{\boldsymbol{\mu},\,\boldsymbol{\Sigma}}(\boldsymbol{x})\mathrm{d}\boldsymbol{x},$$

where Y_{i_k} $(k = 1, \ldots, 7)$ represents the latent emergence time for the kth tooth in one quadrant of the ith child, $\varphi_{\boldsymbol{\mu},\,\boldsymbol{\Sigma}}(\boldsymbol{x})$ represents the multivariate normal density with mean vector $\boldsymbol{\mu}$ and covariance matrix $\boldsymbol{\Sigma}$; and $S = \{(y_{i_1}, y_{i_2}, \ldots, y_{i_7}) \in \mathbb{R}^7 \mid y_{i_1} < y_{i_2} < \ldots < y_{i_7}\}$. This integral was numerically calculated using Quasi-Monte Carlo integration techniques. In contrast to Monte Carlo integration, the technique of Quasi Monte Carlo relies on point sets in which the points are not chosen i.i.d. from the uniform distribution but rather interdependently using pseudo-random sequences, see, e.g., Hickernell et al. (2005). From the Bayesian computations, the most prevalent emergence rankings of permanent teeth in Flanders were obtained and then reported in Cecere et al. (2006), and Leroy et al. (2008, 2009).

Apart from multivariate parametric modelling, a parametric shared frailty approach could be used bringing the multivariate problem essentially back to a univariate problem conditional on the frailty term. When more flexible modelling is required, the frailty approaches in Section 13.1.2 can be of help. However, the dependence structure assumed in the frailty models may be too simple and the conditional interpretation of the regression coefficients may not be preferable. A possible way out is the multivariate semiparametric approach suggested in Jara et al. (2010) and to make use of the R package DPpackage.

13.3 Doubly interval censoring

In Section 8.1 we have reviewed non- and semiparametric frequentist approaches for doubly interval (DI)-censored observations. Recall that DI-censored observations are in fact duration times (gap times) with interval-censored start and end time points. De Gruttola and Lagakos (1989) showed that for DI censoring the distribution of the beginning and end of the interval must be modelled. Hence, in general it is not sufficient to act as if the gap time is interval-censored and model it univariately. This makes DI censoring a special case of bivariate interval-censoring with interval-censored times for the ith subject: T_i^O (onset) and T_i^E (event). The difference with the general bivariate case is that T_i^O must happen earlier than T_i^E. Thus, we have now bivariate interval-censored responses $\boldsymbol{T}_i = (T_i^O, T_i^E)'$, but with $T_i^O \leq T_i^E$.

To illustrate modelling DI-censored data, we have used in Section 8.1 the data obtained from a survey on mobile phone purchases. We mentioned in that section that frequentist semiparametric software dealing appropriately with DI-censoring is not generally available. That is why we have used in that section a self-written R program. In a Bayesian context, the parametric approach to modelling is usually the first option to consider. This is illustrated here with a program using JAGS. A SAS program was attempted but the current version of the package, and especially of procedure MCMC, showed numerical difficulties. Flexibly modelling DI-censored data is exemplified with the R package bayesSurv. This package provides several tools to fit (multivariate) DI-censored observations. The R package DPpackage is used to illustrate Bayesian semiparametric modelling.

13.3.1 Parametric modelling of univariate DI-censored data

We now assume that the two interval-censored outcomes that determine the gap time, i.e., the onset and the event time, have a specific parametric distribution. Note that the true onset and gap times are assumed to be statistically independent, but because they are both interval-censored we need to model them jointly. We make use of JAGS via runjags to model the mobile phone data for which the time in-between two purchases of a mobile phone is DI-censored. Note that a similar program in WinBUGS and OpenBUGS crashed.

The model in Section 8.1 is based on the PH assumption, while here we have chosen for an AFT model. This allows us to make only qualitative comparisons between the two solutions regarding the importance of covariates on the distribution of the gap time.

Example 13.6 Survey on mobile phone purchases
T_i^O, T_i^E $(i = 1, \ldots, n)$ represent the interval-censored times of purchase of the previous and current mobile phone, respectively. Of interest is to know what factors determine the distribution of the time lapse to buy a new mobile phone, i.e., the gap time $G_i = T_i^E - T_i^O$. In the database, the following possibly determinant factors were recorded: *gender* (female = 0, male = 1), *age group* (1 = 15 − 24 yrs, 2 = 25 − 34 yrs, 3 = 35 − 44 yrs, 4 = 45 − 54 yrs, 5 = 55 − 64 yrs, 6 = 65+), *size of household* (1 = 1 person, ..., 4 = 4 persons, 5 = 5 persons or more). In the analyses below, we have included *age* as a continuous variable, but also as a categorical variable (age group 1 as reference class) and *household size* as a categorical variable (size = 1 as reference class). We did not include *tax income* because of 60 missing values. We assumed that T_i^O and G_i have a log-normal distribution, i.e., $\log(T_i^O) \sim \mathcal{N}(\mu_0, \sigma_0^2)$ and $\log(G_i) \sim \mathcal{N}(\mu_{G,i}, \sigma_G^2)$, where $\mu_{G,i} = \boldsymbol{X}_i^{G'} \boldsymbol{\beta}^G$ for $\boldsymbol{X}_i^G$ a vector of the covariates and $\boldsymbol{\beta}^G$ a vector of unknown regression coefficients. We let only the distribution of the gap time depend on covariates, since the time of first purchase was only vaguely remembered.

□

13.3.1.1 JAGS solution

Example 13.6 in JAGS
Full R *script which prepares the data and calls* JAGS *via the* run.jags *function is available in the supplementary materials. We evaluated four models that differ with respect to the covariates that are included in the gap time model (vectors* $\boldsymbol{X}_i^G$*): (1) only gender, (2) only household size, (3) gender, (continuous) age and household size and (4) gender and (categorical) age. The posterior summary measures were obtained from three chains, each with 11 000 burn-in and additional 30 000 (×100) iterations. A thinning factor of 100 was used. Convergence was confirmed by visual inspection of the trace plots, and classical tools such as Brooks-Gelman-Rubin and Geweke diagnostics. We show below part of the* JAGS *model based on gender and (categorical) age. The vector* lS *represents the logarithmically transformed purchase time of the first mobile phone, and* lG *is the log(time) of the second purchase.* rlG *is the difference between the true gap time and its mean, hence it expresses the error gap time.*

```
model{
  for(i in 1:n){
    mulS[i] <- beta0S                    # mean of log(onset time) dist
```

```
    ## log(onset) distribution
    ## l1, r1 = log of left, right endpoint
    ##          purchase previous mobile phone
    ylS[i] ~ dinterval(lS[i], c(l1[i], r1[i]))
    lS[i]  ~ dnorm(mulS[i], taulS)

    ## Gender (x1),
    ## categorical age (x7-x11) in model for log(gap time)
    mulG[i] <- beta[1] + beta[2]*x1[i] + beta[3]*x7[i] +
      beta[4]*x8[i] + beta[5]*x9[i] + beta[6]*x10[i] + beta[7]*x11[i]

    ## Gap time = end - onset time
    ## l2, r2 = log of left, right endpoint
    ##          purchase current mobile phone
    ## Left, right endpoint defined on log-scale
    lg[i] <- log(exp(l2[i]) - exp(lS[i]))
    rg[i] <- log(exp(r2[i]) - exp(lS[i]))

    ## log(gap time) distribution
    ylG[i] ~ dinterval(lG[i], c(lg[i], rg[i]))
    lG[i]  ~ dnorm(mulG[i], taulG)

    ## Error for gap time distribution
    rlG[i] <- lG[i] - mulG[i]
  }
  ...
}
```

The results of the first analysis show that gender is not important with the posterior mean of the regression coefficient (95% CI) being -0.035 (-0.018, 0.11), thereby confirming the result in Section 8.1. Recall that to compare qualitatively the regression coefficient of an AFT model to that of a PH model, one must change signs. We also found that subjects from a family of three tend to buy quicker a new mobile phone with the posterior mean of the regression coefficient (95% CI) being -0.49 (-0.70, -0.25). When these covariates are combined with (continuous) age, the importance of household size disappeared. While gender was not important, we decided to keep gender in the final model but removed household. Model 1 in Table 13.4 is based on gender and (categorical) age. From this model we conclude that, overall, older individuals will wait longer to buy a new mobile phone, with the age class 25–34 being the fastest.

TABLE 13.4: Mobile study. Posterior summary measures (mean, SD, 95% equal-tail CI) from two models that predict the time to buy a new mobile phone based on gender and (categorical) age. Model 1 is based on a log-normal distribution for the gap time and makes use of runjags. Model 2 is based on the function bayessurvreg2 from the package bayesSurv.

Covariate	**Parametric Model 1**		**Flexible Model 2**	
	Est. (SD)	**95% CI**	**Est. (SD)**	**95% CI**
Intercept_{Onset}	2.70 (0.017)	(2.66, 2.70)	2.66 (0.019)	(2.62, 2.69)
σ_{Onset}	0.34 (0.011)	(0.32, 0.36)	0.38 (0.035)	(0.32, 0.45)
Intercept_{Gap}	0.87 (0.12)	(0.64, 1.10)	0.98 (0.11)	(0.77, 1.18)
Gender	−0.003 (0.071)	(−0.14, 0.14)	−0.002 (0.072)	(−0.15, 0.14)
Age-group (yrs)				
25−34	−0.11 (0.14)	(−0.37, 0.17)	−0.18 (0.13)	(−0.43, 0.07)
35−44	0.13 (0.14)	(−0.15, 0.41)	0.04 (0.13)	(−0.22, 0.31)
45−54	0.22 (0.14)	(−0.04, 0.49)	0.17 (0.13)	(−0.08, 0.42)
55−64	0.41 (0.14)	(0.15, 0.67)	0.28 (0.13)	(0.04, 0.53)
65+	0.36 (0.14)	(0.10, 0.64)	0.29 (0.13)	(0.04, 0.54)
σ_{Gap}	0.67 (0.025)	(0.62, 0.72)	0.77 (0.037)	(0.71, 0.85)

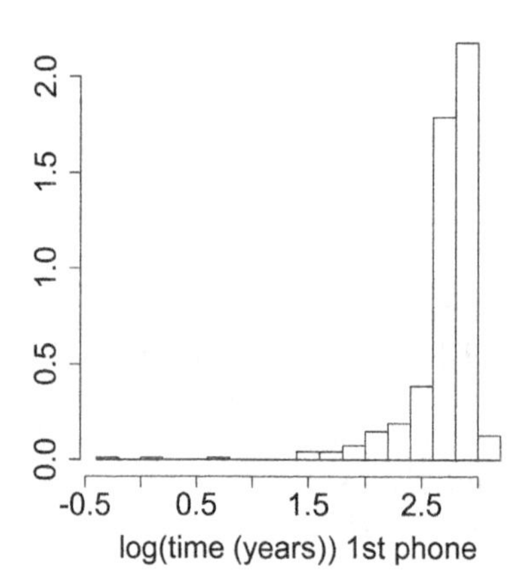

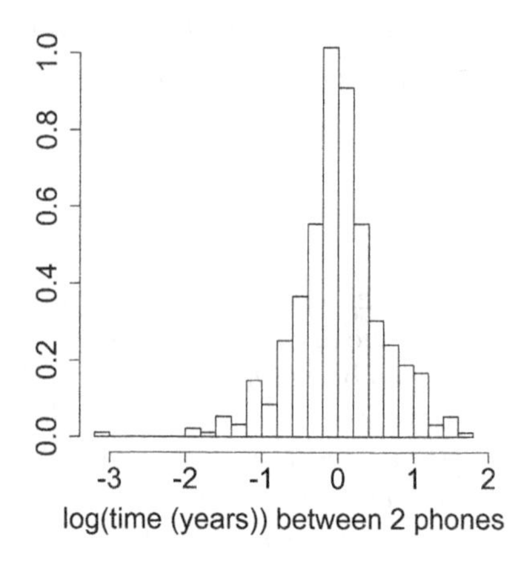

FIGURE 13.7: Mobile study. Histogram log(time) purchase 1st mobile phone ($\log(T^O)$, left) and log(gap time) error between 2 purchases ($\log(T^E - T^O)$, right) fitted using runjags.

The left histogram in Figure 13.7 is based on the mean predicted time of the previous purchase, while the right histogram is based on the error gap time, both on the log-scale. The time of the previous purchase does not appear to be log-normally distributed, but the log-normal distribution does seem to work for the error gap time. To improve the fit, one might choose another distribution or go for a more flexible approach. This is what will be done in the next section.

□

13.3.2 Flexible modelling of univariate DI-censored data

Doubly interval-censored observations can be addressed with the R package bayesSurv, more in particular with the functions bayessurvreg2 and bayessurvreg3 and bayesBisurvreg. In case of univariate DI-censored data, both bayessurvreg2 and bayessurvreg3 lead to the same model, namely, the AFT models for both the onset and the gap times. That is, the following is assumed

$$\begin{aligned} \log(T_i^O) &= \boldsymbol{X}_i^{O'}\boldsymbol{\beta}^O + \varepsilon_i^O, \\ \log(T_i^E - T_i^O) = \log(G_i) &= \boldsymbol{X}_i^{G'}\boldsymbol{\beta}^G + \varepsilon_i^G \qquad (i = 1, \ldots, n), \end{aligned} \tag{13.7}$$

where $\boldsymbol{X}_i^O$ and $\boldsymbol{\beta}^O$ are the covariate vectors and unknown regression coefficients related to the onset time and $\boldsymbol{X}_i^G$ and $\boldsymbol{\beta}^G$ are the covariate vectors and regression coefficients related to the gap time which is usually of primary interest. Flexibility in the distributional assumptions is achieved by assuming Bayesian univariate penalized Gaussian mixtures (PGM, see Section 12.3) for distribution of the error terms ε_i^O in the onset time model as well as the error terms ε_i^G in the gap time model.

The function bayesBisurvreg allows to fit a variant of the PGM AFT model (explained in Section 13.2.3) also to bivariate DI-censored data. This reflects a situation when two onset and two event/gap times are observed on each unit and it is of interest to model the two gap times jointly, e.g., because of interest in their mutual association. We do not cover this here and use the Model (13.7) to analyze the mobile phone data.

Example 13.7 Survey on mobile phone purchases
Model (13.7) with $\boldsymbol{X}_i^O \equiv 1$ $(i = 1, \ldots, n)$, $\boldsymbol{\beta}^0 \equiv \mu_0$ was used to replay the parametric analysis of the mobile data from Section 13.3.1, now assuming the PGM for the logarithms of both the event and the gap time. Table 13.4 (Model 2) shows posterior summary statistics for the parameters of the model where gender and categorical age was included in the covariate vectors $\boldsymbol{X}_i^G$ for the gap time. From Table 13.4 one can observe that the parameter estimates from the flexible fit are similar to those of the parametric fit.

Figure 13.8 shows the estimated baseline densities of the onset time and the gap time, both on log-scale. A comparison with Figure 13.7 shows that the histogram of predicted onset and gap times from the parametric fit look similar to the flexible solution obtained now.

Finally, in the left panel of Figure 13.9 we show the incidence curves for buying a new mobile phone. In other words, this figure shows the estimated cdf of the gap time. It can be again concluded from this figure that older subjects tend to wait longer to buy a new mobile phone, but also that the age class of 25 – 34 years renews its mobile phone fastest.

□

13.3.2.1 R solution

As already mentioned above, the AFT Model (13.7) with the PGM error distributions can be fitted by functions bayessurvreg2 or bayessurvreg3 from the R package bayesSurv. The R script to run the analysis of Example 13.7 (function bayessurvreg2 was used) is available in the supplementary materials. But, since we have illustrated the use of bayesSurv in various sections before, no description of the R commands is given here.

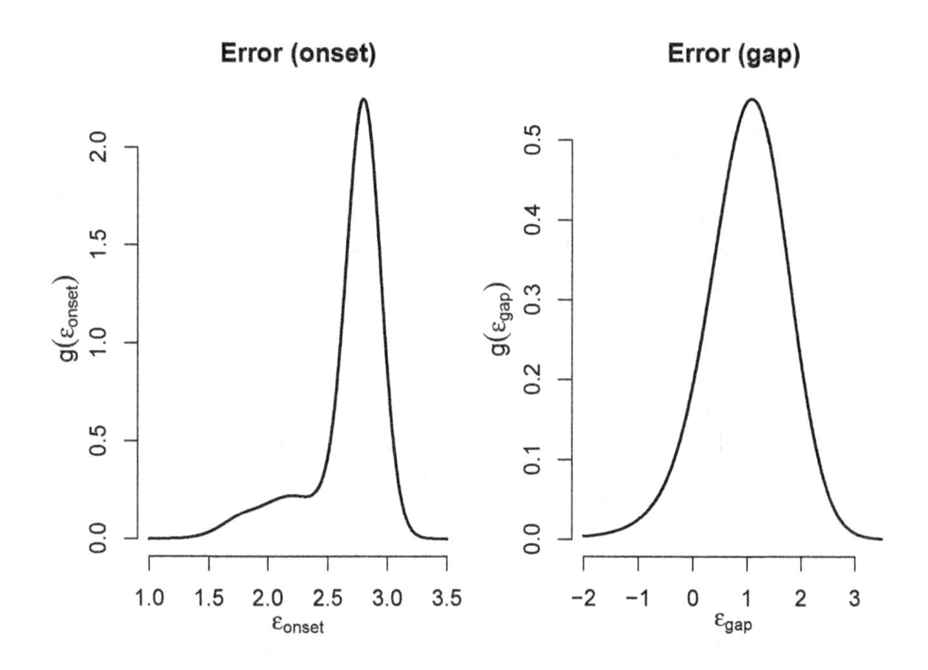

FIGURE 13.8: Mobile study. Estimated error density for log time of first purchase (T^O) and estimated error density for log gap time ($T^E - T^O$) both obtained from the R package bayesSurv .

13.3.3 Semiparametric modelling of univariate DI-censored data

Jara et al. (2010) proposed a Bayesian semiparametric approach for the analysis of multivariate DI-censored data, which was afterward implemented in the function LDPDdoublyint of the R package DPpackage. Here we focus on the univariate case. The approach is an extension of the Dirichlet Process prior introduced in Section 9.2.2 to dependent nonparametric priors. We have seen that $S \sim \mathcal{DP}(cS^*)$ implies a random survival curve that varies around a template S^* with variability that depends on c. When covariates are involved, the question is how to generalize the DP prior such that for each $\boldsymbol{X}_i$ the survival function has a DP prior and when considering $i = 1, \ldots, n$ these priors are linked. To this end Jara et al. (2010) proposed the *Linear Dependent Poisson-Dirichlet Process Mixture of Survival Models* and this approach was implemented in the function LDPDdoublyint to model the onset and gap time of several endpoints on the log-scale. The construction goes briefly as follows.

Let $\boldsymbol{Y}_i \equiv \log(\boldsymbol{T}_i) = (\log(T_i^O),\ \log(G_i))'$ $(i = 1, \ldots, n)$ be the bivariate vector of the onset time and the gap time on the log-scale. Let also $\boldsymbol{X}_i^O$ and $\boldsymbol{X}_i^G$ be the covariate vectors for the onset time and the gap time of dimensions q_O and q_G, respectively and combined into $\boldsymbol{X}_i = (\boldsymbol{X}_i^{O'},\ \boldsymbol{X}_i^{G'})'$. The following

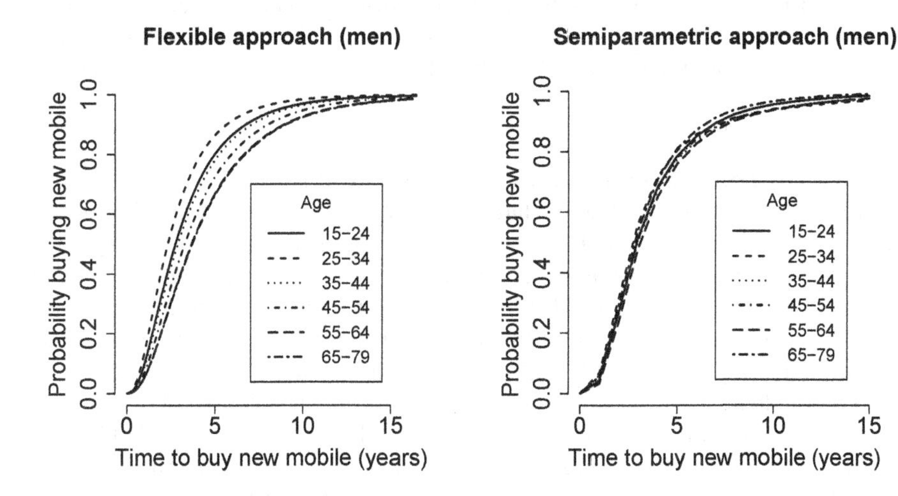

FIGURE 13.9: Mobile study. Estimated incidence function for men to buy a new mobile phone split up into age groups obtained from R packages bayesSurv (left panel) and DPpackage (right panel).

bivariate distribution of $\boldsymbol{Y}_i$ as a function of the covariates is proposed:

$$\begin{aligned}\boldsymbol{Y}_i \mid \boldsymbol{\beta}_i,\, \boldsymbol{\Sigma} &\sim \mathcal{N}_2(\boldsymbol{X}_i'\boldsymbol{\beta}_i,\, \boldsymbol{\Sigma}) \qquad (i = 1, \ldots, n),\\ \boldsymbol{\beta}_1, \ldots, \boldsymbol{\beta}_n \mid F_{\boldsymbol{\beta}} &\sim F_{\boldsymbol{\beta}},\\ F_{\boldsymbol{\beta}} \mid a,\, b,\, F_{\boldsymbol{\beta},0} &\sim \mathcal{PD}(a,\, b,\, F_{\boldsymbol{\beta},0}),\end{aligned}$$

with $\mathcal{PD}(a,\, b,\, F_{\boldsymbol{\beta},0})$ a Poisson-Dirichlet Process prior around the baseline distribution $F_{\boldsymbol{\beta},0}$ assumed to be a $(q_O + q_G)$-dimensional normal distribution $F_{\boldsymbol{\beta},0} = \mathcal{N}_{(q_O+q_G)}(\boldsymbol{m}_{\boldsymbol{\beta}},\, \boldsymbol{S}_{\boldsymbol{\beta}})$ for with mean $\boldsymbol{m}_{\boldsymbol{\beta}}$ and a covariance matrix $\boldsymbol{S}_{\boldsymbol{\beta}}$ which are both random hyperparameters being again assigned a suitable prior. The parameters a and b are involved in the stick-breaking representation of $\mathcal{PD}(a,\, b,\, F_{\boldsymbol{\beta},0})$. It suffices here to mention that $\mathcal{PD}(a,\, b,\, F_{\boldsymbol{\beta},0})$ boils down to the DP prior for special values of a and b. More details can be found in the manual of the package and Jara et al. (2010).

The advantage of this approach is that no independence of the onset and the gap time is assumed. This assumption was made in all above approaches, and may not always be appropriate. Another advantage of the approach is that it does not make restrictive assumptions as in AFT and PH models. For example, the approach allows for crossing survival curves. An obvious disadvantage of the method is its advanced character making it hard to specify reasonable prior values for some of the parameters and to interpret the results.

The use of the function LDPDdoublyint is exemplified below on the mobile data set.

Example 13.8 Survey on mobile phone purchases

Again it is of interest to know what factors determine the distribution of the time lapse to buy a new mobile phone, i.e., the gap time $G_i = T_i^E - T_i^O$. We check the effect of gender and age class, but now using the semiparametric method proposed by Jara et al. (2010). It is not possible to summarize their effect as regression coefficients, but must be evaluated graphically. The result of our analysis is shown in the right panel of Figure 13.9. Now the conclusions are not so clear. In effect, there seems to be not much difference between the age classes. We can also observe that the incidence curves cross. While the effect of age is relatively small, it can be clearly seen that again the age class 25−34 renews its mobile phone first. Although difficult to see, the effect of age seen before is partially confirmed here, i.e., the older the individual the longer (s)he will wait with renewing his or her mobile phone except for the oldest individuals who start off relatively slow in renewing their phone, but then catch up later.

□

13.3.3.1 R solution

The methodology has been implemented in the function LDPDdoublyint from the R package DPpackage.

Example 13.8 in R

As before, the complete R script to run the analysis described here is available in the supplementary materials. In Section 13.2.4.1, we have partially described the R code to use this function. The program to analyze DI-censored observations does not deviate much from that program, so we will highlight only some relevant parts here.

We remind the reader that the covariance matrix $\boldsymbol{S}_{\boldsymbol{\beta}}$ from the normal template $F_{0,\boldsymbol{\beta}} = \mathcal{N}(\boldsymbol{m}_{\boldsymbol{\beta}}, \boldsymbol{S}_{\boldsymbol{\beta}})$ of the Poisson Dirichlet process is assigned an inverse Wishart prior with $\nu_{\boldsymbol{\beta}}$ degrees of freedom (argument nub *in the prior specification when calling the* LPDPdoublyint *function). With a low value for* nub*, say 10, convergence could not be established. Recall that this value was used in Section 13.2.4.1 for the analysis of the Signal Tandmobiel data. The low value of* nub *aims for a vague prior on $\boldsymbol{S}_{\boldsymbol{\beta}}$. This could be a difficulty for convergence for reasons that are beyond the scope of this book. It is sufficient to say that in case of convergence problems* nub *needs to be increased. Here we increased it to 15.*

With 3 000 burn-in iterations and 15 000 iterations to base the posterior summary measures convergence was obtained. Apart from the above change in nub *and the obvious changes in the model parameters, the program used for calculations with the mobile phone data in Example 13.8 is quite similar to that used to analyze the dental data in Example 13.5. Namely, now there is only one onset and event time and as before only covariates are given for the event time, here gender and age class. Some extra calls (hence not using the plot statement provided with the package) produce the right panel of Figure 13.9.*

□

13.3.4 Modelling of multivariate DI-censored data

Multivariate DI-censored observations with dimension $d > 2$ are not common, except in oral health research. That is why especially the dental data, e.g., collected in the Signal Tandmobiel, were the trigger to develop a number of flexible and semiparametric Bayesian methods and software for multivariate DI-censored observations. In particular, the software BITE (Härkänen, 2003) was developed for the analysis of caries on permanent teeth. The software was also applied to analyze the risk factors for caries experience on the permanent first molars (teeth 14, 24, 34 and 44) of the Signal Tandmobiel study in Komárek et al. (2005b). The approach is based on a frailty PH model for both the emergence and caries process with piece-wise constant baseline hazard functions. The emergence and caries processes are linked via a gamma distributed frailty term. The levels of the piece-wise constant functions are assumed to have a gamma distribution while the prior of the jump points is a homogeneous Poisson process. As a result, the posterior predictive survival and hazard functions look smooth. The method is therefore coined as a nonparametric Bayesian approach. The BITE software can be consulted at `http://blogs.helsinki.fi/bayesian-intensity/`.

Alternatively, the functions bayessurvreg2 and bayessurvreg3 of the package bayesSurv can be used for a flexible analysis of multivariate DI-censored data by means of the AFT model with random effects. Illustrations for the bivariate case were given in previous sections. For additional illustrations, also with $d > 2$-dimensional responses, we refer to Komárek and Lesaffre (2006, 2008). In these publications, the aim was to find risk factors for caries experience on permanent teeth.

The semiparametric approach suggested in Section 13.3.3 was originally developed for multivariate DI-censored observations. Indeed, the function LDPDdoublyint of the package DPpackage was developed for this purpose. An illustration of its use on the caries process of permanent teeth of data collected in the Signal Tandmobiel study can be found in Jara et al. (2010).

13.4 Concluding remarks

We have reviewed some more complex problems that can be addressed with generally available software. Surely, there are other important problems that we have ignored in this chapter. An overview of other, somewhat more specialized, topics is discussed in Chapter 14.

Part IV

Concluding remarks

Chapter 14

Omitted topics and outlook

In this closing chapter we review topics that were not covered up to now. It is impossible to give a comprehensive overview since the number of potential extensions to deal with interval censoring case is, virtually, endless. Over the last decade one can observe an increasing interest in developing dedicated methods for interval-censored observations in a great variety of settings. In fact, each statistical approach can be extended to cover interval-censored observations. Therefore, and inevitably, also this final review of omitted topics will be incomplete. Frequentist and Bayesian approaches will now be mixed in each section.

14.1 Omitted topics

We omitted in this book the extra complication caused by *truncation* (see Section 1.2.5). This is not to claim that it can be ignored in practice. But the authors have little to no experience with studies affected by truncation. In addition, (frequentist) software that allows for truncation is somewhat limited. In case truncation is an issue, the following frequentist software can be consulted: SAS procedure ICPHREG (option ENTRY=, see SAS manual for details) allows left-truncation. To our knowledge, there is currently no R package that allows the combination of interval-censored and truncated data. With respect to Bayesian software, OpenBUGS and JAGS allow for truncation. Note that WinBUGS does not differentiate properly censoring from truncation.

For the same reasons we have omitted the discussion of the *proportional odds (PO)* model. The PO model is a regression model that models the conditional survival function given explanatory variables $\boldsymbol{X}$ assuming that

$$\frac{S(t \mid \boldsymbol{X})}{1 - S(t \mid \boldsymbol{X})} = \mathrm{e}^{-\boldsymbol{X}'\boldsymbol{\beta}} \frac{S_0(t)}{1 - S_0(t)}, \tag{14.1}$$

or

$$\mathrm{logit}\{S(t \mid \boldsymbol{X})\} = \mathrm{logit}\{S_0(t)\} - \boldsymbol{X}'\boldsymbol{\beta},$$

where $S_0(t)$ denotes the baseline survival function, that is the survival function for an individual whose explanatory variables all take the value zero and $\mathrm{logit}(s) = \log\{s/(1-s)\}$.

As with the PH model, Model (14.1) assumes that the effect of the explanatory variables is multiplicative, but now applied on the odds of the survival function instead of the hazard function. The logarithm of the odds of survival beyond t for the ith individual, relative to an individual for whom the explanatory variables are all equal to zero, is therefore just $\boldsymbol{X}'\boldsymbol{\beta}$. The model is therefore a linear model for the log-odds ratio.

The PO model has an important property regarding the hazard function. In the two-sample situation where $X = 0$ or 1, one has

$$\frac{\hbar(t \mid X = 1)}{\hbar(t \mid X = 0)} = \frac{1}{1 + \{\exp(-\beta) - 1\} S_0(t)},$$

which is a monotonic function of t and converges to 1 as $t \to \infty$. This means that, in contrast to the PH model, the ratio of hazards changes with time. More specifically, they converge to 1 with time. This property might be useful in applications for which a treatment effect diminishes over time.

A special case of the PO model is the log-logistic proportional odds model. For this model, it is assumed that $S_0(t)$ is a log-logistic distribution. The model is also an accelerated failure time model (see Section 2.3.2). In fact, it is the only distribution to share both the accelerated failure time property and the proportional odds property. The following frequentist software supports the PO model: function ic_par from R package icenReg (see Section 5.1.2) and SAS macro %SNPmacroint implementing the SemiNonParametric estimated baseline hazard approach of Zhang and Davidian (2008) (see Section 5.2.2). As for the Bayesian software, again BUGS-like software can be used.

We now review some specialized topics involving interval-censored observations. The fact that these developments are only briefly touched upon here has only to do with space limitations and not with their importance. In fact, we believe that some of the sections below, e.g., on informative censoring, should and will receive more attention in the future.

14.1.1 Competing risks and multistate models

Two related survival models were not treated in this book: the competing risks model and the multistate model. Both are extensions of the standard survival model. As in any survival model, also here interval-censoring may play a role.

Survival models have been applied to a great variety of endpoints. Examples in clinical research are overall mortality, cardiac mortality, mortality due to cancer, or just any event of interest such as time to re-operation, time to HIV or to AIDS, etc. In all these cases censoring plays a role. The standard approach is to assume that censoring is independent of the event of interest. This assumption is probably never entirely valid in practice, but often a reasonable start. For short-term cardiac studies, e.g., when evaluating the effect of thrombolytics on patients from acute myocardial infarction, death is likely attributable to a failing heart. On the other hand, in long-term cardiac studies, subjects may and will die also from other causes such as cancer. The risk factors for one cause of death often also have impact on other causes of death. Evaluating the risk factors for one cause of death using a standard survival analysis which censors a patient who died from another cause may be therefore questionable.

A *competing risks model* extends the classical survival model by taking into account a collection of mutually exclusive potential event types and their occurrence over time. In addition, the impact of different factors on the event types is examined. When browsing through recent issues of major clinical journals it is clear that a competing risks approach is seldom used even with multiple causes of death. Possibly the difficulty to interpret a competing risk analysis plays a role. But, the same is true for a standard survival analysis in a competing risk setting. An excellent introduction to competing risks models is given by Geskus (2015), but interval-censoring is not treated. For applications of competing risks models with interval censoring, we refer to some papers in Section 14.1.6 where joint modelling and competing risks are combined with interval censoring.

Standard survival models and competing risks models involve only two states: disease-free and diseased, alive and dead, etc. However, there are many occasions wherein one is interested in the way a subject evolves from healthy to diseased, or in general from one state to another. The statistical model that describes the transition from one state to another is called the *multistate model*. In a multistate analysis one models the transition probabilities as a function of covariates. An important example of a multistate model is the illness-death model. The illness-death cancer model assumes that a healthy patient can die in only two ways, either via cancer or directly. One says that the patient who is in the healthy state arrives at the absorbing state (death) either via the cancer state or directly. Multistate models have been shown to be an extremely rich class of models, with broad applications to many important questions in medicine, sociology, biology, ecology, etc. In most applications the transition between states can happen at irregular time points often unknown to the researcher. In that case, the time-intervals between states are interval-censored. Interval-censoring complicates matters especially for non- and semiparametric multistate models. The reader is referred to van den Hout (2016) for a review of multistate analyses for interval-censored data illustrated with a variety of examples. Other applications can be found in Sutradhar et al. (2011), Pak et al. (2017) and Boumezoued et al. (2017).

14.1.2 Survival models with a cured subgroup

In a classical survival model, it is implicitly assumed that eventually all subjects will experience the event, if they are followed-up long enough. This is often a too strong assumption. Namely, fifty years ago basically all cancer patients died from cancer. Nowadays, for several types of cancer there is a nonzero probability that subjects survive from the disease especially when detected early. In a mixture cure model, one assumes that a proportion of subjects $1 - \pi$ will eventually be cured, while the remaining portion of subjects will experience the event when the subjects are monitored long enough. The *mixture cure rate survival model* is formally defined as:

$$S(t) \;=\; \pi S_u(t) \;+\; (1 - \pi),$$

with $S_u(t)$ the survival function of the uncured population. As usual, the survival function may depend on covariates, but also π may depend on (possibly different) covariates. Most of the developments in mixture cure rate survival models assume a PH or AFT model for $S_u(t)$ combined with right-censoring, see Scolas (2016) for a history and review. In the same PhD thesis, mixture cure rate models for interval-censored survival times are reviewed and extensions to a flexible $S_u(t)$ are proposed. Model selection for mixture cure models is treated in Scolas et al. (2016a), while model checking is explored in Scolas et al. (2016b). Finally, mixture cure rate models have been extended to correlated interval-censored responses in Li and Ma (2010) and Xiang et al. (2011).

The cure fraction is by definition the proportion of subjects not experiencing the event in the long run. This requires in principle a long-term study, but in long-term studies subjects may die from other causes (cfr. competing risks). This may complicate the estimation of the cure fraction for the disease of interest even more so with interval-censored observations, which renders mixture cure rate models challenging in practice.

14.1.3 Multilevel models

The extension of frailty models to more than two levels is straightforward. Rabe-Hesketh et al. (2001) discussed the multilevel model for interval-censored data. However, their example on age of onset of smoking has only two levels. A discrete survival model was fitted to these data by considering the interval-censored outcomes as grouped data. The advantage of this approach is that it fits into a classical modelling framework. However, as seen in this book, in general a more sophisticated approach is required. While it is easy to imagine multilevel interval-censored responses, no other mention of such data is made in the literature.

14.1.4 Informative censoring

The methods discussed in this book are based on the noninformative censoring assumption allowing to neglect the distribution of the left and right endpoints and to work on a simplified likelihood function. For well-controlled studies, such as randomized clinical trials (RCTs), this is a reasonable assumption if the patients are compliant with prespecified examination visits. However, this assumption is likely violated when patients miss visits or drop out because of reasons related to the event under study. Pantazis et al. (2013)

argued that "individuals with poorer compliance to their treatment (...) and hence lower probabilities of virologic response, may also have a higher tendency to skip or delay clinical visits. Then the censoring intervals will be associated with the actual and unobserved time-to-event and the noninformative condition will no longer hold." This triggered a simulation study to check the performance of parametric and nonparametric models for interval-censored data when individuals miss some of the prescheduled visits in one- and two group situations. In the simulation study, a subject-specific probability of missing one or more visits was assumed, which could be common for all (missing completely at random (MCAR)), depending on a baseline covariate (missing at random (MAR)) or depending on the subject-specific time-to-virologic response (missing not at random (MNAR)). The correct approaches were compared with single imputation approaches such as the right endpoint and mid-point approach. The simulation results showed that violating the noninformative censoring assumption leads to biased estimators with the direction and the magnitude of the bias depending on the direction and the strength of the association between the probability of missing visits and the actual time-to-event. The conclusion of the exercise was that in practice statisticians should be cautious when frequency and timing of the examinations is highly variable as this may indicate violation of the noninformative assumption.

Various approaches have been suggested to address informative censoring. Just to mention a few early approaches with right censoring: Emoto and Matthews (1990) assumed a bivariate Weibull model for failure and censoring times while Zheng and Klein (1995) used a copula model. Huang and Wolfe (2002) allowed informative censoring on the cluster level in frailty models and suggested an EM procedure and in Huang et al. (2004) they suggested a test for informative censoring in clustered data. For independent subjects with an interval-censored response, Zhang et al. (2007) assumed a multivariate survival model with response (T_i, L_i, W_i) for the ith subject, with T_i the unobserved survival time, L_i the lower end of the interval $[L_i, U_i]$ with $L_i \leq T_i \leq U_i$ and $W_i = U_i - L_i$ the gap time. In fact, the method makes use of a PH frailty model building on the approach suggested in Huang and Wolfe (2002). More specifically, the authors assumed three hazard functions for the ith subject:

$$\begin{aligned} \hbar_i^{(T)}(t \mid \boldsymbol{X}_i, \boldsymbol{b}_i) &= \hbar_{t0}(t) \exp(\boldsymbol{\beta}_t' \boldsymbol{X}_i + b_{1i}), \\ \hbar_i^{(U)}(t \mid \boldsymbol{X}_i, \boldsymbol{b}_i) &= \hbar_{u0}(t) \exp(\boldsymbol{\beta}_u' \boldsymbol{X}_i + b_{2i}), \\ \hbar_i^{(W)}(t \mid \boldsymbol{X}_i, \boldsymbol{b}_i) &= \hbar_{w0}(t) \exp(\boldsymbol{\beta}_w' \boldsymbol{X}_i + b_{3i}), \end{aligned}$$

with unknown baseline hazard functions $\hbar_{t0}(t), \hbar_{u0}(t), \hbar_{w0}(t)$ and $\boldsymbol{\beta}_t, \boldsymbol{\beta}_u, \boldsymbol{\beta}_w$ the unknown regression parameters for T, U, W, respectively. For the latent parameters $\boldsymbol{b}_i = (b_{1i}, b_{2i}, b_{3i})'$ a joint three-variate normal distribution $\mathcal{N}(\mathbf{0}, \boldsymbol{\Sigma})$ is assumed. The method requires discretization of the baseline hazard functions at the observed left- and right-interval endpoints, which renders the method impractical for many distinct time points. Anyway, this approach as other recently developed approaches may be of interest to perform some sensitivity analyses allowing to deviate from the noninformative assumption, similar to what is now common practice in longitudinal studies examining the impact of missing-not-at-random mechanisms on the results of a statistical analysis. Finally, note that Kim et al. (2016) extended this approach to clustered interval-censored responses.

14.1.5 Interval-censored covariates

Up to now we discussed only interval-censored responses with all covariates known exactly. However, also covariates may be interval-censored. A common and simple example occurs with rounded age as covariate in a regression model. Rounding is a trivial example of interval censoring. Biomarker data are often subject to interval censoring because of limits of detection or quantification and regularly used as a regressor in all kinds of models. Thus, statistical models are needed that allow for interval-censored covariates. Note that the case of interval-censored covariates is closely related to covariates measured with error.

In a series of papers, Gómez and colleagues developed methods to deal with interval-censored covariates in a frequentist and a Bayesian context. In Gómez et al. (2003), linear regression is extended to allow for a discrete interval-censored covariate. A likelihood maximization algorithm is suggested to estimate the regression coefficients parametrically and the distribution of the covariate nonparametrically. The authors showed a much worse performance in mean squared error (MSE) in estimating the regression coefficients when instead the mid-point approach was employed. An extension of the computational procedure to the exponential family was further proposed. In Langohr et al. (2004), the methodology was extended to a parametric AFT model with a doubly censored response and a discrete interval-censored covariate. Diagnostics in the presence of an interval-censored covariate were looked at in this paper and in Topp and Gómez (2004). In Calle and Gómez (2005) a Bayesian approach is suggested, now lifting the restriction of discreteness for the interval-censored covariate. A parametric and a nonparametric approach for estimating the distribution of the interval-censored covariate were considered. Recall, however, that in Section 13.2.1 we showed that BUGS and SAS software allow for interval-censored covariates as long they are considered stochastic with a distribution that can be sampled from. In contrast, Calle and Gómez (2005) proposed a nonparametric approach based on a mixture of Dirichlet processes allowing all the components in the model to be specified parametrically, except for the distribution of the interval-censored covariate. A simulation study illustrated that the results are sensitive to the parametric assumption, showing thereby the better performance of the nonparametric approach. In Langohr and Gómez (2014), an R program is described to fit a linear regression model with an interval-censored covariate. The supplementary materials of the paper contain R commands. Finally, the R package SurvRegCensCov (Hubeaux and Rufibach, 2014) allows fitting a Weibull regression model with a right-censored endpoint and an interval-censored covariate.

14.1.6 Joint longitudinal and survival models

Up to now all examples only involved baseline covariates, called time-independent covariates. Survival models can be extended to allow for time varying covariates, such as blood pressure, body-mass index, income, etc. The classical Cox model with time varying covariates, called the extended Cox model, assumes the following hazard function

$$\hbar(t \mid \boldsymbol{Z}) = \hbar_0(t)\exp\{\boldsymbol{Z}(t)'\boldsymbol{\beta}\},$$

with $\boldsymbol{Z}(t)' = (Z_1(t), \ldots, Z_d(t), X_1, \ldots, X_p)$ where the first d covariates are time-dependent and the remaining covariates are time-independent. To estimate the model parameters with right-censored data, one can still use partial likelihood. An alternative popular approach consists of jointly analyzing the repeated measures and the survival outcomes, initiated in the area of HIV/AIDS research. The method can better deal with time-dependent covariates which are most often measured only periodically and with error. The method can also be seen as an approach to model informative dropouts. See Rizopoulos (2012) for a comprehensive treatment of the topic.

Here is an example of a joint model. Suppose that continuous time-dependent covariate $Z(t)$ takes the value Z_{ij} at time t_{ij} for the ith subject, with $i = 1, \ldots, n; j = 1, \ldots, n_i$. Suppose also that a linear mixed model represents the evolution over time of $Z(t)$ as follows:

$$Z_{ij} = \boldsymbol{X}'_{1i}\boldsymbol{\beta}_1 + b_{0i} + b_{1i}t_{ij} + \varepsilon_{ij} \qquad (i = 1, \ldots, n; j = 1, \ldots, n_i),$$

where $\boldsymbol{X}_{1i}$ representing the time-independent covariates of the ith subject affecting the time varying covariate. Often it is assumed that $\varepsilon_{ij} \sim \mathcal{N}(0, \sigma^2)$ independent of the random effects b_{0i}, b_{1i}. The second part of the joint model consists of a model for the true survival time T_i for the ith subject $(i = 1, \ldots, n)$. Often T_i is right-censored and assumed to have a parametric distribution, such as a Weibull distribution with shape parameter ϕ and scale parameter ψ_i. Suppose now that the scale parameter depends on covariates as follows:

$$\log(\psi_i) = \boldsymbol{X}'_{2i}\boldsymbol{\beta}_2 + \gamma_1 b_{0i} + \gamma_2 b_{1i}, \qquad (i = 1, \ldots, n),$$

with $\boldsymbol{X}_{2i}$ representing the time-independent covariates of the ith subject affecting the survival distribution and γ_1 and γ_2 expressing the link with the longitudinal process via the shared random effects b_{0i} and b_{1i}. These random effects describe both the subject-specific (linear) evolution of $Z(t)$ over time and affect the survival distribution. Finally, it is assumed that survival and longitudinal processes are conditionally independent given the random effects

b_{0i}, b_{1i}. That is, denoting the vector of random effects for the ith subject as $\boldsymbol{b}_i$ and the time vector as $\boldsymbol{t}_i$, it is assumed that:

$$p(\boldsymbol{Z}_i, T_i \mid \boldsymbol{X}_{1i}, \boldsymbol{X}_{2i}, \boldsymbol{t}_i, \boldsymbol{b}_i, \boldsymbol{\beta}_1, \boldsymbol{\beta}_2, \sigma^2) = p(\boldsymbol{Z}_i \mid \boldsymbol{X}_{1i}, \boldsymbol{t}_i, \boldsymbol{b}_i, \boldsymbol{\beta}_1, \sigma^2)\, p(T_i \mid \boldsymbol{X}_{2i}, \boldsymbol{b}_i, \boldsymbol{\beta}_2).$$

This conditional independence allows for a relatively easy calculation of the marginal likelihood, especially if one makes use of a data-augmentation algorithm. When the survival time is actually the time that a patient is in a study, i.e., the dropout time, the model provides one type of a missing-not-at-random mechanism, called the shared frailty model. This approach is one of the options to correct for bias in the estimates of the repeated measures outcome.

The majority of the developments and software in joint modelling is based on right-censored time-to-event data. When dealing with interval-censored event times, one approach might be to replace the intervals by singly imputed best guesses, such as the mid-point approach. This was done in Lee et al. (2011) on modelling growth of trees and tree mortality. The exact timing of when the tree died was, however, only known to occur in-between two inspection times. An alternative approach is of course to make full use of the interval-censored survival times, as advocated in this book.

Now follows some recent published examples of joint modelling with an interval-censored response. Sparling et al. (2006) present a parametric family of regression models for interval-censored survival times that accommodates time-independent and time-dependent covariates. The authors suggested a three-parameter family of survival distributions that includes the Weibull, negative binomial, and log-logistic distribution and developed a SAS macro, available from the website of the journal. Gueorguieva et al. (2012) were involved in a RCT comparing the effect of five antipsychotic medications. The authors noticed that patients dropped out of their treatment because of a variety of reasons, such as inefficacy and side effects. Some of these reasons may be informative and others not. Following earlier approach to account for cause-specific dropout, Gueorguieva et al. (2012) proposed a joint model for longitudinal outcome and competing risk, cause-specific dropouts that are interval-censored. A linear mixed model was chosen for the longitudinal part and Weibull or loglogistic cause-specific hazards. Interval-censoring was dealt with in a classical parametric manner, while the marginal likelihood is computed with Gaussian quadrature implemented in the SAS procedure NLMIXED. Rouanet et al. (2016) considered a joint latent class model for longitudinal data and interval-censored semicompeting events. Matters are more complicated when dealing with non- or semiparametric approaches. Hudgens et al. (2007) suggested a frequentist nonparametric joint modelling approach and applied it in a phase III HIV vaccine efficacy trial. Su and Hogan (2011) applied a Bayesian semiparametric approach to examine the impact of hepatitis C virus (HCV) on the effect of HAART treatment on HIV patients. This involved analyzing a doubly interval-censored response.

14.1.7 Spatial-temporal models

Spatial epidemiology is a rapidly growing field of research focused on analyzing various aspects of the geographical spread of diseases. The literature is vast and has seen a tremendous expansion in the last two decades especially. The reader is referred to Lawson (2006) and Lawson (2013) for comprehensive overviews. A spatial survival model combines survival data with spatial information. Pan et al. (2014) suggested a spatial-temporal model whereby the response is interval-censored. Their model is an extension of the model described in Section 11.2.1. More specifically, suppose that there are I regions with the ith region containing n_i subjects. Let the true failure time for the jth subject in the ith region be T_{ij} but only observed in the interval $(L_{ij}, U_{ij}]$. Further, let the $\boldsymbol{X}_{ij}$ be the covariate vector for the jth subject in the ith region, then the authors assumed the following model for the survival function:

$$S(t \mid \boldsymbol{X}_{ij}, \phi_i) = \exp\big\{-\Lambda_0(t)\exp(\boldsymbol{X}_{ij}'\boldsymbol{\beta} + \phi_i)\big\},$$

where $\Lambda_0(t)$ is the baseline cumulative hazards function and ϕ_i the spatial frailty of the ith region. A smooth function for $\Lambda_0(t)$ is assumed using integrated splines as for the R package ICBayes. Similar to what is described in Section 11.2.1 data augmentation is used to facilitate the MCMC procedure. A conditional autoregressive (CAR) prior is given on the vector of ϕ's, i.e., $p(\boldsymbol{\phi}) \propto \exp\Big\{-\frac{1}{2}\boldsymbol{\phi}'\boldsymbol{Q}\boldsymbol{\phi}\Big\}$, whereby $\boldsymbol{Q}$ is $I \times I$ positive definite matrix. A special case of this prior is the intrinsic conditional autoregressive prior whereby $\boldsymbol{Q}$ is based on the neighbouring information of the regions, more details can be found in Pan et al. (2014). Upon request to the first author, the R implementation of this approach can be obtained.

14.1.8 Time points measured with error

In all above statistical techniques we assumed that no errors are made in determining the endpoints of the interval. But this may not be the case in practice. Each time a decision needs to be made about the status of a patient judgment, errors are unavoidable. For instance, in oncology an important clinical endpoint is progression-free survival (PFS) defined as the time needed to see a progression of the disease in the patient, say the occurrence of a new tumor lesion. This endpoint is typically determined every 3 or 6 months with X-rays and judged by the local investigator and/or a team of experts. Whenever progression is detected, a patient might be re-allocated to a so-called rescue medication. However, it is known that progression is not always obvious to detect and that errors in diagnosis are likely to happen. A recent illustration of this problem can be found in Lesaffre et al. (2017)

where a futility analysis in an oncology trial was based on the judgment of PFS by the local investigator. However, it turned out that there was about 20% misclassification of progression when the expert panel diagnosis is taken as gold standard. This misclassification was partly the cause that the study was stopped due to futility.

Misclassification of the outcome is not uncommon and is likely to occur in many clinical and other research areas. Misclassification was also an issue in the Signal Tandmobiel study, first introduced in Chapter 1, where caries on individual teeth was examined by sixteen examiners on an annual basis. The comparison of their scoring with a benchmark scorer in validation studies revealed considerable misclassification. The longitudinal analysis of caries experience (CE), which is caries in the past or the present, can be done with a survival model whereby the response is interval-censored. Based on developments in García-Zattera et al. (2010) and García-Zattera et al. (2012), García-Zattera et al. (2016) proposed a flexible AFT model for misclassified clustered interval-censored data and applied it to analyze CE data of four permanent molars in the Signal Tandmobiel study. A Bayesian approach making use of Gibbs sampling delivers the estimates of the model parameters corrected for misclassification. In contrast to many epidemiological studies, the model does not require any internal nor external validation studies to estimate separately the misclassification correction terms. An option in the R package bayesSurv allows for misclassification of the response in interval-censored and DI-censored observations.

14.1.9 Quantile regression

In linear regression, the aim is to estimate the mean of a response given a covariate value. With a non-Gaussian response or when the aim is the estimate another measure than a central location, quantile regression proposed by Koenker and Bassett (1978) is preferable. Most quantile regression models deal with observed responses. Extensions for right-censored responses were also developed, but only recently quantile regression for interval-censored responses was suggested by Zhou et al. (2017).

14.2 Outlook

Despite the availability of statistical software, interval-censored data are often not analyzed in an appropriate manner. Already in 1998, Lindsey and Ryan (1998) argued that there is no reason anymore that the interval-censored character of the data is not taken into account, since commercial software is capable of dealing with interval censoring. Still too often, ad hoc approaches, such as the mid-point approach, are used in practice. This has been pointed out repeatedly in the literature. For the analysis of PFS, Stone et al. (2011) reported on the research outcomes and recommendations from a working group in the Pharmaceutical Research and Manufacturers Association (PhRMA). Among other things, the report stipulates that, "Despite the presence of multiple published techniques to perform analyses of such interval-censored data in practice, clinical trials are not routinely analysed using interval-censored analysis (ICA) approaches." One of the possible reasons for not addressing interval censoring in RCTs is the belief that the balanced nature of clinical trial data renders them robust for the impact of interval censoring on the results of the statistical analysis. This may well be the case in many situations, but it is never known in advance. Pantazis et al. (2013) concluded from their simulation study that the single imputation methods can yield misleading results even when the censoring intervals depend only on a baseline covariate, say treatment. This was also warned against in Stone et al. (2011). MacKenzie and Peng (2013) examined single imputation methods for interval-censoring, i.e., replacing the intervals with begin, middle and end of the interval, in the context of longitudinal RCTs with two groups. Analytical and simulation research applied to exponential and Weibull survival models both PH and non-PH data revealed that the ad hoc methods are artificially too precise. In addition the ad hoc methods are not consistent, e.g., the mid-point method shows between 1% to 5% bias. For non-PH data, the ICA approach may not always lead to more conservative standard errors, hence the power may be decreased when using the ad hoc methods. To conclude, when appropriate techniques are available there seems no reason anymore not to use ICA approaches. This book aims to fill in the gap in the literature to bring together in one volume the different software tools that can be used to analyze a variety of problems involving interval-censored data. So there is no excuse anymore not to use appropriate ICA techniques.

When decided for an ICA approach, the question remains whether to choose for a parametric or a non-/semiparametric method. From a very limited simulation study, Lindsey and Ryan (1998) found the results quite robust with respect to the choice of a parametric survival function. But they did not go that far claiming that the choice of parametric survival does not matter. They also argued that nonparametric models are highly over-rated, although they can have some role in checking the fit of parametric models. Gong and Fang (2013) examined the performance of parametric PH models for three baseline hazard functions (exponential, Weibull and piecewise exponential) when the baseline hazard function is mis-specified. The conclusion of their simulation study was that mis-specification of the baseline hazard had minimal impact on bias and MSE. In practice we recommend, as in each and every statistical analysis, to perform a sensitivity analysis varying distributional assumptions in a reasonable manner and evaluating whether the conclusions materially change.

Stone et al. (2011) reported on the importance to keep in mind that informative censoring might take place in a study. The impact of violation of the noninformative censoring assumption is clearly illustrated in Pantazis et al. (2013). Some recent models to allow for informative censoring have been proposed and reviewed in Section 14.1. But more research is needed to appreciate their performance, and equally important is to have software tools to perform a sensitivity analysis which allows for informative censoring. Basically, what is needed is a sensible primary analysis of the data and additional sensible analyses. This is the recommendation nowadays for RCTs plagued with missing data.

Part V

Appendices

Appendix A

Data sets

A.1 Homograft study

The homograft study is described in Section 1.6.1. The data are too large to be provided within the book. They are available in the supplementary materials and in the R package icensBKL. Data about the first homograft in 272 patients are provided. The list of variables with some summary statistics are given below:

1. Patient identification number
2. Follow-up time since the operation in years (min = 0, Q1 = 2.6, median = 5.9, mean = 5.8, Q3 = 8.9, max = 14.0)
3. Indicator for homograft failure (0 = Censored (N = 213), 1 = Failure (N = 59))
4. Age of the patient in years (min = 0, Q1 = 5.1, median = 5.9, mean = 13.9, Q3 = 24.4, max = 69.1, Nmiss = 1)
5. Type of graft (PH = Pulmonary donor homograft (N = 230), AH = Aortic donor homograft (N = 42))

A.2 AIDS clinical trial

The AIDS clinical trial study is described in Section 1.6.3. Table A.1 displays the data. The variables L_C and U_C contain the lower and upper limit of the interval that contains time of CMV shedding in months. The variables L_M and U_M contain the lower and upper limit of the interval that contains time of MAC colonization in months. A dot in L_C or L_M represents a left-censored observation and a dot in U_C or U_M represents a right-censored observation. The intervals are given in closed form, that is $[L, R]$. The column # represents the multiplicity of each observation. The data set contains in total 204 observations. The data are available in the supplementary materials and in the R package icensBKL.

TABLE A.1: Data of AIDS trial ACTG 181. Observed intervals of months to CMV shedding ($[L_C, U_C]$) and MAC colonization ($[L_M, U_M]$) with multiplicity of observations (#).

L_C	U_C	L_M	U_M	#	L_C	U_C	L_M	U_M	#	L_C	U_C	L_M	U_M	#
0	3	0	.	3	9	12	9	.	1	18	.	0	.	1
0	3	3	.	1	9	12	12	.	3	18	.	6	.	1
0	3	6	.	3	9	9	15	.	1	18	.	9	.	1
0	6	6	.	1	9	12	24	.	1	18	.	12	.	1
0	3	9	.	1	9	9	27	.	1	18	.	15	.	3
0	3	12	.	5	12	12	0	.	1	18	.	18	.	6
0	3	15	.	5	12	15	0	.	1	21	.	15	.	1
0	6	15	.	1	12	15	6	.	1	.	0	0	.	9
3	3	3	.	1	12	15	15	.	1	.	0	3	.	3
3	3	6	.	1	12	15	21	.	1	.	0	6	.	10
3	3	9	.	3	0	.	0	.	6	.	0	9	.	6
3	6	9	.	2	3	.	0	.	2	.	0	12	.	8
3	6	12	.	3	6	.	0	.	1	.	0	15	.	5
3	3	15	.	2	6	.	3	.	2	.	0	18	.	4
3	6	15	.	2	6	.	6	.	3	.	0	21	.	1
3	6	18	.	1	6	.	9	.	1	0	.	0	3	1
3	3	21	.	1	9	.	0	.	2	6	.	0	6	1
6	6	0	.	2	9	.	9	.	3	6	.	6	6	1
6	9	0	.	1	9	.	12	.	1	12	.	0	3	1
6	9	9	.	1	12	.	0	.	5	12	.	0	6	1
6	6	12	.	1	12	.	6	.	1	15	.	0	3	1
6	9	12	.	2	12	.	9	.	4	21	.	15	15	1
6	6	15	.	1	12	.	12	.	10	3	.	.	0	1
6	9	15	.	1	15	.	0	.	3	9	.	.	0	1
6	6	18	.	1	15	.	3	.	1	12	.	.	0	1
6	9	18	.	2	15	.	6	.	1	0	3	0	6	1
9	9	0	.	1	15	.	9	.	2	3	6	6	12	1
9	12	0	.	2	15	.	12	.	8	9	9	9	9	1
9	9	9	.	2	15	.	15	.	9	.	0	.	0	1

A.3 Survey on mobile phone purchases

The survey on mobile phone purchases is described in Section 1.6.5. The following variables are available for the 478 respondents used for analysis:

1. Lower limit purchase date previous phone
2. Upper limit purchase date previous phone
3. Lower limit purchase date current phone
4. Upper limit purchase date current phone
5. Gender (0 = Male (N = 258), 1 = Female (N = 220))
6. Age group (1 = "15–24 years" (N = 57), 2 = "25–34 years" (N = 83), 3 = "35–44 years" (N = 66), 4 = "45–54 years" (N = 81), 5 = "55–64 years" (N = 95), 6 = "65–79 years" (N = 96))
7. Size of household (1 = "1 person" (N = 103), 2 = "2 persons" (N = 213), 3 = "3 persons" (N = 58), 4 = "4 persons" (N = 57), 5 = "5 persons or more" (N = 47))
8. Household income before taxes (1 = "30 000 Eur or less" (N = 118), 2 = "30 001–50 000 Eur" (N = 152), 3 = "50 001–70 000 Eur" (N = 87), 4 = "> 70 000 Eur" (N = 61), 5 = "No answer" (N = 60))

The data are available in the supplementary materials and in the R package icensBKL.

A.4 Mastitis study

The mastitis study is described in Section 1.6.6. Regarding the parity of the 100 cows, 42 cows had one calving, 44 cows had 2 to 4 calvings and 14 cows had more than 4 calvings. In total, 317 of 400 udder quarters were infected during the lactation period. Hence, 83 observations are right-censored. In addition, 26 udder quarters were already infected at the first visit and result therefore in left-censored observations. The data are too large to be provided within the book. The data are available in the supplementary materials and in the R package icensBKL.

A.5 Signal Tandmobiel study

The Signal Tandmobiel study is described in Section 1.6.7. The data are too large to be provided within the book. They are available in the supplementary materials and in the R package icensBKL. The list of variables with some summary statistics are given below:

1. Patient identification number
2. Gender of the child (0 = boy (N = 256), 1 = girl (N = 244))
3. dmft score at baseline around the age of 7 years (min = 0, Q1 = 0, median = 1, mean = 2.4, Q3 = 4, max = 16)
4. Occlusal plaque status of tooth 16, 26, 36 and 46 (4 variables)

Tooth	**16**	**26**	**36**	**46**
0 = No plaque	291	305	295	300
1 = Pits/fissures	113	102	104	96
2 = Total	14	14	16	20
Missing	82	79	85	84

5. Brushing frequency at baseline (0 = Less than daily (N = 67), 1 = At least once a day (379), Nmiss = 54)
6. Sealing present on tooth 16, 26, 36 and 46 (4 variables)

Tooth	**16**	**26**	**36**	**46**
0 = No	404	415	400	395
1 = Yes	20	20	25	23
Missing	76	65	75	82

7. dmft score of teeth 54, 55, 64, 65, 74, 75, 84 and 85 (8 variables)

Tooth	**54**	**55**	**64**	**65**	**74**	**75**	**84**	**85**
0 = Sound	393	352	381	358	345	365	343	343
1 = Caries experience	102	147	114	141	154	133	153	157
Missing	5	1	5	1	1	2	4	0

8. Lower and upper limit of interval (Lxy, Rxy] that contains the emergence time of tooth xy with x = 1, 2, 3, 4 and y = 4, 5, 6 (24 variables)
9. Age at visit selected to determine current status (mean = 8.9, SD = 1.65)

10. Current status of tooth 14, 24, 34 and 44 (4 variables)

Tooth	**14**	**24**	**34**	**44**
0 = Not emerged	377	372	383	375
1 = Emerged	123	128	117	125

11. Lower and upper limit of interval (CLxy, CRxy] that contains the time to caries of tooth xy with x = 1, 2, 3, 4 and y = 6 (8 variables)

Appendix B

Distributions

This appendix complements the book by a list of distributions used often in survival parametric models. Popular parametric survival distributions are the log-normal, log-logistic, Weibull and gamma distribution. For each of these distributions we provide the density, survival and hazard function, the mean, variance and median. Each distribution is parametrized by a 2-dimensional parameter vector $\boldsymbol{\theta} = (\gamma, \alpha)'$, $\gamma, \alpha > 0$ such that the mean and the variance of T are proportional to α and α^2, respectively. Hence α is often called *scale* of the event time distribution. Parameter γ is often referred to as the *shape* of the event time distribution. Further, we mention the one-parameter exponential and Rayleigh distributions which are special cases of the Weibull distribution.

The chosen parametrizations lead to the distributions of the logarithm of the event time coming from the location and scale family with location $\mu \in \mathbb{R}$ and scale $\sigma > 0$. That is, the density g of $Y = \log(T)$ can be written as

$$g(y) = \frac{1}{\sigma}\, g^*\Big(\frac{y-\mu}{\sigma}\Big), \tag{B.1}$$

where g^* is the standardized density of a particular family of distributions. Further, location μ and scale σ of the log-event time are linked to parameters γ and α of the event time distribution by the relationships

$$\mu = \log(\alpha), \qquad \sigma = \frac{1}{\gamma}. \tag{B.2}$$

By no means the survival distributions mentioned in this appendix should be considered as the only useful distributions for survival analysis. Our choice simply reflects that they can be fitted directly in R, SAS and BUGS-like software.

B.1 Log-normal $\mathcal{LN}(\gamma,\,\alpha)$

$$\left.\begin{aligned} f(t) &= \frac{\gamma}{t\sqrt{2\pi}}\exp\biggl[-\frac{\bigl\{\gamma\log(\frac{t}{\alpha})\bigr\}^2}{2}\biggr],\\ S(t) &= 1-\Phi\Bigl(\gamma\log\Bigl(\frac{t}{\alpha}\Bigr)\Bigr),\\ \hbar(t) &= \frac{\gamma}{t\bigl\{1-\Phi\bigl(\gamma\log(\frac{t}{\alpha})\bigr)\bigr\}\sqrt{2\pi}}\exp\biggl[-\frac{\bigl\{\gamma\log(\frac{t}{\alpha})\bigr\}^2}{2}\biggr] \end{aligned}\right\}\quad t>0, \tag{B.3}$$

$$\left.\begin{aligned} \mathsf{E}(T) &= \alpha\,\exp\Bigl(\frac{1}{2\gamma^2}\Bigr),\\ \mathsf{var}(T) &= \alpha^2\Bigl\{\exp\Bigl(\frac{3}{2\gamma^2}\Bigr)-\exp\Bigl(\frac{1}{2\gamma^2}\Bigr)\Bigr\},\\ \mathsf{med}(T) &= \alpha. \end{aligned}\right\} \tag{B.4}$$

The distribution of $Y=\log(T)$ is normal $\mathcal{N}(\mu,\,\sigma^2)$ with a density g and the first two moments given by

$$\left.\begin{aligned} g(y) &= \frac{1}{\sigma\sqrt{2\pi}}\exp\Bigl\{-\frac{1}{2}\Bigl(\frac{y-\mu}{\sigma}\Bigr)^2\Bigr\},\\ \mathsf{E}(Y) &= \mu,\qquad \mathsf{var}(Y)=\sigma^2. \end{aligned}\right\} \tag{B.5}$$

B.2 Log-logistic $\mathcal{LL}(\gamma,\,\alpha)$

$$\left.\begin{aligned} f(t) &= \frac{\frac{\gamma}{\alpha}\,(\frac{t}{\alpha})^{\gamma-1}}{\bigl\{1+(\frac{t}{\alpha})^{\gamma}\bigr\}^2},\qquad S(t)=\frac{1}{1+(\frac{t}{\alpha})^{\gamma}},\\ \hbar(t) &= \frac{\frac{\gamma}{\alpha}\,(\frac{t}{\alpha})^{\gamma-1}}{1+(\frac{t}{\alpha})^{\gamma}}. \end{aligned}\right\}\quad t>0. \tag{B.6}$$

$$\left.\begin{aligned} \mathsf{E}(T) &= \alpha \cdot \frac{\pi}{\gamma \sin\left(\frac{\pi}{\gamma}\right)}, && \gamma > 1, \\ \mathsf{var}(T) &= \alpha^2 \left\{ \frac{2\pi}{\gamma \sin\left(\frac{2\pi}{\gamma}\right)} - \frac{\pi^2}{\gamma^2 \sin^2\left(\frac{\pi}{\gamma}\right)} \right\}, && \gamma > 2, \\ \mathsf{med}(T) &= \alpha, && \gamma > 0. \end{aligned}\right\} \tag{B.7}$$

The distribution of $Y = \log(T)$ is logistic $\mathcal{L}(\mu, \sigma)$ with a density g, cumulative distribution function G and the first two moments given by

$$g(y) = \frac{\exp\left(\frac{y-\mu}{\sigma}\right)}{\sigma \left\{1 + \exp\left(\frac{y-\mu}{\sigma}\right)\right\}^2}, \qquad G(y) = \frac{\exp\left(\frac{y-\mu}{\sigma}\right)}{1 + \exp\left(\frac{y-\mu}{\sigma}\right)},$$

$$\mathsf{E}(Y) = \mu, \qquad \mathsf{var}(Y) = \frac{\pi^2}{3}\,\sigma^2.$$

B.3 Weibull $\mathcal{W}(\gamma, \alpha)$

$$\left.\begin{aligned} f(t) &= \frac{\gamma}{\alpha}\left(\frac{t}{\alpha}\right)^{\gamma-1} \exp\left\{-\left(\frac{t}{\alpha}\right)^{\gamma}\right\}, \\ S(t) &= \exp\left\{-\left(\frac{t}{\alpha}\right)^{\gamma}\right\}, \\ \hbar(t) &= \frac{\gamma}{\alpha}\left(\frac{t}{\alpha}\right)^{\gamma-1}, \end{aligned}\right\} \quad t > 0. \tag{B.8}$$

$$\left.\begin{aligned} \mathsf{E}(T) &= \alpha\,\Gamma\left(1 + \frac{1}{\gamma}\right), \\ \mathsf{var}(T) &= \alpha^2 \left\{\Gamma\left(1 + \frac{2}{\gamma}\right) - \Gamma^2\left(1 + \frac{1}{\gamma}\right)\right\}, \\ \mathsf{med}(T) &= \alpha\left(\log 2\right)^{\frac{1}{\gamma}}. \end{aligned}\right\} \tag{B.9}$$

Since $\log\{-\log\{S(t)\}] = -\gamma\log(\alpha) + \gamma\,\log(t)$, the empirical check for the Weibull distribution is obtained by a plot of $\log[-\log\{\widehat{S}(t)\}]$, where $\widehat{S}$ is the NPMLE of S, against $\log(t)$ which should resemble approximately a straight line.

The distribution of $Y = \log(T)$ is the type I least extreme value distribution $\mathcal{EV}(\mu, \sigma)$ with a density g, cumulative distribution function G and the first two moments given by

$$g(y) = \frac{1}{\sigma}\,\exp\Bigl\{\frac{y-\mu}{\sigma} - \exp\Bigl(\frac{y-\mu}{\sigma}\Bigr)\Bigr\},$$

$$G(y) = 1 - \exp\Bigl\{-\exp\Bigl(\frac{y-\mu}{\sigma}\Bigr)\Bigr\},$$

$$\mathsf{E}(Y) = \mu - \nu\sigma, \qquad \mathsf{var}(Y) = \frac{\pi^2}{6}\,\sigma^2,$$

where $\nu = 0.57721\ldots$ is Euler's constant. Finally, we remark that the distribution of $-Y = -\log(T)$ is the type I greatest extreme value distribution, referred to as the Gumbel distribution.

B.4 Exponential $\mathcal{E}(\alpha)$

The exponential distribution $\mathcal{E}(\alpha)$ is a special case of the Weibull distribution and it is obtained as $\mathcal{W}(1,\,\alpha)$. Hence,

$$\left.\begin{array}{c} f(t) = \dfrac{1}{\alpha}\,\exp\Bigl(-\dfrac{t}{\alpha}\Bigr), \qquad S(t) = \exp\Bigl(-\dfrac{t}{\alpha}\Bigr), \\[2ex] \hbar(t) = \dfrac{1}{\alpha}, \end{array}\right\} \quad t > 0.$$

$$\mathsf{E}(T) = \alpha, \qquad \mathsf{var}(T) = \alpha^2, \qquad \mathsf{med}(T) = \alpha\,\log 2.$$

B.5 Rayleigh $\mathcal{R}(\alpha)$

Also the Rayleigh distribution $\mathcal{R}(\alpha)$ is a special case of the Weibull distribution and it is obtained as $\mathcal{W}(2,\,\alpha)$. Hence,

$$\left.\begin{array}{c} f(t) = \dfrac{2}{\alpha^2}\,t\,\exp\Bigl(-\dfrac{t^2}{\alpha^2}\Bigr), \qquad S(t) = \exp\Bigl(-\dfrac{t^2}{\alpha^2}\Bigr), \\[2ex] \hbar(t) = 2\,\dfrac{t}{\alpha^2}, \end{array}\right\} \quad t > 0.$$

$$\mathsf{E}(T) = \frac{\sqrt{\pi}}{2}\,\alpha, \qquad \mathsf{var}(T) = \alpha^2\,\Bigl(1 - \frac{\pi}{4}\Bigr), \qquad \mathsf{med}(T) = \alpha\,\sqrt{\log 2}.$$

B.6 Gamma(γ, α)

$$\left.\begin{aligned} f(t) &= \frac{1}{\Gamma(\gamma)\alpha^\gamma} t^{\gamma-1} \exp\Big(-\frac{t}{\alpha}\Big), \\ S(t) &= 1 - \frac{1}{\Gamma(\gamma)} \Gamma_{\frac{t}{\alpha}}(\gamma), \\ \hbar(t) &= \frac{1}{\Big\{\Gamma(\gamma) - \Gamma_{\frac{t}{\alpha}}(\gamma)\Big\}\alpha^{\gamma-1}} t^{\gamma-1} \exp\Big(-\frac{t}{\alpha}\Big) \end{aligned}\right\} \quad t > 0, \qquad \text{(B.10)}$$

where $\Gamma(s) = \int_0^\infty t^{s-1} e^{-t}\,\mathrm{d}t$ is the gamma function and $\Gamma_x(s) = \int_0^x t^{s-1} e^{-t}\,\mathrm{d}t$ is the lower incomplete gamma function.

$$\mathsf{E}(T) = \gamma\alpha, \qquad \mathsf{var}(T) = \gamma\alpha^2, \qquad \mathsf{med}(T) \text{ has no simple closed form.}$$

B.7 R solution

The recommended package stats provides the density, the cumulative distribution function, the quantile function and a random number generator for survival distributions: functions dlnorm, plnorm, qlnorm, rlnorm (log-normal distribution), dweibull, pweibull, qweibull, rweibull (Weibull distribution), dexp, pexp, qexp, rexp (exponential distribution), dgamma, pgamma, qgamma, rgamma (gamma distribution), respectively. In the package icensBKL, we have added functions dllogis, pllogis, qllogis, rllogis for the log-logistic distribution. Parameters of the assumed distributions are supplied in all functions as arguments:

Log-normal distribution

- meanlog: mean $\mu = \log(\alpha)$ of the logarithm of the event time;
- sdlog: standard deviation $\sigma = \gamma^{-1}$ of the logarithm of the event time.

Log-logistic, Weibull, gamma distribution

- shape: shape γ of the event time distribution;

- scale: scale α of the event time distribution.

Exponential distribution

- rate: inverted scale (here equal to the inverted mean) α^{-1} of the event time distribution.

For example, the values of the densities, survival and hazard functions of the three two-parametric survival distributions with parameter values taken from Table 3.2 evaluated at $t = 7, 7.01, 7.02, \ldots, 15$ are computed using the following code.

```
> library("icensKBL")              ### needed for function dllogis
> t <- seq(7, 15, by=0.01)
```

Densities:

```
> ft.lnorm   <- dlnorm(t, meanlog = 2.352, sdlog = exp(-2.175))
> ft.llogis <- dllogis(t, shape = 15.222, scale = 10.518)
> ft.weibull <- dweibull(t, shape = 11.050, scale = 11.014)
```

Survival functions:

```
> St.lnorm   <- plnorm(t, meanlog = 2.352, sdlog = exp(-2.175),
+                        lower.tail = FALSE)
> St.llogis <- pllogis(t, shape = 15.222, scale = 10.518,
+                        lower.tail = FALSE)
> St.weibull <- pweibull(t, shape = 11.050, scale = 11.014,
+                          lower.tail = FALSE)
```

Hazard functions:

```
> ht.lnorm   <- ft.lnorm / St.lnorm
> ht.llogis  <- ft.llogis / St.llogis
> ht.weibull <- ft.weibull / St.weibull
```

Calculated functional values are plotted on Figure B.1 which was drawn using

```
> plot(t, ft.lnorm, type = "l", xlab = "Time", ylab = "Density",
+      ylim = c(0, 0.37))
> lines(t, ft.llogis, lty = 2)
> lines(t, ft.weibull, lty = 3)

> plot(t, St.lnorm, type = "l", xlab = "Time", ylab = "Survival function",
+      ylim = c(0, 1))
> lines(t, St.llogis, lty = 2)
```

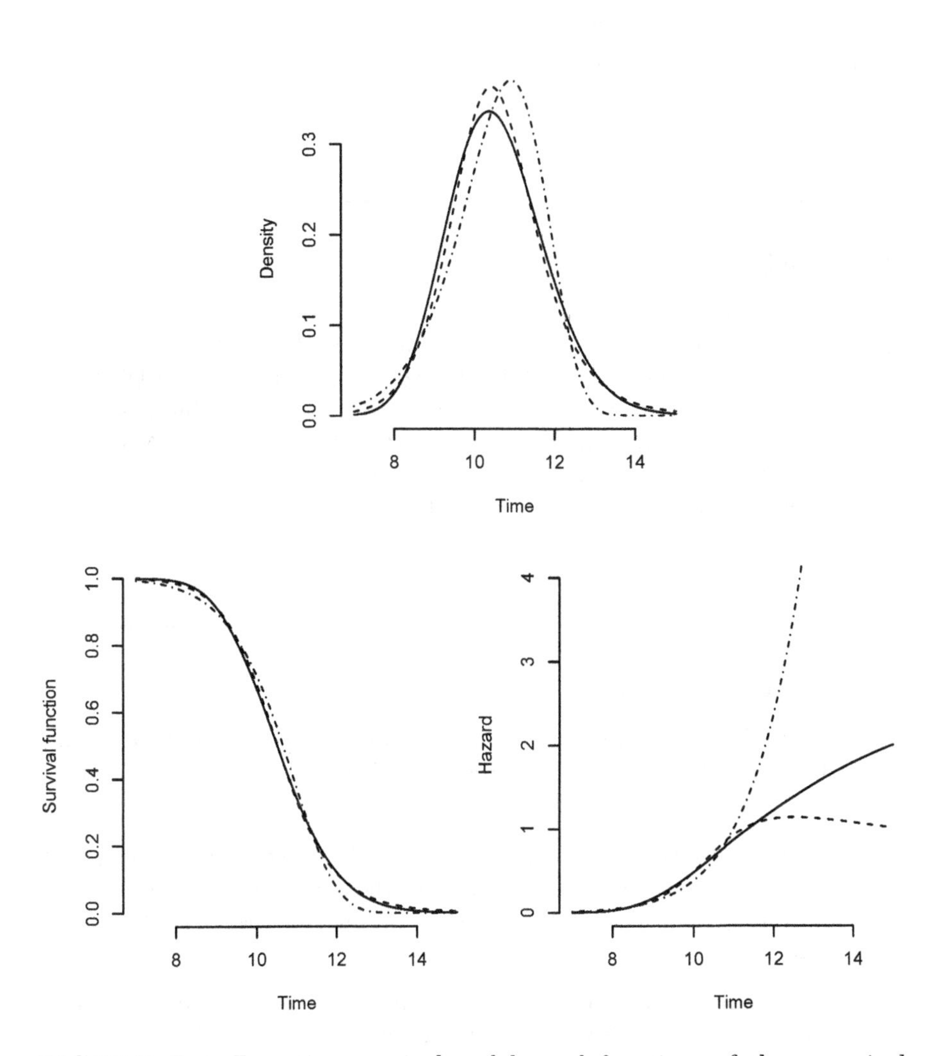

FIGURE B.1: Densities, survival and hazard functions of three survival distributions. Solid line: $\mathcal{LN}(8.806, 10.506)$, dashed line: $\mathcal{LL}(15.222, 10.518)$, dotted-dashed line: $\mathcal{W}(11.050, 11.014)$.

```
> lines(t, St.weibull, lty = 4)

> plot(t, ht.lnorm, type = "l", xlab = "Time", ylab = "Hazard",
+       ylim = c(0, 4))
> lines(t, ht.llogis, lty = 2)
> lines(t, ht.weibull, lty = 4)
```

B.8 SAS solution

Parametric models can be fitted in SAS using the using the SAS/STAT procedure LIFEREG or the SAS/QC procedure RELIABILITY. Using a combination of the options DISTRIBUTION=<distribution-name> (or DIST= or D=) and NOLOG in the model statement a whole list of distributions can be fitted. An overview is given in Table E.1. Note that the keyword GAMMA does not refer to the standard gamma distribution but to the generalized gamma distribution. For more details we refer to the SAS documentation.

B.9 BUGS solution

In WinBUGS, OpenBUGS, as well as in JAGS, log-normal, Weibull and exponential distributions can easily be used either in the likelihood part of the model or as prior distributions as follows.

Log-normal distribution is implemented by the function dlnorm(mu, tau), where mu $= \mu = \log(\alpha)$ is the mean of the logarithm of the event time and tau $= \sigma^{-2} = \gamma^2$ is the inverted variance of the logarithm of the event time.

Weibull distribution is provided by the function dweib(v, lambda), where somehow an alternative parametrization compared to (B.8) is considered. Namely, if v $= \upsilon$ and lambda $= \lambda$, the following density is assumed

$$f(t) = \upsilon\,\lambda\,t^{\upsilon-1}\exp\bigl(-\lambda\,t^{\upsilon}\bigr), \qquad t > 0. \tag{B.11}$$

Comparison of (B.11) with (B.8) shows that

$$\upsilon = \gamma = \frac{1}{\sigma}, \qquad \lambda = \Bigl(\frac{1}{\alpha}\Bigr)^{\gamma} = \exp\Bigl(-\frac{\mu}{\sigma}\Bigr).$$

That is, parameter υ of the BUGS parametrization is the inverted scale of the logarithm of the event time distribution, whereas $-\log(\lambda)$ is the ratio between location and scale parameters of the logarithm of the event time distribution.

Exponential distribution is given by the function dexp(lambda), where lambda $= \alpha^{-1}$ is the inverted scale (here equal to the inverted mean) of the event time distribution.

The **log-logistic** distribution is not directly available in neither WinBUGS/OpenBUGS nor JAGS. Nevertheless, models involving the log-logistic distribution can be used while using the available logistic distribution in combination with the log transformation of the respective random variables. The logistic distribution is provided by the function dlogis(mu, tau). If the logistic distribution is specified as the distribution of the logarithm of the event time then mu $= \mu = \log(\alpha)$ is the location parameter of the log-event time and tau $= \sigma^{-2} = \gamma^2$ is the inverted squared scale parameter of the log-event time.

B.10 R and BUGS parametrizations

To help the reader to find a link between different parametrizations and different syntax used by R and BUGS for the distributions outlined in this appendix, an overview is provided in Tables B.1 and B.2. For R, we list the name of the function that calculates a density of the respective distribution along with its arguments related to parameters of a particular distribution while not mentioning remaining arguments. The distribution specific arguments are also used in the same way inside the function calculating the cumulative distribution, the quantile function and the function for generating random numbers from a particular distribution. For BUGS, we mention the name of the function (along with its arguments) that is used inside the BUGS model to specify that a particular quantity follows a given distribution. First, Table B.1 lists those distributions of the event time T that are mentioned in this appendix and are directly available in R and BUGS. Second, Table B.2 shows distributions of $Y = \log(T)$ that are directly available in the mentioned software packages. The meaning of parameters denoted by α, γ, μ and σ is maintained. That is, for all distributions

$$\mathsf{E}(T) \propto \alpha, \qquad \mathsf{SD}(T) \propto \alpha, \qquad \mathsf{var}(T) \propto \alpha^2,$$

and parameter γ is the shape parameter of the event time distribution. Parameters μ and σ are then location and scale parameters of the distribution of $Y = \log(T)$. At the same time, the parameters γ and α relate to parameters μ and σ by Expression (B.2).

TABLE B.1: Parametrizations of the distributions of the event time T in R and BUGS.

R	BUGS
Log-normal distribution, $T \sim \mathcal{LN}(\gamma,\, \alpha)$	
dlnorm(meanlog, sdlog)	dlnorm(mu, tau)
meanlog $= \mu = \log(\alpha) = \mathsf{E}(Y)$	mu $= \mu = \log(\alpha) = \mathsf{E}(Y)$
sdlog $= \sigma = \gamma^{-1} = \mathsf{SD}(Y)$	tau $= \sigma^{-2} = \gamma^{2} = \big\{\mathsf{var}(Y)\big\}^{-1}$
meanlog $=$ mu	
sdlog $= \dfrac{1}{\sqrt{\text{tau}}}$	
Weibull distribution, $T \sim \mathcal{W}(\gamma,\, \alpha)$	
dweibull(shape, scale)	dweib(v, lambda)
shape $= \gamma = \sigma^{-1}$	v $= \gamma = \sigma^{-1}$
scale $= \alpha = \exp(\mu)$	lambda $= \Big(\dfrac{1}{\alpha}\Big)^{\gamma} = \exp\Big(-\dfrac{\mu}{\sigma}\Big)$
shape $=$ v	
$\Big(\dfrac{1}{\text{scale}}\Big)^{\text{shape}}$ $=$ lambda	
Exponential distribution, $T \sim \mathcal{E}(\alpha)$	
dexp(rate)	dexp(lambda)
rate $= \dfrac{1}{\alpha} = \exp(-\mu) = \dfrac{1}{\mathsf{E}(T)}$	lambda $= \dfrac{1}{\alpha} = \exp(-\mu) = \dfrac{1}{\mathsf{E}(T)}$
rate $=$ lambda	

TABLE B.2: Parametrizations of the distributions of the log-event time Y in R and BUGS.

R	BUGS
Normal distribution, $Y \sim \mathcal{N}(\mu, \sigma^2)$	
dnorm(mean, sd)	dnorm(mu, tau)
$\mathsf{mean} = \mu = \log(\alpha) = \mathsf{E}(Y)$	$\mathsf{mu} = \mu = \log(\alpha) = \mathsf{E}(Y)$
$\mathsf{sd} = \sigma = \gamma^{-1} = \mathsf{SD}(Y)$	$\mathsf{tau} = \sigma^{-2} = \gamma^2 = \{\mathsf{var}(Y)\}^{-1}$
$\mathsf{mean} = \mathsf{mu}$	
$\mathsf{sd} = \dfrac{1}{\sqrt{\mathsf{tau}}}$	
Logistic distribution, $Y \sim \mathcal{L}(\mu, \sigma)$	
dlogis(location, scale)	dlogis(mu, tau)
$\mathsf{location} = \mu = \log(\alpha) = \mathsf{E}(Y)$	$\mathsf{mu} = \mu = \log(\alpha) = \mathsf{E}(Y)$
$\mathsf{scale} = \sigma = \gamma^{-1} = \dfrac{\sqrt{3}}{\pi}\,\mathsf{SD}(Y)$	$\mathsf{tau} = \sigma^{-2} = \gamma^2 = \dfrac{\pi^2}{3}\,\{\mathsf{var}(Y)\}^{-1}$
$\mathsf{location} = \mathsf{mu}$	
$\mathsf{scale} = \dfrac{1}{\sqrt{\mathsf{tau}}}$	

Appendix C

Prior distributions

In this appendix we provide some technical details on prior distributions used in this book. The following distributions were used in this book as priors for the Bayesian models: beta, gamma, inverse gamma, inverse Wishart and Dirichlet. Many other priors are in use in Bayesian applications but are not described here. We also give further details on the link between the Dirichlet and the Dirichlet Process prior. Further, we provide the commands in R, WinBUGS/OpenBUGS, JAGS and the SAS procedure MCMC to invoke the distributions. The symbol "NA" indicates that the distribution is not available in the software. In the remainder of this section θ represents the parameter, and $\boldsymbol{\theta}$ the parameter vector. Covariance parameters are denoted as $\boldsymbol{\Sigma}$.

In the final subsection, we establish the link between the beta and Dirichlet prior and the Dirichlet Process prior.

C.1 Beta prior: Beta(α_1, α_2)

For θ a continuous parameter in the interval $(0,1)$, the beta prior $\text{Beta}(\theta \mid \alpha_1, \alpha_2)$ is given by:

$$p(\theta) = \frac{1}{B(\alpha_1,\alpha_2)}\,\theta^{\alpha_1-1}(1-\theta)^{\alpha_2-1},$$

with $B(\alpha_1,\alpha_2) = \dfrac{\Gamma(\alpha_1)\Gamma(\alpha_2)}{\Gamma(\alpha_1+\alpha_2)}$.

The model parameters α_1 and α_2 satisfy $\alpha_1 > 0, \alpha_2 > 0$ and regulate the shape of the beta distribution. Some special cases:

- $\alpha_1 = \alpha_2 = 1$: uniform prior;
- $\alpha_1 = \alpha_2 = 0.5$: Jeffreys' prior;
- $\alpha_1 = \alpha_2 > 1$: symmetrical informative prior;
- $\alpha_1 \neq \alpha_2 \geq 1$: asymmetrical skewed prior.

Location and variance	Program calls	
$\mathsf{E}(\theta)$: $\dfrac{\alpha_1}{(\alpha_1+\alpha_2)}$	R:	`dbeta(theta, alpha1, alpha2)`
$\mathsf{var}(\theta)$:	WB/JAGS:	`theta ~ dbeta(alpha1, alpha2)`
$\dfrac{\alpha_1\,\alpha_2}{(\alpha_1+\alpha_2)^2(\alpha_1+\alpha_2+1)}$	SAS:	`theta ~ beta(alpha1, alpha2)`

Use in practice: The beta prior is a popular prior for the probability parameter of the (negative) binomial likelihood, because of its conjugacy property. Popular vague priors are Beta(1, 1) (uniform prior) and Beta(0.5, 0.5) (Jeffreys' prior).

C.2 Dirichlet prior: $\mathcal{D}ir(\boldsymbol{\alpha})$

Suppose $\boldsymbol{\theta} = (\theta_1, \ldots, \theta_J)'$ is a J-dimensional vector of continuous parameters satisfying $\sum_{j=1}^J \theta_j = 1$ and $\theta_j > 0$ $(j = 1, \ldots, J)$, then the Dirichlet prior $\mathcal{D}ir(\boldsymbol{\alpha})$ is given by:

$$p(\boldsymbol{\theta}) = \frac{\Gamma(\sum_{j=1}^J \alpha_j)}{\prod_{j=1}^J \Gamma(\alpha_j)} \prod_{j=1}^J \theta_j^{\alpha_j - 1},$$

with $\boldsymbol{\alpha} = (\alpha_1, \ldots, \alpha_J)'$ and $\alpha_j > 0$ $(j = 1, \ldots, J)$.

Some special cases:

- $\alpha_1 = \ldots = \alpha_J = 1$: uniform prior on $(0,1)^J$;
- $\alpha_1 = \ldots = \alpha_J$: Jeffreys' prior on $(0,1)^J$.

Location and (co)variance	Program calls
$\mathsf{E}(\theta_j)$: $\alpha_j / \sum_{j=1}^J \alpha_j$	`R:` `ddirichlet(vtheta, valpha)`
$\mathsf{var}(\theta_j)$: $\dfrac{\alpha_j(\sum_m \alpha_m - \alpha_j)}{(\sum_m \alpha_m)^2(\sum_m \alpha_m + 1)}$	`WB/JAGS:` `vtheta[] ~ ddirich(valpha[])`
$\mathsf{cov}(\theta_j, \theta_k)$: $-\dfrac{\alpha_j \alpha_k}{(\sum_m \alpha_m)^2(\sum_m \alpha_m + 1)}$	`SAS:` `vtheta ~ dirich(valpha)`

Use in practice: The Dirichlet prior is a popular prior for the parameters of the multinomial likelihood, because of its conjugacy property. Popular vague priors are $\mathcal{D}ir(1, \ldots, 1)$ (uniform prior) and Jeffreys' prior $\mathcal{D}ir(0.5, \ldots, 0.5)$.

Some further properties:

- For $J = 2$ the Dirichlet distribution becomes the beta distribution;
- Marginal distributions of a Dirichlet distribution are beta distributions.

C.3 Gamma prior: $\mathcal{G}(\alpha,\ \beta)$

For θ a continuous parameter in the interval $[0, \infty)$, the gamma prior $\mathcal{G}(\theta \mid \alpha, \beta)$ is given by:

$$p(\theta) = \frac{\beta^{\alpha}}{\Gamma(\alpha)}\, \theta^{(\alpha-1)}\, \mathrm{e}^{-\beta\theta},$$

with α the shape parameter and β the rate parameter. An alternative formulation is $p(\theta) = \dfrac{\beta^{-\alpha}}{\Gamma(\alpha)}\, \theta^{(\alpha-1)}\, \mathrm{e}^{-\theta/\beta}$, with β the scale parameter.

The model parameters α and β satisfy $\alpha > 0,\ \beta > 0$. Some special cases:

- $\alpha = 1$: exponential prior;
- $\alpha = \nu/2,\ \beta(\text{rate}) = 0.5$: χ^2_ν prior with ν degrees of freedom.

Location and variance	**Program calls**
	β = **rate**
$\mathsf{E}(\theta)$: $\dfrac{\alpha}{\beta}$	`R:`
	`dgamma(theta, alpha, rate = beta)`
$\mathsf{var}(\theta)$: $\dfrac{\alpha}{\beta^2}$	`SAS:`
	`theta ~ gamma(alpha, iscale=beta)`
	β = **scale**
$\mathsf{E}(\theta)$: $\alpha\beta$	`R:`
	`dgamma(theta, alpha, scale = beta)`
	`WB/JAGS: theta ~ dgamma(alpha, beta)`
$\mathsf{var}(\theta)$: $\alpha\beta^2$	`SAS:`
	`theta ~ gamma(alpha, scale = beta)`

Use in practice: The gamma prior is a popular prior for the mean parameter of the Poisson likelihood, because of its conjugacy property. Popular vague priors are $\mathcal{G}(\alpha,\ \varepsilon)$ with ε small.

C.4 Inverse gamma prior: $\mathcal{IG}(\alpha,\ \beta)$

For θ a continuous parameter in the interval $(0, \infty)$, the inverse gamma prior $\mathcal{IG}(\theta \mid \alpha, \beta)$ is given by:

$$p(\theta) = \frac{\beta^\alpha}{\Gamma(\alpha)}\, \theta^{-(\alpha+1)}\, \mathrm{e}^{-\beta/\theta},$$

with α the shape parameter and β the rate parameter. An alternative formulation is $p(\theta) = \dfrac{1}{\beta^\alpha \Gamma(\alpha)}\, \theta^{-(\alpha+1)}\, \mathrm{e}^{-1/(\theta\beta)}$, with β the scale parameter.

The model parameters α and β satisfy $\alpha > 0, \beta > 0$. Some special cases:

- $\alpha = \nu/2, \beta\,(\text{rate}) = 1/2$: Inv-$\chi^2_\nu$ prior with ν degrees of freedom;
- $\alpha = \nu/2, \beta\,(\text{rate}) = \nu s^2/2$: Scale-inv-$\chi^2_{\nu,s^2}$ prior with ν degrees of freedom and s^2 scale parameter;
- $\alpha = \beta \to 0$: $p(\theta) \propto 1/\theta^2$, which is Jeffreys' prior for σ^2 for Gaussian likelihood $\mathcal{N}(\mu, \sigma^2)$, with μ fixed.

Location and variance	**Program calls**
	β = **rate**
E(θ): $\dfrac{\beta}{\alpha-1}$	R: `dgamma(1/theta, alpha, rate = beta)/theta^2`
var(θ): $\dfrac{\beta^2}{(\alpha-1)^2(\alpha-2)}$	SAS: `theta ~ igamma(alpha, iscale = beta)`
	β = **scale**
E(θ): $\dfrac{1}{(\alpha-1)\beta}$	R: `dgamma(1/theta, alpha, scale = beta)/theta^2`
var(θ): $\dfrac{1}{(\alpha-1)^2(\alpha-2)\beta^2}$	WB/JAGS: `theta <- 1/itheta; itheta ~ dgamma(alpha, beta)`
	SAS: `theta ~ igamma(alpha, scale = beta)`

Use in practice: The inverse gamma prior is a popular prior for the variance σ^2 of a Gaussian likelihood $\mathcal{N}(\mu, \sigma^2)$, with μ fixed (conditional conjugacy property). Popular vague priors for variance parameters are $\mathcal{IG}(\varepsilon, \varepsilon)$ with ε small.

C.5 Wishart prior: Wishart($\boldsymbol{R}$, k)

Suppose $\boldsymbol{\Sigma}$ is a d-dimensional stochastic symmetric and positive definite matrix, then the Wishart($\boldsymbol{R}$, k) prior with $\boldsymbol{R}$ positive definite and $k > 0$ degrees of freedom is given by:

$$p(\boldsymbol{\Sigma}) = c \det(\boldsymbol{R})^{-k/2} \det(\boldsymbol{\Sigma})^{(k-d-1)/2} \exp\left\{-\frac{1}{2}\mathrm{tr}\left(\boldsymbol{R}^{-1}\boldsymbol{\Sigma}\right)\right\},$$

with $c^{-1} = 2^{kd/2}\pi^{d(d-1)/4}\prod_{j=1}^{d}\Gamma\left(\frac{k+1-j}{2}\right)$ and $k \geq d+1$. Further, let $\boldsymbol{\Sigma} = (\sigma_{ij})_{ij}$, and $\boldsymbol{R} = (r_{ij})_{ij}$.

Location and (co)variance	Program calls
$\mathsf{E}(\boldsymbol{\Sigma})$: $k\,\boldsymbol{R}$	`R: dWISHART(W = Sigma, df = k, S = R)` (in mixAK)
	`R: dwish(Sigma, k, Rinv)` (Rinv = $\boldsymbol{R}^{-1}$ in MCMCpack)
$\mathsf{var}(\sigma_{ij})$: $k\,(r_{ij}^2 + r_{ii}r_{jj})$	`WB/JAGS:` `Sigma[,] ~ dwish(R[,], k)`
$\mathsf{cov}(\sigma_{ij}, \sigma_{kl})$: $k(r_{ik}r_{jl} + r_{il}r_{jk})$	`SAS: NA`

Use in practice: The Wishart prior is a popular prior for the precision matrix of $\boldsymbol{\Sigma}^{-1}$ of the Gaussian likelihood $\mathcal{N}(\boldsymbol{\mu}, \boldsymbol{\Sigma})$ with $\boldsymbol{\mu}$ fixed, because of its conditional conjugacy property.

Remark:

- WinBUGS uses an alternative expression of the Wishart distribution: In the above expression $\boldsymbol{R}$ is replaced by $\boldsymbol{R}^{-1}$ and hence represents a covariance matrix in WinBUGS.

C.6 Inverse Wishart prior: Wishart($\boldsymbol{R}$, k)

Suppose $\boldsymbol{\Sigma}$ is a d-dimensional stochastic symmetric and positive definite matrix, then the Inverse Wishart($\boldsymbol{R}$, k) prior with $\boldsymbol{R}$ positive definite and $k > 0$ degrees of freedom is given by:

$$p(\boldsymbol{\Sigma}) = c\det(\boldsymbol{R})^{k/2}\det(\boldsymbol{\Sigma})^{-(k+d+1)/2}\exp\left\{-\frac{1}{2}\mathrm{tr}\left(\boldsymbol{\Sigma}^{-1}\boldsymbol{R}\right)\right\},$$

with $c^{-1} = 2^{kd/2}\pi^{d(d-1)/4}\prod_{j=1}^{d}\Gamma\left(\frac{k+1-j}{2}\right)$ and $k \geq d+1$.

A special case:

- $k \to 0$: $p(\boldsymbol{\Sigma}) \propto \det(\boldsymbol{\Sigma})^{-(d+1)/2}$, which is Jeffreys' prior for $\boldsymbol{\Sigma}$ for Gaussian likelihood $\mathcal{N}(\boldsymbol{\mu}, \boldsymbol{\Sigma})$ with $\boldsymbol{\mu}$ fixed.

Location	Program calls
$\mathsf{E}(\boldsymbol{\Sigma})$: $\boldsymbol{R}/(k-d-1)$ (if $k > d+1$)	`R: diwish(Sigma, k, Rinv)` (Rinv = $\boldsymbol{R}^{-1}$ in MCMCpack)
	`WB/JAGS: NA`
	`SAS: Sigma ~ iwishart(k, R)`

Use in practice: The Inverse Wishart prior is a popular prior for the covariance matrix $\boldsymbol{\Sigma}$ of the Gaussian likelihood $\mathcal{N}(\boldsymbol{\mu}, \boldsymbol{\Sigma})$ with $\boldsymbol{\mu}$ fixed, because of its conditional conjugacy property. Jeffreys' prior $p(\boldsymbol{\Sigma}) \propto \det(\boldsymbol{\Sigma})^{-(d+1)/2}$ reduces to $p(\sigma^2) \propto 1/\sigma^2$.

C.7 Link between Beta, Dirichlet and Dirichlet Process prior

The beta distribution can be viewed as a distribution over distributions. Let the stochastic variable X take two values, i.e., 1 and 2. Denote the space of possible outcomes by $\mathbb{X} = \{1, 2\}$. Then assume X is a Bernoulli random variable with distribution Bern(π_1), with $\pi_1 = \mathsf{P}(X = 1)$ and $\pi_2 = 1 - \pi_1$. Hence, two (related) probabilities, π_1 and π_2 characterize the Bernoulli distribution. Averaged over the beta distribution, the mean distribution could be characterized as

$$(\overline{\pi}_1, \overline{\pi}_2) = \left(\frac{\alpha_1}{\alpha(\mathbb{X})}, \frac{\alpha_2}{\alpha(\mathbb{X})}\right),$$

with $\alpha(\mathbb{X}) = \sum_{X_j \in \mathbb{X}} \alpha_j$. The posterior distribution for the two parameters after having observed i.i.d. Bernoulli random variables $X_1, \ldots, X_n$ is given by

$$p(\pi_1 \mid X_1, \ldots, X_n) \;=\; \text{Beta}\left(\alpha_1 + \sum_{i=1}^{n} \delta_{X_i,1},\; \alpha_2 + \sum_{i=1}^{n} \delta_{X_i,2}\right),$$

with $\delta_{X,a} = 1$ if $X = a$, or zero otherwise.

The Dirichlet distribution is the generalization of the beta distribution to J possible values of X, e.g., $\mathbb{X} = \{1, 2, \ldots, J\}$. The expression of the Dirichlet distribution is given above. Here we generalize some properties of the beta distribution to J possible outcomes. Suppose X has a multinomial distribution with probabilities $\pi_1, \ldots, \pi_J$ and $\sum_{j=1}^{J} \pi_j = 1$, then the Dirichlet distribution $\mathcal{D}ir(\alpha_1, \ldots, \alpha_J)$ has the following properties:

- The expected value of each $(\pi_1, \ldots, \pi_J)$ is:

$$(\overline{\pi}_1, \ldots, \overline{\pi}_J) = \left(\frac{\alpha_1}{\alpha(\mathbb{X})}, \ldots, \frac{\alpha_J}{\alpha(\mathbb{X})}\right).$$

- The posterior distribution of $(\pi_1, \ldots, \pi_J)$ given i.i.d. multinomial distributed $X_1, \ldots, X_n$ is:

$$p(\pi_1, \ldots, \pi_J \mid X_1, \ldots, X_n) = \mathcal{D}ir\left(\alpha_1 + \sum_{i=1}^{n} \delta_{X_i,1}, \ldots, \alpha_J + \sum_{i=1}^{n} \delta_{X_i,J}\right).$$

- Further, when $\mathbb{B}_1, \ldots, \mathbb{B}_J$ is any partition of $\mathbb{X}$, then:

$$p(\pi_{\mathbb{B}_1}, \ldots, \pi_{\mathbb{B}_J}) = \mathcal{D}ir\big(\alpha(\mathbb{B}_1), \ldots, \alpha(\mathbb{B}_J)\big),$$

  with $\pi_{\mathbb{B}_j} = p(X \in \mathbb{B}_j)$ and $\alpha(\mathbb{B}_j) = \sum_{X_i \in \mathbb{B}_j} \alpha_i$.

To make the transition from the Dirichlet distribution to the Dirichlet Process prior, we assume that X takes values in $\mathbb{R}$ (or more generally in $\mathbb{R}^k$, with k an integer). Further, assume that α measures the "size" of each subset of $\mathbb{X}$. More mathematically spoken, one takes α to be a measure on $(\mathbb{X}, \Omega)$, where Ω contains all measurable subsets of $\mathbb{X}$ (actually a σ-algebra). Then Ferguson (1973) showed that there is a unique probability measure $\mathcal{DP}(\alpha)$ on the space of probability measures over $\mathbb{X}$, called the Dirichlet process with parameter α satisfying:

$$\big(p(\mathbb{B}_1), \ldots, p(\mathbb{B}_J)\big) \;\sim\; \mathcal{D}ir\big(\alpha(\mathbb{B}_1), \ldots, \alpha(\mathbb{B}_J)\big),$$

where $\mathbb{B}_1, \ldots, \mathbb{B}_J$ is any partition of $\mathbb{X}$ by Borel sets.

Some properties of the Dirichlet process are generalizations of properties seen above for the Dirichlet distribution and are given below. To shorten notation, we denote $(\mathbb{B}_1, \ldots, \mathbb{B}_J)$ as $(\cdot)$:

- The expected value of each distribution p applied to each partition $\mathbb{B}_1, \ldots, \mathbb{B}_J$ of Ω is:

$$\overline{p}(\cdot) \;=\; \frac{\alpha(\cdot)}{\alpha(\mathbb{X})}.$$

- The posterior distribution of $p(\cdot)$ given a sample of i.i.d. distributed $X_1, \ldots, X_n$ is also a Dirichlet process:

$$p\big(p(\cdot) \,|\, X_1, \ldots, X_n\big) \;=\; \mathcal{DP}\left(\alpha(\cdot) + \sum_{i=1}^{n} \delta_{X_i}\right).$$

- The mean distribution is then given by:

$$\overline{p}\big(p(\cdot) \,|\, X_1, \ldots, X_n\big) \;=\; \frac{\alpha(\mathbb{X})}{\alpha(\mathbb{X}) + n}\left(\frac{\alpha(\cdot)}{\alpha(\mathbb{X})}\right) + \frac{n}{\alpha(\mathbb{X}) + n}\left(\frac{\sum_{i=1}^{n} \delta_{X_i}}{n}\right).$$

  This property shows that the Bayesian estimate is a weighted combination of the "prior guess" (or "template") and the "empirical estimate."

Appendix D

Description of selected R packages

In this appendix, we briefly describe the syntax and explain the meaning of the main arguments for selected R functions which appeared in examples presented throughout the book. Nevertheless, the selection of functions and their arguments being described in this appendix is somehow subjective and is by no means comprehensive, nor covering all capabilities of the included methods. It was also not possible to describe in detail all R functions that appeared throughout the book. On the other hand, most R packages are accompanied by very well-written help pages, numerous vignettes or even accompanied by scientific papers, published in *Journal of Statistical Software* (`https://www.jstatsoft.org`) and *R News*, recently superseded by *The R Journal* (`https://journal.r-project.org`) or other journals where the reader can acquire all necessary information.

D.1 icensBKL package

The icensBKL package accompanies this book and primarily contains the data sets introduced in Section 1.6 and Appendix A. On top of that, few methods described in this book have been implemented and included in the package.

Nonparametric K-sample tests based on the $G^{\varrho,\gamma}$ family of weighted log-rank tests introduced in Subsection 4.1.4 was implemented using guidelines given in Gómez et al. (2009) as function kSampleIcens. The inference is based on the asymptotical permutation distribution of the weighted log-rank vector $\boldsymbol{U}$.

```
kSampleIcens(A, group, icsurv, rho = 0, lambda = 0)
```

The arguments have the following meaning.

- A: $n \times 2$ matrix containing the observed intervals having the same form as A argument of function PGM from package Icens (see Appendix D.2).
- group: factor vector of covariate which indicates pertinence to groups under comparison.

- icsurv: optionally, object which represents a NPMLE of the survival function based on a pooled sample. The pooled NPMLE is then not recalculated inside the function which considerably shortens the computation time if the user wants to perform several K-sample tests for different values of ϱ and γ. It must be an object obtained by using the PGM function from package Icens (see Appendix D.2).
- rho, lambda: parameters which specify the $G^{\varrho,\gamma}$ family such that $\varrho =$ rho and $\gamma =$ lambda.

Additionally, the package icensBKL contains the following functions implementing the methods discussed in the book:

fit.copula: fitting copula models to interval-censored data, see Section 7.3.

icbiplot: biplot for interval-censored data, see Section 8.3.

NPbayesSurv: nonparametric Bayesian estimation of a survival curve with right-censored data based on the Dirichlet Process prior, see Section 9.4.2.

NPICbayesSurv: nonparametric Bayesian estimation of a survival curve with interval-censored data based on the Dirichlet Process prior, see Section 10.3.1.

D.2 Icens package

NPMLE of the survival function under interval censoring was discussed in Section 3.1. For practical calculations, we mainly used the R package Icens (Gentleman and Vandal, 2016). In particular, we distinguish the following functions:

```
EM(A, pvec, maxiter = 500, tol = 1e-12)

PGM(A, pvec, maxiter = 500, tol = 1e-07, told = 2e-05,
    tolbis = 1e-08, keepiter = FALSE)

VEM(A, pvec, maxiter = 500, tol = 1e-07, tolbis = 1e-07,
    keepiter = FALSE)

ISDM(A, pvec, maxiter = 500, tol = 1e-07, tolbis = 1e-08,
     verbose = FALSE)

EMICM(A, EMstep = TRUE, ICMstep = TRUE, keepiter = FALSE, tol = 1e-07,
      maxiter = 1000)
```

The function EM implements the original self-consistency algorithm of Turnbull (1976). The projected gradient method (Wu, 1978) is provided by the function PGM. The vertex exchange method (Böhning, 1986; Böhning et al., 1996) is implemented in the function VEM. The function ISDM provides the NPMLE using the intra-simplex direction method proposed in Lesperance and Kalbfleisch (1992). Finally, the hybrid EM-iterative convex minorant estimator (Wellner and Zahn, 1997) is computed by the function EMICM. The important argument of these functions is:

- A: $n \times 2$ matrix containing the observed intervals. The intervals are assumed to be half-open, i.e., $(L, R]$. Special care must be given to left- and right-censored observations because the usual representation with a missing value is not allowed for this argument. Left-censored data must contain a zero in the lower endpoint. The upper endpoints of right-censored data must be set to an arbitrary fixed value beyond the greatest observed upper endpoint.

All above functions return objects of class icsurv containing the relevant information. A plot method also exists for an object of this class and can be used to visualize the NPMLE. Our package icensBKL further provides the function icsurv2cdf which converts the object of class icsurv into a simple data.frame with columns labeled 'time' and 'cdf'. As was shown in the example in Section 3.1.3, plots of the estimated cdf F and survival function S can then easily be drawn.

D.3 interval package

The interval package was used in Section 3.1 to calculate the NPMLE of the survival function and mainly in Section 4.1 to compare several survival distributions under interval censoring. For a comprehensive description of the package capabilities, we refer to Fay and Shaw (2010) which accompanies the package. The main functions include icfit which calculates the NPMLE of the survival function using the original self-consistency algorithm of Turnbull (1976), and a function ictest which implements various nonparametric K-sample tests, some of which were discussed in Section 4.1. For both icfit and ictest, there are two methods implemented: the default one where the observed intervals are supplied in the form of vectors of lower and upper endpoints and a method for class formula where the observed intervals are supplied in a form of Surv object.

For icfit, the two versions have the following syntax

```
icfit(L, R, initfit = NULL, control = icfitControl(),
      Lin = NULL, Rin = NULL, ...)
```

```
icfit(formula, data, ...)
```

The important arguments of the default version are:

- L, R: vectors of lower (left) and upper (right) limits of observed intervals. By default, it is assumed that intervals are half-open taking $(L,\, R]$ unless $L = R$ (exact observation) or $R = \infty$ (right-censored observation). As L and R arguments, it is also possible to use the first and second columns, respectively, of matrix A which was used as primary argument to functions from package Icens (see Appendix D.2).
- Lin, Rin: logical values which may change the default assumption concerning the half-open observational intervals. For example, by setting Lin = TRUE and Rin = FALSE, observed intervals will be assumed to be $[L,\, R)$.

The method for class formula differs by the mean in which data are supplied. Argument formula specifies the response in the form SURV ~ 1, where SURV is the survival object as explained in Subsection 1.7.1. Furthermore, if GROUP is a categorical covariate (factor), formula can take a form SURV ~ GROUP in which case separate NPMLEs are calculated in strata defined by levels of the categorical covariate. Note that also the formula method accepts other arguments like Lin and Rin to change the default function behavior. The function icfit returns an object of class icfit (which is in fact the same as class icsurv returned by functions from the Icens package). A plot method also exists for an object of this class and can be used to visualize the NPMLE.

The two versions of ictest have the following syntax

```
ictest(L, R, group,
       scores = c("logrank1", "logrank2", "wmw", "normal","general"),
       rho = NULL, alternative = c("two.sided", "less", "greater"),
       icFIT = NULL, initfit = NULL, icontrol = icfitControl(),
       exact = NULL, method = NULL, methodRule = methodRuleIC1,
       mcontrol = mControl(),
       Lin = NULL, Rin = NULL,
       dqfunc = NULL, ...)

ictest(formula, data, subset, na.action, ...)
```

Arguments L, R, Lin and Rin have the same meaning as for the function icfit. Specification of the groups to compare is done either by using the argument group = GROUP or in a formula way as SURV ~ GROUP, where GROUP is a categorical covariate which specifies the group pertinence. Other important arguments include:

- scores: specification of scores c_i $(i = 1, \ldots, n)$ to be used to calculate the test statistic (4.4). Default value scores = "logrank1" leads to the test statistic introduced by Sun (1996), which is an alternative generalization of a classical (unweighted) log-rank test for interval-censored data rather than the generalization characterized by the scores (4.9). The latter is introduced in Section 4.1 and obtained by setting scores = "logrank2". Value scores = "wmw" leads to Peto-Prentice-Wilcoxon interval-censored generalization of the WLRT with scores (4.11). Further, scores = "normal" leads to the WLRT generated by Model (4.5) with normally distributed error terms. Finally, by scores = "general" the user may specify his/her own distribution for these error terms.

- method: specification of the inferential method to calculate P-values. By setting method = "pclt", the P-value is based on an asymptotical permutation test (default in most situations). Value method = "scoretest" leads to usage of standard maximum likelihood theory to get the P-value (see Section 4.1.5 for details).

The function ictest returns an object of class ictest which is a list containing all important elements needed to calculate the test statistic, the value of the test statistic and the P-value. A print method allows to see the most important results in a comprehensive form.

D.4 survival package

The survival package (Therneau, 2017) contains the core survival analysis R routines, mainly suitable in case of right censoring. One of the functions included in this package is the survreg function which can also be used with interval-censored data to fit parametric (AFT) models as exemplified in Sections 3.2 and 6.1. The syntax of the function is the following:

```
survreg(formula, data, weights, subset,
        na.action, dist = "weibull", init = NULL, scale = 0,
        control, parms = NULL, model = FALSE,
        x = FALSE, y = TRUE, robust = FALSE, score = FALSE, ...)
```

The function survreg is primarily designed to fit the accelerated failure time model which is discussed in Section 6.1. However, one can use this function to fit simple parametric models described in Section 3.2 as well.

For the purpose of fitting simple parametric models of Section 3.2, the most important arguments include:

- formula: specification of the response in the form SURV $\sim$ 1, where SURV is the survival object as explained in Section 1.7.1.

- dist: a character string which determines the assumed survival distribution, possible values are "weibull", "exponential", "rayleigh", "lognormal", "loglogistic".
- scale: an optional fixed value of the scale parameter σ. If it is set to 0 or a negative value (default), σ is estimated (except in the case of an exponential or Rayleigh distribution, where σ is equal to 1 or 0.5, respectively).

When using the survreg function to fit the parametric AFT model of Section 6.1, the meaning of the most important arguments is the following:

- formula: specification of the response and the right-hand side of the AFT model in the form SURV $\sim$ MODEL, whereas above, SURV is a survival object and MODEL a symbolic description of the model being the same as with other R regression functions.
- dist: a character string which determines the assumed survival distribution of the response variable T. The same values are possible as above, that is, "lognormal", "loglogistic", and "weibull" leading to the AFT models with the normal, logistic, and type I extreme value, respectively (see Equations (6.5), (6.6) and (6.7)) error distributions. Furthermore, the dist values "exponential" and "rayleigh" provide AFT models with the exponential and Rayleigh distribution of T, respectively. These are obtained by taking the standard type I extreme value distribution (6.7) for the AFT error terms and fixing the scale parameter σ to 1 and 0.5, respectively.

D.5 logspline package

In Section 3.3.1 the function oldlogspline of the package logspline (Kooperberg, 2016) was used for the logspline density estimation. The function has the following syntax:

```
oldlogspline(uncensored, right, left, interval, lbound, ubound,
    nknots, knots, penalty, delete = TRUE)
```

Arguments of the function have the following meaning:

- uncensored, right, left: numeric vectors with uncensored, lower limits of right-censored and upper limits of left-censored observations.
- interval: a two-column matrix containing interval-censored observations.
- lbound, ubound: lower and upper limits of the support of the estimated distribution. For survival analysis, lbound=0 is usually required, whereas ubound is left unspecified corresponding to $T_{max} = \infty$.

- nknots: number of knots to start the stepwise knot deletion or fixed number of knots if a model with prespecified number of knots is required. If left unspecified, the function uses an automatic rule to specify the number of knots.
- knots: sequence of knots $\mu_1 < \cdots < \mu_K$ to start the stepwise knot deletion or fixed number of knots if a model with prespecified number of knots is required. If left unspecified, the function uses an automatic rule to specify the number of knots.
- penalty: argument which specifies the type of the information criterion to select the number of knots. Its default value of $\log(n)$ leads to BIC. Setting penalty=2 leads to AIC.
- delete: By setting delete = FALSE, a model with fixed (number of) knots given by arguments nknots and/or knots can be fitted.

D.6 smoothSurv package

The maximum likelihood based estimation of penalized Gaussian mixture models (Sections 3.3.3 and 6.2) is implemented in the R package smoothSurv (Komárek et al. (2005a)). The main fitting function smoothSurvReg has the following syntax:

```
smoothSurvReg(formula = formula(data), logscale = ~1,
    data = parent.frame(), subset, na.action = na.fail,
    init.beta, init.logscale, init.c, init.dist = "best",
    update.init = TRUE, aic = TRUE, lambda = exp(2:(-9)),
    model = FALSE, control = smoothSurvReg.control(), ...)
```

Important arguments of the function are:

- formula: specification of the response in the form SURV ~ FORMULA, where SURV is the survival object as explained in Section 1.7.1.
- lambda: the values of λ/n for which the penalized log-likelihood is optimized and which are used in the grid search for the optimal value of AIC.

Additionally, the basis functions (knots $\boldsymbol{\mu}$, basis standard deviation σ_0) and the order of the penalty s can be changed from their default values ($\boldsymbol{\mu} = (-6.0, -5.7, \ldots, 5.7, 6.0)$, $\sigma_0 = 0.2$, $s = 3$) if we specify also arguments which make a part of the additional arguments denoted by ... in R. These are

- dist.range: a two-component vector, e.g., c(-6, 6), giving the range of the knots.

- by.knots: distance δ between two consecutive knots.
- sdspline: basis standard deviation σ_0.
- difforder: order of the penalty s.

The functional call results in an object of class smoothSurvReg for which a set of supporting methods exist. These are especially:

- confint calculates confidence intervals for regression coefficients and scale parameters.
- plot draws estimated density of $Y = \log(T)$, that is, a density $g(y)$ and compares it to three common parametric densities, namely normal, logistic, and Gumbel (extreme value), e.g., see Figure 3.10.
- fdensity computes (and draws) estimated survival density $f(t)$.
- survfit computes (and draws) estimated survival function $S(t)$.
- hazard computes (and draws) estimated hazard function $\hbar(t)$.
- estimTdiff estimates the mean survival time $\mathsf{E}(T)$ including the confidence interval derived from using the pseudo-covariance matrix (3.24).

D.7 mixAK package

The Bayesian estimation of the survival distribution modelled by a classical Gaussian mixture and estimated by the (RJ-)MCMC simulation has been discussed in Section 10.2.1. The R implementation of the method is provided by the function NMixMCMC function from the R package mixAK (Komárek, 2009).

The syntax of the function NMixMCMC, which provides the MCMC sampling from the posterior distribution, is the following:

```
NMixMCMC(y0, y1, censor, scale, prior,
    init, init2, RJMCMC,
    nMCMC = c(burn = 10, keep = 10, thin = 1, info = 10),
    PED, keep.chains = TRUE, onlyInit = FALSE, dens.zero = 1e-300)
```

Important arguments of the function have the following meaning.

- y0: numeric vector with uncensored observations, lower limits of interval-censored and right-censored observations, and upper limits of left-censored observations.
- y1: numeric vector having the same length as y0 containing upper limits of interval-censored observations and arbitrary values for uncensored, left-, and right-censored observations.

- censor a vector with censoring indicators δ as described in Table 1.1.
- scale a list which allows the user to specify the shift parameter α and the scale parameter τ. If these are not given, the function chooses the values of α and τ such that the shifted and scaled data have approximately zero mean and unit variance.
- prior a list which can specify parameters of the prior distribution. For most parameters, the function is able to choose reasonable values leading to weakly informative prior distributions.
- init, init2 lists which can specify initial values for two instances of the RJ-MCMC sampling algorithm. The function is able to initialize automatically up to two chains.
- nMCMC a vector which specifies the length of the MCMC simulation.

The call of the function results in an object of class NMixMCMC which contains sampled values of model parameters and several basic posterior summary statistics. The chains can further be processed in a standard way using, e.g., the coda package. Additionally, the resulting object can be processed by a set of other functions which compute several quantities of primary interest. The most important supporting functions are:

- NMixChainsDerived computes the chain for the survival mean $\mathsf{E}(T)$ (Equation (10.5)).
- NMixPredDensMarg computes the values of the posterior predictive density of Y (analogous to Equation (10.7)).
- NMixPredCDFMarg computes the values of the posterior predictive cumulative distribution function of Y.
- Y2T converts computed posterior predictive density/cdf of Y into posterior predictive density/cdf of T (Expression (10.7), one minus Expression (10.8)).

There are plot methods implemented to visualize computed posterior predictive density, cdf, survival function.

D.8 bayesSurv package

In Sections 12.2, 12.3, 13.1 – 13.3, several variants of the accelerated failure time model were introduced where either the classical or the penalized Gaussian mixture was assumed for the random elements of the model. Bayesian inference for the proposed models have been implemented in the bayesSurv package (García-Zattera et al., 2016; Komárek and Lesaffre, 2006, 2007, 2008; Komárek et al., 2007). The following five functions included in the package implement a primary MCMC sampling for different variants of the AFT model.

bayesHistogram: estimation of a univariate or a bivariate density based on interval-censored data. A penalized Gaussian mixture (PGM) is used as a flexible model for an unknown density. Usage of this function was not explicitly discussed in this book. Nevertheless, one could use it as a Bayesian counterpart of the frequentist methods disussed in Section 3.3.3. In particular, either the AFT model of Section 12.3 or the bivariate model of Section 13.2.3, both involving no covariates, can be fitted using this function.

bayessurvreg1: AFT model with a classical Gaussian mixture assumed for the distribution of the error terms (Section 12.2). The function can also be used to fit a more general, shared frailty model with normally distributed frailties being combined with covariates (not discussed in this book) suitable to fit clustered data in a similar way as was shown in Section 13.1.2. See Komárek and Lesaffre (2007) for more details. It is also possible to fit another generalization towards doubly interval-censored data.

bayessurvreg2: AFT model with a penalized Gaussian mixture assumed for the error distribution (Section 12.3) which can be extended to include the normally distributed shared frailty terms in combination with covariates (not covered by this book) and allows for doubly interval-censored data (model without frailties for DI-censored data is discussed in Section 13.3.2). See Komárek et al. (2007) for more details.

bayessurvreg3: AFT model with a penalized Gaussian mixture assumed for the error distribution (Section 12.3). The model can be extended to include also the shared frailty terms whose distribution is also modelled by the penalized Gaussian mixture (Section 13.1.2). On top of that, doubly interval-censored data are allowed (model without frailties for DI-censored data is discussed in Section 13.3.2). See Komárek and Lesaffre (2008) for more details. Finally, the function is capable of fitting misclassified interval-censored data briefly mentioned in Section 14.1.8, see García-Zattera et al. (2016) for details.

bayesBisurvreg: AFT model for bivariate (doubly) interval-censored data (Section 13.2.3). We refer to Komárek and Lesaffre (2006) for more details.

A philosophy of using the above functions is the same for all of them. Below, we briefly discuss usage of the bayessurvreg1 function in context of the CGM AFT model introduced in Section 12.2. The full syntax is the following:

```
bayessurvreg1(formula, random,
   data = parent.frame(), subset,
   na.action = na.fail,
   x = FALSE, y = FALSE, onlyX = FALSE,
   nsimul = list(niter = 10, nthin = 1, nburn = 0,
```

```
                nnoadapt = 0, nwrite = 10),
prior = list(kmax = 5, k.prior = "poisson", poisson.k = 3,
             dirichlet.w = 1,
             mean.mu = NULL, var.mu = NULL,
             shape.invsig2 = 1.5,
             shape.hyper.invsig2 = 0.8, rate.hyper.invsig2 = NULL,
             pi.split = NULL, pi.birth = NULL,
             Eb0.depend.mix = FALSE),
prior.beta, prior.b, prop.revjump,
init = list(iter = 0, mixture = NULL, beta = NULL,
            b = NULL, D = NULL,
            y = NULL, r = NULL, otherp = NULL, u = NULL),
store = list(y = TRUE, r = TRUE, b = TRUE, u = TRUE,
             MHb = FALSE, regresres = FALSE),
dir = getwd(),
toler.chol = 1e-10, toler.qr = 1e-10, ...)
```

Important arguments of the function related to the model of Section 12.2 have the following meaning:

- formula: specification of the response in the form SURV ~ 1, where SURV is the survival object as explained in Section 1.7.1.
- nsimul: a list which specifies the length of the MCMC simulation. The most important elements are niter (total number of generated values after exclusion of the thinned values nevertheless including the burn-in period), nthin (thinning interval), nburn (length of the burn-in period after exclusion of the thinned values), nwrite (an interval with which the sampled values are written to the TXT files).
- prior: a list which specifies the prior distribution of the parameters related to the error distribution.
- prior.beta: a list which specifies the prior hyperparameters of a normal prior for the regression coefficients $\boldsymbol{\beta}$.
- init: a list specifying the initial values of the model paremeters to start the MCMC simulation.
- dir: a character string specifying the path to a directory in which the posterior samples are stored in a form of the TXT files.

The call of the function bayessurvreg1 starts an MCMC simulation. The sampled values are gradually written to the TXT files stored in a directory specified by the argument dir. The files can then be processed by the following set of functions:

- files2coda reads sampled values from the TXT files and prepares the mcmc objects to be able to process the sampled chains using the functions from the coda package.
- bayesDensity calculates the values of the posterior predictive density of the (standardized) error terms of the AFT model. Analogous function bayesGspline calculates different types of the posterior predictive densities for models fitted primarily by the functions bayesHistogram, bayessurvreg2, bayessurvreg3 and bayesBisurvreg.
- predictive calculates the values of the quantities related to the posterior predictive survival distributions for given combinations of covariates. For models fitted by the functions bayessurvreg2, bayessurvreg3 and bayesBisurvreg, the posterior predictive survival distributions are analogously calculated by the function predictive2.

For additional capabilities of the bayesSurv package, we refer to the examples used throughout the book and to the help pages and vignettes of the package.

D.9 DPpackage package

In Sections 10.3.2, 12.4, 13.2.4 and 13.3.3 nonparametric as well as semiparametric Bayesian models were presented. The methods introduced in those sections have been implemented in the R package DPpackage (Jara, 2007; Jara et al., 2011). In the book we have shown only a small fraction of the package capabilities implemented in the following functions:

DPMdencens function was used in Section 10.3.2 to estimate a survival distribution of interval-censored data while using the Dirichlet Process Mixture approach.

DPsurvint function was exploited in Section 12.4 to fit a semiparametric AFT model based on the Mixture of Dirichlet processes.

LDPDdoublyint function was used in Sections 13.2.4 and 13.3.3 to estimate regression models for multivariate (doubly) interval-censored data. The methodology was built upon the linear dependent Poisson-Dirichlet process mixture of survival models.

For more possibilities of the DPpackage we refer to Jara (2007); Jara et al. (2011), references therein and to the help pages of the package.

D.10 Other packages

On top of the previously listed R packages Icens, interval, survival, logspline, smoothSurv, the following packages implement the methods discussed in the frequentist Part II of the book:

FHtest (Oller and Langohr, 2015) to test equality of two or more survival distributions (Sections 2.2 and 4.1);

glrt (Zhao and Sun, 2015) to compare two or more survival distributions (Section 4.1) and to calculate the score tests in Section 5.3;

icenReg (Anderson-Bergman, 2017) to fit a PH model with parametrically specified baseline hazard function (Section 5.1) as well as a semiparametric PH model (Section 5.3);

frailtypack (Rondeau et al., 2015) to fit a PH model with parametrically specified baseline hazard function (Section 5.1), piecewise exponential baseline distribution (Section 5.2.1) as well as with the spline-based smoothed baseline hazard (Section 5.2.3) with possibility to include the frailty terms in the model expression (Section 8.2.1);

MIICD (Delord, 2017) to fit the PH model using the multiple imputation approaches (Section 5.4);

MLEcens (Maathuis, 2013) to calculate the bivariate NPMLE of the survival distribution (Section 7.1);

cubature (Narasimhan and Johnson, 2017) to perform numerical integration towards association measures in Section 7.5.

The general Bayesian R calculations would not have been so handy as they were without the help of the following R packages:

R2WinBUGS, R2OpenBUGS (Sturtz et al., 2005) and BRugs (Thomas et al., 2006) to call WinBUGS/OpenBUGS from R;

runjags (Denwood, 2016) to call JAGS from R;

coda (Plummer et al., 2006) to perform convergence diagnostics of the MCMC sample.

Next to the R packages mixAK, bayesSurv and DPpackage, the following two R packages were used in the Bayesian Part III of the book:

ICBayes (Pan et al., 2015) to fit a Bayesian PH model with a smooth baseline hazard (Section 11.2.1);

dynsurv (Wang et al., 2017) to fit a Bayesian PH model with piecewise constant baseline hazard (Section 11.2.2).

Appendix E

Description of selected SAS procedures

In this appendix, we briefly describe the syntax and useful options for the most important SAS procedures which appeared in the examples presented throughout the book. Nevertheless, the selection of options and their arguments being described is by no means comprehensive, nor covering all capabilities of the selected procedures. To explore the full capabilities of the described SAS procedures, we refer the reader to the corresponding SAS documentation.

E.1 PROC LIFEREG

This section describes some more useful options of the SAS/STAT LIFEREG procedure in the context of analyzing interval-censored observations. For a complete description of the procedure, we refer to the SAS/STAT 14.1 User's Guide (SAS Institute Inc., 2015b).

The LIFEREG procedure has the ability to fit accelerated failure time models (see Sections 3.2 and 6.1) and estimate the NPMLE of the cdf (see Section 3.1). The basic syntax for fitting interval-censored data with the LIFEREG procedure is given below. The symbols $<$ and $>$ are used to indicate an optional argument. The symbols themselves should be omitted in the program if the option is selected.

```
PROC LIFEREG <DATA=SAS-data-set>;
<CLASS variables;>
MODEL (variable1,variable2)=<variables> / <D=distribution>;
PROBPLOT / <MAXITEM=(n_1, n_2)>
           <ITPRINTEM>
           <PRINTPROBS>
           <PPOUT>;
RUN;
```

The DATA= argument specifies the SAS data set that contains the neces-

TABLE E.1: Supported distribution in PROC LIFEREG.

Keyword for DISTRIBUTION=	**NOLOG** specified?	**Fitted distribution**
EXPONENTIAL	No	Exponential
	Yes	One-parameter extreme value
GAMMA	No	Generalized gamma
	Yes	Generalized gamma with untransformed responses
LOGISTIC	No	Logistic
	Yes	Logistic
LLOGISTIC	No	Log-Logistic
	Yes	Logistic
LNORMAL	No	Log-normal
	Yes	Normal
NORMAL	No	Normal
	Yes	Normal
WEIBULL	No	Weibull
	Yes	Extreme value

sary data for your analysis. If omitted, SAS uses the last created data set to work on.

The optional CLASS statement specifies if there are variables that should be treated as categorical rather than continuous variables in the MODEL statement. It should be specified before the MODEL statement.

The variables variable1 and variable2 in the MODEL statement are the variables that contain the left and right endpoints of the interval-censored observations, respectively.

You specify the distribution to be fitted using the option DISTRIBUTION= or its abbreviations DIST= or D=. When no distribution is specified, the procedure fits a Weibull distribution on the logarithm of T. Table E.1 shows the possible distributions that can be fitted using the different keywords that can be used in conjunction with the NOLOG option. This option suppresses the default applied natural logarithmic transformation to the response. However, note that the NOLOG option has no effect when LOGISTIC or NORMAL is given as a keyword for the distribution. In addition, procedure LIFEREG fits a generalized gamma distribution (Lawless, 2003). The standard two-parameter gamma distribution is not available in PROC LIFEREG. There are some relations among the available distributions:

- The gamma with Shape=1 is a Weibull distribution.
- The gamma with Shape=0 is a log-normal distribution.
- The Weibull with Scale=1 is an exponential distribution.

If necessary, you can fix the scale parameter with the options NOSCALE and SCALE=. If no SCALE= value is specified, the scale parameter is fixed at the value 1. Note that if the log transformation has been applied to the response, the effect of the scale parameter is a power transformation of the original response. You also can fix the first shape parameter of the generalized gamma distribution using the options NOSHAPE1 and SHAPE1=. If no SHAPE1= value is specified, the shape parameter is fixed at the value 1.

By default, the LIFEREG procedure computes initial values for the parameters by using ordinary least squares and ignoring censoring. This might cause sometimes overflow or convergence problems if for example there are extreme values in your data. Using the INITIAL=, INTERCEPT=, SCALE= and SHAPE1= options, you can override the default starting values. The INITIAL= option sets initial values for the regression parameters excluding the intercept which can be initialized using the INTERCEPT= option. For further details about specifying the initial values, we refer to the SAS documentation manual.

E.2 PROC RELIABILITY

This section describes some more useful options of the SAS/QC RELIABILITY procedure in the context of analyzing interval-censored observations. For a complete description of the procedure, we refer to the SAS/QC 14.1 User's Guide (SAS Institute Inc., 2015a).

The RELIABILITY procedure has the ability to fit accelerated failure time models (see Sections 3.2 and 6.1) and estimate the NPMLE of the cdf (see Section 3.1).

By default, the initial estimate for the self-consistency algorithm assigns equal probabilities to each interval. Initial values can be provided with the PROBLIST= option. Convergence is obtained if the change in the log-likelihood between two successive iterations is less than delta (10^{-8} by default). A different value for delta is specified with the TOLLIKE= option.

For instance, MAXITEM=(10000,500) requests a maximum of 10 000 iterations and every 500 iterations, the iteration history will be displayed if it is requested with the ITPRINTEM option.

E.3 PROC ICLIFETEST

This section describes some more useful options of the SAS/STAT ICLIFETEST procedure. The procedure is available since SAS/STAT 13.1

in SAS 9.4. For a complete description of the procedure, we refer to the SAS/STAT 14.1 User's Guide (SAS Institute Inc., 2015b).

The ICLIFETEST computes nonparametric estimates of the survival function for interval-censored data (see Section 3.1.4) and tests the equality of different survival survival functions (see Section 4.1.7). The basic syntax for the ICLIFETEST procedure is given below. The symbols $<$ and $>$ are used to indicate an optional argument. The symbols themselves should be omitted in the program if the option is selected.

```
PROC ICLIFETEST <DATA=SAS-data-set> <OUTSURV=SAS-data-set>
                <BOOTSTRAP> <IMPUTE(SEED=n)> <METHOD=>
                <PLOTS<(global options)>=(plot-request(s))>;
TIME (variable1,variable2);
STRATA variables;
TEST variable /<WEIGHT=test-request>
               <TREND><DIFF=keyword>
               <ADJUST=method><OUTSCORE=SAS-data-set>;
RUN;
```

The DATA= argument specifies the SAS data set that contains the necessary data for your analysis. If omitted, SAS uses the last created data set to work on.

The OUTSURV= argument specifies the SAS data set in which survival estimates and confidence limits are stored.

By default, standard errors and the covariance matrix of the generalized log-rank statistic are calculated using a multiple imputation method according to Sun (1996). Using SEED= guarantees reproducible results. The number of imputations (1 000 by default) for computing the standard errors and covariance matrix are controlled by the options NIMSE= and NIMTEST=, respectively. Alternatively, for the standard errors a bootstrap method is available too using the keyword BOOTSTRAP with options NBOOT= to control the number of samples to be generated (1 000 by default) and SEED= to generate reproducible results. The standard method to compute the survival estimates is the EMICM algorithm but also the Turnbull method (METHOD=TURNBULL) or iterative convex minorant algorithm (METHOD=ICM) are available. The maximum number of iterations is controlled by MAXITER=. The EMICM method uses by default at most 200 iterations.

The PLOTS option may produce plots of the survival (or failure) curve versus time (SURVIVAL), the negative log of the estimated survival function versus time (LOGSURV), the log of the negative log of the estimated survival function versus the log of time (LOGLOGS) or the kernel smoothed hazard function versus time (HAZARD). Global plot options are ONLY, which specifies to plot only the requested plots (otherwise, the default survival plot is also generated) or ALL which produces all the above-mentioned plots. For the specific options for the kernel smoothing hazard function, we refer to the SAS documentation. For the survival plot the following options are available: 1) CL which displays pointwise confidence limits for the survival function; 2) FAILURE to display the failure function instead of the survival function; 3) NODASH which suppresses the plotting of the dashed lines a linear interpolation for the Turnbull intervals; 4) STRATA=INDIVIDUAL, OVERLAY or PANEL to control how the curves are plotted for multiple strata; 5) TEST to display the P-value for the K-sample test that is specified in the TEST statement.

The variables variable1 and variable2 in the TIME statement are the variables that contain the left and right endpoints of the interval-censored observations, respectively.

The STRATA statement is used to obtain separate survival estimates for different groups formed by one or more variables that you specify. By default, strata are formed according to the nonmissing values of these variables. However, missing values are treated as a valid stratum level if you specify the MISSING option in the PROC ICLIFETEST statement.

The TEST statement is used to test whether survival functions are the same between the groups. When you specify both the TEST and STRATA statements, the ICLIFETEST procedure calculates a stratified test in which the comparisons are conditional on the strata. The variables that you specify in the STRATA statement must be different from the variable that you specify in the TEST statement. Note that this setup differs from that of the LIFETEST procedure, in which survival comparisons are handled implicitly by the STRATA statement. The type of generalized log-rank test can be chosen by specifying the WEIGHT= option. The following tests are available using Fay's weights (FAY, based on a proportional odds model (Fay, 1999)), using Finkelstein's weights (FINKELESTEIN, assuming a PH model (Finkelstein, 1986)), using Sun's weights (SUN, default option (Sun, 1996)) or using Harrington and Fleming's weights (FLEMING(ϱ, γ), Harrington and Fleming (1982)). For ordered groups, also a trend test may be requested by adding the keyword TREND. The test is specifically designed to detect ordered alternatives such as $\mathrm{H}_1 : S_1(t) \geq S_2(t) \geq \cdots \geq S_K(t)$, with at least one inequality.

In case of a significant result, post-hoc pairwise testing may be requested by adding the keyword DIFF=ALL or CONTROL to perform all pairwise tests or all comparisons with a control group, respectively. To specify the control group, you specify the quoted string of the formatted value representing the control group in parentheses after the keyword. By default, the unadjusted and Tukey-Kramer (Kramer, 1956) adjusted P-values are reported. Other options for adjusting are possible with the specification of the keyword ADJUST=. The Bonferroni correction (BON), Scheffé correction (SCHEFFE), *Šidák* correction (SIDAK), the studentized maximum modulus correction (SMM), Tukey's correction (TUKEY), simulation based corrections (SIMULATE) or Dunnett's two-tailed comparison of a control group with all other groups (DUNNETT) are available.

The OUTSCORE= option allows to output scores that are derived from a permutation form of the generalized log-rank statistic. These scores can be post-processed with the NPAR1WAY procedure to provide a permutation test.

E.4 PROC ICPHREG

This section describes some more useful options of the SAS/STAT ICPHREG procedure. The procedure is available since SAS/STAT 13.2 in SAS 9.4. For a complete description of the procedure, we refer to the SAS/STAT 14.1 User's Guide (SAS Institute Inc., 2015b).

The ICPHREG procedure is designed to fit PH models to interval-censored data. It has the ability to model the baseline hazard function as a piecewise constant model (see Section 5.2.1) or a cubic spline model (see Section 5.2.3). The ICPHREG procedure maximizes the full likelihood instead of the Cox partial likelihood to obtain the regression coefficients and baseline parameters. Standard errors of the estimates are obtained by inverting the observed information matrix.

The ICPHREG procedure has to ability to include an offset variable in the model, weight the observations in the input data set, test linear hypotheses about the regression coefficients, compute hazard ratios, estimate and plot the survival function and the cumulative hazard function for a new set of covariates.

The basic syntax for fitting interval-censored data with the ICPHREG procedure is given below. The symbols $<$ and $>$ are used to indicate an optional argument. The symbols themselves should be omitted in the program if the option is selected.

```
PROC ICPHREG <DATA=SAS-data-set> <NAMELEN=n>
             <PLOTS<(global options)>=(plot-request(s))>;
<CLASS variables;>
MODEL (variable1,variable2)=<variables> / <BASE=options>;
TEST <model-effect> / <options>;
HAZARDRATIO <'label'> variable / <AT(variable=option)> <DIFF=option>;
BASELINE <OUT=SAS-data-set> <COVARIATES=SAS-data-set>
         <TIMELIST=list> /
         <GROUP=variable> <ROWID=variable>
         <keyword options for variables e.g. SURVIVAL=, CUMHAZ=, ...>;
RUN;
```

The DATA= argument specifies the SAS data set that contains the necessary data for your analysis. If omitted, SAS uses the last created data set to work on.

By default, the length of effect names in tables and output data sets are limited to 20 characters. Using the NAMELEN= option, the user may augment it up to 200.

Using the PLOTS option, the survival function and/or the cumulative hazard function for a new set of covariates may be plotted. ODS GRAPHICS must be turned on in order to produce any plot. The user may request the estimated cumulative hazard (CUMHAZ) or survival plot (SURVIVAL) to be plotted for each set of covariates in the COVARIATES= data set specified in the BASELINE statement. If the COVARIATES= data set is not specified, the plots are created for the reference set of covariates, which consists of reference levels for the CLASS variables and average values for the continuous variables. Global plot options include: 1) CL which generates pointwise confidence limits; 2) OVERLAY=GROUP or INDIVIDUAL which either overlays on the same plot all functions that are plotted for the covariate sets that have the same GROUP= value in the COVARIATES= data set or creates a separate plot for each covariate set, respectively; 3) RANGE=(min, max) which specifies the range of values on the time axis to clip the display. By default, min = 0 and max is the largest boundary value.

The optional CLASS statement, to be specified before the MODEL statement, specifies if there are variables that should be treated as categorical rather than continuous variables in the MODEL statement. Several options are available to choose the reference category. A useful global option is the PARAM=REF combined with REF=FIRST or LASTwhich allows you to choose the first or last ordered level as a reference category, respectively. The individual variable option REF='formatted level' can be used to deviate from this global option. For instance,

```
CLASS gender (REF='Male') educ weight / PARAM=REF REF=LAST;
```

The variables variable1 and variable2 in the MODEL statement are the variables that contain the left and right endpoints of the interval-censored observations, respectively. The left endpoints may contain either a zero or a missing.

By default, the model fits a piecewise constant model (Friedman, 1982) for the baseline hazard (BASE=PIECEWISE). Using the options NINTERVAL= or INTERVAL= the user controls the number of intervals to be used or the partition of the time axis, respectively. The option NINTERVAL= specifies in how many intervals PROC ICPHREG should divide the time axis such that each interval contains an approximate equal number of unique boundary values and imputed mid-points. Alternatively, using the INTERVAL= option, a list of numbers that partition the time axis into disjoint intervals that have a constant hazard rate in each interval can be provided. For example, INTERVALS=(10,20,30) specifies a model that has a constant hazard in the intervals $[0, 10), [10, 20), [20, 30)$ and $[30, \infty)$. If neither option is specified, 5 intervals are used by default. PROC ICPHREG by default does not transform the baseline parameters and names them Haz1, Haz2,..., and so on. As these parameters are bounded below by 0, applying a logarithmic transformation removes this constraint. Specifying HAZSCALE=LOG uses log-transformed baseline parameters which are named LogHaz1, LogHaz2,..., and so on. By default, the ICPHREG procedure "polishes" the hazard estimates by fixing these parameters at the lower bound value and refitting the model. The lower bound values are set to 0 for modelling on the original scale and −10 for modelling on the log scale. Adding the option NOPOLISH suppresses polishing of the parameter estimates of the baseline function.

A cubic splines model (Royston and Parmar, 2002) for the cumulative baseline hazard is obtained by specifying BASE=SPLINES. The only available option is to set the degrees of freedom by adding DF=m (with m an integer > 0) in parentheses. The number of knots equals $m+1$. By default, two degrees of freedom are used.

The TEST statement allows to perform Wald tests for the model effects. By default, type III tests are applied but also type I and II tests may be obtained by specifying the option HTYPE=1 or 2, respectively.

The HAZARDRATIO statement enables the user to request hazard ratios for any variable in the model. If the variable is continuous, the hazard ratio compares the hazards for an increase of 1 unit in the variable. A different unit may be specified with the UNITS= option. For a CLASS variable, a hazard ratio compares the hazards of two levels of the variable. By default, all comparisons of the distinct combinations of pairs are reported. This means that, e.g., A1 vs. A2 is reported but not A2 vs. A1. Using DIFF=, one can change this behavior using the keyword PAIRWISE to request all possible comparisons or the keyword REF to request only the comparisons between the reference level and all other levels of the CLASS variable. If the variable is part of an interaction term in the model, the option AT controls at which level of the interacting variables the hazard ratio is calculated. If the interacting variable is continuous and a numeric list is specified after the equal sign, hazard ratios are computed for each value in the list. If the interacting variable is a CLASS variable, a list of quoted strings that correspond to various levels of the CLASS variable or the keyword ALL or REF may be specified. Hazard ratios are then computed at each value of the list, at each level of the interacting variable or at the reference level of the interacting variable, respectively. If you do not specify the AT option, PROC ICPHREG finds all the variables that interact with the variable of interest. If an interacting variable is a CLASS variable, variable=ALL is the default; if the interacting variable is continuous, variable=m is the default, where m is the average of all the sampled values of the continuous variable. As an example, for a model where treatment (variable tx) has an interaction with age (variable age), the code below requests the hazard ratios for treatment for patients who are aged 40, 50 or 60 years old.

```
HAZARDRATIO tx / AT (age=40,50,60);
```

Several HAZARDRATIO statements may be included and the optional label can be used to identify the different requests.

The BASELINE statement creates a SAS data set that contains predicted values of several statistics at specified times for a set of covariates. The output data set is named in the OUT= option. If omitted, a data set named according to the DATAn convention will be created. The predicted values to be outputted are specified by adding after the slash the specific keywords followed by equal sign and a variable name that indicates under which name the statistic is stored in the output data set. Available statistics include among others the cumulative hazard estimate (CUMHAZ=), the survival estimate (SURVIVAL=), the standard error of the survival function (STDERR=) and cumulative hazard (STDCUMHAZ=), the lower and upper pointwise confidence limits for the survival function (LOWER= and UPPPER=) and cumulative hazard (LOWERCUMHAZ= and UPPERCUMHAZ=). These predicted values are calculated for each set of covariates specified in the COVARIATES= data set and at all the times that partition the time axis (by default) or the values specified in the TIMELIST= option. If no COVARIATES= data set is specified, the ICPHREG procedure uses the reference set of covariates described above. The options GROUP= and ROWID= name the grouping variable for the plots and the identifying variable to be used in the legend of these plots, respectively.

Appendix F

Technical details

In this appendix, we briefly explain some technical details about concepts and algorithms that are useful for a better understanding of the method but were too technical to be included in the main text.

F.1 Iterative Convex Minorant (ICM) algorithm

The objective of the Iterative Convex Minorant algorithm (ICM)-algorithm is to maximize the log-likelihood

$$\ell(F_0, \boldsymbol{\beta}) = \sum_{i=1}^{n} \log\left[\{1 - F_0(l_i-)\}^{\exp(\boldsymbol{X}_i'\boldsymbol{\beta})} - \{1 - F_0(u_i)\}^{\exp(\boldsymbol{X}_i'\boldsymbol{\beta})}\right]$$

by a modified Newton-Raphson algorithm. The gradients needed for the maximization are $\nabla_1 \ell(F_0, \boldsymbol{\beta}) = \frac{\partial \ell(F_0, \boldsymbol{\beta})}{\partial F_0}$ and $\nabla_2 \ell(F_0, \boldsymbol{\beta}) = \frac{\partial \ell(F_0, \boldsymbol{\beta})}{\partial \boldsymbol{\beta}}$. The baseline distribution function F_0 is considered to be piecewise constant and thus can be represented by a finite k-dimensional vector which is parametrized by the finite steps of the cumulative baseline hazard function. Let $s_1, \ldots, s_k$ denote the points at which F_0 may have jumps and $\mathcal{R} = \{F : 0 \leq F(s_1) \leq \ldots \leq F(s_k) \leq 1\}$. The derivative with respect to F_0 is the gradient of the log-likelihood with respect to the vector of the baseline cumulative distribution function values. The derivative with respect to $\boldsymbol{\beta}$ is the usual derivative of the log-likelihood with respect to the components of $\boldsymbol{\beta}$. The full Hessian in the original Newton-Raphson algorithm is replaced by the diagonal matrices of the negative second derivatives $G_1(F_0, \boldsymbol{\beta})$ and $G_2(F_0, \boldsymbol{\beta})$.

Denote estimates from the ith iteration as $F^{(i)}$ and $\boldsymbol{\beta}^{(i)}$. The update from $F^{(m)}$ to $F^{(m+1)}$ and $\boldsymbol{\beta}^{(m)}$ to $\boldsymbol{\beta}^{(m+1)}$ are done iteratively with control of the step size. The starting point is always a step size of $\alpha = 1$. The new candidates for F_0 and $\boldsymbol{\beta}$ result from

$$F_0^{(m+1)} = \mathrm{Proj}[F_0^{(m)} + \alpha G_1(F_0^{(m)}, \boldsymbol{\beta}^{(m)})^{-1} \nabla_1 \ell(F_0^{(m)}, \boldsymbol{\beta}^{(m)}), G_1(F_0^{(m)}, \boldsymbol{\beta}^{(m)}), \mathcal{R}],$$

$$\boldsymbol{\beta}^{(m+1)} = \boldsymbol{\beta}^{(m)} + \alpha G_2(F_0^{(m)}, \boldsymbol{\beta}^{(m)})^{-1} \nabla_2 \ell(F_0^{(m)}, \boldsymbol{\beta}^{(m)}).$$

A projection into the restricted range $\mathcal{R}$ weighted by G is used to assure that $F_0^{(m+1)}$ is again a distribution function:

$$\text{Proj}[y, G, \mathcal{R}] = \arg\min_{\boldsymbol{x}} \left\{ \sum_{i=1}^{k} (y_i - s_i)^2 G_{ii} \, : \, 0 \leq s_1 \leq \ldots \leq s_k \leq 1 \right\}.$$

In case of $\ell(F^{(m+1)}) < \ell(F^{(m)})$, α is halved and the step is reiterated. A numerical procedure for the restricted projection is the pool adjacent violators algorithm (PAVA), which is described in Robertson et al. (1988). Starting values can be calculated by treating the data as right censored and using the classical proportional hazards model. In order to do so, left- or genuine interval-censored observations are interpreted as right-censored at their upper limit. Right-censored observations are used as such. The Breslow-estimator is used to get a starting value for the baseline hazard $H_0(t)$.

F.2 Regions of possible support for bivariate interval-censored data

As described in Section 7.1, the first step in finding the NPMLE for bivariate interval-censored data is determining the regions of possible support. In addition to the simple but unpractical algorithm of Betensky and Finkelstein (1999b) for finding the regions of possible support in moderate to large data sets, Gentleman and Vandal (2001), Bogaerts and Lesaffre (2004) and Maathuis (2005) proposed more efficient algorithms.

F.2.1 Algorithm of Gentleman and Vandal (2001)

Using graph theoretic concepts, Gentleman and Vandal (2001) proposed a more efficient algorithm based on a marginal approach. To understand the algorithm we need some graph theory terminology. Let each (bivariate) interval-censored observation correspond to a vertex. Two vertices are joined by an edge if the corresponding observations intersect. Figure F.1 depicts the graph corresponding to our example data presented in Figure 7.3.

Let V denote the set of vertices. A clique C is a subset of V such that every member of C is connected by an edge to every other member of C. A clique is maximal if it is not a proper subset of any other clique. In the context of interval-censored data, a clique can be viewed as a set of those observations whose regions intersect. The regions of possible support are the maximal cliques. In our example, we observe 4 maximal cliques $\{4\}$, $\{2, 3, 5\}$, $\{1, 2\}$ and $\{5, 6\}$. The algorithm of Gentleman and Vandal can be described as follows: Search (marginally) in each dimension for the regions of possible support using, e.g., Turnbull's algorithm (see Section 3.1). For each interval of possible support, determine the bivariate observations (i.e., rectangles) that contain that specific interval. The collection of these observations for that interval constitute a maximal clique. The final regions of possible support are a subset of the intersections of all these maximal cliques from both dimensions. The algorithm determines the intersections in a straightforward manner. By approaching the problem first in a marginal sense and then determining the intersections, the number of intersections to be established is greatly reduced compared to the algorithm of Betensky and Finkelstein (1999b). More details on the algorithm and the terminology from graph theory can be found in the original publication (Gentleman and Vandal, 2001). The algorithm can be quite easily extended to higher dimensions.

F.2.2 Algorithm of Bogaerts and Lesaffre (2004)

Bogaerts and Lesaffre (2004) described a more efficient algorithm by directly searching the regions of possible mass. The algorithm can be briefly described as follows: For each distinct endpoint (left or right) in the first dimension (say $t = j$), determine the regions of possible support (e.g., with Turnbull's algorithm) on the intervals that are obtained by intersecting the observed rectangles with the vertical line $T = j$. This results in a set of endpoints (l, u) with possible non-zero mass for each left or right endpoint in the first dimension. Do this similarly for the left and right endpoints in the second dimension. The regions of possible mass for the bivariate case are determined by rectangles that have on each side of the rectangle a region of possible mass from the univariate stratified process and have no other such rectangle completely inside it. Finally, note that the original algorithm is based on closed

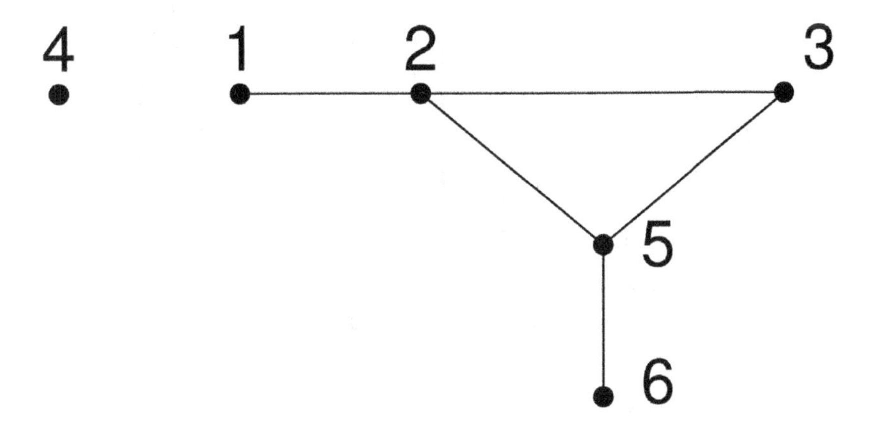

FIGURE F.1: Graph corresponding to the data presented in Figure 7.3

intervals. In case semi-open intervals are given, the algorithm can still be applied by adding a small enough value to the open endpoint turning it into a closed interval and such that no other endpoint lies between the original endpoint and the newly constructed one. This value must be removed at the end of the algorithm. The algorithm can also be extended to higher dimensions but the extension is less transparent, certainly for more than 3 dimensions.

F.2.3 Height map algorithm of Maathuis (2005)

Maathuis (2005) published the most efficient algorithm up to now for determining the possible regions of support. In a first step, the observed rectangles $\{R_1, \ldots, R_n\}$ are transformed into canonical rectangles with the same inter-

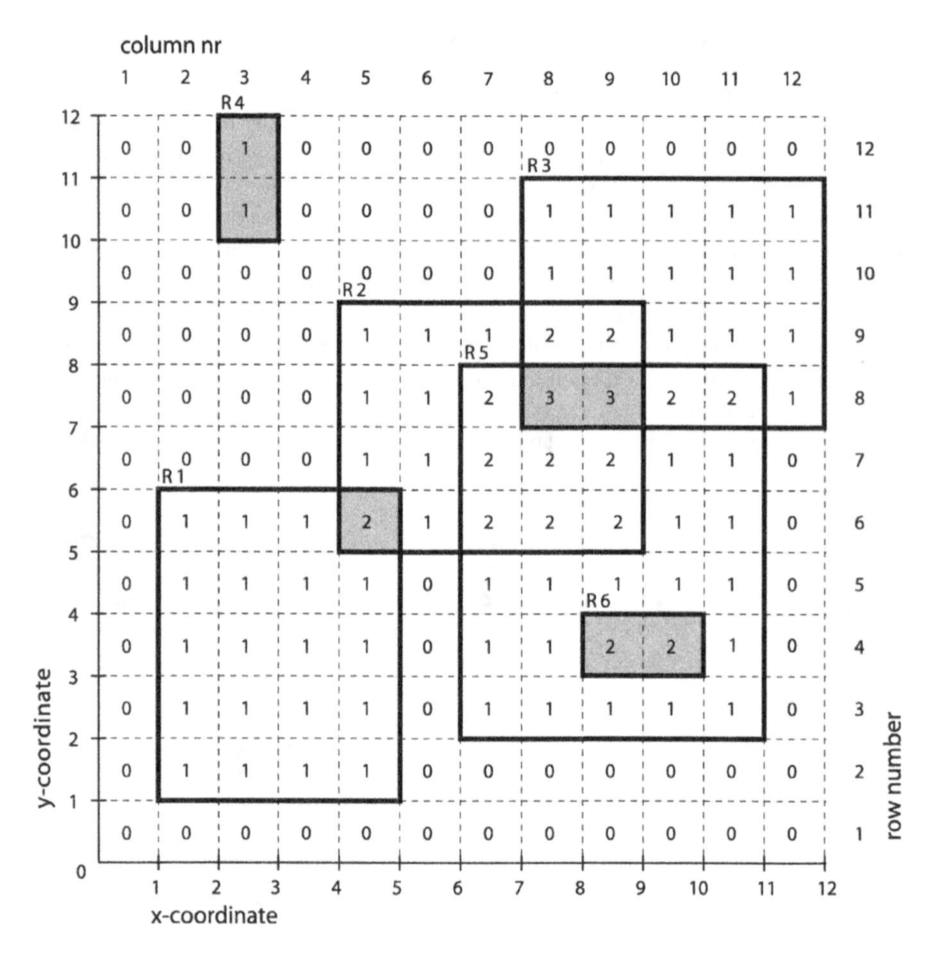

FIGURE F.2: Height map corresponding to the example data presented in Figure 7.3. Regions of possible support are indicated in gray.

section structure. A set of n rectangles are called canonical if all x-coordinates are different and all y-coordinates are different, and if they take on values in the set $\{1, 2, \ldots, 2n\}$. These steps force in the very beginning to deal with ties and with the fact whether endpoints are open or closed. As a consequence, no attention should be paid to this in the actual algorithm. In addition, it simplifies the reduction algorithm, since the column and row numbers in the height map directly correspond to the x- and y-coordinates of the canonical rectangles. As for the algorithm of Gentleman and Vandal (2001), in the second step the algorithm searches for the maximal cliques. Now the observed rectangles are first turned into canonical rectangles to define a height map of the observed rectangles. The height map is a function $h : \mathbb{R}^2 \to \mathbb{N}$, where $h(x, y)$ is the number of observation rectangles that contain the point (x, y). Figure F.2 depicts the height map for the example. A sweeping technique (commonly used in computational geometry, see Lee, 1983) is applied to efficiently determine the regions of possible support which are formed by the regions of maximal height. In Figure F.2 we observe 4 such regions indicated in gray. Also, this algorithm can be quite easily extended to higher dimensions.

F.3 Splines

In this section we give a brief overview of the smoothing splines techniques used in the book. An excellent introduction to this increasingly popular topic can be found in Fahrmeir et al. (2013). Smoothing splines techniques have been implemented on the hazard function or its logarithm, the cumulative hazard function and the density function or its logarithm. Smoothing methods have been applied in a great variety of applications. In a regression context, these methods are examples of nonparametric regression. In this brief review, we assume that smoothing is applied on a variable t representing time to an event. We limit ourselves to univariate smoothing. To achieve some generality we denote the function for which a smooth representation is required by $f(t)$. This function can be a hazard function, a log-hazard function, a cumulative hazard function, a density function or a log-density function. Often the spline smoothing approach may be applicable to all of such functions, in other cases the approach may only be applied on one type of function.

For all methods we assume that the variable t is defined on the real axis, but observed in a data set in an interval $[a, b]$. In survival applications, this interval will be $[0,\ b]$, with b the largest observed censored or uncensored survival time.

F.3.1 Polynomial fitting

The classical polynomial of degree $l \geq 0$ (or order = degree + 1) in the variable t is given by

$$f(t) = \gamma_0 + \gamma_1 t + \ldots + \gamma_l t^l.$$

The function $f(t)$ could represent (or better, aims to approximate) a hazard function, a log-hazard function, etc. The regression coefficients $\gamma_0, \gamma_1, \ldots, \gamma_l$ then become model parameters to estimate from the data. In a frequentist context these are estimated using maximum likelihood, while in a Bayesian context MCMC computations will deliver the posterior estimates. However, it is known that classical polynomials of a relatively high degree may have a wild behavior in the neighborhood of a and b, but also when used for interpolation especially when data are sparse in a particular interval. In addition, classical polynomials may not provide a good fit to certain patterns in the data, e.g., they may not fit well an asymptote in the data.

F.3.2 Polynomial splines

To make the polynomial model more flexible, we could partition the domain of t into intervals and estimate separate polynomials in each interval. Thus, we fit several locally defined polynomials instead of one global model. The major disadvantage of these piecewise polynomial models is that they do not lead to an overall smooth function. A polynomial spline or regression spline connects smoothly the adjacent local polynomials. The polynomial spline of degree l is defined as a real function $f(t)$ on $[a, b]$ with knots s_j $(j = 1, \ldots, m)$ satisfying $a = s_0 < s_1 < \ldots < s_m < s_{m+1} = b$ if it fulfills the following conditions:

- $f(t)$ is $(l-1)$-times continuously differentiable. The special case of $l = 1$ corresponds to $f(t)$ being continuous (but not differentiable). There are no smoothness requirements for $f(t)$ when $l = 0$.
- $f(t)$ is a polynomial of degree l on the intervals $[s_j, s_{j+1})$ $(j = 0, \ldots, m)$.

The knots $s_1 < s_2 < \ldots < s_m$ are called the inner knots. Note that the definition of a polynomial spline is not specific enough. There are several implementations of this concept suggested in the literature, these will be reviewed in the next subsections. We consider natural cubic splines, truncated power series, B-splines, M-splines and I-splines.

For all the above-defined systems of smoothing splines, the number and the position of the knots needs to be specified by the user. In general, one observes that for a small number of knots the fitted function is smooth but biasedly estimates the true function, while for a large number of knots the fitted function becomes (quite) wiggly. Also, the position of the knots needs to be determined by the user. One generally distinguishes: equidistant knots, quantile-based knots and visually chosen knots. The number and position of the knots can be chosen by comparing different choices via a model selection criterion such as AIC in the frequentist context or DIC in a Bayesian context. In the ICBayes package the log pseudo-marginal likelihood (LPML) was chosen as criterion (larger is better), see Section 9.1.5.

Alternatively, one could use P-splines, which are discussed in the last subsection below. P-splines are based on a large number of knots but with γ-values that are discouraged to vary wildly. In that case, equidistant knots may be chosen, in the other case (without penalization) quantile-based knots or visually chosen knots may be preferred to avoid placing knots in areas where data are sparse.

F.3.3 Natural cubic splines

The function $f(t)$ is a natural cubic spline based on the knots $a = s_0 < s_1 < \ldots < s_m < s_{m+1} = b$, if:

- $f(t)$ is a cubic polynomial spline for the given knots;
- $f(t)$ satisfies the boundary conditions: the second derivatives at the boundaries of $[a, b]$ are zero, i.e., $f''(a) = f''(b) = 0$. Hence, $f(t)$ is linear at the boundaries.

F.3.4 Truncated power series

A truncated polynomial spline of degree l is defined as the sum of a global polynomial of degree l and local polynomial emanating at the knots, i.e.,

$$f(t) = \gamma_0 + \gamma_1 t + \ldots + \gamma_l t^l + \gamma_{l+1}(t - s_1)_+^l + \ldots + \gamma_{l+m}(t - s_m)_+^l$$

with

$$(t - s_j)_+^l = \begin{cases} (t - s_j)^l, & \text{if } t \geq s_j, \\ 0, & \text{otherwise.} \end{cases}$$

The basis functions of the truncated polynomial spline of degree l therefore are the global basis functions: $1, t, \ldots, t^l$ and the local basis functions $(t - s_1)_+^l, \ldots, (t - s_m)_+^l$. In Figure F.3 the basis functions of a truncated polynomial spline of degree 1 on $[0, 10]$ are shown, with inner knots at $1, \ldots, 9$.

F.3.5 B-splines

The B-spline basis consists of locally defined basis functions. Namely, a B-spline basis function, or shorter a B-spline, of degree l is built up from $(l+1)$ polynomial pieces of degree l joined in an $(l-1)$ times continuously differentiable way. With a complete basis $f(t)$ can be represented as a linear combination of $d = m + l + 1$ basis functions, i.e., $f(t) = \sum_{j=0}^{d-1} \gamma_j B_j(t)$. The

B-splines have a local support and are defined recursively. For the B-spline of degree $l = 0$:

$$B_j^0(t) = I(s_j \le t < s_{j+1}) = \begin{cases} 1, & s_j \le t < s_{j+1}, \\ 0, & \text{otherwise.} \end{cases} \qquad (j = 0, \dots, d-1),$$

with $I(\cdot)$ denoting the indicator function. The B-spline of degree $l = 1$ is then obtained by:

$$B_j^1(t) = \frac{t - s_{j-1}}{s_j - s_{j-1}} I(s_{j-1} \le t < s_j) + \frac{s_{j+1} - t}{s_{j+1} - s_j} I(s_j \le t < s_{j+1}).$$

Hence each basis function is defined by two linear segments on the intervals $[s_{j-1}, s_j)$ and $[s_j, s_{j+1})$, continuously combined at knot s_j. For the general case, the B-spline of degree l is obtained from the B-spline of order $(l-1)$ by:

$$B_j^l(t) = \frac{t - s_{j-l}}{s_j - s_{j-l}} B_{j-1}^{l-1}(t) + \frac{s_{j+1} - t}{s_{j+1} - s_{j+1-l}} B_j^{l-1}(t).$$

The recursive definition of B-splines requires $2l$ outer knots outside the domain $[a, b]$, and m interior knots. This produces an expanded knots sequence $s_{1-l}, s_{1-l+1}, \dots, s_0, s_1, \dots, s_m, s_{m+1}, \dots, s_{m+l-1}, s_{m+l}$. In practice one most often uses equidistant knots, and in that case one can show that B-splines can be computed from truncated polynomials. B-splines form a local basis. Namely, each basis function is positive in an interval formed by $l + 2$ adjacent

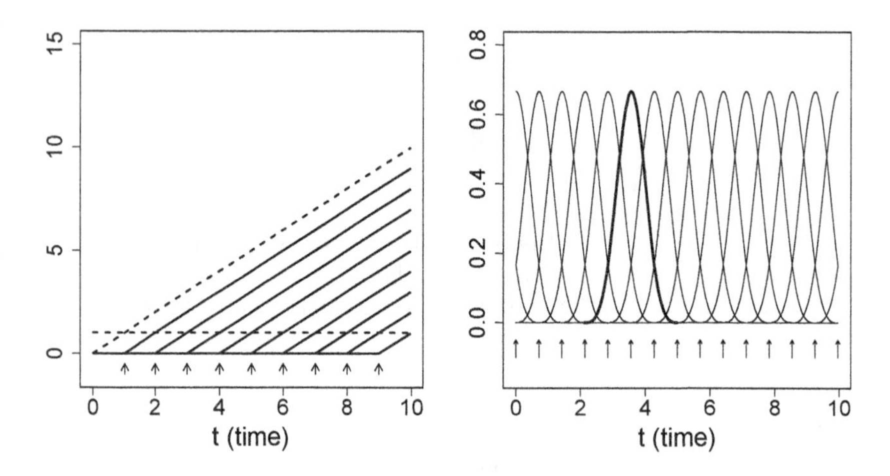

FIGURE F.3: Left: Truncated polynomial of degree 1 on [0, 10]. Global components of degree 0 and 1 (dashed), local components of degree 1 (solid) emanate at inner knots indicated by arrows. Right: B-splines of degree 3 defined on indicated knots.

knots and overlaps within $[a, b]$ with exactly $2l$ adjacent basis functions. In addition, all basis functions have the same form and are only shifted along the X-axis. At any point $t \in [a, b]$, $(l+1)$ basis functions are positive and $\sum_{j=0}^{d-1} B_j(t) = 1$.

F.3.6 M-splines and I-splines

A related construct to B-splines are M-splines introduced by Ramsay (1988). M-splines of order l (degree $l-1$) are constructed recursively in l steps and involve the choice of m inner knots $s_1 < \ldots < s_m$ in a prespecified interval $[a, b]$. More specifically for order $= 1$, let $s_0 = a < s_1 < \ldots < s_m < s_{m+1} = b$, then the $(m+1)$ M-splines of order 1, $M_j(t \mid 1)$, are defined for $j = 0, \ldots, m$ as:

$$M_j(t \mid 1) = \begin{cases} \dfrac{1}{s_{j+1} - s_j}, & \text{for } s_j \leq t < s_{j+1}, \\ 0, & \text{elsewhere.} \end{cases}$$

With the following recursive algorithm all $(m+l)$ M-splines of order $l \geq 2$ are defined:

$$M_j(t \mid l) = \begin{cases} \dfrac{(t - s_{j-l})M_j(t \mid l-1) + (s_j - t)M_{j+1}(t \mid l-1)}{s_j - s_{j-l}} \cdot \dfrac{l}{l-1} & \\ \qquad\qquad\qquad\qquad\qquad\qquad \text{for } s_{j-l} \leq t < s_j & \\ 0 & \\ \qquad\qquad\qquad\qquad\qquad\qquad \text{elsewhere,} & \end{cases}$$

with now $s_{-l+1} = \ldots = s_0 = a < s_1 < \ldots < s_m < s_{m+1} = \ldots = s_{m+l} = b$ and $j = 1, \ldots, m+l$. In Figure F.4, we show that each $M_j(t \mid l)$ is a piecewise polynomial non-zero only within $[s_{j-l}, s_j]$.

The M-splines are then integrated over time from a to t and in this way yield $m+l$ monotone increasing functions $I_j(t \mid l) = \int_a^t M_j(u \mid l)\, du$, $(j = 1, \ldots, m+l)$, which can be used as a tool to represent cumulative baseline hazard functions in a smooth way. Namely, the baseline cumulative hazard function $\Lambda_0(t)$ can be written as $\Lambda_0(t) = \sum_{j=1}^{m+l} \gamma_j I_j(t \mid l)$, with $I_j(t \mid l)$ a monotone spline basis function of order l and γ_j nonnegative spline functions to ensure that $\Lambda_0(t)$ is nondecreasing. Monotone splines are also called integrated (I-) splines.

F.3.7 Penalized splines (P-splines)

The quality of a nonparametric function estimated by the above-defined smoothing splines depends importantly on the number of knots. Alternatively, one could choose a large number of knots and penalize for roughness. This leads to penalized splines (P-splines). For the truncated power series, the penalty term on coefficients $(\gamma_{l+1}, \ldots, \gamma_{l+m})'$ is given by $P_{TP}(\boldsymbol{\gamma}) = \sum_{j=l+1}^{l+m} \gamma_j^2$. For B-splines, a popular and simple penalty term was introduced by Eilers and Marx (1996). They suggested to penalize solutions with too different adjacent γ values (for B-splines, $\boldsymbol{\gamma} = (\gamma_0, \ldots, \gamma_{d-1})'$), i.e., to take as penalty term $P_{BS} = \sum_{j=r}^{d-1} (\Delta^r \gamma_j)^2$, where Δ^r denotes rth-order differences, recursively defined by

$$\begin{aligned}
\Delta^1 \gamma_j &= \gamma_j - \gamma_{j-1}, \\
\Delta^2 \gamma_j &= \Delta^1 \Delta^1 \gamma_j = \Delta^1 \gamma_j - \Delta^1 \gamma_{j-1} = \gamma_j - 2\gamma_{j-1} + \gamma_{j-2}, \\
&\vdots \\
\Delta^r \gamma_j &= \Delta^{r-1} \gamma_j - \Delta^{r-1} \gamma_{j-1}.
\end{aligned}$$

If $\ell(\boldsymbol{\theta}, \boldsymbol{\gamma} \mid \boldsymbol{\mathcal{D}})$ denotes the (unpenalized) log-likelihood depending on the spline coefficients $\boldsymbol{\gamma}$ and other model parameters $\boldsymbol{\theta}$, then the penalized log-likelihood is given by

$$\ell_P(\boldsymbol{\theta}, \boldsymbol{\gamma} \mid \boldsymbol{\mathcal{D}}) = \ell(\boldsymbol{\theta}, \boldsymbol{\gamma} \mid \boldsymbol{\mathcal{D}}) - \frac{1}{2} \lambda P(\boldsymbol{\gamma}),$$

with P equal to P_{TP} or P_{BS} for the truncated power series and the B-splines, respectively. The penalized MLEs are obtained by maximizing $\ell_P(\boldsymbol{\theta}, \boldsymbol{\gamma} \mid \boldsymbol{\mathcal{D}})$. The parameter λ controls the smoothness of the penalized solution, with $\lambda = 0$ implying no penalty correction and $\lambda \to \infty$ a parametric fit completely determined by the penalty term. In a Bayesian context, since the penalty term is a quadratic function in the elements of $\boldsymbol{\gamma}$, the product $\lambda\times$ penalty term can be considered as the logarithm of a Gaussian prior on $\boldsymbol{\gamma}$. This is similar to a ridge prior for regression coefficients in a Bayesian regression problem.

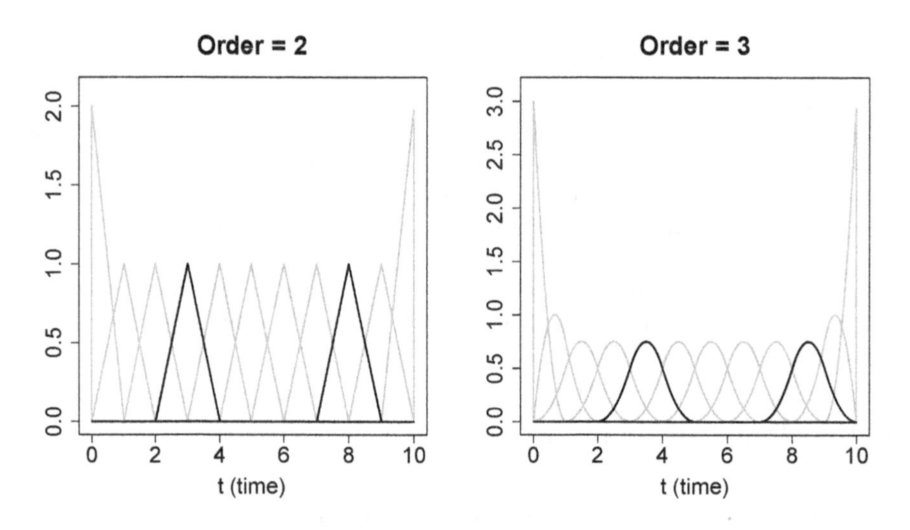

FIGURE F.4: M-splines of order 2 and 3 on [0, 10] with each time two M-splines highlighted.

Remaining is the choice of λ. In a classical context, the optimal λ is chosen using a cross-validation argument. But, because exact cross-validation is too computer intensive, approximate cross-validation is used in practice. In a classical regression context, a popular version of the cross-validation approach is generalized cross-validation, see Fahrmeir et al. (2013). In a Bayesian context the parameter λ can be estimated via its (marginal) posterior distribution.

References

Aalen, O. (1978). Nonparametric inference for a family of counting processes. *The Annals of Statistics*, *6*(4), 701–726.

Adkinson, R. W., Ingawa, K. H., Blouin, D. C., and Nickerson, S. C. (1993). Distribution of clinical mastitis among quarters of the bovine udder. *Journal of Dairy Science*, *76*(11), 3453–3459.

Aerts, M., Claeskens, G., and Hart, J. D. (1999). Testing the fit of a parametric function. *Journal of the American Statistical Association*, *94*(447), 869–879.

Akaike, H. (1974). A new look at the statistical model identification. *IEEE Transactions on Automatic Control*, *AC-19*(6), 716–723.

Altshuler, B. (1970). Theory for the measurement of competing risks in animal experiments. *Mathematical Biosciences*, *6*(1), 1–11.

Andersen, P. K., Borgan, Ø., Gill, R. D., and Keiding, N. (1993). *Statistical models based on counting processes.* New York: Springer-Verlag.

Andersen, P. K., and Gill, R. D. (1982). Cox's regression model for counting processes: A large sample study. *The Annals of Statistics*, *10*(4), 1100–1120.

Andersen, P. K., and Rønn, B. B. (1995). A nonparametric test for comparing two samples where all observations are either left- or right-censored. *Biometrics*, *51*(1), 323–329.

Anderson-Bergman, C. (2017). icenReg: Regression models for interval censored data in R. *Journal of Statistical Software*(Accepted).

Antoniak, C. E. (1974). Mixtures of Dirichlet processes with applications to Bayesian nonparametric problems. *Annals of Statistics*, *2*(6), 1152–1174.

Barlow, W. E., and Prentice, R. L. (1988). Residuals for relative risk regression. *Biometrika*, *75*(1), 65–74.

Beadle, G. F., Silver, B., Botnick, L., Hellman, S., and Harris, J. R. (1984). Cosmetic results following primary radiation therapy for early breast cancer. *Cancer*, *54*(12), 2911–2918.

Besag, J., Green, P., Higdon, D., and Mengersen, K. (1995). Bayesian computation and stochastic systems (with Discussion). *Statistical Science*, *10*(1), 3–66.

Betensky, R. A., and Finkelstein, D. M. (1999a). An extension of Kendall's coefficient of concordance to bivariate interval censored data. *Statistics in Medicine*, *18*(22), 3101–3109.

Betensky, R. A., and Finkelstein, D. M. (1999b). A non-parametric maximum likelihood estimator for bivariate interval censored data. *Statistics in Medicine*, *18*(22), 3089–3100.

Billingsley, P. (1995). *Probability and measure* (Third ed.). New York: John Wiley & Sons.

Bogaerts, K. (2007). *Statistical modelling of tooth emergence based on multivariate interval-censored data* (Unpublished doctoral dissertation). Katholieke Universiteit Leuven, Leuven, Belgium.

Bogaerts, K., Leroy, R., Lesaffre, E., and Declerck, D. (2002). Modelling tooth emergence data based on multivariate interval-censored data. *Statistics in Medicine*, *21*(24), 3775–3787.

Bogaerts, K., and Lesaffre, E. (2003, August 3 to 7). A smooth estimate of the bivariate survival density in the presence of left, right and interval censored data. In *Proceedings of the joint statistical meetings, biometrics section [cd-rom]* (pp. 633–639). Alexandria, VA: American Statistical Association.

Bogaerts, K., and Lesaffre, E. (2004). A new, fast algorithm to find the regions of possible support for bivariate interval-censored data. *Journal of Computational and Graphical Statistics*, *13*(2), 330–340.

Bogaerts, K., and Lesaffre, E. (2008a). Estimating local and global measures of association for bivariate interval censored data with a smooth estimate of the density. *Statistics in Medicine*, *27*(28), 5941–5955.

Bogaerts, K., and Lesaffre, E. (2008b). Modeling the association of bivariate interval-censored data using the copula approach. *Statistics in Medicine*, *27*(30), 6379–6392.

Böhning, D. (1986). A vertex-exchange method in *D*-optimal design theory. *Metrika*, *33*(1), 337–347.

Böhning, D., Schlattmann, P., and Dietz, E. (1996). Interval censored data: A note on the nonparametric maximum likelihood estimator of the distribution function. *Biometrika*, *83*(2), 462–466.

Borgan, Ø., and Liestøl, K. (1990). A note on confidence intervals and bands for the survival curves based on transformations. *Scandinavian Journal of Statistics*, *17*, 35–41.

Boumezoued, A., El Karoui, N., and Loisel, S. (2017). Measuring mortality heterogeneity with multi-state models and interval-censored data. *Insurance: Mathematics and Economics*, *72*, 67–82.

Breslow, N. E. (1974). Covariance analysis of censored survival data. *Biometrics*, *30*(1), 89–99.

Brooks, S. P., and Gelman, A. (1998). General methods for monitoring convergence of iterative simulations. *Journal of Computational and Graphical Statistics*, *7*(4), 434–455.

Buckley, J., and James, I. (1979). Linear regression with censored data. *Biometrika*, *66*(3), 429–436.

Cai, T., and Cheng, S. (2004). Semiparametric regression analysis for doubly censored data. *Biometrika*, *91*(2), 277–290.

Calle, M. L., and Gómez, G. (2001). Nonparametric Bayesian estimation from interval-censored data using Monte Carlo methods. *Journal of Statistical Planning and Inference*, *98*(1–2), 73–87.

Calle, M. L., and Gómez, G. (2005). A semi-parametric hierarchical method for a regression model with an interval-censored covariate. *Australian & New Zealand Journal of Statistics*, *47*(3), 351–364.

Carlin, B. P., and Louis, T. A. (2008). *Bayes and empirical Bayes methods for data analysis* (Third ed.). Boca Raton: CRC Press.

Carpenter, B., Gelman, A., Hoffman, M. D., Lee, D., Goodrich, B., Betancourt, M., Brubaker, M. A., Guo, J., Li, P., and Riddell, A. (2017). Stan: A probabilistic programming language. *Journal of Statistical Software*, *76*, 1–32.

Carvalho, J. C., Ekstrand, K. R., and Thylstrup, A. (1989). Dental plaque and caries on occlusal surfaces of first permanent molars in relation to stage of eruption. *Journal of Dental Research*, *68*(5), 773–779.

Casella, G., and Berger, R. L. (2002). *Statistical inference* (Second ed.). Pacific Grove: Duxbury.

Çetinyürek Yavuz, A., and Lambert, P. (2016). Semi-parametric frailty model for clustered interval-censored data. *Statistical Modelling*, *16*(5), 360–391.

Cecere, S., Groenen, P. J. F., and Lesaffre, E. (2013). The interval-censored biplot. *Journal of Computational and Graphical Statistics*, *22*(1), 123–134.

Cecere, S., Jara, A., and Lesaffre, E. (2006). Analyzing the emergence times of permanent teeth: an example of modeling the covariance matrix with interval-censored data. *Statistical Modelling*, *6*(4), 337–351.

Cecere, S., and Lesaffre, E. (2008, July 7 to 11). On the Bayesian 2-stage procedure for parameter estimation in Copula models. In H. C. Eilers (Ed.), *Proceedings of the 23rd international workshop on statistical modelling* (pp. 163–168). Utrecht.

Celeux, G., and Diebolt, J. (1985). The SEM algorithm: A probabilistic teacher algorithm derived from the EM algorithm for the mixture problem. *Computational Statistics Quarterly*, *2*, 73–82.

Chen, K., and Zhou, M. (2003). Non-parametric hypothesis testing and confidence intervals with doubly censored data. *Lifetime Data Analysis*, *9*(1), 71–91.

Chen, M.-H., Shao, Q.-M., and Ibrahim, J. G. (2000). *Monte carlo methods in bayesian computation.* New York: Springer-Verlag.

Christensen, R., and Johnson, W. (1988). Modelling accelerated failure time with a Dirichlet process. *Biometrika*, *75*(4), 693–704.

Clayton, D. G. (1978). A model for association in bivariate life-tables and its application in epidemiological studies of familial tendency in chronic disease incidence. *Biometrika*, *65*(1), 141–151.

Clayton, D. G. (1991). A Monte-Carlo method for Bayesian inference in frailty models. *Biometrics*, *47*(2), 467–485.

Collett, D. (2003). *Modelling survival data in medical research* (Second ed.). Boca Raton: Chapman & Hall/CRC.

Cox, D. R. (1972). Regression models and life-tables (with Discussion). *Journal of the Royal Statistical Society, Series B*, *34*(2), 187–220.

Cox, D. R. (1975). Partial likelihood. *Biometrika*, *62*(2), 269–276.

Cox, D. R., and Hinkley, D. V. (1974). *Theoretical statistics.* London: Chapman & Hall.

Cox, D. R., and Snell, E. J. (1968). A general definition of residuals (with Discussion). *Journal of the Royal Statistical Society, Series B*, *30*(2), 248–275.

De Gruttola, V., and Lagakos, S. W. (1989). Analysis of doubly-censored survival data, with application to AIDS. *Biometrics*, *45*(1), 1–11.

Dejardin, D., and Lesaffre, E. (2013). Stochastic EM algorithm for doubly interval-censored data. *Biostatistics*, *14*(4), 766-778.

Delord, M. (2017). MIICD: Data augmentation and multiple imputation for interval censored data [Computer software manual]. Retrieved from `http://CRAN.R-project.org/package=MIICD` (R package version 2.4)

Dempster, A. P., Laird, N. M., and Rubin, D. B. (1977). Maximum likelihood from incomplete data via the EM algorithm. *Journal of the Royal Statistical Society, Series B*, *39*(1), 1–38.

Denwood, M. J. (2016). runjags: An R package providing interface utilities, model templates, parallel computing methods and additional distributions for MCMC models in JAGS. *Journal of Statistical Software*, *71*(9), 1–25. doi: 10.18637/jss.v071.i09

Diamond, I. D., McDonald, J. W., and Shah, I. H. (1986). Proportional hazards models for current status data: Application to the study of differentials in age at weaning in Pakistan. *Demography*, *23*, 607–620.

Duchateau, L., and Janssen, P. (2008). *The frailty model.* New York: Springer Science+Business Media.

Eckart, C., and Young, G. (1936). The approximation of one matrix by another of lower rank. *Psychometrika*, *1*(3), 211-218.

Efron, B. (1977). The efficiency of Cox's likelihood function for censored data. *Journal of the American Statistical Association*, *72*(359), 557–565.

Eilers, P. H. C., and Marx, B. D. (1996). Flexible smoothing with B-splines and penalties (with Discussion). *Statistical Science*, *11*(1), 89–121.

Embrechts, P., McNeil, A. J., and Strauman, D. (1999). Correlation: pitfalls and alternatives. *Risk Magazine*, *12*, 69–71.

Emoto, S. E., and Matthews, P. C. (1990). A Weibull model for dependent censoring. *Annals of Statistics*, *18*(4), 1556–1577.

Fahrmeir, L., Kneib, T., Lang, S., and Marx, B. (2013). *Regression. models, methods and applications.* New York: Springer.

Fang, H.-B., Sun, J., and Lee, M.-L. T. (2002). Nonparametric survival comparisons for interval-censored continuous data. *Statistica Sinica*, *12*(4), 1073–1083.

Farrington, C. P. (1996). Interval censored survival data: A generalized linear modelling approach. *Statistics in Medicine*, *15*(3), 283–292.

Farrington, C. P. (2000). Residuals for proportional hazards models with interval-censored survival data. *Biometrics*, *56*(2), 473–482.

Fay, M. P. (1996). Rank invariant tests for interval censored data under grouped continuous model. *Biometrics*, *52*(3), 811–822.

Fay, M. P. (1999). Comparing several score tests for interval censored data. *Statistics in Medicine*, *18*(3), 273–285.

Fay, M. P. (2014). interval: Weighted logrank tests and NPMLE for interval censored data [Computer software manual]. Retrieved from `https://CRAN.R-project.org/package=interval` (R package version 1.1-0.1)

Fay, M. P., and Shaw, P. A. (2010). Exact and asymptotic weighted logrank tests for interval censored data: The interval R package. *Journal of Statistical Software*, *36*(2), 1–34. Retrieved from `"http://www.jstatsoft.org/v36/i02"`

Fay, M. P., and Shih, J. H. (1998). Permutation tests using estimated distribution functions. *Journal of the American Statistical Association*, *93*(441), 387–396.

Ferguson, T. S. (1973). A Bayesian analysis of some nonparametric problems. *The Annals of Statistics*, *1*(2), 209–230.

Finkelstein, D. M. (1986). A proportional hazards model for interval-censored failure time data. *Biometrics*, *42*(4), 845–854.

Finkelstein, D. M., and Wolfe, R. A. (1985). A semiparametric model for regression analysis of interval-censored failure time data. *Biometrics*, *41*(4), 933–945.

Fleming, T. R., and Harrington, D. P. (1984). Nonparametric estimation of the survival distribution in censored data. *Communications in Statistics – Theory and Methods*, *13*, 2469–2486.

Fleming, T. R., and Harrington, D. P. (1991). *Counting processes and survival analysis*. New York: John Wiley & Sons.

Friedman, M. (1982). Piecewise exponential models for survival data with covariates. *The Annals of Statistics*, *10*(1), 101–113.

Gabriel, K. R. (1971). The biplot graphic display of matrices with application to principal component analysis. *Biometrika*, *58*(3), 453–467.

Galassi, M., Davies, J., Theiler, J., Gough, B., Jungman, G., Alken, P., Booth, M., and Rossi, F. (2009). Gnu scientific library reference manual (Third ed.) [Computer software manual]. Retrieved from `http://www.gnu.org/software/gsl/`

Gallant, A. R., and Nychka, D. W. (1987). Seminonparametric maximum likelihood estimation. *Econometrica*, *55*(2), 363–390.

Gamerman, D., and Lopes, H. F. (2006). *Markov chain monte carlo: Stochastic simulation for bayesian inference* (Second ed.). Boca Raton: Chapman & Hall/CRC.

García-Zattera, M. J., Jara, A., and Komárek, A. (2016). A flexible AFT model for misclassified clustered interval-censored data. *Biometrics*,

72(2), 473–483.

García-Zattera, M. J., Jara, A., Lesaffre, E., and Marshall, G. (2012). Modeling of multivariate monotone disease processes in the presence of misclassification. *Journal of the American Statistical Association*, *107*(499), 976–989.

García-Zattera, M. J., Mutsvari, T., Jara, A., Declerck, D., and Lesaffre, E. (2010). Correcting for misclassification for a monotone disease process with an application in dental research. *Statistics in Medicine*, *29*(30), 3103–3117.

Gehan, E. A. (1965a). A generalized two-sample Wilcoxon test for doubly censored data. *Biometrika*, *52*(3/4), 650–653.

Gehan, E. A. (1965b). A generalized Wilcoxon test for comparing arbitrarily singly-censored samples. *Biometrika*, *52*(1/2), 203–223.

Gelfand, A. E., and Smith, A. F. M. (1990). Sampling-based approaches to calculating marginal densities. *Journal of the American Statistical Association*, *85*(410), 398–409.

Gelman, A. (1996). Inference and monitoring convergence. In W. R. Gilks, S. Richardson, and D. J. Spiegelhalter (Eds.), *Markov chain Monte Carlo in practice* (pp. 131–143). London: Chapman & Hall.

Gelman, A., Carlin, J. B., Stern, H. S., Dunson, D. B., Vehtari, A., and Rubin, D. B. (2013). *Bayesian data analysis* (Third ed.). Boca Raton: Chapman & Hall/CRC.

Gelman, A., Meng, X. L., and Stern, H. (1996). Posterior predictive assessment of model fitness via realized discrepancies. *Statistica Sinica*, *6*(4), 733–807.

Gelman, A., and Rubin, D. B. (1992). Inference from iterative simulation using multiple sequences. *Statistical Science*, *7*(4), 457–472.

Gelman, A., Van Mechelen, I., Verbeke, G., Heitjan, D. F., and Meulders, M. (2005). Multiple imputation for model checking: completed-data plots with missing and latent data. *Biometrics*, *61*(1), 74–85.

Geman, S., and Geman, D. (1984). Stochastic relaxation, Gibbs distributions and the Bayes restoration of image. *IEEE Transactions on Pattern Analysis and Machine Intelligence*, *6*, 721–741.

Gentleman, R., and Geyer, C. J. (1994). Maximum likelihood for interval censored data: Consistency and computation. *Biometrika*, *81*(3), 618–623.

Gentleman, R., and Vandal, A. (2016). Icens: NPMLE for censored and truncated data [Computer software manual]. Retrieved from `http://bioconductor.org/packages/Icens/` (R package version 1.46.0)

Gentleman, R., and Vandal, A. C. (2001). Computational algorithms for censored-data problems using intersection graphs. *Journal of Computational and Graphical Statistics*, *10*(3), 403–421.

Gentleman, R., and Vandal, A. C. (2002). Nonparametric estimation of the bivariate CDF for arbitrarily censored data. *The Canadian Journal of*

Statistics, *30*(4), 557–571.

Geskus, R. (2015). *Data analysis with competing risks and intermediate states.* Boca Raton: Chapman & Hall/CRC.

Geweke, J. (1992). Evaluating the accuracy of sampling-based approaches to the calculation of posterior moments. In J. M. Bernardo et al. (Eds.), *Bayesian statistics* (Vol. 4, pp. 169–193). Oxford: Clarendon Press.

Geyer, C. J. (1992). Practical Markov chain Monte Carlo (with Discussion). *Statistical Science*, *7*(4), 473–511.

Ghidey, W., Lesaffre, E., and Eilers, P. (2004). Smooth random effects distribution in a linear mixed model. *Biometrics*, *60*(4), 945–953.

Gilks, W. R., Richardson, S., and Spiegelhalter, D. J. (Eds.). (1996). *Markov chain monte carlo in practice.* London: Chapman & Hall.

Godambe, V. P. (Ed.). (1991). *Estimating functions.* Oxford University Press.

Goethals, K., Ampe, B., Berkvens, D., Laevens, H., Janssen, P., and Duchateau, L. (2009). Modeling interval-censored, clustered cow udder quarter infection times through the shared gamma frailty model. *Journal of Agricultural, Biological, and Environmental Statistics*, *14*(1), 1–14.

Goggins, W. B., Finkelstein, D. M., Schoenfeld, D. A., and Zaslavsky, A. M. (1998). A Markov chain Monte Carlo EM algorithm for analyzing interval-censored data under the Cox proportional hazards model. *Biometrics*, *54*(4), 1498-1507.

Goggins, W. B., Finkelstein, D. M., and Zaslavsky, A. M. (1999). Applying the Cox proportional hazards model for analysis of latency data with interval censoring. *Statistics in Medicine*, *18*(20), 2737–2747.

Gómez, G., and Calle, M. L. (1999). Non-parametric estimation with doubly censored data. *Journal of Applied Statistics*, *26*(1), 45–58.

Gómez, G., Calle, M. L., and Oller, R. (2004). Frequentist and Bayesian approaches for interval-censored data. *Statistical Papers*, *45*(2), 139–173.

Gómez, G., Calle, M. L., Oller, R., and Langohr, K. (2009). Tutorial on methods for interval-censored data and their implementation in R. *Statistical Modelling*, *9*(4), 259–297.

Gómez, G., Espinal, A., and Lagakos, S. W. (2003). Inference for a linear regression model with an interval-censored covariate. *Statistics in Medicine*, *22*(3), 409-425.

Gómez, G., and Lagakos, S. W. (1994). Estimation of the infection time and latency distribution of AIDS with doubly censored data. *Biometrics*, *50*(1), 204–212.

Gong, Q., and Fang, L. (2013). Comparison of different parametric proportional hazards models for interval-censored data: A simulation study. *Contemporary Clinical Trials*, *36*, 276–283.

Goodall, R. L., Dunn, D. T., and Babiker, A. G. (2004). Interval-censored survival time data: Confidence intervals for the non-parametric survivor function. *Statistics in Medicine*, *23*(7), 1131–1145.

Gower, J. C., and Hand, D. J. (1996). *Biplots.* London: Chapman & Hall.

Grambsch, P. M., and Therneau, T. M. (1994). Proportional hazards tests and diagnostics based on weighted residuals. *Biometrika, 81*(3), 515–526.

Gray, R. J. (1992). Flexible methods for analyzing survival data using splines, with application to breast cancer prognosis. *Journal of the American Statistical Association, 87*(420), 942–951.

Green, P. J. (1995). Reversible jump Markov chain computation and Bayesian model determination. *Biometrika, 82*(4), 711–732.

Greenwood, M. (1926). The natural duration of cancer. *Reports on Public Health and Medical Subjects, 33*, 1–26.

Groeneboom, P., Jongbloed, G., and Wellner, J. A. (2008). The support reduction algorithm for computing non-parametric function estimates in mixture models. *Scandinavian Journal of Statistics, 35*(3), 385–399.

Groeneboom, P., and Wellner, J. A. (1992). *Information bounds and non-parametric maximum likelihood estimation.* Basel: Birkhäuser-Verlag.

Grummer-Strawn, L. M. (1993). Regression analysis of current-status data: An application to breast-feeding. *Journal of the American Statistical Association, 88*(3), 758–765.

Gueorguieva, R., Rosenheck, R., and Lin, H. (2012). Joint modelling of longitudinal outcome and interval-censored competing risk dropout in a schizophrenia clinical trial. *Journal of the Royal Statistical Society, Series A, 175*(2), 417–433.

Hájek, J., Šidák, Z., and Sen, P. K. (1999). *Theory of rank tests* (Second ed.). San Diego: Academic Press.

Hannan, E. J., and Quinn, B. G. (1979). The determination of the order of an autoregression. *Journal of the Royal Statistical Society, Series B, 41*(2), 190–195.

Hanson, T., and Johnson, W. O. (2004). A Bayesian semiparametric AFT model for interval-censored data. *Journal of Computational and Graphical Statistics, 13*(2), 341–361.

Hanson, T. E., and Jara, A. (2013). Surviving fully Bayesian nonparametric regression models. In P. Damien et al. (Eds.), *Bayesian theory and applications* (p. 593-615). Oxford, UK: Oxford University Press.

Härkänen, T. (2003). BITE: A Bayesian intensity estimator. *Computational Statistics, 18*, 565–583.

Harrington, D. P., and Fleming, T. R. (1982). A class of rank test procedures for censored survival data. *Biometrika, 69*(3), 553–566.

Hastings, W. K. (1970). Monte Carlo sampling methods using Markov chains and their applications. *Biometrika, 57*(1), 97–109.

Heinz, G., Gnant, M., and Schemper, M. (2003). Exact log-rank tests for unequal follow-up. *Biometrics, 59*(3), 1151–1157.

Held, L. (2004). Simultaneous posterior probability statements from Monte Carlo output. *Journal of Computational and Graphical Statistics, 13*(1), 20–35.

Hens, N., Wienke, A., Aerts, M., and Molenberghs, G. (2009). The correlated and shared gamma frailty model for bivariate current status data: An illustration for cross-sectional serological data. *Statistics in Medicine*, *28*(22), 2785–2800.

Henschel, V., Engel, J., Hölzel, D., and Mansmann, U. (2009). A semiparametric Bayesian proportional hazards model for interval censored data with frailty effects. *BMC Medical Research Methodology*, *9*, 9.

Hickernell, F. J., Lemieux, C., and Owen, A. B. (2005). Control variates for Quasi-Monte Carlo (with Discussion). *Statistical Science*, *20*(1), 1–31.

Hougaard, P. (1999). Fundamentals of survival data. *Biometrics*, *55*(1), 13–22.

Hougaard, P. (2000). *Analysis of multivariate survival data.* New York: Springer-Verlag.

Hough, G. (2010). *Sensory shelf life estimation of food products.* Boca Raton: CRC Press.

Huang, J., Lee, C., and Yu, Q. (2008). A generalized log-rank test for interval-censored failure time data via multiple imputation. *Statistics in Medicine*, *27*(17), 3217–3226.

Huang, X., and Wolfe, R. A. (2002). A frailty model for informative censoring. *Biometrics*, *58*(3), 510–520.

Huang, X., Wolfe, R. A., and Hu, C. (2004). A test for informative censoring in clustered survival data. *Statistics in Medicine*, *23*(13), 2089–2107.

Hubeaux, S., and Rufibach, K. (2014). *SurvRegCensCov: Weibull regression for a right-censored endpoint with interval-censored covariate.* Retrieved from `https://arxiv.org/pdf/1402.0432v2.pdf`

Hudgens, M. G., Maathuis, M. H., and Gilbert, P. B. (2007). Nonparametric estimation of the joint distribution of a survival time subject to interval censoring and a continuous mark variable. *Biometrics*, *63*(2), 372–380.

Huster, W. J., Brookmeyer, R., and Self, S. G. (1989). Modelling paired survival data with covariates. *Biometrics*, *45*(1), 145–156.

Ibrahim, J. G., Chen, M.-H., and Sinha, D. (2001). *Bayesian survival analysis.* New York: Springer. (Corrected printing 2005)

Institute of Food Science and Technology (U.K.). (1993). *Shelf life of foods: Guidelines for its determination and prediction.* London: Institute of Food Science & Technology (UK).

Jara, A. (2007). Applied Bayesian non- and semiparametric inference using DPpackage. *R News*, *7*(3), 17–26. Retrieved from `http://CRAN.R-project.org/doc/Rnews/`

Jara, A., Hanson, T. E., Quintana, F. A., Müller, P., and Rosner, G. L. (2011). DPpackage: Bayesian semi- and nonparametric modeling using R. *Journal of Statistical Software*, *40*(5), 1–30. Retrieved from `http://www.jstatsoft.org/v40/i05/`

Jara, A., Lesaffre, E., De Iorio, M., and Quintana, F. (2010). Bayesian semiparametric inference for multivariate doubly-interval-censored data. *The Annals of Applied Statistics*, *4*(4), 2126–2149.

Jasra, A., Holmes, C. C., and Stephens, D. A. (2005). Markov chain Monte

Carlo methods and the label switching problem in Bayesian mixture modeling. *Statistical Science*, *20*(1), 50–67.

Jin, Z., Lin, D. Y., Wei, L. J., and Ying, Z. (2003). Rank-based inference for the accelerated failure time model. *Biometrika*, *90*(2), 341–353.

Joe, H. (1997). *Multivariate models and dependence concepts* (No. 73). Chapman & Hall Ltd.

Johnson, W., and Christensen, R. (1989). Nonparametric Bayesian analysis of the accelerated failure time model. *Statistics & Probability Letters*, *8*, 179–184.

Joly, P., Commenges, D., and Letenneur, L. (1998). A penalized likelihood approach for arbitrarily censored and truncated data: Application to age-specific incidence of dementia. *Biometrics*, *54*(1), 185–194.

Kalbfleisch, J. D. (1978). Non-parametric Bayesian analysis of survival time data. *Journal of the Royal Statistical Society, Series B*, *40*(2), 214–221.

Kalbfleisch, J. D., and MacKay, R. J. (1979). On constant-sum models for censored survival data. *Biometrika*, *66*(1), 87–90.

Kalbfleisch, J. D., and Prentice, R. L. (2002). *The statistical analysis of failure time data* (Second ed.). Chichester: John Wiley & Sons.

Kaplan, E. L., and Meier, P. (1958). Nonparametric estimation from incomplete observations. *Journal of the American Statistical Association*, *53*(282), 457–481.

Karvanen, J., Rantanen, A., and Luoma, L. (2014). Survey data and bayesian analysis: a cost-efficient way to estimate customer equity. *Quantitative Marketing and Economics*, *12*(3), 305–329.

Kim, H. Y., Williamson, J. M., and Lin, H. M. (2016). Power and sample size calculations for interval-censored survival analysis. *Statistics in Medicine*, *35*(8), 1390–1400.

Kim, J., Kim, Y. N., and Kim, S. W. (2016). Frailty model approach for the clustered interval-censored data with informative censoring. *Journal of the Korean Statistical Society*, *45*(1), 156–165.

Kim, M. Y., De Gruttola, V. G., and Lagakos, S. W. (1993). Analyzing doubly censored data with covariates, with application to AIDS. *Biometrics*, *49*(1), 13–22.

Kimeldorf, G., and Sampson, A. (1975). Uniform representations of bivariate distributions. *Communication in Statistics*, *4*(7), 617–627.

Klein, J. P., and Moeschberger, M. L. (2003). *Survival analysis: Techniques for censored and truncated data.* New York: Springer-Verlag Inc.

Koenker, R., and Bassett, G. J. (1978). Regression quantiles. *Econometrica*, *46*(1), 33–50.

Komárek, A. (2009). A new R package for Bayesian estimation of multivariate normal mixtures allowing for selection of the number of components and interval-censored data. *Computational Statistics and Data Analysis*, *53*(12), 3932–3947.

Komárek, A. (2015). smoothSurv: Survival regression with smoothed error distribution [Computer software manual]. Retrieved from http://CRAN.R-project.org/package=smoothSurv (R package version 1.6)

Komárek, A., and Komárková, L. (2014). Capabilities of R package mixAK for clustering based on multivariate continuous and discrete longitudinal data. *Journal of Statistical Software*, *59*(12), 1–38. Retrieved from http://www.jstatsoft.org/v59/i12/

Komárek, A., and Lesaffre, E. (2006). Bayesian semi-parametric accelerated failure time model for paired doubly-interval-censored data. *Statistical Modelling*, *6*(1), 3–22.

Komárek, A., and Lesaffre, E. (2007). Bayesian accelerated failure time model for correlated censored data with a normal mixture as an error distribution. *Statistica Sinica*, *17*(2), 549–569.

Komárek, A., and Lesaffre, E. (2008). Bayesian accelerated failure time model with multivariate doubly-interval-censored data and flexible distributional assumptions. *Journal of the American Statistical Association*, *103*(482), 523–533.

Komárek, A., and Lesaffre, E. (2009). The regression analysis of correlated interval-censored data: illustration using accelerated failure time models with flexible distributional assumptions. *Statistical Modelling*, *9*(4), 299–319.

Komárek, A., Lesaffre, E., Härkänen, T., Declerck, D., and Virtanen, J. I. (2005b). A Bayesian analysis of multivariate doubly-interval-censored data. *Biostatistics*, *6*(1), 145–155.

Komárek, A., Lesaffre, E., and Hilton, J. F. (2005a). Accelerated failure time model for arbitrarily censored data with smoothed error distribution. *Journal of Computational and Graphical Statistics*, *14*(3), 726–745.

Komárek, A., Lesaffre, E., and Legrand, C. (2007). Baseline and treatment effect heterogeneity for survival times between centers using a random effects accelerated failure time model with flexible error distribution. *Statistics in Medicine*, *26*(30), 5457–5472.

Kooperberg, C. (2016). logspline: Logspline density estimation routines [Computer software manual]. Retrieved from http://CRAN.R-project.org/package=logspline (R package version 2.1.9)

Kooperberg, C., and Stone, C. J. (1992). Logspline density estimation for censored data. *Journal of Computational and Graphical Statistics*, *1*(4), 301–328.

Kramer, C. Y. (1956). Extension of multiple range tests to group means with unequal numbers of replications. *Biometrics*, *12*(3), 307–310.

Lagakos, S. W. (1981). The graphical evaluation of explanatory variables in proportional hazard regression models. *Biometrika*, *68*(1), 93–98.

Laird, N. M., and Ware, J. H. (1982). Random-effects models for longitudinal data. *Biometrics*, *38*(4), 963–974.

Lang, S., and Brezger, A. (2004). Bayesian P-splines. *Journal of Computational and Graphical Statistics*, *13*(1), 183–212.

Langohr, K., Gómez, G., and Muga, R. (2004). A parametric survival model with an interval-censored covariate. *Statistics in Medicine*, *23*(20), 3159-3175.

Langohr, K., and Gómez, G. M. (2014). Estimation and residual analysis with R for a linear regression model with an interval-censored covariate. *Biometrical Journal*, *56*(5), 867–885.

Law, C. G., and Brookmeyer, R. (1992). Effects of mid-point imputation on the analysis of doubly censored data. *Statistics in Medicine*, *11*(12), 1569–1578.

Lawless, J. F. (2003). *Statistical models and methods for lifetime data.* New York: John Wiley & Sons.

Lawless, J. F., and Babineau, D. (2006). Models for interval censoring and simulation-based inference for lifetime distributions. *Biometrika*, *93*(3), 671–686.

Lawson, A. B. (2006). *Statistical methods in spatial epidemiology* (Second ed.). Hoboken: John Wiley & Sons, Inc.

Lawson, A. B. (2013). *Bayesian disease mapping: Hierarchical modeling in spatial epidemiology* (Second ed.). Boca Raton: Chapman & Hall/CRC.

Lee, D. T. (1983). Maximum clique problem of rectangle graphs. In F. P. Preparata (Ed.), *Advances in computing research* (Vol. 1, pp. 91–107). Greenwich, CT: JAI Press.

Lee, T. C. K., Zeng, L., Thompson, D. J. S., and Dean, C. B. (2011). Comparison of imputation methods for interval censored time-to-event data in joint modelling of tree growth and mortality. *The Canadian Journal of Statistics*, *39*(3), 438–457.

Lehmann, E. L., and Casella, G. (1998). *Theory of point estimation* (Second ed.). New York: Springer-Verlag.

Leroy, R., Bogaerts, K., Lesaffre, E., and Declerck, D. (2003). The emergence of permanent teeth in Flemish children. *Community Dentistry and Oral Epidemiology*, *31*(1), 30–39.

Leroy, R., Cecere, S., Declerck, D., and Lesaffre, E. (2008). Variability in permanent tooth emergence sequences in Flemish children. *European Journal of Oral Sciences*, *116*(1), 11–17.

Leroy, R., Cecere, S., Lesaffre, E., and Declerck, D. (2009). Caries experience in primary molars and its impact on the variability in permanent tooth emergence sequences. *Journal of Dentistry*, *37*(11), 865–871.

Lesaffre, E., Edelman, M. J., Hanna, N. H., Park, K., Thatcher, N., Willemsen, S., Gaschler-Markefski, B., Kaiser, R., and Manegold, C. (2017). Futility analyses in oncology lessons on potential pitfalls from a randomised controlled trial. *Annals in Oncology*, *28*(7), 1419-1426.

Lesaffre, E., Komárek, A., and Declerck, D. (2005). An overview of methods for interval-censored data with an emphasis on applications in dentistry. *Statistical Methods in Medical Research*, *14*(6), 539–552.

Lesaffre, E., and Lawson, A. (2012). *Bayesian methods in biostatistics.* New York: John Wiley & Sons.

Lesperance, M. L., and Kalbfleisch, J. D. (1992). An algorithm for computing the nonparametric MLE of a mixing distribution. *Journal of the American Statistical Association*, *87*(417), 120–126.

Li, J., and Ma, S. (2010). Interval-censored data with repeated measurements and a cured subgroup. *Journal of the Royal Statistical Society, Series C*, *59*(4), 693–705.

Liang, K. Y., and Zeger, S. L. (1986). Longitudinal data analysis using generalized linear models. *Biometrika*, *73*(1), 13–22.

Lim, H.-J., and Sun, J. (2003). Nonparametric tests for interval-censored failure time data. *Biometrical Journal*, *45*(3), 263–276.

Lin, D. Y., Wei, L. J., and Ying, Z. (1993). Checking the Cox model with cumulative sums of martingale-based residuals. *Biometrika*, *80*(3), 557–572.

Lin, X., Cai, B., Wang, L., and Zhang, Z. (2015). A Bayesian proportional hazards model for general interval-censored data. *Lifetime Data Analysis*, *21*(3), 470–490.

Lindsey, J. C., and Ryan, L. M. (1998). Methods for interval-censored data. *Statistics in Medicine*, *17*(2), 219-238.

Liu, L., and Yu, Z. (2008). A likelihood reformulation method in non-normal random effects models. *Statistics in Medicine*, *27*(16), 3105–3124.

Lunn, D., Spiegelhalter, D., Thomas, A., and Best, N. (2009). The BUGS project: Evolution, critique and future directions. *Statistics in Medicine*, *28*(25), 3049–3082.

Lunn, D. J., Thomas, A., Best, N., and Spiegelhalter, D. (2000). WinBUGS – A Bayesian modelling framework: Concepts, structure, and extensibility. *Statistics and Computing*, *10*(4), 325–337.

Lyles, R. H., Lin, H. M., and Williamson, J. M. (2007). A practical approach to computing power for generalized linear models with nominal, count, or ordinal responses. *Statistics in Medicine*, *26*(7), 1632–1648.

Maathuis, M. (2013). MLEcens: Computation of the MLE for bivariate (interval) censored data [Computer software manual]. Retrieved from `http://CRAN.R-project.org/package=MLEcens` (R package version 0.1-4)

Maathuis, M. H. (2005). Reduction algorithm for the NPMLE for the distribution function of bivariate interval censored data. *Journal of Computational and Graphical Statistics*, *14*(2), 362–362.

MacKenzie, G., and Peng, D. (2013). Interval-censored parametric regression survival models and the analysis of longitudinal trials. *Statistics in Medicine*, *32*(16), 2804–2822.

Mantel, N. (1966). Evaluation of survival data and two new rank order statistics arising in its consideration. *Cancer Chemotherapy Reports*, *50*, 163–170.

Mantel, N. (1967). Ranking procedures for arbitrarily restricted observations. *Biometrics*, *23*(1), 65–78.

McLachlan, G. J., and Basford, K. E. (1988). *Mixture models: Inference and applications to clustering.* New York: Marcel Dekker, Inc.

McNeil, A. J., Frey, R., and Embrechts, P. (2005). *Quantitative risk management: Concepts, techniques and tools.* Princeton: Princeton University Press.

Meeker, W. Q., and Escobar, L. A. (1998). *Statistical methods for reliability data.* John Wiley & Sons.

Meng, X. L. (1994). Posterior predictive p-values. *The Annals of Statistics, 22*(3), 1142–1160.

Metropolis, N., Rosenbluth, A. W., Rosenbluth, M. N., and Teller, A. H. (1953). Equations of state calculations by fast computing machines. *Journal of Chemical Physics, 21*, 1087–1091.

Meyns, B., Jashari, R., Gewillig, M., Mertens, L., Komárek, A., Lesaffre, E., Budts, W., and Daenen, W. (2005). Factors influencing the survival of cryopreserved homografts. The second homograft performs as well as the first. *European Journal of Cardio-thoracic Surgery, 28*(2), 211–216.

Miller, R. G. (1976). Least squares regression with censored data. *Biometrika, 63*(2), 449–464.

Mitra, R., and Müller, P. (Eds.). (2015). *Nonparametric Bayesian inference in biostatistics.* Cham: Springer.

Molenberghs, G., and Verbeke, G. (2005). *Models for discrete longitudinal data.* New York: Springer-Verlag.

Mostafa, A. A., and Ghorbal, A. B. (2011). Using WinBUGS to Cox model with changing from the baseline hazard function. *Applied Mathematical Sciences, 5*(45), 2217–2240.

Müller, P., Quintana, F. A., Jara, A., and Hanson, T. (2015). *Bayesian nonparametric data analysis.* New York: Springer-Verlag.

Nair, V. N. (1984). Confidence bands for survival functions with censored data: A comparative study. *Technometrics, 26*(3), 265–275.

Narasimhan, B., and Johnson, S. G. (2017). cubature: Adaptive multivariate integration over hypercubes [Computer software manual]. Retrieved from `https://CRAN.R-project.org/package=cubature` (R package version 1.3-8)

Nelsen, R. B. (1998). *An introduction to copulas* (No. 139). New York: Springer-Verlag.

Nelson, K. P., Lipsitz, S. R., Fitzmaurice, G. M., Ibrahim, J., Parzen, M., and Strawderman, R. (2006). Use of the probability integral transformation to fit nonlinear mixed-effects models with nonnormal random effects. *Journal of Computational and Graphical Statistics, 15*(1), 39–57.

Nelson, W. (1969). Hazard plotting for incomplete failure data. *Journal of Quality Technology, 1*(1), 27–52.

Nelson, W. (1972). Theory and applications of hazard plotting for censored failure data. *Technometrics, 14*, 945–966.

Ng, M. P. (2002). A modification of Peto's nonparametric estimation of survival curves for interval-censored data. *Biometrics, 58*(2), 439–442.

Nysen, R., Aerts, M., and Faes, C. (2012). Testing goodness-of-fit of parametric models for censored data. *Statistics in Medicine*, *31*(21), 2374–2385.

Oakes, D. (1982). A concordance test for independence in the presence of censoring. *Biometrics*, *38*(2), 451–455.

Oakes, D. (1989). Bivariate survival models induced by frailties. *Journal of the American Statistical Association*, *84*(406), 487–493.

Oller, R., and Gómez, G. (2012). A generalized Fleming and Harrington's class of tests for interval-censored data. *The Canadian Journal of Statistics*, *40*(3), 501–516.

Oller, R., Gómez, G., and Calle, M. L. (2004). Interval censoring: model characterization for the validity of the simplified likelihood. *The Canadian Journal of Statistics*, *32*, 315–326.

Oller, R., and Langohr, K. (2015). FHtest: Tests for right and interval-censored survival data based on the Fleming-Harrington class [Computer software manual]. Retrieved from `http://CRAN.R-project.org/package=FHtest` (R package version 1.3)

Pak, D., Li, C., Todem, D., and Sohn, W. (2017). A multistate model for correlated interval-censored life history data in caries research. *Journal of the Royal Statistical Society, Series C*, *66*(2), 413–423.

Pan, C., Cai, B., Wang, L., and Lin, X. (2014). Bayesian semiparametric model for spatially correlated interval-censored survival data. *Computational Statistics and Data Analysis*, *74*, 198-208.

Pan, C., Cai, B., Wang, L., and Lin, X. (2015). ICBayes: Bayesian semiparametric models for interval-censored data [Computer software manual]. Retrieved from `https://CRAN.R-project.org/package=ICBayes` (R package version 1.0)

Pan, W. (1999). Extending the iterative convex minorant algorithm to the Cox model for interval-censored data. *Journal of Computational and Graphical Statistics*, *8*(1), 109–120.

Pan, W. (2000a). A multiple imputation approach to Cox regression with interval-censored data. *Biometrics*, *56*(1), 199–203.

Pan, W. (2000b). A two-sample test with interval censored data via multiple imputation. *Statistics in Medicine*, *19*(1), 1–11.

Pan, W. (2001). A multiple imputation approach to regression analysis for doubly censored data with application to AIDS studies. *Biometrics*, *57*(4), 1245–1250.

Panageas, K. S., Ben-Porat, L., Dickler, M. N., Chapman, P. B., and Schrag, D. (2007). When you look matters: the effect of assessment schedule on progression-free survival. *Journal of the National Cancer Institute*, *99*(6), 428–432.

Pantazis, N., Kenward, M. G., and Touloumi, G. (2013). Performance of parametric survival models under non-random interval censoring: A simulation study. *Computational Statistics and Data Analysis*, *63*, 16–30.

Pepe, M. S., and Fleming, T. R. (1989). Weighted Kaplan-Meier statistics: a class of distance tests for censored survival data. *Biometrics*, *45*(2),

497–507.

Pepe, M. S., and Fleming, T. R. (1991). Weighted Kaplan-Meier statistics: large sample and optimality considerations. *Journal of the Royal Statistical Society, Series B*, *53*(2), 341–352.

Peto, R. (1973). Experimental survival curves for interval-censored data. *Applied Statistics*, *22*(1), 86–91.

Peto, R., and Peto, J. (1972). Asymptotically efficient rank-invariant test procedures (with Discussion). *Journal of the Royal Statistical Society, Series A*, *135*(2), 185–206.

Petroni, G. R., and Wolfe, R. A. (1994). A two-sample test for stochastic ordering with interval-censored data. *Biometrics*, *50*(1), 77–87.

Pierce, D. A., Stewart, W. H., and Kopecky, K. J. (1979). Distribution-free regression analysis of grouped survival data. *Biometrics*, *35*(4), 785–793.

Pinheiro, J. C., and Bates, D. M. (1995). Approximations to the loglikelihood function in the nonlinear mixed effects model. *Journal of Computational and Graphical Statistics*, *4*, 12–35.

Plackett, R. L. (1965). A class of bivariate distributions. *Journal of the American Statistical Association*, *60*(310), 516–522.

Plummer, M. (2016). rjags: Bayesian graphical models using MCMC [Computer software manual]. Retrieved from `https://CRAN.R-project.org/package=rjags` (R package version 4-6)

Plummer, M., Best, N., Cowles, K., and Vines, K. (2006). CODA: Convergence diagnosis and output analysis for MCMC. *R News*, *6*(1), 7–11. Retrieved from `https://journal.r-project.org/archive/`

Pourahmadi, M. (1999). Joint mean-covariance models with applications to longitudinal data: unconstrained parametrisation. *Biometrika*, *86*(3), 677–690.

Prentice, R. L. (1978). Linear rank tests with right censored data. *Biometrika*, *65*(1), 167–179.

Prentice, R. L., and Gloeckler, L. A. (1978). Regression analysis of grouped survival data with application to breast cancer data. *Biometrics*, *34*(1), 57–67.

Quintero, A., and Lesaffre, E. (2017). Comparing latent variable models via the deviance information criterion. *submitted*.

R Core Team. (2016). R: A language and environment for statistical computing [Computer software manual]. Vienna, Austria. Retrieved from `http://www.R-project.org`

Rabe-Hesketh, S., Yang, S., and Pickles, A. (2001). Multilevel models for censored and latent responses. *Statistical Methods in Medical Research*, *10*(6), 409–427.

Ramsay, J. O. (1988). Monotone regression splines in action. *Statistical Science*, *3*(4), 425–441.

Ren, J.-J. (2003). Goodness of fit tests with interval censored data. *Scandi-*

navian Journal of Statistics, *30*(1), 211–226.

Richardson, S., and Green, P. J. (1997). On Bayesian analysis of mixtures with unknown number of components (with Discussion). *Journal of the Royal Statistical Society, Series B*, *59*(4), 731–792.

Rizopoulos, D. (2012). *Joint models for longitudinal and time-to-event data: With applications in R*. Boca Raton: Chapman & Hall/CRC.

Robert, C. P. (2007). *The Bayesian choice: From decision-theoretic foundations to computational implementation* (Second ed.). New York: Springer-Verlag.

Robert, C. P., and Casella, G. (2004). *Monte carlo statistical methods* (Second ed.). New York: Springer-Verlag.

Robertson, T., Wright, F. T., and Dykstra, R. (1988). *Order restricted statistical inference.* New York: John Wiley & Sons.

Romeo, J. S., Tanaka, N. I., and Pedroso-de Lima, A. C. (2006). Bivariate survival modeling: a Bayesian approach based on Copulas. *Lifetime Data Analysis*, *12*(2), 205–222.

Rondeau, V., Commenges, D., and Joly, P. (2003). Maximum penalized likelihood estimation in a Gamma-frailty model. *Lifetime Data Analysis*, *9*(2), 139–153.

Rondeau, V., and Gonzalez, J. R. (2005). frailtypack: A computer program for the analysis of correlated failure time data using penalized likelihood estimation. *Computer Methods and Programs in Biomedicine*, *80*(2), 154–164.

Rondeau, V., Gonzalez, J. R., Mazroui, Y., Mauguen, A., Krol, A., Diakite, A., Laurent, A., and Lopez, M. (2015). frailtypack: General frailty models: Shared, joint and nested frailty models with prediction [Computer software manual]. Retrieved from `http://CRAN.R-project.org/package=frailtypack` (R package version 2.10.4)

Rondeau, V., Marzroui, Y., and Gonzalez, J. R. (2012). frailtypack: An R package for the analysis of correlated survival data with frailty models using penalized likelihood estimation or parametrical estimation. *Journal of Statistical Software*, *47*(4), 1–28. Retrieved from `http://www.jstatsoft.org/v47/i04"`

Rouanet, A., Joly, P., Dartigues, J.-F., Proust-Lima, C., and Jacqmin-Gadda, H. (2016). Joint latent class model for longitudinal data and interval-censored semi-competing events: Application to dementia. *Biometrics*, *72*(4), 1123–1135.

Royall, R. M. (1986). Model robust confidence intervals using maximum likelihood estimators. *International Statistical Review*, *54*, 221–226.

Royston, P., and Parmar, M. K. (2002). Flexible parametric proportional-hazards and proportional-odds models for censored survival data, with application to prognostic modelling and estimation of treatment effects. *Statistics in Medicine*, *21*(15), 2175–2197.

RStudio Team. (2016). Rstudio: Integrated development environment for r [Computer software manual]. Boston, MA. Retrieved from `http://www.rstudio.com/`

Rubin, D. B. (1984). Bayesian justifiable and relevant frequency calculations for the applied statistician. *The Annals of Statistics*, *12*(4), 1151–1172.

Rubin, D. B. (1987). *Multiple imputation for nonresponse in surveys.* New York: John Wiley & Sons.

SAS Institute Inc. (2015a). SAS/QC® 14.1 User's Guide [Computer software manual]. Cary, NC, SAS Institute Inc.

SAS Institute Inc. (2015b). SAS/STAT® 14.1 User's Guide [Computer software manual]. Cary, NC, SAS Institute Inc.

Schick, A., and Yu, Q. (2000). Consistency of the GMLE with mixed case interval-censored data. *Scandinavian Journal of Statistics*, *27*(1), 45–55.

Schwarz, G. (1978). Estimating the dimension of a model. *The Annals of Statistics*, *6*(2), 461–464.

Scolas, S. (2016). *Modelling interval-censored event times in the presence of a cure fraction: From building to refining* (Unpublished doctoral dissertation). Université Catholique de Louvain, Louvain-la-Neuve, Belgium.

Scolas, S., El Ghouch, A., Legrand, C., and Oulhaj, A. (2016a). Variable selection in a flexible parametric mixture cure model with interval-censored data. *Statistics in Medicine*, *35*(7), 1210-1225.

Scolas, S., Legrand, C., Oulhaj, A., and El Ghouch, A. (2016b). Diagnostic checks in mixture cure models with interval-censoring. *Statistical Methods in Medical Research.* Retrieved from `https://doi.org/10.1177/0962280216676502`

Self, S. G., and Grossman, E. A. (1986). Linear rank tests for interval-censored data with application to PCB levels in adipose tissue of transformer repair workers. *Biometrics*, *42*(3), 521–530.

Sethuraman, J. (1994). A constructive definition of Dirichlet priors. *Statistica Sinica*, *4*, 639–650.

Shih, J. H., and Louis, T. A. (1995). Inferences on the association parameter in copula models for bivariate survival data. *Biometrics*, *51*(4), 1384–1399.

Silva, R. S., and Lopes, H. F. (2008). Copula, marginal distributions and model selection: a Bayesian note. *Statistics and Computing*, *18*(3), 313–320.

Silverman, B. W. (1985). Some aspects of the spline smoothing approach to non-parametric regression curve fitting. *Journal of the Royal Statistical Society, Series B*, *47*(1), 1–52.

Sinha, D., Ibrahim, J. G., and Chen, M.-H. (2003). A Bayesian justification of Cox's partial likelihood. *Biometrika*, *90*(3), 629–641.

Sklar, A. (1959). Fonctions de répartition à n-dimensions et leurs marges. *Publications de l'Institut Statistique de l'Université de Paris*, *8*, 229–231.

Sparling, Y. H., Younes, N., Lachin, J. M., and Bautista, O. M. (2006). Parametric survival models for interval-censored data with time-dependent covariates. *Biostatistics*, *7*(4), 599.

Spiegelhalter, D. J., Best, N. G., Carlin, B. P., and van der Linde, A. (2002). Bayesian measures of model complexity and fit (with Discussion). *Journal of the Royal Statistical Society, Series B*, *64*(4), 583–639.

Spiegelhalter, D. J., Best, N. G., Carlin, B. P., and van der Linde, A. (2014). The deviance information criterion: 12 years on (with Discussion). *Journal of the Royal Statistical Society, Series B*, *76*(3), 485–493.

Spiegelhalter, D. J., Thomas, A., Best, N. G., and Gilks, W. R. (1996). Bugs examples volume 1, version 0.5 (version ii) [Computer software manual].

Stone, A. M., Bushnell, W., Denne, J., Sargent, D. J., Amit, O., Chen, C., Bailey-Iacona, R., Helterbrand, J., and Williams, G. (2011). Research outcomes and recommendations for the assessment of progression in cancer clinical trials from a PhRMA working group. *European Journal of Cancer*, *47*(12), 1763–1771.

Stone, C. J., Hansen, M. H., Kooperberg, C., and Truong, Y. K. (1997). Polynomial splines and their tensor products in extended linear modeling (with Discussion). *The Annals of Statistics*, *25*(4), 1371–1470.

Sturtz, S., Ligges, U., and Gelman, A. (2005). R2WinBUGS: A package for running WinBUGS from R. *Journal of Statistical Software*, *12*(3), 1–16. Retrieved from `http://www.jstatsoft.org`

Su, L., and Hogan, J. W. (2011). HIV dynamics and natural history studies: Joint modelling with doubly interval-censored event time and infrequent longitudinal data. *The Annals of Applied Statistics*, *5*(1), 400–426.

Su, Y.-S., and Yajima, M. (2015). R2jags: Using R to run JAGS [Computer software manual]. Retrieved from `https://CRAN.R-project.org/package=R2jags` (R package version 0.5-7)

Sun, J. (1995). Empirical estimation of a distribution function with truncated and doubly interval-censored data and its application to AIDS studies. *Biometrics*, *51*(3), 1096–1104.

Sun, J. (1996). A non-parametric test for interval-censored failure time data with application to AIDS studies. *Statistics in Medicine*, *15*(13), 1387–1395.

Sun, J. (2001). Variance estimation of a survival function for interval-censored survival data. *Statistics in Medicine*, *20*(8), 1249-1257.

Sun, J. (2006). *The statistical analysis of interval-censored failure time data.* New York: Springer Science + Business Media.

Sun, J., Liao, Q., and Pagano, M. (1999). Regression analysis of doubly censored failure time data with applications to AIDS studies. *Biometrics*, *55*(3), 909–914.

Sun, J., Zhao, Q., and Zhao, X. (2005). Generalized log-rank tests for interval-censored failure time data. *Scandinavian Journal of Statistics*, *32*(1), 49–57.

Sun, L., Kim, Y.-J., and Sun, J. (2004). Regression analysis of doubly censored failure time data using the additive hazards model. *Biometrics*, *60*(3), 637–643.

Sun, L., Wang, L., and Sun, J. (2006). Estimation of the association for bivariate interval-censored failure time data. *Scandinavian Journal of Statistics*, *33*(4), 637–649.

Sun, X., and Chen, C. (2010). Comparison of Finkelsteins's method with the convential approach for interval censored data analysis. *Statistics in Biopharmaceutical Research*, *2*(1), 97–108.

Susarla, V., and Van Ryzin, J. (1976). Nonparametric Bayesian estimation of survival curves from incomplete observations. *Journal of the American Statistical Association*, *71*(356), 897–902.

Susarla, V., and Van Ryzin, J. (1978). Large sample theory for a Bayesian nonparametric survival curve estimator based on censored samples. *The Annals of Statistics*, *6*(4), 755–768.

Sutradhar, R., Barbera, L., Seow, H., Howell, D., Husain, A., and Dudgeon, D. (2011). Multistate analysis of interval-censored longitudinal data: Aapplication to a cohort study on performance status among patients diagnosed with cancer. *American Journal of Epidemiology*, *173*(4), 468–475.

Tanner, M. A., and Wong, W. H. (1987). The calculation of posterior distributions by data augmentation (with Discussion). *Journal of the American Statistical Association*, *82*(398), 528–550.

Tarone, R. E., and Ware, J. (1977). On distribution-free tests for equality of survival distributions. *Biometrika*, *64*(1), 156–160.

Therneau, T. (2017). A package for survival analysis in S [Computer software manual]. Retrieved from `http://CRAN.R-project.org/package=survival` (R package version 2.41-3)

Therneau, T. M., Grambsch, P. M., and Fleming, T. R. (1990). Martingale-based residuals for survival models. *Biometrika*, *77*(1), 147–160.

Thomas, A., O'Hara, B., Ligges, U., and Sturtz, S. (2006). Making BUGS open. *R News*, *6*(1), 12–17. Retrieved from `http://cran.r-project.org/doc/Rnews/`

Tierney, L. (1994). Markov chains for exploring posterior distributions (with Discussion). *The Annals of Statistics*, *22*(4), 1701–1762.

Titterington, D. M., Smith, A. F. M., and Makov, U. E. (1985). *Statistical analysis of finite mixture distributions*. Chichester: John Wiley & Sons.

Topp, R., and Gómez, G. (2004). Residual analysis in linear regression models with an interval-censored covariate. *Statistics in Medicine*, *23*(21), 3377-3391.

Turnbull, B. W. (1974). Nonparametric estimation of a survivorship functions with doubly-censored data. *Journal of the American Statistical Association*, *69*(345), 169–173.

Turnbull, B. W. (1976). The empirical distribution function with arbitrarily grouped, censored and truncated data. *Journal of the Royal Statistical*

Society, Series B, *38*(3), 290–295.

van den Hout, A. (2016). *Multi-state survival models for interval-censored data.* Boca Raton: Chapman & Hall/CRC.

van der Vaart, A., and Wellner, J. A. (2000). Preservation theorems for Glivenko-Cantelli and uniform Glivenko-Cantelli class. In E. Gine, D. M. Mason, and J. A. Wellner (Eds.), *High dimensional probability II* (p. 115-133). Birkhäuser Verlag.

Wahba, G. (1983). Bayesian "confidence intervals" for the cross–validated smoothing spline. *Journal of the Royal Statistical Society, Series B*, *45*(1), 133–150.

Wang, W., and Ding, A. A. (2000). On assessing the association for bivariate current status data. *Biometrika*, *87*(4), 879–893.

Wang, X., Chen, M.-H., Wang, W., and Yan, J. (2017). dynsurv: Dynamic models for survival data [Computer software manual]. Retrieved from `https://CRAN.R-project.org/package=dynsurv` (R package version 0.3-5)

Wang, X., Chen, M.-H., and Yan, J. (2013). Bayesian dynamic regression models for interval censored survival data with application to children dental health. *Lifetime Data Analysis*, *19*(3), 297–316.

Wang, X., Sinha, A., Yan, J., and Chen, M.-H. (2012). Bayesian inference of interval-censored survival data. In D.-G. Chen, J. Sun, and K. E. Peace (Eds.), *Interval-censored time-to-event data: Methods and applications* (pp. 167–195). Boca Rotan, US: Chapman and Hall/CRC.

Watanabe, S. (2010). Asymptotic equivalence of Bayes cross validation and widely applicable information criterion in singular learning theory. *Journal of Machine Learning Research*, *11*, 3571–3594.

Wei, G. C. G., and Tanner, M. A. (1991). Applications of multiple imputation to the analysis of censored regression data. *Biometrics*, *47*(4), 1297–1309.

Weller, J. I., Saran, A., and Zeliger, Y. (1992). Genetic and environmental relationships among somatic cell count, bacterial infection, and clinical mastitis. *Journal of Dairy Science*, *75*(9), 2532–2540.

Wellner, J. A., and Zahn, Y. (1997). A hybrid algorithm for computation of the nonparametric maximum likelihood estimator from censored data. *Journal of the American Statistical Association*, *92*(439), 945–959.

Wienke, A. (2010). *Frailty models in survival analysis.* Chapman & Hall/CRC.

Williams, J. S., and Lagakos, S. W. (1977). Models for censored survival analysis: Constant-sum and variable-sum models. *Biometrika*, *64*(2), 215–224.

Wong, G. Y. C., and Yu, Q. (1999). Generalized MLE of a joint distribution function with multivariate interval-censored data. *Journal of Multivariate Analysis*, *69*(2), 155–166.

Wu, C.-F. (1978). Some algorithmic aspects of the theory of optimal designs. *The Annals of Statistics*, *6*(6), 1286–1301.

Xiang, L., Ma, X., and Yau, K. K. W. (2011). Mixture cure model with random effects for clustered interval-censored survival data. *Statistics in Medicine*, *30*(9), 995–1006.

Yashin, A. I., Vaupel, J. W., and Iachine, I. A. (1995). Correlated individual frailty: An advantageous approach to survival analysis of bivariate data. *Mathematical Population Studies*, *5*(2), 145–159.

Yu, Q., Li, L., and Wong, G. Y. C. (2000). On consistency of the self-consistent estimator of survival functions with interval-censored data. *Scandinavian Journal of Statistics*, *27*(1), 35–44.

Yu, S., Yu, Q., and Wong, G. Y. C. (2006). Consistency of the generalized MLE of a joint distribution function with multivariate interval-censored data. *Journal of Multivariate Analysis*, *97*(3), 720–732.

Yuen, K.-C., Shi, J., and Zhu, L. (2006). A k-sample test with interval censored data. *Biometrika*, *93*(2), 315–328.

Zhang, M., and Davidian, M. (2008). "Smooth" semiparametric regression analysis for arbitrarily censored time-to-event data. *Biometrics*, *64*(2), 567–576.

Zhang, Y., Liu, W., and Wu, H. (2003). A simple nonparametric two-sample test for the distribution function of event time with interval censored data. *Journal of Nonparametric Statistics*, *15*(6), 643–652.

Zhang, Y., Liu, W., and Zhan, Y. (2001). A nonparametric two-sample test of the failure function with interval censoring case 2. *Biometrika*, *88*(3), 677–686.

Zhang, Z., and Sun, J. (2010). Interval censoring. *Statistical Methods in Medical Research*, *19*(1), 53–70.

Zhang, Z., Sun, L., Sun, J., and Finkelstein, D. M. (2007). Regression analysis of failure data with informative interval censoring. *Statistics in Medicine*, *26*(12), 2533–2546.

Zhao, Q., and Sun, J. (2004). Generalized log-rank test for mixed interval-censored failure time data. *Statistics in Medicine*, *23*(10), 1621–1629.

Zhao, Q., and Sun, J. (2015). glrt: Generalized logrank tests for interval-censored failure time data [Computer software manual]. Retrieved from `http://CRAN.R-project.org/package=glrt` (R package version 2.0)

Zhao, X., Zhao, Q., Sun, J., and Kim, J. S. (2008). Generalized log-rank tests for partly interval-censored failure time data. *Biometrical Journal*, *50*(3), 375–385.

Zheng, M., and Klein, J. P. (1995). Estimates of marginal survival for dependent competing risks based on an assumed copula. *Biometrika*, *82*(1), 127–138.

Zhou, M., and Jin, Z. (2009). rankreg: Rank regression for censored data AFT model. [Computer software manual]. Retrieved from `http://CRAN.R-project.org/package=rankreg` (R package version 0.2-2)

Zhou, X., Feng, Y., and Du, X. (2017). Quantile regression for interval censored data. *Communications in Statistics - Theory and Methods*, *46*(8), 3848–3863.

Author Index

Subject Index